Molecular Oral Microbiology

Edited by Anthony Rogers

Caister Academic Press

Caister Academic Press
Norfolk, UK

www.caister.com

British Library Cataloguing-in-Publication Data
A catalogue record for this book is available from the British Library

ISBN: 978-1-904455-24-0

Printed and bound in Great Britain

Contents

Contributors

Sirkka Asikainen
Section of Oral Microbiology
Institute of Odontology
Umea University
Umea
Sweden

sirkka.asikainen@odont.umu.se

Mohsen Amin
CIHR Group in Matrix Dynamics
Faculty of Dentistry
University of Toronto
Toronto, ON
Canada

mohsen.amin@utoronto.ca

Ching Seng Ang
Cooperative Research Centre for Oral Health Sciences
School of Dental Science and Bio 21 Institute
University of Melbourne
Parkville, Vic.
Australia

c.ang@postgrad.unimelb.edu.au

Gilad Bachrach
Institute of Dental Sciences
The Hebrew University-Hadassah
School of Dental Medicine
Jerusalem
Israel

bgilad@md.huji.ac.il

Alex B. Berezow
Department of Periodontics
School of Dentistry
University of Washington
Seattle, WA
USA

alexbb@u.washington.edu

Richard Darveau
Department of Periodontics
School of Dentistry
University of Washington
Seattle, WA
USA

rdarveau@u.washington.edu

Stuart G. Dashper
Cooperative Research Centre for Oral Health Sciences
School of Dental Science and Bio 21 Institute
University of Melbourne
Parkville, Vic.
Australia

stuartgd@unimelb.edu.au

David D. Davies
Department of Biological Sciences
SUNY at Binghamton
Binghamton, NY
USA

dgdavies@binghampton.edu

Margaret J. Duncan
Department of Molecular Genetics
Forsyth Dental Center
Boston, MA
USA

mduncan@forsyth.org

Richard P. Ellen
CIHR Group in Matrix Dynamics
Faculty of Dentistry
University of Toronto
Toronto, ON
Canada

richard.ellen@dentistry.utoronto.ca

Daniel H. Fine
Department of Oral Biology
Centre for Oral Infectious Diseases
University of Medicine and Dentistry of New Jersey
Newark, NJ
USA

finedh@umdnj.edu

David H. Figurski
Department of Microbiology
School of Physicians and Surgeons
Columbia University
New York, NY
USA

figurski@cancercenter.columbia.edu

George Hajishengallis
Centre for Oral Health and Systemic Disease
University of Louisville School of Dentistry
Louisville, KY
USA

g0haji01@louisville.edu

Lijian Jin
Department of Periodontology
School of Dentistry
University of Hong Kong
Hong Kong

ljjin@hkusua.hku.hk

Marie-Claude Jobin
CIHR Group in Matrix Dynamics
Faculty of Dentistry
University of Toronto
Toronto, ON
Canada

marieclaude.jobin@utoronto.ca

Scott C. Kachlany
Department of Oral Biology
Centre for Oral Infectious Diseases
University of Medicine and Dentistry of New Jersey
Newark, NJ
USA

kachlasc@umdnj.edu

Jeffrey B. Kaplan
Department of Oral Biology
Centre for Oral Infectious Diseases
University of Medicine and Dentistry of New Jersey
Newark, NJ
USA

kaplanjb@umdnj.edu

Maribasappa Karched
Section of Oral Microbiology
Institute of Odontology
Umea University
Umea
Sweden

maribasappa.karched@odont.umu.se

Janina P. Lewis
Philips Institute of oral and Craniofacial Molecular Biology
Virginia Commonwealth University
Richmond, VA
USA

jplewis@vcu.edu

Alex Mira
Evolutionary Genomics Group
Division of Microbiology
Miguel Hernandez University
San Juan
Alicante
Spain

alex.mira@umh.es

Eric C. Reynolds
Cooperative Research Centre for Oral Health Sciences
School of Dental Science and Bio 21 Institute
University of Melbourne
Parkville, Vic.
Australia

e.reynolds@unimelb.edu.au

Alexander H. Rickard
Department of Biological Sciences
SUNY at Binghamton
Binghamton, NY
USA

arickard@binghamton.edu

Michael W. Russell
Departments of Microbiology and Immunology, and Oral Biology
Witebsky Center for Microbial Pathogenesis and Immunology
University at Buffalo
Buffalo, NY
USA

russellm@buffalo.edu

Roy R.B. Russell
School of Dental Sciences
University of Newcastle
Newcastle-upon-Tyne
UK

R.R.Russell@newcastle.ac.uk

Paul D. Veith
Cooperative Research Centre for Oral Health Sciences
School of Dental Science and Bio 21 Institute
University of Melbourne
Parkville, Vic.
Australia

pdv@unimelb.edu.au

Peter S. Zilm
Microbiology Laboratory
Dental School
The University of Adelaide
Adelaide, SA
Australia

peter.zilm@adelaide.edu.au

Preface

Molecular microbiology may be defined as studies that deepens our understanding of the molecular principles underlying physiological processes involved in the life cycle of pro- and eukaryotic microorganisms. These processes include gene expression, regulation and transfer; the synthesis of macromolecules; sub-cellular organization; cell–cell communication; and pathogenicity and virulence. Put more succinctly, it involves studying the interaction between various cell systems and attempting to understand their regulation. The aim of the present book is to illustrate how molecular biological techniques are being used to study the fundamental biology of some of the microorganisms thought to be involved in the two major dental diseases. These diseases, namely caries and periodontal disease, result from the interaction between the resident oral bacteria themselves and the host. Indeed, it seems that ecological imbalance within the oral microbiota is the fundamental cause of both diseases. Accordingly, in recent years, research has focused on factors which might modulate such imbalances (H.K. Kuramitsu and R.P. Ellen, 2000, *Oral Bacterial Ecology: The Molecular Basis*. Wymondham, UK: Horizon Scientific Press). Apart from molecular studies on specific pathogens, the present book deals, at a molecular level, with the interactions of oral microorganisms with one another and with host; the innate defence mechanisms of the host; and the development of vaccines against oral diseases. In the context of attacking potentially pathogenic bacteria, and aside from characterizing possible virulence factors as targets for vaccine development, molecular techniques can, for example, identify genes encoding proteins essential for bacterial cell viability or proliferation. Any that do not have a human homologue are potential targets for anti-microbial agents. Perhaps these techniques will also lead to the development of useful disease-predictive tests.

The topics addressed in the present book are intertwined to varying degrees and so overlap, in terms of some of the subject matter, is unavoidable. Hopefully, the book will be of value, not only to under- and post-graduate students and researchers in the field of oral biology and oral microbiology, in particular, but also to microbiologists in general. It will have served a useful purpose if it poses more questions than it answers.

I thank Mr Hugh Griffin of Horizon Press for his encouragement and assistance in the production of this book. I am also grateful to my colleague, Peter Zilm, for his logistical help in getting the book into a form suitable for submission to the publishers.

Among married couples, particularly those who are retired, one often hears the term 'golfing widows'. My wife, Dorothy, is a member of an international band of 'academic widows'. I thank her for her patience and tolerance, which are (almost) boundless.

Anthony Rogers
September 2007

Other Books of Interest

Helicobacter pylori: Molecular Genetics and Cellular Biology	2008
Corynebacteria: Genomics and Molecular Biology	2008
Staphylococcus: Molecular Genetics	2008
Leishmania: After The Genome	2008
Archaea: New Models for Prokaryotic Biology	2008
RNA and the Regulation of Gene Expression	2008
Legionella Molecular Microbiology	2008
Epigenetics	2008
Animal Viruses: Molecular Biology	2008
Segmented Double-Stranded RNA Viruses	2008
Acinetobacter Molecular Biology	2008
Pseudomonas: Genomics and Molecular Biology	2008
Microbial Biodegradation: Genomics and Molecular Biology	2008
The Cyanobacteria: Molecular Biology, Genomics and Evolution	2008
Coronaviruses: Molecular and Cellular Biology	2007
Real-Time PCR in Microbiology: From Diagnosis to Characterisation	2007
Bacteriophage: Genetics and Molecular Biology	2007
Candida: Comparative and Functional Genomics	2007
Bacillus: Cellular and Molecular Biology	2007
AIDS Vaccine Development: Challenges and Opportunities	2007
Alpha Herpesviruses: Molecular and Cellular Biology	2007
Pathogenic *Treponema*: Molecular and Cellular Biology	2007
PCR Troubleshooting: The Essential Guide	2006
Influenza Virology: Current Topics	2006
Microbial Subversion of Immunity: Current Topics	2006
Cytomegaloviruses: Molecular Biology and Immunology	2006
Papillomavirus Research: From Natural History To Vaccines and Beyond	2006
Epstein Barr Virus	2005
HIV Chemotherapy: A Critical Review	2005
Probiotics and Prebiotics: Scientific Aspects	2005
Foodborne Pathogens: Microbiology and Molecular Biology	2005

Caister Academic Press www.caister.com

Molecular Techniques in Oral Microbial Taxonomy, Identification and Typing

1

Sirkka Asikainen and Maribasappa Karched

Good opening Chapter introducing the molecular techniques that are currently used in Oral Microbiology

Abstract

The rapid development of molecular techniques during the past decade has revolutionized the field of microbiology. Two issues are of profound importance. First, the discovery of phylogenetically informative DNA sequences, such as the 16S rRNA gene, radically changed the concept of bacterial relatedness and provided a universal system for bacterial identification and categorization. Second, it became possible to detect, identify, and type bacteria independent of their cultivability and, by these new means, to elucidate the diversity and spatial organization of complex oral bacterial communities. Of considerable benefit has been the fact that the same nucleic acid-based molecular approaches can be applied in all microbial environments, ranging from the oral cavity, to the surfaces of historical monuments, to the depths of open oceans. This has led to development of versatile PCR- and hybridization-based techniques that allow a rapid and convenient analysis of the bacterial contents of oral samples and offer previously unattainable possibilities for expanding studies on bacterial epidemiology and characterization. In particular, the emerging new microarray technology will facilitate great strides in understanding the structure and dynamics of oral bacterial communities and bacteria-host interactions, and will form the basis for developing novel diagnostics for oral infections.

Introduction

Current understanding of oral microbiology is based on the knowledge obtained from culture-based studies during the past 100 years. Culture does not, however, give a full picture of the diversity of the complex host-associated or environmental bacterial communities. It has been estimated that 40–50% of the bacteria in the human oral cavity (Kroes *et al.*, 1999; Paster *et al.*, 2001; Kumar *et al.*, 2005), 20–40% of intestinal bacteria (Rondon *et al.*, 1999), and only 1% of environmental bacteria (Amann *et al.*, 1995; Oren, 2004) have been cultured. Current concepts about dental diseases are thus based upon microbiological knowledge obtained from only part of the oral bacterial populations by using methods that are biased towards bacteria that survive transportation from sampling site to the laboratory, and are easily or rapidly grown under laboratory conditions. Therefore, the possibilities remain that the bacteria detected by culture-dependent methods are neither dominant in the oral sites sampled, nor functionally significant in their natural environment.

The rapid development of molecular biological technology during the past 15 years has led to an increased use of culture-independent methods to supplement or replace culture in the detection, identification, and typing of bacteria. An example of the recent molecular advances is the improved and more cost-effective DNA sequencing technology that has made possible the dramatic progress in bacteriology; particularly in bacterial identification. This progress is illustrated by the ever increasing number of sequenced bacterial genomes since the first genome sequence (*Haemophilus influenzae*) was published in 1995 (Fleischmann *et al.*, 1995). As of April 2007, 449 sequenced bacterial genomes have been published and 1091 are in progress (http://www.genomesonline.org/gold.cgi). The advent of phylogenetically informative DNA sequences (Young *et al.*, 1979; Woese, 1987; Maiden *et al.*, 1998), together with advances in sequencing technology have revolutionized the process of identifying bacteria; this, in turn, promotes elucidation of the composition and diversity of complex bacterial communities. These innovations also accelerate the study of the phylogeny of cultivable and not-yet-cultured bacteria. Other notable molecular biological achievements include the establishment of continuously growing and diversifying public databases in which enormous amounts of bacterial sequence data have been collected and which can be exploited by researchers. Furthermore, actively developing bioinformatics plays a central role in the path from data generation to biological interpretation (Goodman, 2002).

As discussed below, a variety of molecular tools is currently available for studying detailed questions that range from bacterial evolutionary relationships to monitoring and visualizing the dynamics of complex bacterial communities, as well as detection and identification of selected single or multiple bacteria, their clones or nucleotide sequences of interest.

Bacterial taxonomy, species and phylotype

A knowledge of the members of the oral microbiota is a prerequisite for understanding its functions and role, not only in disease, but also in maintaining health. The process of 'inventoring' oral bacteria on the basis of phylogenetically significant sequences (e.g. 16S rDNA), is closely related to bacterial taxonomy. Taxonomy means classification of organisms in an ordered system, a hierarchy that shows their evolutionary relationships (Table 1.1). The aims of taxonomy are to reconstruct evolution and organize biological information for improved understanding of the functional roles of bacteria in their environments, their biodiversity and community structure (Fenchel and Finlay, 2006).

The basic official taxonomic group in bacterial systematics is the species (Amann *et al.*, 1995; Konstantinidis *et al.*, 2006; Staley, 2006). In ascending taxonomic order, each species belongs to a genus, family, order, class, phylum (or division), and domain. The current bacterial species definition derives from the recommendations of

Table 1.1 Definitions of common terms in bacterial systematics

Taxonomy	The science of classification
Classification	The ordering of organisms into groups (taxa) on the basis of their phylogenetic relationships
Phylogeny	The process in which lineages of organisms evolved by separation from a common ancestor
Identification	The act and result of assigning unidentified organisms to a particular rank in a previously made classification
Species	A taxonomic group below the genus level in the taxonomic hierarchy

Adapted from Busse *et al.* (1996).

an *ad hoc* re-evaluation committee of the species definition in bacteriology in 2002 (Staley, 2006). The definition was a consensus conclusion considered to cause the fewest contradictions with the traditional species classification (e.g. with regard to human pathogens). Accordingly, the bacterial species is defined using a polyphasic approach that takes into account a combination of phenotypic and genotypic characteristics of the bacterium. A species is 'a category that circumscribes a (preferably) genomically coherent group of individual isolates/strains sharing a high degree of similarity in (many) independent features, comparatively tested under highly standardized conditions'. For practical purposes, 'species' is considered to be 'a group of strains (including the type strain) that are characterized by a certain degree of phenotypic consistency, showing 70% of DNA–DNA binding and over 97% of 16S ribosomal RNA (rRNA) gene-sequence identity' (Vandamme *et al.*, 1996; Gevers *et al.*, 2005).

A pragmatic polyphasic approach in bacterial classification starts with the screening of isolates using different phenotypic and genotypic techniques. Isolate clusters are formed by distinguishing isolates that are more related from those less related to each other. Application of new genomic methods, such as multilocus sequence typing (MLST), DNA profiling with e.g. repetitive element-based PCR (rep-PCR), PCR-restriction fragment length polymorphism (PCR-RFLP), and the DNA microarray technology, is recommended for the further development of bacterial classification (Stackebrandt *et al.*, 2002). This recommendation is due partly to the procedural difficulties associated with the DNA–DNA reassociation hybridization method and its unsuitability for classification of uncultivable bacteria, or of single bacteria because of the required pair-wise comparisons.

It is notable that there is no official bacterial classification available, but the Taxonomic Outline of the Prokaryotes, published in Bergey's Manual of Systematic Bacteriology serves among bacteriologists as the most widely accepted bacterial classification. On the other hand, bacterial nomenclature or naming is controlled by the International Code of Nomenclature of Bacteria. As maintained by this Code, each bacterial species is named using the binomial system (genus and species) initiated by Carl von Linné nearly 300 years ago. The lowest rank accepted by the Code is the subspecies. A 'candidatus' status can be given to incompletely described bacteria.

The term phylotype signifies the phylogenetic position of a bacterium, usually as identified by comparative 16S rDNA sequence analysis. Phylotypes that can be further identified to the species level are shown in phylogenetic trees by their official names, whereas the phylotypes not identifiable to the species level are usually assigned according to the genus of the closest relative, followed by the phylotype code. Examples of detailed phylogenetic trees constructed from oral bacterial community samples can be found in several original publications (Kroes *et al.*, 1999; Sakamoto *et al.*, 2000; Paster *et al.*, 2001; Becker *et al.*, 2002; Kazor *et al.*, 2003; Aas *et al.*, 2005).

Molecular techniques: a preamble

Molecular techniques generally involve methods that are used to study cellular macromolecules, such as nucleic acids, DNA and RNA, and proteins. This discussion focuses on nucleic acid-based techniques employed for detection, identification, and typing of oral bacteria. The techniques selected in this context are divided into three groups, each of which is discussed in detail below:

(i) Detection and identification of oral bacteria
Techniques for individual or multiple bacteria already known or recently identified by molecular methods, such as PCR-based methods (conventional PCR, real-time PCR [qRT-PCR]) and hybridization-based techniques (species-specific oligonucleotide probes, whole-genomic checkerboard DNA–DNA hybridization, reverse-capture oligonucleotide hybridization, DNA microarray, and fluorescence *in-situ* hybridization [FISH]).
(ii) Bacterial community diversity in the oral cavity
Techniques for studying bacterial community diversity (i.e. detection and identification of all oral bacteria), such as 16S rRNA gene cloning and sequencing, denaturing gradient gel electrophoresis (DGGE), and terminal restriction fragment length polymorphism (T-RFLP).
(iii) Bacterial typing
Techniques for determining intraspecies or interspecies genetic differences among bacterial strains, such as DNA sequence-based typing (DNA microarray-based genomic profiling and typing, and MLST) and gel electrophoresis-based DNA fingerprinting (random primed polymorphic DNA/arbitrarily primed-PCR [RAPD/AP-PCR], rep-PCR, PCR-RFLP, MLST, and ribosomal intergenic spacer analysis [RISA]).

Most of the above-mentioned techniques involve PCR amplification of the target DNA. PCR is often used at the first step of the bacterial analysis and is followed by characterization of the amplicons by sequencing or by visualization with electrophoretic separation according to their size and/or sequence polymorphism. Due to the central role of PCR in molecular analyses, its principles, advantages, and limitations, as relevant to the subsequent discussion on molecular techniques and their applications in oral microbiology, are summarized in the following paragraphs.

The greatest advantage of the PCR technique, developed in the early 1980s by Kari Mullis (Mullis, 1990), is that minute amounts of DNA can be multiplied up to a billion times within a few hours. This is a major advance, since the early DNA-based research was complicated by the fact that the time-consuming cloning procedure had to be used to increase the amount of DNA for the subsequent analyses. The PCR amplification technique is comprised of repeated thermal cycling of the reaction mixture containing the DNA template to be amplified. The cycling can be conveniently and precisely done in an automatic thermocycler. The first step in PCR amplification involves denaturation of the target DNA at high temperature (*ca.* 94°C) to obtain single-stranded template DNA. The primers anneal at a lower temperature (50–60°C) to their specific binding sites and initiate the synthesis of the new DNA strand. At *ca.* 72°C, a thermostable DNA polymerase adds free deoxynucleotide triphosphates one-by-one to the 3′ end of the primer according to the order of nucleotides in the template. This leads to elongation of the primer and finally to completion of a new DNA strand that is complementary to the template strand. Since each new DNA strand synthesized in each cycle can act as a new template in the next cycle, amplification of the target sequence proceeds exponentially.

Besides the fact that PCR has obvious advantages, which have made it the most commonly used molecular method in laboratories worldwide, it has certain limitations that can cause biased amplification of sequences from heterogeneous bacterial samples (von Wintzingerode *et al.*, 1997), such as those from the oral cavity. The first source of error derives from the extraction of DNA. Due to differences in the susceptibility of bacterial cells to lysis, the methods optimal for DNA extraction from Gram-negative bacteria may extract an insufficient amount of DNA from sturdier Gram-positive bacteria. Conversely, harsh extraction methods may cause excessive shearing of DNA from Gram-negative bacteria and lead to amplification bias and artefacts like chimeric products. Oral samples often contain PCR amplification inhibitors (e.g. blood). The inhibition mechanism involves blocking of the active site of DNA polymerase by haem (Abu Al-Soud and Radstrom, 1998). Other possible PCR inhibition mechanisms include polymerase degradation by proteinases or denaturation by phenol and detergents. Furthermore, DNA sequence differences among bacteria in a community sample may lead to unequal DNA denaturation and primer hybridization and cause biased amplification of the target DNA segments. Noteworthy is the fact that, due to its sensitivity, PCR amplifies contaminant DNA in addition to the target DNA segment from the sample, resulting in erroneous results.

Detection and identification of oral bacteria

Since the beginning of the 1990s, molecular methods have been increasingly used in oral bacterial epidemiology. The majority of these studies have aimed at detecting bacterial species that had previously been delineated as 'pathogens' by culture methods. In periodontitis, these bacterial species are mainly Gram-negative anaerobic rods and the most frequently studied species include *Porphyromonas gingivalis, Aggregatibacter (Actinobacillus) actinomycetemcomitans, Tannerella forsythia, Prevotella intermedia,* and *Treponema denticola*. In caries, these species are mainly Gram-positive facultatively anaerobic bacteria and the most frequently studied are *Streptococcus mutans, Streptococcus sobrinus*, and other oral streptococci. The molecular methods used for the detection and identification of oral bacteria are divided into two groups in the following discussion: the PCR-based methods and hybridization-based methods (Fig. 1.1).

PCR-based methods

Conventional PCR

Due to its simplicity, sensitivity, and cost-effectiveness conventional PCR detection of a single species at a time has been the most commonly used strategy in molecular-based oral bacterial epidemiology (Fig. 1.1). The majority of the studies have employed species-specific oligonucleotide primers targeting the 16S rRNA gene in the sample (Slots *et al.*, 1995; Ashimoto *et al.*, 1996; Matto *et al.*, 1998; Rudney and Larson, 1999; Tang *et al.*, 2003; Umeda *et al.*, 2004; Rocas and Siqueira, 2005). However, apart from 16S rDNA, a number of studies have amplified other species-specific genes, such as the *A. actinomycetemcomitans* leukotoxin gene (Goncharoff *et al.*, 1993; Fujise *et al.*, 1995; Yuan *et al.*, 2001), *P. gingivalis* fimbrillin gene (Amano *et al.*, 1999), *S. mutans spaP* gene, encoding the surface protein antigenI/II (Ono *et al.*, 1994; Allaker *et al.*, 1998), and glucosyltransferase genes (*gtf*) of oral streptococci, such as *S. sanguinis, S. gordonii, S. mutans, S. sobrinus*, and *S. salivarius* (Hoshino *et al.*, 2004; Nakano *et al.*, 2006). Furthermore, the 16S-23S rRNA gene intergenic spacer region has been used to identify *A. actinomycetemcomitans, P. gingivalis, T. forsythia* (Griffen *et al.*, 1999; Leys *et al.*, 2002), and *Streptococcus mitis* (Barsotti *et al.*, 2002).

Whereas conventional PCR involves amplification of a single genetic locus at a time by using two primers, multiplex PCR is a variation that amplifies several genetic loci in the same reaction (Chamberlain *et al.*, 1988) (Fig. 1.1). Amplification of more than one locus at a time means that the number of primer pairs is equivalent to the number of loci to be amplified. The presence of several primers in the same reaction may, however, lead to non-specific amplification products and formation of primer dimers (Brownie *et al.*, 1997; Vallone and Butler, 2004).

In oral microbiology, multiplex PCR was first used in the simultaneous detection of *A. actinomycetemcomitans* and *P. gingivalis* in periodontal samples (Tran and Rudney, 1996). From endodontal samples, multiplex PCR succeeded in detecting *Dialister pneumosintes* and *Filifactor alocis* (Siqueira and Rocas, 2004), and pigmented Gram-negative rods, such as *Porphyromonas endodontalis, P. gingivalis, Prev. intermedia, Prev. nigrescens*, and *Prev. tannerae* (Seol *et al.*, 2006). The multiplex PCR method has also been tested with samples from various oral surfaces for the simultaneous detection of *Prev. melaninogenica* and *Prev. loescheii* together with *Prev. intermedia* and *Prev. nigrescens* (Yoshida *et al.*, 2005).

Real-time PCR

The quantitative real-time PCR (qRT-PCR) technique is another variation utilizing the principles of the PCR technique (Fig. 1.1). The

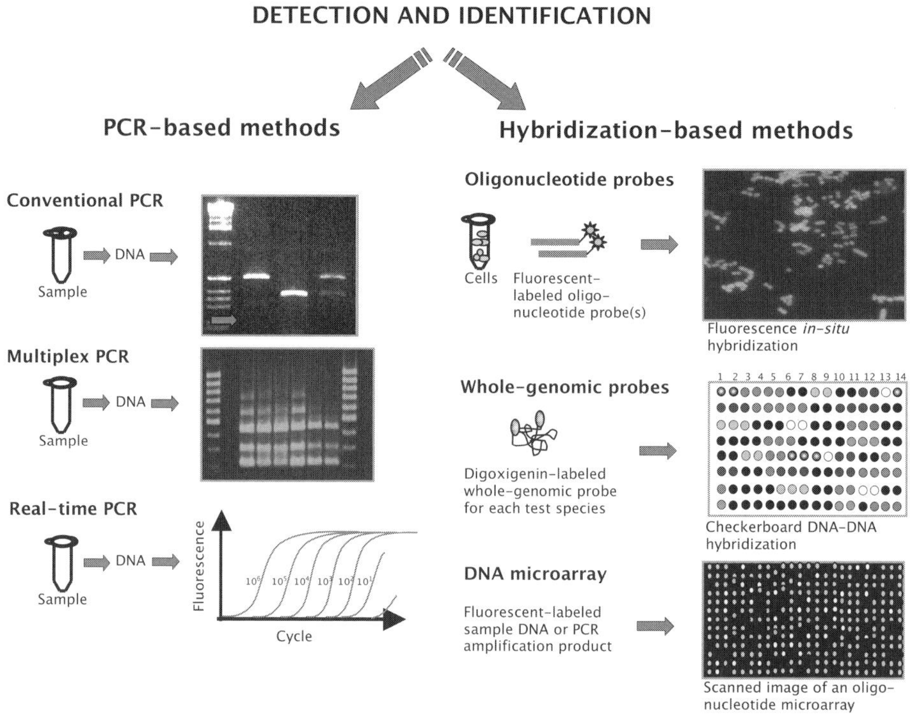

Figure 1.1 Examples of molecular techniques used for the detection and identification of oral bacteria from clinical samples. A colour version of this figure is located in the plate section at the back of the book.

qRT-PCR is also known as RT-PCR, which creates confusion, since in the literature the acronym RT-PCR also refers to reverse transcription PCR. Furthermore, since both methods are often used together, the terminology becomes even more confusing. Here the acronym qRT-PCR is used for the real-time PCR (Zhang and Fang, 2006).

Conventional PCR suffers from the drawback that the amount of amplified PCR product correlates poorly with the amount of DNA template originally present in the sample. The qRT-PCR technique was developed to overcome this limitation (Higuchi *et al.*, 1993). Two findings were used to produce qRT-PCR; namely, the 5′–3′-exonuclease activity of *Taq* DNA polymerase and the construction of dual-labelled oligonucleotide probes that emit a fluorescence signal only on cleavage (Giulietti *et al.*, 2001). In contrast to conventional PCR, which allows monitoring the PCR product after the amplification procedure, qRT-PCR monitors the formation of the product in real time. The monitoring is accomplished using various types of fluorescent dyes.

The most widely used fluorescent reagents in qRT-PCR are SYBR Green I, TaqMan probe, and molecular beacons (Zhang and Fang, 2006). The SYBR Green I binds to double-stranded DNA formed at the end of each PCR cycle, thereby increasing the fluorescence intensity cycle by cycle. The TaqMan probe and molecular beacons have different structures, but both are oligonucleotides with a 5′ end fluorophor and 3′ end quencher. Despite the differences, the outcome of the qRT-PCR amplification is the same. The emitted fluorescence intensifies when the distance between the fluorophor and quencher increases during the PCR reaction. The fluorescence increase is proportional to the PCR amplicon concentration in real time.

The qRT-PCR technique offers attractive advantages over conventional PCR. These include sensitivity, specificity, and reduced risk of releasing PCR amplicons to the laboratory environment, since both amplification and analysis steps are performed in a closed system. During the past few years, qRT-PCR has been extensively tested for its suitability to detect, identify, and quantify bacteria in oral bacterial samples. So far, most studies have looked for traditional periodontal pathogens in subgingival samples (Suzuki *et al.*, 2005; Boutaga *et al.*, 2006), saliva (Sakamoto *et al.*, 2004), or mouth wash samples (Boutaga *et al.*, 2007). Interestingly, recent studies have also found representatives of the domain *Archaea* in periodontal (Lepp *et al.*, 2004) and endodontal infections (Vianna *et al.*, 2006; Vickerman *et al.*, 2007).

Hybridization-based techniques

The discovery of the hybridization reaction (i.e. spontaneous pairing of two strands of a nucleic acid double helix), has made possible most of the molecular biological advances during the past 50 years. A plethora of techniques has been developed, based on the hybridization of nucleic acid strands that are either DNA or RNA. For bacterial detection and identification purposes, DNA probes, that are most commonly oligonucleotide probes comprising phylogenetically informative sequences or whole-genomic probes originating in cultivable species, have been used to probe bacterial DNA preparations from the oral cavity (Fig. 1.1). The hybridization-based methods have detection threshold of 10^4–10^5 DNA or RNA copies, which is higher than that achieved by PCR-based methods. Some hybridization-based methods, such as oligonucleotide probes and whole genomic checkerboard hybridization, have gained extensive use in oral microbiology. Emerging techniques, using the hybridization reaction for detection of bacteria, include FISH and DNA microarray (Fig. 1.1).

Oligonucleotide probes

Oligonucleotide probes are short (usually 15 to 30 bases) single-stranded DNA sequences complementary to their target sequence. Oligonucleotide probes were initially developed to identify genes of interest in basic molecular biology studies. Soon after, oligonucleotides were found to be applicable tools for identifying and detecting oral microorganisms (Moncla *et al.*, 1990; Dix *et al.*, 1990; Tsai *et al.*, 2003). Probes are highly selective under stringent conditions and precisely hybridize to their complementary target sequences which are usually distributed in complex mixtures of heterogeneous DNA of bacterial and human origin. Hybridization is visualized by markers that are either radioactive or fluorescent molecules tagged to the probes (Fig. 1.1). Phylogenetically informative oligonucleotide probes (e.g. 16S rRNA gene) enable precise taxonomic positioning of bacteria and are, therefore, widely used in medicine and microbial ecology. As hybridization probes, the oligonucleotides are specific but may suffer the problem of low sensitivity when the target bacteria are present at low levels in the bacterial sample.

Whole-genomic checkerboard DNA–DNA hybridization

The whole-genomic checkerboard DNA–DNA hybridization technique was first described by Socransky and colleagues (1994). It was developed from a previous checkerboard methodology for the detection of antigen-antibody cross-reactions (Kazemi and Finkelstein, 1990). In checkerboard DNA–DNA hybridization, bacterial whole genomic DNA is labelled with digoxigenin and then used as a hybridization probe (Fig. 1.1). The labelled probes from a variety of species are hybridized to a large number of denatured DNA samples on a nylon membrane. The hybridization is detected by using anti-digoxigenin antibody – alkaline phosphatase conjugate and incubation with a chemiluminescent substrate. An advantage of the checkerboard hybridization method is that large numbers of samples can be rapidly analysed for high levels of different bacteria, allowing a simultaneous analysis of more than 1600 reactions on each membrane. During the past 15 years the method has been used in large-scale epidemiological studies. The specificity of whole genomic probes is, however, lower than that of oligonucleotide probes due to shared genomic sequences with other bacteria. Additionally, whole-genomic checkerboard strategy requires cultivable bacteria for constructing the probes and is thus not suitable for studying not-yet-cultured bacteria.

It has, however, been extensively used for the screening of selected oral bacteria in supragingival and subgingival plaque, on oral mucosal and tongue surfaces, and in saliva (Socransky and Haffajee, 2005). The results have been compared between patients with and without periodontitis, before and after periodontal treatment, and in relation to geographic locations and various oral and systemic diseases (Mager *et al.*, 2005; Didilescu *et al.*, 2005; Renvert *et al.*, 2006).

Reverse-capture oligonucleotide hybridization

The reverse-capture oligonucleotide hybridization technique (Paster *et al.*, 1998) is a modification of the checkerboard hybridization method and, instead of conventional DNA extraction and whole-genomic probes, it involves PCR amplification of 16S rDNA from the sample and species-specific 16S rDNA oligonucleotide probes. The oligonucleotide probes are bound on a membrane via polythymidine tails leaving the probes available for hybridization to their digoxigenin-labelled target sequences in the sample.

When samples from caries-active and inactive subjects were compared by using a test panel of 82 bacterial species, an overabundance of *Actinomyces* sp., *S. mutans*, and *Lactobacillus* spp. was seen in caries-active subjects (Corby *et al.*, 2005). In another study, using 23 bacterial species in the test panel, *Actinomyces gerencseriae*, *Bifidobacterium* sp., and *S. mutans* were the bacteria suggested as playing a role in caries initiation in children (Becker *et al.*, 2002).

Recently, a modification of the reverse-capture oligonucleotide hybridization technique was introduced (Paster *et al.*, 2006). Instead of the checkerboard hybridization on a nylon membrane, the new method applies a microarray format, where a selection of up to 400 predominant cultivable or not-yet-cultured oral species/phylotypes can be detected simultaneously from bacterial samples.

DNA microarray

In contrast to the other nucleic acid-based techniques, DNA microarray technology provides an outstandingly powerful tool for analysing a massive number of nucleotide sequences in a single experiment (Schena *et al.*, 1995; Ye *et al.*, 2001; Garaizar *et al.*, 2006). DNA microarray has several advantages, such as its wide applicability in different areas of biological research and the enormous amount of information it generates. An extensive number of different bacteria or genetic characters of a test bacterium can be analysed simultaneously (Fig. 1.1). Microarray, equipped with advanced computer-based analysis and visualization tools can detect pathogenic bacterial species to an extent never before achieved with other techniques. Research effort is currently directed at establishing basic knowledge of the sets of gene targets to be used in clinical panel diagnostics (Vernet *et al.*, 2004) containing several diagnostic features in the same array. These could be used, for example, in the diagnosis of multibacterial infections, and for determining antibiotic resistance of bacteria. There are, however, challenges in the microarray-based analysis of complex bacterial communities. For example, there is a multitude of unknown bacteria that still lack probes in the arrays and there is the difficulty in obtaining, without DNA amplification, enough material from target bacteria found at low levels in a background of other bacteria.

DNA microarray consists of DNA hybridization reactions run as in the reverse-capture hybridization method described above. The probes for detection and identification of bacteria are generally genomic DNA or PCR products that provide phylogenetic or functional information. The oligonucleotide probes are fixed onto solid supports, such as nitrocellulose or nylon membranes or glass slides, which are commonly known as 'arrays' or 'chips' (Fig. 1.1). Nitrocellulose or nylon membranes can be used for hundreds of probes to form macroarrays or low-density microarrays. Contrarily, glass slides allow the deposition of tens to hundreds of thousands of probes on a single high-density microarray.

Although the DNA microarray technique was introduced as recently as 1995, it has already been widely applied throughout the biosciences, from microbiology to pharmacology and medicine. As the microarray was originally used as a tool for gene expression profiling and detection of specific mutations in DNA sequences, it is currently also employed in the detection and identification of food borne (Kostrzynska and Bachand, 2006) and water borne (Maynard *et al.*, 2005) pathogens and bacteria in mixed infections or contaminations (Tobler *et al.*, 2006). So far, DNA microarray has had limited diagnostic application in detection and identification of oral bacteria.

Fluorescence in-situ hybridization (FISH)
Fluorescence *in-situ* hybridization (FISH) was developed for the detection and identification of single microbial cells (DeLong *et al.*, 1989) (Fig. 1.1). It enables direct visualization and identification of individual bacteria in their natural habitats or in clinical specimens (Moter and Gobel, 2000; Wagner *et al.*, 2003; Bottari *et al.*, 2006). The oligonucleotide probes used in FISH are covalently linked at their 5′ end with fluorescent dyes. The probes can be designed to target phylogenetic group-specific sequences to distinguish between species, genera, or domains. In most studies, the target for FISH has been the 16S rRNA. Labelling each probe with a separate fluorescent dye allows identification and quantification of cells of several microbial species in the same sample.

In comparison with the afore-mentioned molecular techniques, FISH has so far been less used in oral microbiology. Recently, FISH was coupled with confocal laser scanning microscopy to simultaneously detect and determine the spatial organization of six Gram-positive and Gram-negative bacteria grown as biofilm *in vitro* (Thurnheer *et al.*, 2004) as well as to detect *Prev. intermedia* and *Prev. nigrescens* from subgingival samples (Gmur and Thurnheer, 2002). In periradicular lesions of asymptomatic teeth, FISH revealed polymicrobial bacterial consortia composed of *Treponema* spp, *P. gingivalis, Prev. intermedia, T. forsythia,* and *Streptococcus* spp. (Sunde *et al.*, 2003).

Bacterial community diversity in the oral cavity

Sampling and molecular techniques

Sampling is a crucial step for a rational outcome of microbial community analysis. In the oral cavity, there are various anatomic sites supporting different ecosystems. Thus, it is not likely that sampling a single oral site is representative of the oral microbiota and would reveal the extent and nature of oral microbial diversity. For instance, the availability of host-derived substances differs in various oral sites. In the subgingival area, bacteria are in close contact with nutrients and host defence molecules from blood circulation, whereas bacteria growing on supragingival tooth surfaces have easy access to saliva as the major source of host molecules. Regenerating oral mucosal surfaces provide a different environment for bacterial growth, because epithelial cell shedding affects the development and maturation of bacterial communities. On the other hand, sampling saliva provides other options. Besides the possibility to standardize the quantification of bacteria in the sample, saliva is a practical sampling material that gives an overview of bacteria living in the oral cavity.

Not only do the compositions of bacterial communities differ in various ecological environments in the oral cavity, they also differ among individuals. Factors affecting the individual oral ecosystem include oral and systemic health status, oral hygiene level, diet, medications, and presence or absence of teeth, as well as the genetic makeup of the individual. Thus, to answer the questions concerning the diversity of oral microbiota, not only do we need to obtain samples representative of the different ecosystems in the oral cavity, but a representative number of subjects also needs to be studied.

With the advent of molecular methods, there has been a great enthusiasm among oral microbiology researchers to identify and characterize the hard-to-culture or not-yet-cultured oral bacteria that have, despite considerable efforts, remained elusive for well over a century. The most obvious example is the clear difference in the detection of oral spirochetes and large motile rods by microscopy and bacterial culture. The microscopic findings were originally reported over 300 years ago when Antonie van Leeuwenhoek described the forms and movements of these 'animalcules' in samples scraped from tooth surfaces. Furthermore, as early as the end of the nineteenth century, Willoughby D. Miller, the 'father' of oral microbiology, suggested that these microorganisms, which he could not culture, are related to gum disease. But even though the disease-association of spirochetes has turned out to be increasingly recognized during the ensuing 100 years, most oral spirochetes and large motile rods remain non-cultivable. Thus, it is no wonder that describing the species diversity of oral spirochetes was among the first aims in oral microbiology when the 16S rDNA cloning and sequencing technique became available (Choi *et al.*, 1994). More holistic studies to detect and identify all bacteria in the oral bacterial community then rapidly followed.

Direct cloning or 'shot gun cloning' involves no PCR amplification and thus causes minimum bias in studies of bacterial community composition, since the total community DNA is cloned. However, this method is very labour-intensive and has not been applied in studies of oral bacterial diversity. Conversely, cloning and sequencing of taxonomically useful genes, particularly the 16S rRNA gene, has been widely used for the culture-independent detection and identification of bacteria from oral community samples. Furthermore, as compared with cloning and sequencing, less laborious and costly methods include bacterial community fingerprinting approaches, such as DGGE and T-RFLP (Fig. 1.2, Table 1.2). The DGGE method relies on sequence-specific DNA melting behaviour, whereas the T-RFLP approach is based on comparative sequence analysis and size polymorphism of a PCR-amplified marker sequence. Both techniques are capable of detecting not only cultivable but also not-yet-cultured bacteria. In the following sections, the background and recently reported findings for the cloning and sequencing of the 16S rRNA gene are dealt with, followed by a description of the principles, and some oral microbiology applications of, DGGE and T-RFLP.

16S rRNA gene cloning and sequencing

The 16S rRNA gene has proven to be the most useful phylogenetic marker among the macromolecules employed to identify bacteria and to determine their evolutionary relationships. Ribosomal RNA is essential for life and can be found in all bacteria. It contains nucleic acid sequences with regions that are highly conserved across bacterial species, reflecting absence of evolutionary divergence, and variable regions, which in turn reflect evolutionary divergence among different species. In addition, the 16S rRNA gene is large enough (about 1500 bases) to allow sufficient sequence variability among bacteria, thereby making comparisons possible at different taxonomic levels all the way from bacterial subspecies up to domain.

The practical approach for 16S rRNA gene cloning and sequencing entails isolation of bacterial DNA from a community sample followed by PCR amplification of the 16S rRNA genes

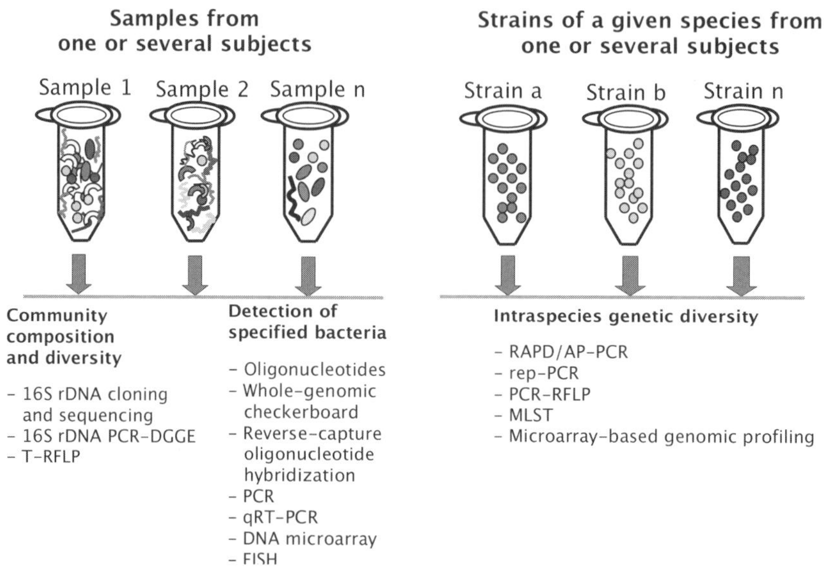

Figure 1.2 Examples of molecular techniques used for studying the composition and diversity of oral bacterial community samples; detection of specified bacteria from clinical samples; and comparison of different strains of the same species. For abbreviations, see Table 1.2. A colour version of this figure is located in the plate section at the back of the book.

Table 1.2 Examples of molecular techniques used for the analyses of oral microbiota

Method	Abbreviation	Principle	Reference
Whole-genomic checkerboard DNA–DNA hybridization	–	Hybridization of a selection of labelled whole-genomic DNA probes to sets of sample DNAs fixed on a membrane	(Socransky *et al.*, 1994)
Reverse-capture oligonucleotide hybridization	–	Hybridization of a selection of labelled PCR-amplified 16S rDNA segments from sets of samples to species-specific oligonucleotide probes fixed on a membrane	(Paster *et al.*, 1998)
Fluorescence *in situ* hybridization	FISH	*In-situ* hybridization of fluorescent-labelled 16S rDNA oligonucleotide probes to bacterial cell rRNA in the sample	(DeLong *et al.*, 1989)
DNA microarray	–	Hybridization of labelled DNA sequences in the sample to target-specific oligonucleotides fixed on a membrane/glass slide	(Schena *et al.*, 1995)
Real-time PCR	qRT-PCR	Quantitative monitoring of the formation of PCR product in real time by using special fluorescent dyes	(Higuchi *et al.*, 1993)
PCR-denaturing gradient gel electrophoresis	PCR-DGGE	Electrophoretic separation of PCR amplified 16S rDNA fragments through polyacrylamide gel with increasing gradient of denaturants	(Muyzer *et al.*, 1993)
Random amplified polymorphic DNA/ Arbitrarily primed-PCR	RAPD/ AP-PCR	PCR amplification of genomic DNA using a single random primer in low-stringency conditions and size separation by gel electrophoresis	(Welsh and McClelland, 1990; Williams *et al.*, 1990)
Repetitive element based-PCR	rep-PCR	PCR amplification using consensus primers for randomly dispersed repetitive palindromic sequences in genomic DNA and size separation by gel electrophoresis	(Versalovic *et al.*, 1991)
Multilocus sequence typing	MLST	Comparison of the sequences of PCR amplified housekeeping genes between test and reference strains	(Maiden *et al.*, 1998)
Ribosomal intergenic spacer analysis	RISA	PCR amplification of the spacer region between 16S and 23S rRNA genes and Southern blot hybridization with probes of interest	(Borneman and Triplett, 1997)
PCR-restriction fragment length polymorphism	PCR-RFLP	Restriction digestion of the PCR amplified target DNA and fragment size separation by gel electrophoresis	(Vaneechoutte *et al.*, 1995)
Pulsed field gel electrophoresis	PFGE	Restriction digestion of genomic DNA with infrequently cutting restriction enzymes followed by gel electrophoresis by alternating the direction of the electric field	(Schwartz and Cantor, 1984)
Terminal restriction fragment length polymorphism	T-RFLP	Restriction digestion of PCR amplified 16S rDNA using fluorescent-labelled forward primer and detection by DNA sequencer	(Liu *et al.*, 1997)

present in the sample by using universal primers targeting all prokaryotes, the domains *Bacteria* or *Archaea*, or specific bacterial groups. The PCR amplicons are purified and ligated into a plasmid vector, and *Escherichia coli* is transfected with the plasmid DNA. The insert 16S rDNA is amplified from randomly chosen clones (usually 50–100 clones) and sequenced (usually 300–1000 bases). Sequence data are compared with the existing information in databases, followed by phylogenetic analysis techniques which rely on tree graphs as the main device for the interpretation of the data. The rapidly growing Ribosomal Database (RDP) (http://rdp.cme.msu.edu/), as of early April, 2007, contains 351 796 16S rRNA gene sequences and seven online analysis tools.

It is important to note that, in using online 16S rRNA gene sequence comparisons, the 16S rRNA gene sequences from all cultivable bacteria have not been deposited in the databases and so non-matching results do not necessarily mean that the bacterium has not been cultured. Moreover, as discussed above, the PCR-based techniques suffer from possible biases due to disproportionate lysis of different types of bacteria, primers used, and differential target-gene amplification. An additional pitfall may arise from the possibility that different cloning systems may lead to different results in bacterial distribution in the same DNA sample (von Wintzingerode *et al.*, 1997). Furthermore, PCR amplification of 16S rDNA from mixed bacterial populations favours bacteria with a higher copy number of rRNA gene operons in their genomes (http://rrndb.cme.msu.edu/). This may have an effect on the identification and quantification of different bacteria in the community samples.

In most studies using the 16S rRNA gene cloning and sequencing technique to investigate oral bacterial community diversity, distinct oral sites have been sampled. These include subgingival (Muyzer *et al.*, 1993; Kroes *et al.*, 1999; Paster *et al.*, 2001; Sakamoto *et al.*, 2002; Kumar *et al.*, 2003; Kumar *et al.*, 2005; Kumar *et al.*, 2006) and supragingival dental surfaces (Becker *et al.*, 2002), tongue surface (Kazor *et al.*, 2003) and root canals (Siqueira and Rocas, 2005). Only rarely have several sites in the same individual or salivary samples been analysed (Aas *et al.*, 2005). Studies also differ from each other with regard to the PCR primer sets chosen for the amplification of the 16S rDNA from the sample, the number of clones sequenced in each sample, and the number of bases in each sequence. The primers have broadly targeted the domain *Bacteria* (Paster *et al.*, 2001; Kumar *et al.*, 2003; Kumar *et al.*, 2005; Kumar *et al.*, 2006), *Archaea* (Lepp *et al.*, 2004; Vickerman *et al.*, 2007) or bacterial groups, such as *Spirochaetales* (Kroes *et al.*, 1999; Paster *et al.*, 2001) and *Bacteroidetes* (Paster *et al.*, 2001).

Although the currently available 16S rDNA sequence data of oral bacteria derive from a relatively limited number of individuals with varying oral status, over 700 different phylotypes have been identified; among them close to 400 phylotypes from subgingival samples (Paster *et al.*, 2006). More than half of the phylotypes represent bacteria that have not previously been cultured or characterized. This indicates that all cultivated oral bacteria (509 taxa) (Moore and Moore, 1994) have not been found by molecular methods and that the human oral microbiota may comprise at least 1000 different bacteria.

The bacterial genera found by comparative 16S rDNA sequence analysis in periodontal samples belong to seven bacterial phyla (Paster *et al.*, 2001; Kumar *et al.*, 2005) and in endodontal samples to five bacterial phyla (Munson *et al.*, 2002; Vickerman *et al.*, 2007), according to the phylum designation in Bergey's Manual of Systematic Bacteriology. In both types of infections, the *Firmicutes* (class *Clostridia*, e.g. *Peptostreptococcus*, *Selenomonas*, *Dialister*, *Filifactor*, and class *Bacilli*, e.g. *Streptococcus*, *Lactobacillus*) was the predominant phylum. Other phyla found in the oral samples included *Proteobacteria* (e.g. *Aggregatibacter*, *Neisseria*, *Haemophilus*, *Campylobacter*), and *Bacteroidetes* (e.g. *Porphyromonas*, *Prevotella*, *Capnocytophaga*), *Actinobacteria* (e.g. *Actinomyces*, *Rothia*), *Spirochaetes* (e.g. *Treponema*), *Fusobacteria* (e.g. *Fusobacterium*), *Synergistes*, and *Deferribacteres*. In addition, sequences of the candidate phyla TM7 and Obsidian pool OP11 can also occur in oral samples. Rather surprisingly, archaeal strains, belonging to a single phylum and genus (*Methanobrevibacter*), have been recently found in periodontal (Lepp *et al.*, 2004) and endodontal samples (Vickerman *et al.*, 2007).

Comparison of the phylum-level diversity between oral bacteria and all bacteria catalogued in designated prokaryotic phyla in Bergey's Manual of Systematic Bacteriology (second edition, 2004) shows that 17 of the 24 phyla ascribed to the domain *Bacteria* have not been found in the oral cavity. Interestingly, the most intensively studied bacterial genera, all of which are important pathogens in human infections, are members of only three bacterial phyla: i.e. *Firmicutes* (*Bacillus*, *Streptococcus*, *Staphylococcus*), *Proteobacteria* (*Escherichia*, *Helicobacter*, *Pseudomonas*, *Salmonella*), and *Actinobacteria* (*Mycobacterium*) (Hugenholtz, 2002).

Relative abundance of bacteria in community samples

As determined by culture studies, the dental diseases are multibacterial infections. These bacteria belong to the indigenous, or 'resident', oral bacteria. Dental disease pathogens can be also found in healthy subjects although usually less frequently and in lower proportions of the cultivable bacterial flora than in disease. Thus, in addition to cataloguing oral bacteria occurring in various clinical conditions it is important to know the relative abundance of individual bacteria in the oral bacterial communities to understand, more fully, their function and role in health and disease.

Accordingly, questions have arisen as to the relative abundance and stability of bacteria in subgingival samples from patients with different periodontal status. Through standardizing various steps in the 16S rDNA cloning and sequencing procedures, an attempt was made to determine the original proportions of bacteria in the various samples (Kumar *et al.*, 2003; Kumar *et al.*, 2005; Kumar *et al.*, 2006). Most (>90%) of the nearly 300 bacteria identified occurred in proportions lower than 1% of the total number of clones studied. Comparisons of the prevalence and proportion of the identified bacteria between the study subjects showed that Gram-positive bacteria predominated in periodontitis and Gram-negative bacteria in health (Kumar *et al.*, 2003; Kumar *et al.*, 2005). The phylum *Firmicutes* (class *Clostridia*) dominated the subgingival flora both in periodontal health and disease. The prominent species/phylotypes, namely, Gram-positive anaerobes *F. alocis* (previously *Fusobacterium alocis*, reclassified to the family *Peptostreptococcaceae*) and *Peptostreptococcus* sp. and a Gram-negative anaerobe *Selenomonas* sp., occurred in higher proportions of the total clones in periodontitis; whereas a Gram-negative anaerobe, *Veillonella* sp., predominated in healthy sites (Kumar *et al.*, 2003; Kumar *et al.*, 2006). In contrast, the results obtained from culture studies have shown quite the opposite, indicating a predominance of Gram-negative bacteria in periodontitis and Gram-positive bacteria in health (Socransky, 1977; Slots, 1979; Socransky and Haffajee, 1992).

Detection of recently identified 'new' bacteria in oral samples

Molecular methods can be applied to study the prevalence and distribution of hard-to-culture or not-yet-cultured bacteria in clinical samples; for example, by constructing PCR primers or hybridization probes according to the 16S rDNA sequence information obtained from 16S rDNA cloning and sequencing studies and deposited in the RDP database (Harper-Owen *et al.*, 1999; Brinig *et al.*, 2003). Two examples are presented below to illustrate the role of various molecular techniques in detecting 'new' disease-associated bacteria.

The clone TM7 (TM for **T**orf, **m**ittlere Schicht = peat, middle layer) was first found in a German peat bog and then later on, in other environmental samples by using the 16S rDNA cloning and sequencing approach (Rheims *et al.*, 1996). As shown by the FISH method, the TM7 sequence derived from a sheathed-filament with Gram-positive ultrastructure (Hugenholtz *et al.*, 2001; Pei *et al.*, 2004). The clone TM7 was regarded uncultivable until micromanipulation of the bacterial filaments resulted in its laboratory growth (Hugenholtz, 2002). Unexpectedly, recent molecular studies revealed TM7 sequences in clinical habitats, such as subgingival sites, tongue surface (Paster *et al.*, 2001; Kumar *et al.*, 2003; Kazor *et al.*, 2003) and oesophagus (Pei *et al.*, 2004). In addition, qRT-PCR and FISH showed that bacteria related to TM7 are widespread in both healthy and diseased oral sites, and that a subgroup of TM7 was more common in samples from periodontitis than healthy sites (Brinig *et al.*, 2003). The above results, indicating that an environmental bacterium can frequently occur in oral samples, differs from the earlier belief that the human oral cavity does not support the growth of environmental bacteria.

The clone PUS9.180 is another example of molecular-based detection of 'new' oral bacteria. This clone was also found by the 16S rDNA cloning and sequencing approach in samples from oral abscesses and it was identified as a new species belonging to the genus *Prevotella* (Dymock *et al.*, 1996; Wade *et al.*, 1997). Later on, bacterial culture of samples from endodontal abscesses produced isolates that were identified by 16S rDNA sequence analysis as representatives of the new *Prevotella* sp. (clone PUS9.180) and were characterized as Gram-negative, non-pigmented anaerobic rods. Furthermore, the isolates exhibited higher sensitivity to oxygen than the phylogenetic relatives *Prevotella oralis* and *Prev. intermedia* (Khemaleelakul *et al.*, 2002). While *Prevotella* and *Streptococcus* were the most frequent genera in the studied abscess samples, the new *Prevotella* sp. (clone PUS9.180) appeared to be one of the most common *Prevotella* species.

Denaturing gradient gel electrophoresis

Denaturing gradient gel electrophoresis (DGGE) was introduced in 1993 for the analysis of bacterial communities (Muyzer *et al.*, 1993). It is a powerful molecular technique that can, at least theoretically, reveal single-base differences between DNA sequences. The DGGE method is based on an electrophoretic separation of PCR amplicons run through polyacrylamide gel containing a linear gradient of DNA denaturants, such as urea and formamide. The greater the GC content of the DNA fragment, the higher denaturant concentration that is needed for it to melt. The resolution efficiency of the technique is enhanced by attaching a long GC-tail (GC clamp) through PCR incorporation to one strand of the test DNA. The GC-rich tail changes the melting conditions of the fragment so that the DNA strands do not totally separate in the GC-clamp end. The partial melting of the DNA strands reduces the migration speed in the gel until the fragments stop migrating and form an individual fingerprint for each sample.

The DGGE technique is suitable for a variety of purposes and the 16S rDNA sequences are the most commonly used targets. Different taxonomic levels, such as species, genus, or domain, can be chosen by selecting primers for the target DNA sequence of interest. The DGGE method has been widely used to study bacterial community diversity, to monitor community dynamics and to identify predominant species in community samples. Taxonomic information can be obtained by excising bands of interest and by re-amplifying, cloning, and sequencing them. Another variation of DGGE involves an additional Southern blot analysis in which the gel bands are transferred to a membrane and then probed for sequences of interest. DGGE gel expansion (DGGEGE) is a recently reported modification to study whether different sequences co-migrate on the gel (Gafan and Spratt, 2005). This method involves a post-DGGE extension step, where individual bands are excised and run through a shorter denaturant gradient.

In the field of oral microbiology, DGGE has been used for bacterial community profiling of supragingival samples (Li *et al.*, 2006; 2007), monitoring interindividual variability and temporal changes in the bacterial composition of saliva and dental plaque (Rasiah *et al.*, 2005; Ledder *et al.*, 2006), detecting traditional periodontal pathogens from subgingival samples (Zijnge *et al.*, 2006) and analysing root canal microbiota after failed endodontal treatment (Siqueira and Rocas, 2005). Interestingly, no difference in the species diversity was found with DGGE analysis between periodontal health and disease (Ledder *et al.*, 2007). Furthermore, the DGGE analysis of salivary and dental plaque bacteria after *in vitro* exposure to antimicrobial agents represents the first study to use culture-independent methods for determining antimicrobial susceptibility of oral bacterial communities (McBain *et al.*, 2003a,b).

Terminal-restriction fragment length polymorphism (T-RFLP)

The terminal-restriction fragment length polymorphism (T-RFLP) method was introduced at the end of 1990s (Liu *et al.*, 1997). The method involves restriction enzyme digestion of PCR-amplified 16S rRNA gene. The PCR amplification entails fluorescent-labelled forward primer, which allows detection of the digests by a DNA sequencer. The method can be used for bacterial community profiling as well as for monitoring changes over time or after known environmental alterations. Since the locations of the restriction sites are not random, but have a phylogenetic background, T-RFLP provides knowledge of the taxonomic position of the test bacteria. However, T-RFLP, like DGGE, may underestimate the species diversity in a community sample because the numerically low bacterial populations are not represented in the PCR product.

So far, there are few studies using T-RFLP in the field of oral microbiology. Sakamoto and collaborators studied the diversity and stability of salivary bacteria in patients with and without periodontitis (Sakamoto *et al.*, 2003); and, subsequently, changes in saliva and subgingival plaque after treatment of periodontitis (Sakamoto *et al.*, 2004). The results showed that the populations of *P. gingivalis*, *Prev. intermedia*, and *Peptostreptococcus* sp. were markedly lower after periodontal treatment. It also appeared that the T-RFLP results were in agreement with the results from qRT-PCR and 16S rDNA cloning and sequencing performed on the same samples. In samples obtained from endodontal infections, T-RFLP demonstrated a greater bacterial diversity than that found by culture. In addition, bacterial diversity was greater in symptomatic than asymptomatic endodontal infections (Sakamoto *et al.*, 2006).

Bacterial typing

In contrast to phenotypic methods, nucleic acid-based genotyping is independent of gene expression and enables typing of all bacterial strains. Genotyping thus provides a more reliable tool than phenotyping in studying bacterial epidemiology, for example. Identification of a specified clonal type, a genotype, facilitates its tracing in human populations, making it possible to determine its prevalence, distribution, and association with disease. Disease or health-associated bacterial clonal types can then be separated from each other and pathogenic characteristics of the disease-associated types defined. In addition to separating strains of the same bacterial species, genotyping can also be used to rapidly distinguish between strains of species that are difficult to classify with conventional methods. Examples include oral streptococci (Alam *et al.*, 1999) and *Prev. intermedia* and *Prev. nigrescens* (Matto *et al.*, 1996).

A variety of genotyping techniques has been developed to distinguish between bacterial strains and species (Olive and Bean, 1999; Bensch and Akesson, 2005) and, during the past decade, these techniques have gained extensive use for typing human pathogens. Selected genotyping methods, based on DNA sequence data and on gel electrophoresis fingerprinting (Fig. 1.3), are introduced in the following sections.

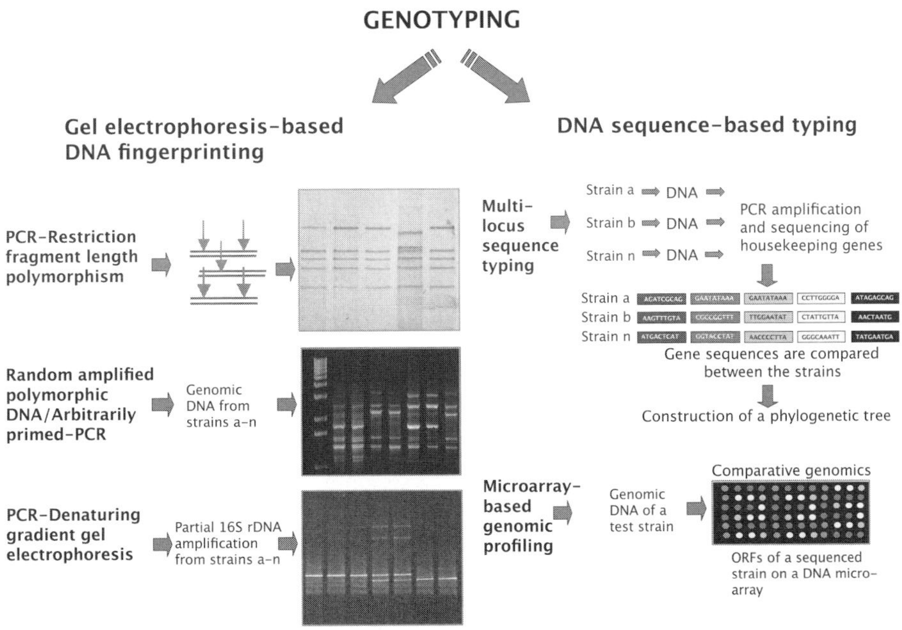

Figure 1.3 Examples of molecular techniques used for the genotyping of bacterial strains. A colour version of this figure is located in the plate section at the back of the book.

DNA sequence-based typing

DNA microarray-based genomic profiling

Although whole-genome sequencing is the ultimate genotyping tool, it is an extensive and costly procedure. Recently, microarray-based genomic profiling has become possible for genome comparisons between strains of the same species without the requirement of whole-genome sequencing of the test strains. Microarray-based comparative genomics has been used to study genetic diversity within human pathogens, such as *E. coli*, *Shigella*, *Helicobacter*, and *Salmonella* (Salama *et al.*, 2000; Fukiya *et al.*, 2004; Anjum *et al.*, 2005), and dental disease-associated bacteria, *P. gingivalis* (Duncan, 2003; Chen *et al.*, 2004b; Chen *et al.*, 2005) and *S. mutans* (Waterhouse *et al.*, 2007). A whole genome microarray is currently available for two additional oral bacteria: *A. actinomycetemcomitans* (HK1651) and *T. denticola* (ATCC 35405) (http://array.brop.org/).

In the field of oral microbiology, the first results from microarray-based comparisons of the genomes of two *P. gingivalis* strains, W83 (virulent) and ATCC 33277 (avirulent), showed that 7% of open reading frames were divergent (Chen *et al.*, 2004a). For instance, a cluster of open reading frames involved in capsular polysaccharide synthesis was absent in the avirulent strain. The finding was interesting, since the presence of capsule has been related to increased *P. gingivalis* virulence. A genomic comparison of nine *S. mutans* test strains with the reference strain (UA159) demonstrated that 20% of open reading frames were absent in one or several test strains, suggesting that one-fifth of the open reading frames of the strain UA159 are not required for the survival of *S. mutans* (Waterhouse *et al.*, 2007).

Multilocus sequence typing (MLST)

Multilocus sequence typing (MLST) (Maiden *et al.*, 1998) is a modification of multilocus enzyme electrophoresis, but is based on the analysis of allelic polymorphism in housekeeping genes instead of the electrophoretic mobility of their gene products. It provides a tool to use phylogenetic markers other than rRNA genes. Commonly, 6–10 housekeeping gene loci are PCR amplified and sequenced (*ca.* 500 bases), and the allelic profiles of the sequences are compared between bacterial strains. Each bacterial isolate is characterized by a series of integers corresponding to the number of genes studied. Since MLST is based on nucleotide sequences the results are more easily compared between laboratories than the results of the gel electrophoresis band-based techniques. A sequence database has been constructed for the online MLST analysis of medically important pathogens (http://www.mlst.net). In oral microbiology, MLST analysis has been used to investigate the population structure of *P. gingivalis* (Koehler *et al.*, 2003; Enersen *et al.*, 2006).

As MLST is used for distinguishing strains of a known species, multilocus sequence analysis (MLSA) is a method used for the genotypic characterization of diverse groups of bacteria, even entire genera (Gevers *et al.*, 2005). The MLSA method is suitable both for establishing the phylogenetic position of new species and for resolving species clusters without trying to infer their phylogenetic relationships (Hanage *et al.*, 2006).

Gel electrophoresis-based DNA fingerprinting

Bacterial DNA fingerprinting usually relies on electrophoretic size separation of the DNA fragments that derive from genomic DNA (e.g. RAPD/AP-PCR or rep-PCR) or locus-specific DNA (e.g. PCR-RFLP and RISA) (Table 1.2, Fig. 1.3). The DNA fragments are produced by PCR amplification or restriction enzyme digestion. Fragments of different sizes form strain-specific band patterns, 'fingerprints', on the gel. Some fingerprinting methods (e.g. rep-PCR and RISA) have proven amenable to standardization and automation.

Fingerprinting techniques allow simultaneous screening and rapid comparison of several test strains. With regard to technical and cost considerations, fingerprinting is a more practical strategy for strain comparisons than is comparative genomics. As MLST requires gene sequence data from the test and reference strains for each species studied, fingerprinting provides a more rapid and quite easy analysis of genotype differences among bacterial strains.

Since the beginning of the 1990s, bacterial genotyping by various fingerprinting methods has been widely used to distinguish between strains of oral bacteria and to determine the extent of genetic diversity of species in individual oral cavities (Loos *et al.*, 1992; Alaluusua *et al.*, 1996; Alpha *et al.*, 2001; Alves *et al.*, 2006); in groups of individuals with and without periodontitis (DiRienzo *et al.*, 1990; Loos *et al.*, 1990; Preus *et al.*, 1993; Slots *et al.*, 1993; Van Steenbergen *et al.*, 1993; Leys *et al.*, 1994; Chen and Slots, 1994; Lotufo *et al.*, 1994; Asikainen *et al.*, 1995; Matto *et al.*, 1996; Macheleidt *et al.*, 1999; Kaplan *et al.*, 2002; Huang *et al.*, 2003) or caries (Saarela *et al.*, 1996; Li *et al.*, 2005), from different geographic locations (Menard and Mouton, 1995; Asikainen *et al.*, 1997; He *et al.*, 1998; Dogan *et al.*, 2003; Toi *et al.*, 2005; Caufield *et al.*, 2007), and ethnic and racial backgrounds (Haubek *et al.*, 1997; Umeda *et al.*, 1998). There is a wide consensus that populations of oral bacterial species exhibit extensive genetic heterogeneity and that, within individual subjects, there is varied heterogeneity among the strains of these species.

According to the infectious disease concept, also oral microbiologists have been interested in finding the source and route of infection for dental pathogens. The family model has mainly been used to study vertical and horizontal person-to-person transmission of dental disease-associated bacteria (Saarela *et al.*, 1993; Van Steenbergen *et al.*, 1993; Matto *et al.*, 1996; Asikainen *et al.*, 1996; von Troil-Linden *et al.*, 1996; van Steenbergen *et al.*, 1997; Fukui *et al.*, 1999; Li *et al.*, 2000; Kononen *et al.*, 2000) and the time (Kononen *et al.*, 1994; Kononen *et al.*, 2000; Hames-Kocabas *et al.*, 2006) and stability (Alaluusua *et al.*, 1994; Saarela *et al.*, 1999; Koehler *et al.*, 2003; Asano *et al.*, 2003) of their colonization in the oral cavity.

In the following sections, selected fingerprinting methods will be described.

Random amplification of polymorphic DNA/ arbitrarily primed PCR (RAPD/AP-PCR)

Two similar approaches for typing bacterial genomic DNA were independently developed by two research groups. One method is called random amplification of polymorphic DNA (RAPD) (Williams *et al.*, 1990) and the other one is known as arbitrarily primed PCR (AP-PCR) (Welsh and McClelland, 1990). In both PCR techniques a short single primer of arbitrary nucleotide sequence is used to amplify genomic DNA of the study strains. Thus, in contrast to many other techniques, RAPD/AP-PCR do not require previous sequence information of the DNA to be typed. In RAPD/AP-PCR the primer-annealing conditions are non-stringent and enable primer annealing to unknown target DNA sequences of low complementarity. PCR products of differing lengths are amplified and separated using gel electrophoresis. As a result, the gel bands form 'fingerprint patterns', specific for bacterial strains or species. The methods are easy to use and a universal set of random primers is available for the genomic analysis of a wide variety of bacteria.

Numerous studies have demonstrated the good applicability of RAPD/AP-PCR in periodontitis-associated bacteria, such as *A. actinomycetemcomitans* (Preus *et al.*, 1993; Slots *et al.*, 1993; He *et al.*, 1998; Saarela *et al.*, 1999; Kaplan *et al.*, 2002; Dogan *et al.*, 2003), *P. gingivalis* (Menard and Mouton, 1993; Chen and Slots, 1994), *Prev. intermedia* and *Prev. nigrescens* (Matto *et al.*, 1996; Fukui *et al.*, 1999; Kononen *et al.*, 2000; Alves *et al.*, 2006), *F. nucleatum* (George *et al.*, 1997), *T. forsythia* (Lotufo *et al.*, 1994; Huang *et al.*, 2003), *Eikenella corrodens* (Lotufo *et al.*, 1994; Huang *et al.*, 2003; Fujise *et al.*, 2004), and *Peptostreptococcus micros* (Alpha *et al.*, 2001). Similarly, good performance of RAPD/AP-PCR has been shown for distinguishing caries-associated species, such as *S. mutans* and *S. sobrinus* (Saarela *et al.*, 1996; Truong *et al.*, 2000; Li *et al.*, 2001; Kamiya *et al.*, 2005), *Actinomyces* spp. (Ruby *et al.*, 2002) and strains of other oral streptococci, such as mitis group (Rudney and Larson, 1999; Pan *et al.*, 2001).

Despite the good intralaboratory reproducibility of RAPD/AP-PCR when standardized protocols are used (Dogan *et al.*, 1999; Paju *et al.*, 2000; Lakio *et al.*, 2002), the reproducibility of the band patterns between laboratories seems to be poor, probably because the low-stringency conditions used in PCR amplification are easily affected by different working environments.

Repetitive element-based PCR (rep-PCR)

Prokaryotic microorganisms have non-coding repetitive palindromic sequences that are randomly dispersed in their genomes. Repetitive extragenic palindromic (REP) elements, or palindromic units, were first utilized for bacterial genotyping by Versalovic and co-workers (1991). Enterobacterial Repetitive Intergenic Consensus (ERIC) sequences (Hulton *et al.*, 1991), not related to the REP elements, comprise another family of repetitive elements in bacteria. Both REP and ERIC elements are efficient binding sites for PCR consensus primers (Versalovic *et al.*, 1991). The practical procedures of the rep-PCR are straight-forward and involve extraction of template DNA, using rapid methods (Woods *et al.*, 1993), from test bacteria, followed by PCR amplification and amplicon separation and visualization by gel electrophoresis. Another benefit of rep-PCR is that the same primer pair can be used for various species, whereas the primers in AP-PCR, for example, have to be optimized separately for each species (Rudney and Larson, 1999). An automated rep-PCR-based bacterial DNA typing system uses fluorescent-labelled primers for amplification, allowing genotype profile analysis by a DNA sequencer (Healy *et al.*, 2005).

In oral bacteria, a high discriminative ability of rep-PCR genotyping has been shown among strains of *P. gingivalis* (Teanpaisan and Douglas, 1999), *F. nucleatum* (Moraes *et al.*, 2002), and various oral streptococcal species (Alam *et al.*, 1999), as well as among *Streptococcus oralis* strains (Alam *et al.*, 2000). As a result, this technique has been used in studies addressing various questions in the fields of oral bacterial epidemiology and micro-ecology. With regard to oral streptococci, numerous temporally unstable *S. oralis* genotypes were found in the oral cavity of each studied subject (Paddick *et al.*, 2003). The genotypes were not shared, except among some co-habiting subjects. Furthermore, the diversity of *S. oralis* was higher in caries-active than caries-free subjects. This supported the notion that an increased environmental stress (caries) leads to a greater genetic heterogeneity within bacterial species involved. Similarly, a positive association was found between the number of *Prev. intermedia* genotypes and periodontal pocket depth, when AP-PCR and heteroduplex methods were used for genotyping (Alves *et al.*, 2006).

Restriction fragment length polymorphism (RFLP)

DNA restriction enzymes catalyse cleavage of double-stranded DNA through the recognition of specific nucleotide sequences (usually 4–6 bp long) called 'restriction sites'. The resulting DNA fragments can be displayed by separating them, according to size, by gel electrophoresis. Unless low frequency cutting restriction enzymes and special running conditions are used, as in pulsed field gel electrophoresis (PFGE), the restriction digests of genomic DNA create a multitude of gel bands, making comparisons between strains difficult. As with other gel electrophoresis-based techniques, the restriction analysis can be complemented by probe hybridization. The gel bands are transferred to a membrane and probed for specific sequences of interest. Ribotyping presents an example in which known nucleotide sequences are used as probes (Bingen *et al.*, 1994). For instance, *E. coli* rRNA genes can be used as a universal probe to study restriction patterns of a variety of bacterial genomes (Grimont and Grimont, 1986; Stull *et al.*, 1988). Because of its discriminative power and consistency of the band patterns, ribotyping has also been used for typing oral bacteria (Bowden *et al.*, 1993; Saarela *et al.*, 1993; Kononen *et al.*, 1994; Fiehn *et al.*, 1995; Saarela *et al.*, 1999; Pearce *et al.*, 2000). However, due to the relatively laborious Southern blot method, PCR-based methods have largely displaced genotyping by hybridization probes.

As mentioned above, PFGE technique is a variation of RFLP (Table 1.2). The method was first described by Schwartz and Cantor (1984). Even though the DNA sequencing-based methods are becoming the methods of choice, PFGE is still regarded as the 'gold standard' in bacterial molecular epidemiology. Contrary to the restriction analyses described above, in PFGE, the genomic DNA is digested using restriction enzymes that cleave DNA at low frequency and thus separate large pieces of the genome (30–10 000 kb). The resolution of the large DNA fragments run through the gel is improved by alternating the direction of the electric field. The method has been extensively used in medical microbiology and also in a number of studies typing oral bacteria, such as *S. mutans* and *S. sobrinus*

(Okahashi *et al.*, 1990; Kiska and Macrina, 1994; Motegi *et al.*, 2006), mitis streptococci (Konig *et al.*, 1998), *P. gingivalis* and *Prev. intermedia* (Lewis and Macrina, 1999; Sumida *et al.*, 2002; Asano *et al.*, 2003), *T. forsythia* (Persson *et al.*, 2000), *F. nucleatum* (George *et al.*, 1997), *A. actinomycetemcomitans* (Valcarcel *et al.*, 1997), and *E. corrodens* (Steffens *et al.*, 1994).

Other RFLP modifications target specific genes or DNA sequences rather than bacterial genomic DNA. In PCR-RFLP, the target DNA is PCR-amplified and the amplicons are digested by selected restriction enzyme(s) followed by size separation of the digests by gel electrophoresis. The PCR-RFLP requires only a small amount of sample DNA and yields easily comparable gel patterns among the bacterial strains. Amplified rDNA restriction analysis is a variation of PCR-RFLP and is based on the restriction patterns of PCR-amplified rDNA; for example, the 16S subunit (Vaneechoutte *et al.*, 1995a) or the 16S-23S intergenic spacer region (Barsotti *et al.*, 2002), which allow phylogenetic inference from the fingerprints.

In the oral microbiology field, the 16S rDNA PCR-RFLP method (*rrn*-ARDRA) has been used, for example, to distinguish between strains of various oral species that are difficult to identify. The benefit of the method is that the sequence data in the 16S rDNA database can be used, in prior testing, for the discrimination capability of different restriction enzymes and for the expected sizes of the restriction fragments. The method was successful in distinguishing between reference strains of five cultivable oral treponemas (Sato and Kuramitsu, 1999), seven mutans streptococcal species (Schlegel *et al.*, 2003), and five *Capnocytophaga* species (Ciantar *et al.*, 2001).

Ribosomal intergenic spacer analysis (RISA)

Between the 16S and 23S rRNA genes, there is a heterogeneous spacer region that contains genes encoding several tRNAs and direct repeat sequences (Bacot and Reeves, 1991). Ribosomal intergenic spacer analysis (RISA) is based on species-specific length and sequence variation in the spacer region (Borneman and Triplett, 1997). PCR amplification is performed using primers chosen from conserved regions flanking the spacer region. Gel electrophoresis separation, according to amplicon size, is followed by Southern blot hybridization with probes for sequences of interest. RISA has been successfully used for distinguishing between closely related bacterial species and for taxonomic studies at the genus or domain level. As of early 2007, there were 2614 ribosomal spacer sequences deposited in the Ribosomal Intergenic Spacer Sequence Collection (RISSC) (http://egg.umh.es/rissc).

A modification of the RISA method has been used to determine the genetic variability among *P. gingivalis* strains directly from subgingival samples. The spacer region between the 16S and 23S ribosomal genes was PCR-amplified using species-specific primers and the amplicons then directly compared, using the heteroduplex analysis, with those of the known *P. gingivalis* strains. When the results were compared with the RISA sequence phylogeny, a relationship was found between disease-association and phylogeny of certain strains (Rumpf *et al.*, 2000).

Future considerations

Molecular techniques have come to stay in most oral microbiology laboratories. Without underestimating the importance of maintaining conventional microbiological methods and skills in the field of oral microbiology, molecular techniques greatly enhance the possibilities of addressing a variety of questions about the composition, structure, and function of complex bacterial populations and their interaction with the host in oral health and disease. In particular, among the methods described above, microarray technology, coupled with advanced computer-based data analysis, has endless potential as a tool for resolving intricate bacteria-host relationships. The newly opened era of molecular oral microbiology could be used for the development and implementation of novel diagnostic tools that simultaneously test for both microbial and host variables, thereby leading to a better understanding of oral infections.

Acknowledgements

We thank Casey Chen for fruitful discussions and useful comments on the manuscript.

References

Aas, J.A., Paster, B.J., Stokes, L.N., Olsen, I., and Dewhirst, F.E. (2005) Defining the normal bacterial flora of the oral cavity. J. Clin. Microbiol. *43*, 5721–5732.

Abu Al-Soud, W., and Radstrom, P. (1998) Capacity of nine thermostable DNA polymerases to mediate DNA amplification in the presence of PCR-inhibiting samples. Appl. Environ. Microbiol. *64*, 3748–3753.

Alaluusua, S., Alaluusua, S.J., Karjalainen, J., Saarela, M., Holttinen, T., Kallio, M., Holtta, P., Torkko, H., Relander, P., and Asikainen, S. (1994) The demonstration by ribotyping of the stability of oral *Streptococcus mutans* infection over 5 to 7 years in children. Arch. Oral Biol. *39*, 467–471.

Alaluusua, S., Matto, J., Gronroos, L., Innila, S., Torkko, H., Asikainen, S., Jousimies-Somer, H., and Saarela, M. (1996) Oral colonization by more than one clonal type of mutans streptococcus in children with nursing-bottle dental caries. Arch. Oral Biol. *41*, 167–173.

Alam, S., Brailsford, S.R., Adams, S., Allison, C., Sheehy, E., Zoitopoulos, L., Kidd, E.A., and Beighton, D. (2000) Genotypic heterogeneity of *Streptococcus oralis* and distinct aciduric subpopulations in human dental plaque. Appl. Environ. Microbiol. *66*, 3330–3336.

Alam, S., Brailsford, S.R., Whiley, R.A., and Beighton, D. (1999) PCR-based methods for genotyping viridans group streptococci. J. Clin. Microbiol. *37*, 2772–2776.

Allaker, R.P., Seddon, S.V., Tredwin, C., and Lynch, E. (1998) Detection of *Streptococcus mutans* by PCR amplification of the *spaP* gene in teeth rendered caries free. J. Dent. *26*, 443–445.

Alpha, C.X., Guthmiller, J.M., Cummings, H.E., Schomberg, L.L., and Noorani, S.M. (2001) Molecular analysis of *Peptostreptococcus micros* isolates from patients with periodontitis. J. Periodontol. *72*, 877–882.

Alves, A.C., Napimoga, M.H., Klein, M.I., Hofling, J.F., and Goncalves, R.B. (2006) Increase in probing depth is correlated with a higher number of *Prevotella intermedia* genotypes. J. Periodontol. *77*, 61–66.

Amann, R.I., Ludwig, W., and Schleifer, K.H. (1995) Phylogenetic identification and *in situ* detection of individual microbial cells without cultivation. Microbiol. Rev. *59*, 143–169.

Amano, A., Nakagawa, I., Kataoka, K., Morisaki, I., and Hamada, S. (1999) Distribution of *Porphyromonas gingivalis* strains with *fimA* genotypes in periodontitis patients. J. Clin. Microbiol. *37*, 1426–1430.

Anjum, M.F., Marooney, C., Fookes, M., Baker, S., Dougan, G., Ivens, A., and Woodward, M.J. (2005) Identification of core and variable components of the *Salmonella enterica* subspecies I genome by microarray. Infect. Immun. *73*, 7894–7905.

Asano, H., Ishihara, K., Nakagawa, T., Yamada, S., and Okuda, K. (2003) Relationship between transmission of *Porphyromonas gingivalis* and *fimA* type in spouses. J. Periodontol. *74*, 1355–1360.

Ashimoto, A., Chen, C., Bakker, I., and Slots, J. (1996) Polymerase chain reaction detection of 8 putative periodontal pathogens in subgingival plaque of gingivitis and advanced periodontitis lesions. Oral Microbiol. Immunol. *11*, 266–273.

Asikainen, S., Chen, C., Saarela, M., Saxen, L., and Slots, J. (1997) Clonal specificity of *Actinobacillus actinomycetemcomitans* in destructive periodontal disease. Clin. Infect. Dis. *25 Suppl 2*, S227–229.

Asikainen, S., Chen, C., and Slots, J. (1995) *Actinobacillus actinomycetemcomitans* genotypes in relation to serotypes and periodontal status. Oral Microbiol. Immunol. *10*, 65–68.

Asikainen, S., Chen, C., and Slots, J. (1996) Likelihood of transmitting *Actinobacillus actinomycetemcomitans* and *Porphyromonas gingivalis* in families with periodontitis. Oral Microbiol. Immunol. *11*, 387–394.

Bacot, C.M., and Reeves, R.H. (1991) Novel tRNA gene organization in the 16S-23S intergenic spacer of the *Streptococcus pneumoniae* rRNA gene cluster. J. Bacteriol. *173*, 4234–4236.

Barsotti, O., Decoret, D., and Renaud, F.N. (2002) Identification of *Streptococcus mitis* group species by RFLP of the PCR-amplified 16S-23S rDNA intergenic spacer. Res. Microbiol. *153*, 687–691.

Becker, M.R., Paster, B.J., Leys, E.J., Moeschberger, M.L., Kenyon, S.G., Galvin, J.L., Boches, S.K., Dewhirst, F.E., and Griffen, A.L. (2002) Molecular analysis of bacterial species associated with childhood caries. J. Clin. Microbiol. *40*, 1001–1009.

Bensch, S., and Akesson, M. (2005) Ten years of AFLP in ecology and evolution: why so few animals? Mol. Ecol. *14*, 2899–2914.

Bingen, E.H., Denamur, E., and Elion, J. (1994) Use of ribotyping in epidemiological surveillance of nosocomial outbreaks. Clin. Microbiol. Rev. *7*, 311–327.

Borneman, J., and Triplett, E.W. (1997) Molecular microbial diversity in soils from eastern Amazonia: evidence for unusual microorganisms and microbial population shifts associated with deforestation. Appl. Environ. Microbiol. *63*, 2647–2653.

Bottari, B., Ercolini, D., Gatti, M., and Neviani, E. (2006) Application of FISH technology for microbiological analysis: current state and prospects. Appl. Microbiol. Biotechnol. *73*, 485–494.

Boutaga, K., Savelkoul, P.H., Winkel, E.G. and van Winkelhoff, A.J. (2007) Comparison of subgingival bacterial sampling with oral lavage for detection and quantification of periodontal pathogens by real-time polymerase chain reaction. J. Periodontol. *78*, 79–86.

Boutaga, K., van Winkelhoff, A.J., Vandenbroucke-Grauls, C.M., and Savelkoul, P.H. (2006) The additional value of real-time PCR in the quantitative detection of periodontal pathogens. J. Clin. Periodontol. *33*, 427–433.

Bowden, G., Johnson, J., and Schachtele, C. (1993) Characterization of *Actinomyces* with genomic DNA fingerprints and rRNA gene probes. J. Dent. Res. *72*, 1171–1179.

Brinig, M.M., Lepp, P.W., Ouverney, C.C., Armitage, G.C., and Relman, D.A. (2003) Prevalence of bacteria of division TM7 in human subgingival plaque and their association with disease. Appl. Environ. Microbiol. *69*, 1687–1694.

Brownie, J., Shawcross, S., Theaker, J., Whitcombe, D., Ferrie, R., Newton, C., and Little, S. (1997) The elimination of primer-dimer accumulation in PCR. Nucleic Acids Res. *25*, 3235–3241.

Busse, H.J., Denner, E.B., and Lubitz, W. (1996) Classification and identification of bacteria: current approaches to an old problem. Overview of methods used in bacterial systematics. J. Biotechnol. *47*, 3–38.

Caufield, P.W., Saxena, D., Fitch, D., and Li, Y. (2007) Population structure of plasmid-containing strains of *Streptococcus mutans*, a member of the human indigenous biota. J. Bacteriol. *189*, 1238–1243.

Chamberlain, J.S., Gibbs, R.A., Ranier, J.E., Nguyen, P.N., and Caskey, C.T. (1988) Deletion screening of the Duchenne muscular dystrophy locus via multiplex DNA amplification. Nucleic Acids Res. *16*, 11141–11156.

Chen, C., and Slots, J. (1994) Clonal analysis of *Porphyromonas gingivalis* by the arbitrarily primed polymerase chain reaction. Oral Microbiol. Immunol. *9*, 99–103.

Chen, T., Abbey, K., Deng, W.J., and Cheng, M.C. (2005) The bioinformatics resource for oral pathogens. Nucleic Acids Res. *33*, W734–740.

Chen, T., Hosogi, Y., Nishikawa, K., Abbey, K., Fleischmann, R.D., Walling, J., and Duncan, M.J. (2004a) Comparative whole-genome analysis of virulent and avirulent strains of *Porphyromonas gingivalis*. J. Bacteriol. *186*, 5473–5479.

Chen, T., Hosogi, Y., Nishikawa, K., Abbey, K., Fleischmann, R.D., Walling, J., and Duncan, M.J. (2004b) Comparative whole-genome analysis of virulent and avirulent strains of *Porphyromonas gingivalis*. J. Bacteriol. *186*, 5473–5479.

Choi, B.K., Paster, B.J., Dewhirst, F.E., and Gobel, U.B. (1994) Diversity of cultivable and uncultivable oral spirochetes from a patient with severe destructive periodontitis. Infect. Immun. *62*, 1889–1895.

Ciantar, M., Spratt, D.A., Newman, H.N., and Wilson, M. (2001) *Capnocytophaga granulosa* and *Capnocytophaga haemolytica*: novel species in subgingival plaque. J. Clin. Periodontol. *28*, 701–705.

Corby, P.M., Lyons-Weiler, J., Bretz, W.A., Hart, T.C., Aas, J.A., Boumenna, T., Goss, J., Corby, A.L., Junior, H.M., Weyant, R.J., and Paster, B.J. (2005) Microbial risk indicators of early childhood caries. J. Clin. Microbiol. *43*, 5753–5759.

DeLong, E.F., Wickham, G.S., and Pace, N.R. (1989) Phylogenetic stains: ribosomal RNA-based probes for the identification of single cells. Science *243*, 1360–1363.

Didilescu, A.C., Skaug, N., Marica, C., and Didilescu, C. (2005) Respiratory pathogens in dental plaque of hospitalized patients with chronic lung diseases. Clin. Oral Investig. *9*, 141–147.

DiRienzo, J.M., Cornell, S., Kazoroski, L., and Slots, J. (1990) Probe-specific DNA fingerprinting applied to the epidemiology of localized juvenile periodontitis. Oral Microbiol. Immunol. *5*, 49–56.

Dix, K., Watanabe, S.M., McArdle, S., Lee, D.I., Randolph, C., Moncla, B., and Schwartz, D.E. (1990) Species-specific oligodeoxynucleotide probes for the identification of periodontal bacteria. J. Clin. Microbiol. *28*, 319–323.

Dogan, B., Antinheimo, J., Cetiner, D., Bodur, A., Emingil, G., Buduneli, E., Uygur, C., Firatli, E., Lakio, L., and Asikainen, S. (2003) Subgingival microflora in Turkish patients with periodontitis. J. Periodontol. *74*, 803–814.

Dogan, B., Asikainen, S., and Jousimies-Somer, H. (1999) Evaluation of two commercial kits and arbitrarily primed PCR for identification and differentiation of *Actinobacillus actinomycetemcomitans, Haemophilus aphrophilus*, and *Haemophilus paraphrophilus*. J. Clin. Microbiol. *37*, 742–747.

Duncan, M.J. (2003) Genomics of oral bacteria. Crit. Rev. Oral Biol. Med. *14*, 175–187.

Dymock, D., Weightman, A.J., Scully, C., and Wade, W.G. (1996) Molecular analysis of microflora associated with dentoalveolar abscesses. J. Clin. Microbiol. *34*, 537–542.

Enersen, M., Olsen, I., van Winkelhoff, A.J., and Caugant, D.A. (2006) Multilocus sequence typing of *Porphyromonas gingivalis* strains from different geographic origins. J. Clin. Microbiol. *44*, 35–41.

Fenchel, T., and Finlay, B.J. (2006) The diversity of microbes: resurgence of the phenotype. Philos. Trans. R. Soc. Lond. B. Biol. Sci. *361*, 1965–1973.

Fiehn, N.E., Bangsborg, J.M., and Colding, H. (1995) Ribotyping on small-sized spirochetes isolated from subgingival plaque. Oral Microbiol. Immunol. *10*, 13–18.

Fleischmann, R.D., Adams, M.D., White, O., Clayton, R.A., Kirkness, E.F., Kerlavage, A.R., Bult, C.J., Tomb, J.F., Dougherty, B.A., Merrick, J.M., and *et al.* (1995) Whole-genome random sequencing and assembly of *Haemophilus influenzae* Rd. Science *269*, 496–512.

Fujise, O., Chen, W., Rich, S., and Chen, C. (2004) Clonal diversity and stability of subgingival *Eikenella corrodens*. J. Clin. Microbiol. *42*, 2036–2042.

Fujise, O., Hamachi, T., Hirofuji, T., and Maeda, K. (1995) Colorimetric microtiter plate based assay for detection and quantification of amplified *Actinobacillus actinomycetemcomitans* DNA. Oral Microbiol. Immunol. *10*, 372–377.

Fukiya, S., Mizoguchi, H., Tobe, T., and Mori, H. (2004) Extensive genomic diversity in pathogenic *Escherichia coli* and *Shigella* strains revealed by comparative genomic hybridization microarray. J. Bacteriol. *186*, 3911–3921.

Fukui, K., Kato, N., Kato, H., Watanabe, K., and Tatematsu, N. (1999) Incidence of *Prevotella intermedia* and *Prevotella nigrescens* carriage among family members with subclinical periodontal disease. J. Clin. Microbiol. *37*, 3141–3145.

Gafan, G.P., and Spratt, D.A. (2005) Denaturing gradient gel electrophoresis gel expansion (DGGEGE)--an attempt to resolve the limitations of co-migration in the DGGE of complex polymicrobial communities. FEMS Microbiol. Lett. *253*, 303–307.

Garaizar, J., Rementeria, A., and Porwollik, S. (2006) DNA microarray technology: a new tool for the epidemiological typing of bacterial pathogens? FEMS Immunol. Med. Microbiol. *47*, 178–189.

George, K.S., Reynolds, M.A., and Falkler, W.A., Jr. (1997) Arbitrarily primed polymerase chain reaction fingerprinting and clonal analysis of oral *Fusobacterium nucleatum* isolates. Oral Microbiol. Immunol. *12*, 219–226.

Gevers, D., Cohan, F.M., Lawrence, J.G., Spratt, B.G., Coenye, T., Feil, E.J., Stackebrandt, E., Van de Peer, Y., Vandamme, P., Thompson, F.L., and Swings, J. (2005) Opinion: Re-evaluating prokaryotic species. Nat Rev Microbiol *3*, 733–739.

Giulietti, A., Overbergh, L., Valckx, D., Decallonne, B., Bouillon, R., and Mathieu, C. (2001) An overview of real-time quantitative PCR: applications to quantify cytokine gene expression. Methods *25*, 386–401.

Gmur, R., and Thurnheer, T. (2002) Direct quantitative differentiation between *Prevotella intermedia* and *Prevotella nigrescens* in clinical specimens. Microbiology *148*, 1379–1387.

Goncharoff, P., Figurski, D.H., Stevens, R.H., and Fine, D.H. (1993) Identification of *Actinobacillus actinomycetemcomitans*: polymerase chain reaction amplification of *lktA*-specific sequences. Oral Microbiol. Immunol. *8*, 105–110.

Goodman, N. (2002) Biological data becomes computer literate: new advances in bioinformatics. Curr. Opin. Biotechnol. *13*, 68–71.

Griffen, A.L., Lyons, S.R., Becker, M.R., Moeschberger, M.L., and Leys, E.J. (1999) *Porphyromonas gingivalis* strain variability and periodontitis. J. Clin. Microbiol. *37*, 4028–4033.

Grimont, F., and Grimont, P.A. (1986) Ribosomal ribonucleic acid gene restriction patterns as potential taxonomic tools. Ann. Inst. Pasteur. Microbiol. *137B*, 165–175.

Hames-Kocabas, E.E., Ucar, F., Kocatas Ersin, N., Uzel, A., and Alpoz, A.R. (2006) Colonization and vertical transmission of *Streptococcus mutans* in Turkish children. Res. Microbiol. May 28;[E-pub ahead of print]PMID:16735109.

Hanage, W.P., Fraser, C., and Spratt, B.G. (2006) Sequences, sequence clusters and bacterial species. Philos. Trans. R. Soc. Lond. B. Biol. Sci. *361*, 1917–1927.

Harper-Owen, R., Dymock, D., Booth, V., Weightman, A.J., and Wade, W.G. (1999) Detection of unculturable bacteria in periodontal health and disease by PCR. J. Clin. Microbiol. *37*, 1469–1473.

Haubek, D., Dirienzo, J.M., Tinoco, E.M., Westergaard, J., Lopez, N.J., Chung, C.P., Poulsen, K., and Kilian, M. (1997) Racial tropism of a highly toxic clone of *Actinobacillus actinomycetemcomitans* associated with juvenile periodontitis. J. Clin. Microbiol. *35*, 3037–3042.

He, T., Hayashi, J., Yamamoto, M., and Ishikawa, I. (1998) Genotypic characterization of *Actinobacillus actinomycetemcomitans* isolated from periodontitis patients by arbitrarily primed polymerase chain reaction. J. Periodontol. *69*, 69–75.

Healy, M., Huong, J., Bittner, T., Lising, M., Frye, S., Raza, S., Schrock, R., Manry, J., Renwick, A., Nieto, R., *et al.* (2005) Microbial DNA typing by automated repetitive-sequence-based PCR. J. Clin. Microbiol. *43*, 199–207.

Higuchi, R., Fockler, C., Dollinger, G., and Watson, R. (1993) Kinetic PCR analysis: real-time monitoring of DNA amplification reactions. Biotechnology. (NY). *11*, 1026–1030.

Hoshino, T., Kawaguchi, M., Shimizu, N., Hoshino, N., Ooshima, T., and Fujiwara, T. (2004) PCR detection and identification of oral streptococci in saliva samples using gtf genes. Diagn. Microbiol. Infect. Dis. *48*, 195–199.

Huang, Y., Umeda, M., Takeuchi, Y., Ishizuka, M., Yano-Higuchi, K., and Ishikawa, I. (2003) Distribution of *Bacteroides forsythus* genotypes in a Japanese periodontitis population. Oral Microbiol. Immunol. *18*, 208–214.

Hugenholtz, P. (2002) Exploring prokaryotic diversity in the genomic era. Genome Biol 3: reviews 0003.1–0003.8.

Hugenholtz, P., Tyson, G.W., Webb, R.I., Wagner, A.M., and Blackall, L.L. (2001) Investigation of candidate division TM7, a recently recognized major lineage of the domain *Bacteria* with no known pure-culture representatives. Appl. Environ. Microbiol. *67*, 411–419.

Hulton, C.S., Higgins, C.F., and Sharp, P.M. (1991) ERIC sequences: a novel family of repetitive elements in the genomes of *Escherichia coli*, *Salmonella typhimurium* and other enterobacteria. Mol. Microbiol. *5*, 825–834.

Kamiya, R.U., Napimoga, M.H., Rosa, R.T., Hofling, J.F., and Goncalves, R.B. (2005) Mutacin production in *Streptococcus mutans* genotypes isolated from caries-affected and caries-free individuals. Oral Microbiol. Immunol. *20*, 20–24.

Kaplan, J.B., Schreiner, H.C., Furgang, D., and Fine, D.H. (2002) Population structure and genetic diversity of *Actinobacillus actinomycetemcomitans* strains isolated from localized juvenile periodontitis patients. J. Clin. Microbiol. *40*, 1181–1187.

Kazemi, M., and Finkelstein, R.A. (1990) Checkerboard immunoblotting (CBIB): an efficient, rapid, and sensitive method of assaying multiple antigen/antibody cross-reactivities. J. Immunol. Methods *128*, 143–146.

Kazor, C.E., Mitchell, P.M., Lee, A.M., Stokes, L.N., Loesche, W.J., Dewhirst, F.E., and Paster, B.J. (2003) Diversity of bacterial populations on the tongue dorsa of patients with halitosis and healthy patients. J. Clin. Microbiol. *41*, 558–563.

Khemaleelakul, S., Baumgartner, J.C., and Pruksakorn, S. (2002) Identification of bacteria in acute endodontic infections and their antimicrobial susceptibility. Oral Surg. Oral Med. Oral Pathol. Oral Radiol. Endod. *94*, 746–755.

Kiska, D.L., and Macrina, F.L. (1994) Genetic analysis of fructan-hyperproducing strains of *Streptococcus mutans*. Infect. Immun. *62*, 2679–2686.

Koehler, A., Karch, H., Beikler, T., Flemmig, T.F., Suerbaum, S., and Schmidt, H. (2003) Multilocus sequence analysis of *Porphyromonas gingivalis* indicates frequent recombination. Microbiology *149*, 2407–2415.

Konig, A., Reinert, R.R., and Hakenbeck, R. (1998) *Streptococcus mitis* with unusually high level resistance to beta-lactam antibiotics. Microb. Drug Resist. *4*, 45–49.

Kononen, E., Saarela, M., Karjalainen, J., Jousimies-Somer, H., Alaluusua, S., and Asikainen, S. (1994) Transmission of oral *Prevotella melaninogenica* between a mother and her young child. Oral Microbiol. Immunol. *9*, 310–314.

Kononen, E., Wolf, J., Matto, J., Frandsen, E.V., Poulsen, K., Jousimies-Somer, H., and Asikainen, S. (2000) The *Prevotella intermedia* group organisms in young children and their mothers as related to maternal periodontal status. J. Periodont. Res. *35*, 329–334.

Konstantinidis, K.T., Ramette, A., and Tiedje, J.M. (2006) The bacterial species definition in the genomic era. Philos. Trans. R. Soc. Lond. B. Biol. Sci. *361*, 1929–1940.

Kostrzynska, M., and Bachand, A. (2006) Application of DNA microarray technology for detection, identification, and characterization of food-borne pathogens. Can. J. Microbiol. *52*, 1–8.

Kroes, I., Lepp, P.W., and Relman, D.A. (1999) Bacterial diversity within the human subgingival crevice. Proc. Natl. Acad. Sci. USA *96*, 14547–14552.

Kumar, P.S., Griffen, A.L., Barton, J.A., Paster, B.J., Moeschberger, M.L., and Leys, E.J. (2003) New bacterial species associated with chronic periodontitis. J. Dent. Res. *82*, 338–344.

Kumar, P.S., Griffen, A.L., Moeschberger, M.L., and Leys, E.J. (2005) Identification of candidate periodontal pathogens and beneficial species by quantitative 16S clonal analysis. J. Clin. Microbiol. *43*, 3944–3955.

Kumar, P.S., Leys, E.J., Bryk, J.M., Martinez, F.J., Moeschberger, M.L., and Griffen, A.L. (2006) Changes in periodontal health status are associated with bacterial community shifts as assessed by quantitative 16S cloning and sequencing. J. Clin. Microbiol. *44*, 3665–3673.

Lakio, L., Kuula, H., Dogan, B., and Asikainen, S. (2002) *Actinobacillus actinomycetemcomitans* proportion of subgingival bacterial flora in relation to its clonal type. Eur. J. Oral Sci. *110*, 212–217.

Ledder, R.G., Gilbert, P., Huws, S.A., Aarons, L., Ashley, M.P., Hull, P.S., and McBain, A.J. (2007) Molecular analysis of the subgingival microbiota in health and disease. Appl. Environ. Microbiol. *73*, 516–523.

Ledder, R.G., Gilbert, P., Pluen, A., Sreenivasan, P.K., De Vizio, W., and McBain, A.J. (2006) Individual microflora beget unique oral microcosms. J. Appl. Microbiol. *100*, 1123–1131.

Lepp, P.W., Brinig, M.M., Ouverney, C.C., Palm, K., Armitage, G.C., and Relman, D.A. (2004) Methanogenic *Archaea* and human periodontal disease. Proc. Natl. Acad. Sci. USA *101*, 6176–6181.

Lewis, J.P., and Macrina, F.L. (1999) Localization of HArep-containing genes on the chromosome of *Porphyromonas gingivalis* W83. Infect. Immun. *67*, 2619–2623.

Leys, E.J., Griffen, A.L., Strong, S.J., and Fuerst, P.A. (1994) Detection and strain identification of *Actinobacillus actinomycetemcomitans* by nested PCR. J. Clin. Microbiol. 32, 1288–1294.

Leys, E.J., Lyons, S.R., Moeschberger, M.L., Rumpf, R.W., and Griffen, A.L. (2002) Association of *Bacteroides forsythus* and a novel *Bacteroides* phylotype with periodontitis. J. Clin. Microbiol. *40*, 821–825.

Li, Y., Caufield, P.W., Emanuelsson, I.R., and Thornqvist, E. (2001) Differentiation of *Streptococcus mutans* and *Streptococcus sobrinus* via genotypic and phenotypic profiles from three different populations. Oral Microbiol. Immunol. *16*, 16–23.

Li, Y., Ge, Y., Saxena, D., and Caufield, P.W. (2007) Genetic profiling of the oral microbiota associated with severe early-childhood caries. J. Clin. Microbiol. *45*, 81–87.

Li, Y., Ku, C.Y., Xu, J., Saxena, D., and Caufield, P.W. (2005) Survey of oral microbial diversity using PCR-based denaturing gradient gel electrophoresis. J. Dent. Res. *84*, 559–564.

Li, Y., Saxena, D., Barnes, V.M., Trivedi, H.M., Ge, Y., and Xu, T. (2006) Polymerase chain reaction-based denaturing gradient gel electrophoresis in the evaluation of oral microbiota. Oral Microbiol. Immunol. *21*, 333–339.

Li, Y., Wang, W., and Caufield, P.W. (2000) The fidelity of mutans streptococci transmission and caries status correlate with breast-feeding experience among Chinese families. Caries Res. *34*, 123–132.

Liu, W.T., Marsh, T.L., Cheng, H., and Forney, L.J. (1997) Characterization of microbial diversity by determining terminal restriction fragment length polymorphisms of genes encoding 16S rRNA. Appl. Environ. Microbiol. *63*, 4516–4522.

Loos, B.G., Mayrand, D., Genco, R.J., and Dickinson, D.P. (1990) Genetic heterogeneity of *Porphyromonas (Bacteroides) gingivalis* by genomic DNA fingerprinting. J. Dent. Res. *69*, 1488–1493.

Loos, B.G., Van Winkelhoff, A.J., Dunford, R.G., Genco, R.J., De Graaff, J., Dickinson, D.P., and Dyer, D.W. (1992) A statistical approach to the ecology of *Porphyromonas gingivalis*. J. Dent. Res. *71*, 353–358.

Lotufo, R.F., Flynn, J., Chen, C., and Slots, J. (1994) Molecular detection of *Bacteroides forsythus* in human periodontitis. Oral Microbiol. Immunol. *9*, 154–160.

Macheleidt, A., Muller, H.P., Eger, T., Putzker, M., and Zoller, L. (1999) Clonal diversity of *Actinobacillus actinomycetemcomitans* isolates from young adults with minimal periodontal disease. J. Periodont. Res. *34*, 179–187.

Mager, D.L., Haffajee, A.D., Devlin, P.M., Norris, C.M., Posner, M.R., and Goodson, J.M. (2005) The salivary microbiota as a diagnostic indicator of oral cancer: A descriptive, non-randomized study of cancer-free and oral squamous cell carcinoma subjects. J. Transl. Med. *3*, 27.

Maiden, M.C., Bygraves, J.A., Feil, E., Morelli, G., Russell, J.E., Urwin, R., Zhang, Q., Zhou, J., Zurth, K., Caugant, D.A., *et al.* (1998) Multilocus sequence typing: a portable approach to the identification of clones within populations of pathogenic microorganisms. Proc. Natl. Acad. Sci. USA *95*, 3140–3145.

Matto, J., Saarela, M., Alaluusua, S., Oja, V., Jousimies-Somer, H., and Asikainen, S. (1998) Detection of *Porphyromonas gingivalis* from saliva by PCR by using a simple sample-processing method. J. Clin. Microbiol. *36*, 157–160.

Matto, J., Saarela, M., von Troil-Linden, B., Alaluusua, S., Jousimies-Somer, H., and Asikainen, S. (1996) Similarity of salivary and subgingival *Prevotella inter-*

media and *Prevotella nigrescens* isolates by arbitrarily primed polymerase chain reaction. Oral Microbiol. Immunol. *11*, 395–401.

Maynard, C., Berthiaume, F., Lemarchand, K., Harel, J., Payment, P., Bayardelle, P., Masson, L., and Brousseau, R. (2005) Waterborne pathogen detection by use of oligonucleotide-based microarrays. Appl. Environ. Microbiol. *71*, 8548–8557.

McBain, A.J., Bartolo, R.G., Catrenich, C.E., Charbonneau, D., Ledder, R.G., and Gilbert, P. (2003a). Effects of a chlorhexidine gluconate-containing mouthwash on the vitality and antimicrobial susceptibility of *in vitro* oral bacterial ecosystems. Appl. Environ. Microbiol. *69*, 4770–4776.

McBain, A.J., Bartolo, R.G., Catrenich, C.E., Charbonneau, D., Ledder, R.G., and Gilbert, P. (2003b). Effects of triclosan-containing rinse on the dynamics and antimicrobial susceptibility of *in vitro* plaque ecosystems. Antimicrob. Agents Chemother. *47*, 3531–3538.

Menard, C., and Mouton, C. (1993) Randomly amplified polymorphic DNA analysis confirms the biotyping scheme of *Porphyromonas gingivalis*. Res. Microbiol. *144*, 445–455.

Menard, C., and Mouton, C. (1995) Clonal diversity of the taxon *Porphyromonas gingivalis* assessed by random amplified polymorphic DNA fingerprinting. Infect. Immun. *63*, 2522–2531.

Moncla, B.J., Braham, P., Dix, K., Watanabe, S., and Schwartz, D. (1990) Use of synthetic oligonucleotide DNA probes for the identification of *Bacteroides gingivalis*. J. Clin. Microbiol. *28*, 324–327.

Moore, W.E., and Moore, L.V. (1994) The bacteria of periodontal diseases. Periodontol. 2000 5, 66–77.

Moraes, S.R., Siqueira, J.F., Jr., Rocas, I.N., Ferreira, M.C., and Domingues, R.M. (2002) Clonality of *Fusobacterium nucleatum* in root canal infections. Oral Microbiol. Immunol. *17*, 394–396.

Motegi, M., Takagi, Y., Yonezawa, H., Hanada, N., Terajima, J., Watanabe, H., and Senpuku, H. (2006) Assessment of genes associated with *Streptococcus mutans* biofilm morphology. Appl. Environ. Microbiol. *72*, 6277–6287.

Moter, A., and Gobel, U.B. (2000) Fluorescence in situ hybridization (FISH) for direct visualization of microorganisms. J. Microbiol. Methods *41*, 85–112.

Mullis, K.B. (1990) The unusual origin of the polymerase chain reaction. Sci. Amer. *262*, 56–61, 64–55.

Munson, M.A., Pitt-Ford, T., Chong, B., Weightman, A., and Wade, W.G. (2002) Molecular and cultural analysis of the microflora associated with endodontic infections. J. Dent. Res. *81*, 761–766.

Muyzer, G., de Waal, E.C., and Uitterlinden, A.G. (1993) Profiling of complex microbial populations by denaturing gradient gel electrophoresis analysis of polymerase chain reaction-amplified genes coding for 16S rRNA. Appl. Environ. Microbiol. *59*, 695–700.

Nakano, K., Inaba, H., Nomura, R., Nemoto, H., Takeda, M., Yoshioka, H., Matsue, H., Takahashi, T., Taniguchi, K., Amano, A., and Ooshima, T. (2006) Detection of cariogenic *Streptococcus mutans* in extirpated heart valve and atheromatous plaque specimens. J. Clin. Microbiol. *44*, 3313–3317.

Okahashi, N., Sasakawa, C., Okada, N., Yamada, M., Yoshikawa, M., Tokuda, M., Takahashi, I., and Koga, T. (1990) Construction of NotI restriction map of the *Streptococcus mutans* genome. J. Gen. Microbiol. *136*, 2217–2223.

Olive, D.M., and Bean, P. (1999) Principles and applications of methods for DNA-based typing of microbial organisms. J. Clin. Microbiol. *37*, 1661–1669.

Ono, T., Hirota, K., Nemoto, K., Fernandez, E.J., Ota, F., and Fukui, K. (1994) Detection of *Streptococcus mutans* by PCR amplification of spaP gene. J. Med. Microbiol. *41*, 231–235.

Oren, A. (2004) Prokaryote diversity and taxonomy: current status and future challenges. Philos. Trans. R. Soc. Lond. B. Biol. Sci. *359*, 623–638.

Paddick, J.S., Brailsford, S.R., Kidd, E.A., Gilbert, S.C., Clark, D.T., Alam, S., Killick, Z.J., and Beighton, D. (2003) Effect of the environment on genotypic diversity of *Actinomyces naeslundii* and *Streptococcus oralis* in the oral biofilm. Appl. Environ. Microbiol. *69*, 6475–6480.

Paju, S., Carlson, P., Jousimies-Somer, H., and Asikainen, S. (2000) Heterogeneity of *Actinobacillus actinomycetemcomitans* strains in various human infections and relationships between serotype, genotype, and antimicrobial susceptibility. J. Clin. Microbiol. *38*, 79–84.

Pan, Y.P., Li, Y., and Caufield, P.W. (2001) Phenotypic and genotypic diversity of *Streptococcus sanguis* in infants. Oral Microbiol. Immunol. *16*, 235–242.

Paster, B.J., Boches, S.K., Galvin, J.L., Ericson, R.E., Lau, C.N., Levanos, V.A., Sahasrabudhe, A., and Dewhirst, F.E. (2001) Bacterial diversity in human subgingival plaque. J. Bacteriol. *183*, 3770–3783.

Paster, B.J., Dewhirst, F.E., Coleman, B.C., Lau, C.N., and Ericson, R.L. (1998) Phylogenetic analysis of cultivable oral treponemes from the Smibert collection. Int. J. Syst. Bacteriol. *48 Pt 3*, 713–722.

Paster, B.J., Olsen, I., Aas, J.A., and Dewhirst, F.E. (2006) The breadth of bacterial diversity in the human periodontal pocket and other oral sites. Periodontol. 2000 *42*, 80–87.

Pearce, M.A., Devine, D.A., Dixon, R.A., and van Steenbergen, T.J. (2000) Genetic heterogeneity in *Prevotella intermedia, Prevotella nigrescens, Prevotella corporis* and related species isolated from oral and nonoral sites. Oral Microbiol. Immunol. *15*, 89–95.

Pei, Z., Bini, E.J., Yang, L., Zhou, M., Francois, F., and Blaser, M.J. (2004) Bacterial biota in the human distal esophagus. Proc. Natl. Acad. Sci. USA *101*, 4250–4255.

Persson, G.R., Schlegel-Bregenzer, B., Chung, W.O., Houston, L., Oswald, T., and Roberts, M.C. (2000) Serum antibody titers to *Bacteroides forsythus* in elderly subjects with gingivitis or periodontitis. J. Clin. Periodontol. *27*, 839–845.

Preus, H.R., Haraszthy, V.I., Zambon, J.J., and Genco, R.J. (1993) Differentiation of strains of *Actinobacillus actinomycetemcomitans* by arbitrarily primed polymerase chain reaction. J. Clin. Microbiol. *31*, 2773–2776.

Rasiah, I.A., Wong, L., Anderson, S.A., and Sissons, C.H. (2005) Variation in bacterial DGGE patterns from human saliva: over time, between individuals and in corresponding dental plaque microcosms. Arch. Oral Biol. *50*, 779–787.

Renvert, S., Pettersson, T., Ohlsson, O., and Persson, G.R. (2006) Bacterial profile and burden of periodontal infection in subjects with a diagnosis of acute coronary syndrome. J. Periodontol. *77*, 1110–1119.

Rheims, H., Sproer, C., Rainey, F.A., and Stackebrandt, E. (1996) Molecular biological evidence for the occurrence of uncultured members of the actinomycete line of descent in different environments and geographical locations. Microbiology *142 (Pt 10)*, 2863–2870.

Rocas, I.N., and Siqueira, J.F., Jr. (2005) Species-directed 16S rRNA gene nested PCR detection of *Olsenella* species in association with endodontic diseases. Lett. Appl. Microbiol. *41*, 12–16.

Rondon, M.R., Goodman, R.M., and Handelsman, J. (1999) The Earth's bounty: assessing and accessing soil microbial diversity. Trends Biotechnol. *17*, 403–409.

Ruby, J.D., Li, Y., Luo, Y., and Caufield, P.W. (2002) Genetic characterization of the oral *Actinomyces*. Arch. Oral Biol. *47*, 457–463.

Rudney, J.D., and Larson, C.J. (1999) Identification of oral mitis group streptococci by arbitrarily primed polymerase chain reaction. Oral Microbiol. Immunol. *14*, 33–42.

Rumpf, R.W., Griffen, A.L., and Leys, E.J. (2000) Phylogeny of *Porphyromonas gingivalis* by ribosomal intergenic spacer region analysis. J. Clin. Microbiol. *38*, 1807–1810.

Saarela, M., Hannula, J., Matto, J., Asikainen, S., and Alaluusua, S. (1996) Typing of mutans streptococci by arbitrarily primed polymerase chain reaction. Arch. Oral Biol. *41*, 821–826.

Saarela, M., von Troil-Linden, B., Torkko, H., Stucki, A.M., Alaluusua, S., Jousimies-Somer, H., and Asikainen, S. (1993) Transmission of oral bacterial species between spouses. Oral Microbiol. Immunol. *8*, 349–354.

Saarela, M.H., Dogan, B., Alaluusua, S., and Asikainen, S. (1999) Persistence of oral colonization by the same *Actinobacillus actinomycetemcomitans* strain(s). J. Periodontol. *70*, 504–509.

Sakamoto, M., Hayashi, H., and Benno, Y. (2003) Terminal restriction fragment length polymorphism analysis for human fecal microbiota and its application for analysis of complex bifidobacterial communities. Microbiol. Immunol. *47*, 133–142.

Sakamoto, M., Huang, Y., Ohnishi, M., Umeda, M., Ishikawa, I., and Benno, Y. (2004) Changes in oral microbial profiles after periodontal treatment as determined by molecular analysis of 16S rRNA genes. J. Med. Microbiol. 53, 563–571.

Sakamoto, M., Huang, Y., Umeda, M., Ishikawa, I., and Benno, Y. (2002) Detection of novel oral phylotypes associated with periodontitis. FEMS Microbiol. Lett. *217*, 65–69.

Sakamoto, M., Rocas, I.N., Siqueira, J.F., Jr., and Benno, Y. (2006) Molecular analysis of bacteria in asymptomatic and symptomatic endodontic infections. Oral Microbiol. Immunol. *21*, 112–122.

Sakamoto, M., Umeda, M., Ishikawa, I., and Benno, Y. (2000) Comparison of the oral bacterial flora in saliva from a healthy subject and two periodontitis patients by sequence analysis of 16S rDNA libraries. Microbiol. Immunol. *44*, 643–652.

Salama, N., Guillemin, K., McDaniel, T.K., Sherlock, G., Tompkins, L., and Falkow, S. (2000) A whole-genome microarray reveals genetic diversity among *Helicobacter pylori* strains. Proc. Natl. Acad. Sci. USA *97*, 14668–14673.

Sato, T., and Kuramitsu, H.K. (1999) Restriction fragment-length polymorphism analysis of 16S ribosomal RNA genes amplified by polymerase chain reaction for rapid identification of cultivable oral treponemes. Oral Microbiol. Immunol. *14*, 117–121.

Schena, M., Shalon, D., Davis, R.W., and Brown, P.O. (1995) Quantitative monitoring of gene expression patterns with a complementary DNA microarray. Science *270*, 467–470.

Schlegel, L., Grimont, F., Grimont, P.A., and Bouvet, A. (2003) Identification of major Streptococcal species by rrn-amplified ribosomal DNA restriction analysis. J. Clin. Microbiol. *41*, 657–666.

Schwartz, D.C., and Cantor, C.R. (1984) Separation of yeast chromosome-sized DNAs by pulsed field gradient gel electrophoresis. Cell *37*, 67–75.

Seol, J.H., Cho, B.H., Chung, C.P., and Bae, K.S. (2006) Multiplex polymerase chain reaction detection of black-pigmented bacteria in infections of endodontic origin. J. Endodont. *32*, 110–114.

Siqueira, J.F., Jr., and Rocas, I.N. (2004) Simultaneous detection of *Dialister pneumosintes* and *Filifactor alocis* in endodontic infections by 16S rDNA-directed multiplex PCR. J. Endodont. *30*, 851–854.

Siqueira, J.F., Jr., and Rocas, I.N. (2005) Exploiting molecular methods to explore endodontic infections: Part 1--current molecular technologies for microbiological diagnosis. J. Endod. *31*, 411–423.

Slots, J. (1979) Subgingival microflora and periodontal disease. J. Clin. Periodontol. *6*, 351–382.

Slots, J., Ashimoto, A., Flynn, M.J., Li, G., and Chen, C. (1995) Detection of putative periodontal pathogens in subgingival specimens by 16S ribosomal DNA amplification with the polymerase chain reaction. Clin. Infect. Dis. *20 Suppl 2*, S304–307.

Slots, J., Liu, Y.B., DiRienzo, J.M., and Chen, C. (1993) Evaluating two methods for fingerprinting genomes of *Actinobacillus actinomycetemcomitans*. Oral Microbiol. Immunol. *8*, 337–343.

Socransky, S.S. (1977) Microbiology of periodontal disease – present status and future considerations. J. Periodontol. *48*, 497–504.

Socransky, S.S., and Haffajee, A.D. (1992) The bacterial etiology of destructive periodontal disease: current concepts. J. Periodontol. *63*, 322–331.

Socransky, S.S., and Haffajee, A.D. (2005) Periodontal microbial ecology. Periodontol. 2000 *38*, 135–187.

Socransky, S.S., Smith, C., Martin, L., Paster, B.J., Dewhirst, F.E., and Levin, A.E. (1994) 'Checkerboard' DNA–DNA hybridization. Biotechniques *17*, 788–792.

Stackebrandt, E., Frederiksen, W., Garrity, G.M., Grimont, P.A., Kampfer, P., Maiden, M.C., Nesme, X., Rossello-Mora, R., Swings, J., Truper, H.G., *et al.* (2002) Report of the ad hoc committee for the re-evaluation of the species definition in bacteriology. Int. J. Syst. Evol. Microbiol. *52*, 1043–1047.

Staley, J.T. (2006) The bacterial species dilemma and the genomic-phylogenetic species concept. Philos. Trans. R. Soc. Lond. B. Biol. Sci. *361*, 1899–1909.

Steffens, L., Franke, S., Nickel, S., Schwarzkopf, A., Flemmig, T.F., and Karch, H. (1994) DNA fingerprinting of *Eikenella corrodens* by pulsed-field gel electrophoresis. Oral Microbiol. Immunol. *9*, 95–98.

Stull, T.L., LiPuma, J.J., and Edlind, T.D. (1988) A broad-spectrum probe for molecular epidemiology of bacteria: ribosomal RNA. J. Infect. Dis. *157*, 280–286.

Sumida, S., Ishihara, K., Kishi, M., and Okuda, K. (2002) Transmission of periodontal disease-associated bacteria from teeth to osseointegrated implant regions. Int. J. Oral Maxillofac. Implants *17*, 696–702.

Sunde, P.T., Olsen, I., Gobel, U.B., Theegarten, D., Winter, S., Debelian, G.J., Tronstad, L., and Moter, A. (2003) Fluorescence in situ hybridization (FISH) for direct visualization of bacteria in periapical lesions of asymptomatic root-filled teeth. Microbiology *149*, 1095–1102.

Suzuki, N., Yoshida, A., and Nakano, Y. (2005) Quantitative analysis of multi-species oral biofilms by TaqMan Real-Time PCR. Clin. Med. Res. 3, 176–185.

Tang, G., Samaranayake, L.P., Yip, H.K., Chu, F.C., Tsang, P.C., and Cheung, B.P. (2003) Direct detection of *Actinomyces* spp. from infected root canals in a Chinese population: a study using PCR-based, oligonucleotide-DNA hybridization technique. J. Dent. *31*, 559–568.

Teanpaisan, R., and Douglas, C.W. (1999) Molecular fingerprinting of *Porphyromonas gingivalis* by PCR of repetitive extragenic palindromic (REP) sequences and comparison with other fingerprinting methods. J. Med. Microbiol. *48*, 741–749.

Thurnheer, T., Gmur, R., and Guggenheim, B. (2004) Multiplex FISH analysis of a six-species bacterial biofilm. J. Microbiol. Meth. *56*, 37–47.

Tobler, N.E., Pfunder, M., Herzog, K., Frey, J.E., and Altwegg, M. (2006) Rapid detection and species identification of *Mycobacterium* spp. using real-time PCR and DNA-microarray. J. Microbiol. Meth. *66*, 116–124.

Toi, C.S., Cleaton-Jones, P., and Fatti, P. (2005) Characterization of *Streptococcus mutans* diversity by determining restriction fragment-length polymorphisms of the gtfB gene of isolates from 5-year-old children and their mothers. Antonie Van Leeuwenhoek *88*, 75–85.

Tran, S.D., and Rudney, J.D. (1996) Multiplex PCR using conserved and species-specific 16S rRNA gene primers for simultaneous detection of *Actinobacillus actinomycetemcomitans* and *Porphyromonas gingivalis*. J. Clin. Microbiol. *34*, 2674–2678.

Truong, T.L., Menard, C., Mouton, C., and Trahan, L. (2000) Identification of mutans and other oral streptococci by random amplified polymorphic DNA analysis. J. Med. Microbiol. *49*, 63–71.

Tsai, C.Y., Wolff, L.F., Germaine, G., and Hodges, J. (2003) A rapid DNA probe test compared to culture methods for identification of subgingival plaque bacteria. J Clin. Periodontol. *30*, 57–62.

Umeda, M., Chen, C., Bakker, I., Contreras, A., Morrison, J.L., and Slots, J. (1998) Risk indicators for harboring periodontal pathogens. J. Periodontol. *69*, 1111–1118.

Umeda, M., Miwa, Z., Takeuchi, Y., Ishizuka, M., Huang, Y., Noguchi, K., Tanaka, M., Takagi, Y., and Ishikawa, I. (2004) The distribution of periodontopathic bacteria among Japanese children and their parents. J. Periodont. Res. *39*, 398–404.

Valcarcel, J., Allardet-Servent, A., Bourg, G., O'Callaghan, D., Michailesco, P., and Ramuz, M. (1997) Investigation of the *Actinobacillus actinomycetemcomitans* genome by pulsed field gel electrophoresis. Oral Microbiol. Immunol. *12*, 33–39.

Vallone, P.M., and Butler, J.M. (2004) AutoDimer: a screening tool for primer-dimer and hairpin structures. Biotechniques *37*, 226–231.

van Steenbergen, T.J., Bosch-Tijhof, C.J., Petit, M.D., and Van der Velden, U. (1997) Intra-familial transmission and distribution of *Prevotella intermedia* and *Prevotella nigrescens*. J. Periodont. Res. 32, 345–350.

Van Steenbergen, T.J., Menard, C., Tijhof, C.J., Mouton, C., and De Graaff, J. (1993) Comparison of three molecular typing methods in studies of transmission of *Porphyromonas gingivalis*. J. Med. Microbiol. *39*, 416–421.

Vandamme, P., Pot, B., Gillis, M., de Vos, P., Kersters, K., and Swings, J. (1996) Polyphasic taxonomy, a consensus approach to bacterial systematics. Microbiol. Rev. *60*, 407–438.

Vaneechoutte, M., Riegel, P., de Briel, D., Monteil, H., Verschraegen, G., De Rouck, A., and Claeys, G. (1995) Evaluation of the applicability of amplified rDNA-restriction analysis (ARDRA) to identification of species of the genus *Corynebacterium*. Res. Microbiol. *146*, 633–641.

Vernet, G., Jay, C., Rodrigue, M., and Troesch, A. (2004) Species differentiation and antibiotic susceptibility testing with DNA microarrays. J. Appl. Microbiol. *96*, 59–68.

Versalovic, J., Koeuth, T., and Lupski, J.R. (1991) Distribution of repetitive DNA sequences in eubacteria and application to fingerprinting of bacterial genomes. Nucleic Acids Res. *19*, 6823–6831.

Vianna, M.E., Conrads, G., Gomes, B.P., and Horz, H.P. (2006) Identification and quantification of archaea involved in primary endodontic infections. J. Clin. Microbiol. *44*, 1274–1282.

Vickerman, M.M., Brossard, K.A., Funk, D.B., Jesionowski, A.M., and Gill, S.R. (2007) Phylogenetic analysis of bacterial and archaeal species in symptomatic and asymptomatic endodontic infections. J. Med. Microbiol. *56*, 110–118.

von Troil-Linden, B., Saarela, M., Matto, J., Alaluusua, S., Jousimies-Somer, H., and Asikainen, S. (1996) Source of suspected periodontal pathogens re-emerging after periodontal treatment. J. Clin. Periodontol. *23*, 601–607.

von Wintzingerode, F., Gobel, U.B., and Stackebrandt, E. (1997) Determination of microbial diversity in environmental samples: pitfalls of PCR-based rRNA analysis. FEMS Microbiol. Rev. *21*, 213–229.

Wade, W.G., Spratt, D.A., Dymock, D., and Weightman, A.J. (1997) Molecular detection of novel anaerobic species in dentoalveolar abscesses. Clin. Infect. Dis. *25 Suppl 2*, S235–236.

Wagner, M., Horn, M., and Daims, H. (2003) Fluorescence in situ hybridisation for the identification and characterisation of prokaryotes. Curr. Opin. Microbiol. *6*, 302–309.

Waterhouse, J.C., Swan, D.C., and Russell, R.R. (2007) Comparative genome hybridization of *Streptococcus mutans* strains. Oral Microbiol. Immunol. *22*, 103–110.

Welsh, J., and McClelland, M. (1990) Fingerprinting genomes using PCR with arbitrary primers. Nucleic Acids Res. *18*, 7213–7218.

Williams, J.G., Kubelik, A.R., Livak, K.J., Rafalski, J.A., and Tingey, S.V. (1990) DNA polymorphisms amplified by arbitrary primers are useful as genetic markers. Nucleic Acids Res. *18*, 6531–6535.

Woese, C.R. (1987) Bacterial evolution. Microbiol. Rev. *51*, 221–271.

Woods, C.R., Versalovic, J., Koeuth, T., and Lupski, J.R. (1993) Whole-cell repetitive element sequence-based polymerase chain reaction allows rapid assessment of clonal relationships of bacterial isolates. J. Clin. Microbiol. *31*, 1927–1931.

Ye, R.W., Wang, T., Bedzyk, L., and Croker, K.M. (2001) Applications of DNA microarrays in microbial systems. J. Microbiol. Meth. *47*, 257–272.

Yoshida, A., Tachibana, M., Ansai, T., and Takehara, T. (2005) Multiplex polymerase chain reaction assay for simultaneous detection of black-pigmented *Prevotella* species in oral specimens. Oral Microbiol. Immunol. *20*, 43–46.

Young, R.A., Macklis, R., and Steitz, J.A. (1979) Sequence of the 16 S-23 s spacer region in two ribosomal RNA operons of *Escherichia coli*. J. Biol. Chem. *254*, 3264–3271.

Yuan, K., Hsu, P.C., Tseng, C.C., Kiang, D., and Wang, J.R. (2001) Detection rate of *Actinobacillus actinomycetemcomitans* on the permanent 1st molars of primary school children in Taiwan by polymerase chain reaction. J. Clin. Periodontol. *28*, 348–352.

Zhang, T., and Fang, H.H. (2006) Applications of real-time polymerase chain reaction for quantification of microorganisms in environmental samples. Appl. Microbiol. Biotechnol. *70*, 281–289.

Zijnge, V., Welling, G.W., Degener, J.E., van Winkelhoff, A.J., Abbas, F., and Harmsen, H.J. (2006) Denaturing gradient gel electrophoresis as a diagnostic tool in periodontal microbiology. J. Clin. Microbiol. *44*, 3628–3633.

Applied Genomics of Some Oral Bacteria

2

Stuart G. Dashper, Paul D. Veith, Ching Seng Ang, Peter S. Zilm and Eric C. Reynolds

Abstract

The simultaneous and interrelated rapid advances in genome sequencing, mass spectrometry and computing power have provided the tools for examining oral bacteria and their interactions with each other and the host at a global molecular level. In this chapter we focus on the development and usage of applied genomics techniques, mainly proteomics, for the study of Gram-negative anaerobic bacterial species found primarily in subgingival plaque. Proteomic analyses have already been applied to the discovery and characterization of proteins that comprise the proteome of *Porphyromonas gingivalis* and are now being applied to aid our understanding of the effects of environmental cues on the proteomes of oral bacteria, including *Fusobacterium nucleatum*, *Tannerella forsythia* and *Treponema denticola*. These species share the same environment and have been shown to interact with each other on a number of levels. We examine the range of currently available proteomic methodologies that can be used to investigate the elaborate interactions between cells in the complex microbial biofilm communities found in the oral cavity and the interaction of these bacteria with the host.

Introduction

In this chapter we examine how the simultaneous and interrelated rapid advances in genome sequencing, mass spectrometry and computing power have provided the tools to examine oral bacteria and their interactions with each other and the host at a global molecular level. Not only will these technologies be useful tools in answering current research questions but, by their very scale and nature, they will also affect the nature of the questions.

The overwhelming majority of oral bacteria exist as components of complex polymicrobial biofilms (dental plaque) accreted onto the surface of the tooth. There is a major division between the ecologies of supragingival plaque found above the gingival margin and subgingival plaque found below the gingival margin. In this chapter we will concentrate on the development and usage of applied genomics techniques, mainly proteomics, for the study of Gram-negative anaerobic bacterial species found primarily in subgingival plaque. These species share the same environment and have been shown to interact with each other on a number of levels. Considerable effort has been expended in sequencing the genomes of six species of subgingival plaque bacteria, thereby allowing comparison and interspecies studies.

One of the imperatives in the study of subgingival plaque bacteria is to determine how periodontal diseases are initiated and progress, with a view to developing new preventive and therapeutic strategies. The most prevalent form of periodontal disease, chronic periodontitis, is associated with a subgingival infection of a consortium of specific Gram-negative bacteria. In a study of 13 261 plaque samples by Socransky *et al.* (1998) the levels of 40 subgingival taxa were determined using checkerboard DNA–DNA hybridisation analysis. Five clusters or complexes of subgingival bacterial species were distinguished and these could be correlated with the periodontal health of the patient. Three species, *Porphyromonas gingivalis*, *Tannerella forsythia* and *Treponema denticola* were found to form a tightly related group, called the 'Red Complex'; this was highly associated with the clinical signs of advanced chronic periodontitis. This confirmed other research demonstrating a strong relationship between disease severity and the presence of *P. gingivalis* with *T. forsythia* (Gmur *et al.*, 1989; Yang *et al.*, 2004) and *P. gingivalis* with *T. denticola* (Simonson *et al.*, 1992) in subgingival plaque. *Spirochaetes* (mainly *T. denticola* and *Treponema socranskii*) can represent up to 50% of the cultivable flora in subgingival plaque of subjects with chronic periodontitis, whereas they account for less than 1% of the flora at healthy sites (Listgarten and Hellden, 1978; Loesche, 1988). A second tightly related group of bacterial species termed the 'Orange Complex' was also found in the Socransky study and included *Fusobacterium nucleatum*, *Prevotella intermedia* and *Campylobacter rectus*. An association between the Red and Orange Complexes was also evident as members of the Red Complex were rarely found in the absence of members of the Orange Complex (Socransky *et al.*, 1998). Thus, it has been proposed that an ecological succession of bacterial species is associated with the development of disease and that the presence of *F. nucleatum* and *P. intermedia* is needed for the establishment of a pathogenic plaque containing Red Complex members. The environmental conditions allowing the proliferation of one or more of the Red Complex species in subgingival plaque, which can lead to the initiation and progression of periodontitis, are still poorly understood. However, there is increasing evidence to indicate that these species act cooperatively or synergistically to cause disease.

Many interactions between these species have already been demonstrated. For example, *P. gingivalis* has been shown to co-aggregate with *T. denticola* (Grenier, 1992; Onagawa *et al.*, 1994) and other oral bacterial species, such as *Fusobacterium* spp. and *Actinomyces* spp., but not with *A. actinomycetemcomitans* (Eke *et al.*, 1989; Ellen and Galimanas, 2005; Kinder and Holt, 1989; Kolenbrander and Andersen, 1989). *T. denticola* co-aggregates with *Fusobacterium* spp. but not with *A. actinomycetemcomitans* and *Actinomyces* spp (Kolenbrander *et al.*, 1995). Bacterial co-aggregations such as these are important in the development of the subgingival plaque biofilm and colonization of tissue surfaces (Bolstad *et al.*, 1996). The virulence of periodontal pathogens has been studied using a number of animal lesion/abscess models. Mono-infection, by subcutaneous injection of *T. forsythia*, *T. denticola*, *A. actinomycetemcomitans* and *P. gingivalis* non-invasive strains, ATCC 33277 and 381, (Neiders *et al.*, 1989; van Steenbergen *et al.*, 1982) produces localized abscesses at the site of injection (Bird *et al.*, 2001; Chen *et al.*, 1995; Ebersole *et al.*, 1997; Kesavalu *et al.*, 1998). However, *P. gingivalis* invasive strains W50 and A7A1-28 (ATCC 53977) produce spreading ulcerative lesions at, and distant from, the site of injection (Neiders *et al.*, 1989; van Steenbergen *et al.*, 1982). In a murine lesion model, co-infection of *P. gingivalis–T. denticola*, *P. gingivalis–T. forsythia*, and *P. gingivalis–F. nucleatum* produced larger lesions and enhanced virulence compared with mono-infection with individual bacterial species (Chen *et al.*, 1996; Ebersole *et al.*, 1997; Kesavalu *et al.*, 1998; Neiders *et al.*, 1989; van Steenbergen *et al.*, 1982; Yoneda *et al.*, 2001). We have recently demonstrated that co-infection of *P. gingivalis-T. denticola* causes significant alveolar bone loss in a mouse periodontal disease model whereas mono-infection with either species, using equivalent cell numbers, failed to result in bone loss (unpublished data).

Growth studies in continuous culture systems have demonstrated that members of the Red Complex will grow to reproducible ratios that are dependent on, amongst other factors, the growth medium, redox potential and pH (Bradshaw and Marsh, 1998; Marsh, 2003). These data indicate that, at least in these model systems, these species do not directly compete with each other and that there is a degree of nutritional cooperativity between them (Marsh, 2003). *T. denticola* and *P. gingivalis* exhibit a mutualistic enhancement of growth *in vitro* as the production of isobutyric acid by *P. gingivalis* enhanced the growth of *T. denticola* while succinic acid, produced by the latter enhanced the growth of *P. gingivalis* (Grenier, 1992). Recent studies have also demonstrated a positive synergy between *F. nucleatum* and *T. forsythia* and between *T. denticola* and *P. gingivalis* in biofilm formation (Sharma *et al.*, 2005; Vesey and Kuramitsu, 2004).

Of the Red Complex species, *P. gingivalis* has been singled out for study as the recovery of this microorganism from periodontitis lesions can be up to 50% of the subgingival anaerobically cultivable flora (Christersson *et al.*, 1989). In animal experiments, subgingival implantation of *P. gingivalis* in mice (Baker *et al.*, 1994), rats (Evans *et al.*, 1992; Katz *et al.*, 1996), and non-human primates (Holt *et al.*, 1988; Persson *et al.*, 1994) has been associated with initiation and progression of periodontal bone resorption. Moreover, immunisation with killed whole *P. gingivalis* cells protected against periodontal bone loss in the rat and non-human primate models (Evans *et al.*, 1992; Persson *et al.*, 1994). Tissue damage in chronic periodontitis largely results from a dysregulated host immune response to specific bacteria in the subgingival plaque biofilm. Our ability to understand the bacterial ecology of subgingival plaque at a molecular level will enable us to determine how certain species of bacteria come to proliferate and are then able to dysregulate the host response.

The determination of the sequence of DNA was pioneered by Sanger *et al.* (1977) who developed the dideoxynucleotide termination method, initially to sequence short pieces of DNA. This technology has formed the basis for the massive expansion in sequencing of whole genomes that has seen the determination of the genomes of representatives of most kingdoms of life. The genome era was launched in 1995 with the publication of the first full DNA sequence of a free-living organism, the bacterium *Haemophilus influenzae* (Fleischmann *et al.*, 1995). The whole genome shotgun (WGS) approach that was used to sequence this bacterium is still the most widely used method of genome sequencing. To date, some 500 bacterial genomes have been published with another 1800 bacterial genomes currently being sequenced (www.genomesonline.org). There has been an almost exponential increase in the number of bacterial species being sequenced in the last ten years and as the sequencing technology becomes even more automated, cheaper and faster, this trend will only continue.

The ability to determine the sequence of a bacterial genome and the accurate prediction of the genes that compose that genome have enabled the comparative study of the relative abundances of mRNA transcripts, using DNA-based microarrays that can contain the entire genome of the bacterium on a single slide. This type of analysis is known as transcriptomics and provides a determination of global gene expression within the studied organism. Transcriptomics is especially applicable to the study of the effects of environmental stimuli on bacteria. The reducing cost of oligonucleotide synthesis, the increased accuracy and efficiency of synthesis of longer oligonucleotides and the commercial and non-commercial availability of microarrays has made possible the transcriptomic analysis of oral bacteria. Transcriptomic analyses have recently been applied to a number of pathogenic and opportunistic bacterial species to determine gene products involved in infection and disease initiation and progression (Boyce *et al.*, 2002; Stintzi *et al.*, 2005). However, a major limitation of this technology is that the level of a mRNA transcript does not necessarily reflect protein abundance (Anderson and Anderson, 1998; Gygi *et al.*, 1999b). Proteins have varying degrees of stability and turnover can be affected by a range of cellular processes, including specific and non-specific proteolytic activity of the bacterium. In addition, post-translational modification and processing of proteins can directly affect protein activity. Therefore, the study of the proteins that make up the proteome provides a more accurate representation of the cellular state, as these are the biomolecules that directly influence cellular activity. The term 'proteomics' was first coined by an Australian scientist, Marc Wilkins from Macquarie University at the 1994 Conference on Genome and Protein Maps (Siena, Italy) and has come to describe the study of all proteins within a particular system, whether that system is a bacterial cell or a subcellular fraction. These analyses largely involve one or more separation steps followed by protein identification using a range of mass spectrometric techniques. Currently, the major limitation of proteomics is the limited dynamic range of detection leading to incomplete identification of the proteome.

Proteomic analyses have already been applied to the discovery and characterization of proteins that comprise the proteome of *P. gingivalis* (Pathirana *et al.*, 2006; Veith *et al.*, 2001; Zhang *et al.*, 2005) and are now being applied to aid our understanding of the effects of environmental cues on the proteomes of this organism (Zhang *et al.*, 2005) and *F. nucleatum* (Zilm *et al.*, 2007). These techniques will no doubt be used in studying the elaborate interactions between cells in the complex microbial biofilm communities found in the oral cavity and the interaction of these bacteria with the host (Lamont *et al.*, 2006).

Oral anaerobic bacterial genomes

The majority of oral bacterial species that have been sequenced to date have, to varying degrees, been implicated in the initiation or progression of forms of periodontal disease. Specifically, the genomes of all of the Red Complex bacterial species (Socransky *et al.*, 1998) as well as *A. actinomycetemcomitans, P. intermedia* and three subspecies of *F. nucleatum*, have been sequenced.

Porphyromonas gingivalis W83

P. gingivalis is an anaerobic, asaccharolytic Gram-negative bacterium that has essential requirements for both iron and protoporphyrin IX, which it preferentially obtains in the form of haem. *P. gingivalis* is an opportunistic oral pathogen linked to chronic periodontitis and was the first of the oral Gram-negative anaerobes to be sequenced by The Institute for Genomic Research (TIGR) in conjunction with the Forsyth Dental Institute. Its annotated sequence was added to the Comprehensive Microbial Database on the TIGR website in June 2001 (Nelson *et al.*, 2003). Interestingly, it was, at the same time, being sequenced by private industry (Commonwealth Serum Laboratories) in Australia as part of a project to develop a vaccine against periodontitis (Ross *et al.*, 2001).

Perhaps surprisingly, only a single strain of *P. gingivalis* has so far been sequenced to date even though a strain-based mode of disease causation has been proposed (Baker *et al.*, 2000; Griffen *et al.*, 1999). Studies of *P. gingivalis* have been largely restricted to two strains W50 (W83) and ATCC33277. The *P. gingivalis* W83 genome is a single, circular chromosome composed of 2 343 479 base pairs. *P. gingivalis* is not known to carry any naturally occurring plasmids and the chromosome is predicted to contain approximately 2000 protein encoding genes out of a total of 2053 predicted open reading frames (ORFs). Approximately 63% of (putative) protein gene products have been assigned to a functional category, whilst ~8% are predicted to have significant similarity to predicted proteins in other species but have no assigned functional role (conserved hypothetical proteins). The function of a further 26% of predicted proteins is unknown and these do not appear to have orthologues in other bacterial species, although this number is likely to decrease as more species are sequenced. Of the predicted proteins, 240 appear to be exported beyond the cytoplasmic membrane, 712 have transmembrane domains, 285 have coil–coil regions and 40 are predicted to be lipoproteins (www.oralgen.lanl.gov/). More detailed information on the molecular biology *P. gingivalis* can be found in Chapter 7 of this book.

Treponema denticola ATCC 35405

Treponema denticola (Chan *et al.*, 1993) is a motile, asaccharolytic, anaerobic spirochete with characteristic helical morphology that is a consequence of its unusual periplasmically located flagellae. *T. denticola* was sequenced by George Weinstock, Steven Norris and colleagues at the Baylor College of Medicine and the University of Texas in collaboration with TIGR. Nearly 2800 ORFs are located on the 2.8 Mb circular chromosome. Of the predicted 2786 protein encoding genes 734 (26%) encode putative proteins that have no significant similarity to any other sequenced protein (Seshadri *et al.*, 2004). Treponemes are evolutionarily quite separate from the majority of Gram-negative eubacteria and have evolved a number of unique characteristics, such as the form of glycoconjugate in the outer leaflet of the outer sheath and the presence of periplasmic flagellae. The ultrastructure of the outer membrane sheath of *T. denticola* is similar to the outer membrane of other Gram-negative bacteria (Wolf *et al.*, 1993), but the lipid composition of the outer sheath is similar to the lipoteichoic acid of the cell surface of Gram-positive bacteria (Schultz *et al.*, 1998). *T. denticola* shares many characteristics with Gram-positive bacteria and recent phylogenetic studies have shown that spirochetes form a single cluster distinct from both Gram-positive and Gram-negative bacteria. It is highly likely therefore that the treponemes as a group will also have evolved a range of other unique characteristics. An example of this is the lack of TonB-linked receptors in the outer membrane of the organism, which could be due to the periplasmic location of the flagellae interfering with the trans-periplasmic TonB protein. The divergence of *T. denticola* from a common treponeme ancestor was an ancient event indicating that in addition to having a range of characteristics unique to treponemes *T. denticola* will have evolved many virulence factors unique at a species level (Seshadri *et al.*, 2004). More detailed information on the molecular biology of this organism can be found in Chapter 8 of this book.

Tannerella forsythia ATCC 43037

Tannerella forsythia is an anaerobic, Gram-negative member of the Cytophaga-Bacteroides family implicated in periodontitis. As with *P. gingivalis* and *T. denticola*, it has significant extracellular proteolytic activity, including trypsin-like serine proteases (Grenier, 1992) and a haemolytic cysteine-protease, designated PrtH (Saito *et al.*, 1997). This organism, originally designated '*Bacteroides forsythus*' (Tanner *et al.*, 1986), has recently been moved into the newly created genus *Tannerella*, based upon its phylogenetic position (Sakamoto *et al.*, 2002). The specific epithet has been revised to '*forsythia*' (Maiden *et al.*, 2003; Olsen and Shah, 2003). Relatively little work has been undertaken on this species because it has been difficult to cultivate in the laboratory. The *T. forsythia* genome was sequenced by Gill and colleagues at TIGR and has been released prior to publication. The *T. forsythia* genome has 3034 predicted ORFs on a single chromosome.

Fusobacterium nucleatum ATCC 25586, 49256 and 10953

Fusobacterium nucleatum is a Gram-negative, saccharolytic, facultative anaerobe that plays a role in the establishment of normal and disease-associated subgingival dental plaque (Kolenbrander and London, 1993). *F. nucleatum* is the type species of the genus *Fusobacterium* and its name characterizes its morphology; it is a spindle (*fusus*)-shaped small rod (*bacterion*) (Bolstad *et al.*, 1996). The organism's ability to ferment carbohydrates (glucose, fructose and galactose) and certain amino acids (Rogers *et al.*, 1991) demonstrates its metabolic versatility and may help explain its prevalence in various sites in the mouth and in systemic infections of the body (Moore and Moore, 1994). Butyric acid, the primary end-product of glutamic acid catabolism by *F. nucleatum*, can effect mitotic arrest in human gingival fibroblasts (Bartold *et al.*, 1991).

Based on the electrophoretic mobility of cellular proteins and DNA homology, the *F. nucleatum* genome was considered to be remarkably heterogeneous and Dzink *et al.* (1990) proposed that the species be divided into three sub-species, *nucleatum*, *polymorphum* and *vincentii*. Of all the subspecies, only *vincentii* is thought to be associated with periodontal disease (Dzink *et al.*, 1990; Gharbia and Shah, 1992). The validity of the three human isolate subspecies and a fourth (subsp. *animalis*) has been controversial, but was confirmed during a meeting in Paris in 2002 by the 'Subcommittee on the taxonomy of Gram-negative anaerobic rods' (Olsen and Shah, 2003). The Type strains for the subsp. *nucleatum*, *vincentii and polymorphum* are ATCC 25586, 49256 and 10953, repectively.

The 2.17 Mb genome sequence of *F. nucleatum* subsp. *nucleatum* was sequenced in 2002 by Integrated Genomics, Chicago (Kapatral *et al.*, 2002). Analysis of 16S rRNA sequence data and a GC content of 27% led to the description of *F. nucleatum* as being 'a Gram-positive organism in Gram-negative clothing' (Sebaihia *et al.*, 2002). Most BLAST hits appear as Gram-positive *Clostridium* species but the outer membrane transport mechanisms and lipid A synthesis pathways suggest that *F. nucleatum* is a sophisticated Gram-negative organism (Mira *et al.*, 2004). The genome consists of 2067 ORFs with 1394 of these predicted to encode proteins which have a putative function, 673 ORFs encode proteins with no known function of which 74 are unique to *F. nucleatum* (Kapatral *et al.*, 2002). The 'patchy' nature of the *F. nucleatum* genome is thought to have resulted from a high degree of horizontal gene transfer of niche-specific genes between bacteria which inhabit dental plaque. This topic is dealt with in greater detail in Chapter 3 of this book.

The draft genome sequence for ATCC 49256 (subsp. *vincentii*) has also been released by Integrated Genomics and the comparison with ATCC 25586 has been published by (Kapatral *et al.*, 2003). The genome contains a total of 2277 ORFs of which 1576 are predicted to encode proteins that have been assigned a function; 570 encode hypothetical proteins and 441 ORFs (20% of the genome) are not present in subsp. *nucleatum*. Some of these genes have functional orthologues in other species but 118 are unique to subsp. *vincentii*. However, they are extremely short (~440 bp) suggesting they do not represent 'real' genes (Mira *et al.*, 2004). In terms of bacterial physiology, subsp. *vincentii* was also found to encode unique peptidases, suggesting that the organism can catabolize a broader range of peptides. The presence of eukaryotic-type phosphatases and serine/threonine kinases also suggests that subsp. *vincentii* possesses unique signal transduction pathways (Kapatral *et al.*, 2003).

The sequence for ATCC 10953 (subsp. *polymorphum*) has recently been completed by The Human Genome Sequencing Centre, Baylor College of Medicine, Texas under the guidance of George Weinstock and publication is expected in 2007. The chromosome is estimated to be 2.4 Mb and differs from ATCC 25586 in that it contains three plasmids (pFn1, pFn2 and pFn3) which have become useful as potential vectors for gene transfer systems (Haake *et al.*, 2000).

A.a. HK1651

A. actinomycetemcomitans is a small, non-motile capnophilic Gram-negative rod that was, until recently, considered to be a member of the *Actinobacillus* genus (Norskov-Lauritsen and Kilian, 2006). *A. actinomycetemcomitans* has been designated an aetiological agent of aggressive forms of periodontal disease that typically affect younger adults. The genome sequence has been made available on the TIGR website, in advance of publication by Dyer, Roe and colleagues at the University of Oklahoma. More than 3000 ORFs are present on the 2.1 Mb circular chromosome. In addition there is an *A. actinomycetemcomitans* plasmid, pVT745 (approximately 925.4 kb) containing 36 ORFs, some of which are orthologous to chromosomal genes (Chen *et al.*, 2002a). More detailed aspects of the molecular biology of this organism are to found in Chapter 6 of this book.

Prevotella intermedia strain 17

P. intermedia (formerly *Bacteroides intermedius*) is a saccharolytic, black-pigmented, Gram-negative anaerobic bacillus believed to be involved in some forms of periodontal disease (Ximenez-Fyvie *et al.*, 2000). The *P. intermedia* genome sequence has been made available prior to publication by John Heidelberg at TIGR and Margaret Duncan at the Forsyth Institute. There are two main DNA molecules in *P. intermedia*; namely, a chromosome of approximately 2120 kb with 2178 predicted ORFs and a smaller chromosome or megaplasmid (pPI1) of 580 kb containing 642 ORFs. In addition, there is also a much smaller (5.1 kb) naturally occurring plasmid (pYHBi) that contains two sequenced genes. Some 48.4% of the predicted *P. intermedia* protein-encoding genes have been assigned a functional category whilst 39% have no known function and are not conserved with any other sequenced species; and 11% are conserved with other species but still do not have an assigned function.

More than a quarter of the genome of each of the species described above encodes for proteins with no significant similarity to other known proteins currently available in the databases. In addition, a further 8–11% of the predicted proteins have some identity with predicted proteins from other species that have no known function and in many cases are hypothetical. Therefore, between one third to a half of all the genes in these species encode proteins that have no known or predicted function. In addition to this, many putative proteins will have significantly different functions to those predicted from sequence data alone. For example, in *P. gingivalis*, a predicted homologue of CbiK, an intracellular enzyme involved in cobalamin synthesis (Roper *et al.*, 2000), was shown to be a lipoprotein located on the outer membrane of *P. gingivalis* and involved in haem capture (Dashper *et al.*, 2000). Proteomic and genomic analyses of these species offer opportunities to determine the environmental conditions under which these proteins are produced which can, in turn, aid in elucidating their function. There will, therefore, be an ongoing need for detailed biochemical characterisation of proteins discovered by genomic and proteomic analyses.

Comparative genome analyses (*in silico* searching)

Comparison of the sequenced *P. gingivalis* W83 strain to the 'avirulent' type strain ATCC 33277 revealed that 7% of genes were absent or highly divergent in strain 33277 (Chen *et al.*, 2004). In strain W83, a large genetic locus containing genes encoding enzymes believed to be involved in the synthesis of capsular polysaccharide was identified; however, it was absent in ATCC 33277. More detailed information on the *P. gingivalis* strain differences can be found in Chapter 7 of this book.

In silico comparisons between sequenced genomes of different species can be highly informative. For example *P. gingivalis* possesses a family of up to 34 cell surface-located outer membrane proteins that have no significant sequence similarities, apart from a conserved C-terminal domain(CTD) of approximately 80 residues (Seers *et al.*, 2006; Veith *et al.*, 2002). This family of proteins includes the gingipains, Lys- and Arg-specific cysteine proteinases and associated adhesins that are considered to be the major virulence factors of this bacterium (Abe *et al.*, 1998; Adamson *et al.*, 1993; Curtis *et al.*, 1999; O'Brien-Simpson *et al.*, 2001; O'Brien-Simpson *et al.*, 2003; Potempa *et al.*, 1995). In addition, many other putative proteinases, adhesins, haem-binding proteins and immunoreactive proteins are predicted to belong to this novel family of proteins. Most of these predicted proteins have typical *P. gingivalis* type I cleavable signal peptides, indicating that they would be exported to the periplasm by the SEC general secretory system. The CTD has been proposed to play a number of roles in the secretion of the proteins across the outer membrane and their attachment to the surface of the cell, probably via glycosylation. BLAST similarity searches with the conserved CTD sequence revealed that some species of the *Bacteroidetes* class, such as *Bacteroides fragilis* and *Bacteroides thetaiotaomicron*, which are closely related to *P. gingivalis*, do not contain genes encoding proteins predicted to contain the CTD. However, other members of the *Bacteroidetes* including *Rhodothermus marinus*, *Cytophaga hutchinsonii*, *Zobellia galactanovorans*, *Flavobacterium columnare* and two oral species, *T. forsythia* and *P. intermedia*, that have also been implicated in the progression of periodontal diseases were predicted to have proteins containing the CTD (Nguyen *et al.*, 2007; Seers *et al.*, 2006).

A recent report has implicated PorT, a protein associated with the *P. gingivalis* periplasm, in translocation of the gingipain and HagA adhesin precursors which belong to the CTD family. A PorT mutant accumulated unprocessed gingipains and HagA within the periplasm (Saito *et al.*, 1997). BLAST searches of sequenced bacterial genomes identified PorT homologues in *P. intermedia, T. forsythia, C. hutchinsonii* and other species containing CTD family proteins (Nguyen *et al.*, 2007; Seers *et al.*, 2006). PorT homologues were not found in species that did not have predicted CTD family proteins. This is good evidence that PorT plays a role in the export and/or attachment of the CTD family proteins.

An adhesin of *A. actinomycetemcomitans* (*A.a*) was recently identified using the sequence of a known autotransporter adhesin, Hap, from the related bacterium *H. influenzae* (Rose *et al.*, 2003). A search of the *A.a.* genome revealed a homologue with significant similarity in the pore-forming domain; indeed, inactivation of this gene was shown to reduce the binding of *A.a.* to epithelial cells.

Qualitative proteomics

P. gingivalis outer membrane proteome analyses using 2D-PAGE separation followed by mass spectrometry

Commensal and pathogenic bacterial cells do not grow in isolation, having many interactions with the other bacterial cells that share their environment. Additionally, they have a range of interactions with the host, the outcomes of which can lead to disease. The proteins that are located on (or in) the surface of the bacterial cell are those that are most likely to mediate interactions between other cells and the host immune response. They are also responsible for the sensory capacity of the organism. Consequently, of this the description of the outer membrane proteome of Gram-negative parasitic bacteria is of great importance in furthering our understanding of the disease process and for the development of vaccines (Ross *et al.*, 2001). This is especially pertinent for the study of oral subgingival bacteria such as the members of the Red Complex which have a significant amount of extracellular (usually cell surface-associated) proteolytic activity. In the case of *P. gingivalis*, the gingipains, extracellular cysteine proteinases and associated adhesins are recognized as the major virulence factors, as well as functioning to provide the cell with peptides and essential micronutrients, such as iron and protoporphyrin IX and playing a role in protection against oxidative stress. Three genes encode the gingipains and these have been designated *rgpA*, *rgpB* and *kgp* (Curtis *et al.*, 1999). The Arg-gingipain (RgpA) is a polyprotein that is proteolytically processed to produce an N-terminal Arg-Xaa-specific cysteine proteinase ($RgpA_{cat}$) and several adhesins that are designated $RgpA_{A1}$, $RgpA_{A2}$, $RgpA_{A3}$ and $RgpA_{A4}$, whilst the Kgp polyprotein is processed to produce an N-terminal Lys-Xaa-specific cysteine proteinase (Kgp_{cat}) and the adhesins Kgp_{A1}, Kgp_{A2}, Kgp_{A3}, Kgp_{A4} and Kgp_{A5} (O'Brien-Simpson *et al.*, 2003). The RgpA and Kgp proteinase and adhesin domains occur together as a surface associated complex (O'Brien-Simpson *et al.*, 2003; Slakeski *et al.*, 1999), with relatively little of the complex found free in the extracellular milieu. The second Arg-Xaa-specific cysteine proteinase, RgpB, has high sequence similarity to RgpA but lacks the adhesin domains and is not associated with RgpA and Kgp on the cell surface. A considerable amount of the current information regarding these proteinase and associated adhesins and their processing has been obtained by proteomic analyses of the *P. gingivalis* outer membrane proteome, using a two dimensional polyacrylamide gel electrophoresis (2D-PAGE) separation and visualization step followed by identification of the proteins using mass spectrometry techniques (Veith *et al.*, 2004; Veith *et al.*, 2002).

2D-PAGE is an ideal method for visualizing bacterial subcellular proteomes due to its high resolving power and provision of both the molecular weights and isoelectric points of the separated proteins. There are a number of forms of 2D-PAGE. The conventional method was first introduced in 1975 and utilized iso-electric focussing (IEF) for the first dimension followed by SDS-PAGE in the second dimension (Klose, 1975; O'Farrell, 1975). Conventional 2D-PAGE can be carried out using two different IEF methodologies. IEF is electrophoresis within a pH gradient in which amphoteric compounds, such as proteins, will migrate the until each one reaches its iso-electric point (pI). Initially, IEF involved the use of a mixture of small amphoteric molecules termed carrier ampholytes (CA), with high conductivity and buffering capacity at their pI. Immobilized pH gradients (IPG) in which the pH gradient-forming molecules, termed 'immobilines', are covalently bonded to the acrylamide gel enabling higher reproducibility and increased sample capacity (Bjellqvist *et al.*, 1982), have recently become available.

In a 2D-PAGE experiment, utilizing IPGs, the proteins to be analysed are solubilized in high concentrations of chaotropes such as urea and thiourea, and non-ionic or zwitterionic detergents such as CHAPS and triton X-100. The IPG strip is purchased dry, and must first be rehydrated before application of the sample using the 'cup' method (Görg *et al.*, 1995). Alternatively, the IPG strips may be rehydrated in the sample directly (Rabilloud *et al.*, 1994). IEF is conducted at voltages usually in excess of 3000 V to achieve maximum focussing power, after which the IPGs are equilibrated in an SDS-containing buffer and placed on top of an SDS-PAGE gel. In the past, 2D-PAGE was often considered an onerous task but, with the advent of IPGs, modern equipment and solution kits, the technique is considerably more user-friendly and less time consuming. Complete 2D-PAGE experiments using a small (approx. 7 × 7 cm) 2D gel format can be performed within one day.

Although limited in its capacity to resolve highly hydrophobic proteins, 2D-PAGE can be very successfully applied to the separation of outer membrane proteins of Gram-negative bacteria, as these are generally much less hydrophobic than cytoplasmic membrane proteins. Much detailed information on proteins, including post-translational modifications and proteolytic processing, can be obtained from 2D gels when combined with mass spectrometric identification. It is a powerful tool with which to explore complex protein mixtures, as can be demonstrated using the outer membrane of *P. gingivalis* W50 as an example (see below).

Proteins are identified by mass spectrometry (MS) using either peptide mass fingerprinting (PMF) or peptide fragmentation data. In both cases, gel-separated proteins need first to be digested with a specific endoproteinase, such as trypsin. Digestion of proteins within gels tends to work well as the proteins are generally fully denatured (by SDS), concentrated and easy to wash or derivatise, since the protein is trapped within the gel matrix. As most complex protein mixtures contain cysteine residues, it is highly advisable to derivatise (reduce and alkylate) these residues as part of the 2D PAGE procedure, or during the in-gel digestion procedure. This aids digestion and simplifies the set of digested peptides obtained. For most proteins, a quality digest can be obtained by simple diffusion of the digested peptides out of the gel into the surrounding aqueous solution during digestion (Mortz *et al.*, 2001). However, large or hydrophobic peptides may be extracted more efficiently by repeated extraction of the digested peptides (from the gel pieces) with organic solvents (Shevchenko *et al.*, 1996).

PMF involves measuring the mass of the resultant peptides on a mass spectrometer, and submitting the peptide mass list to a computer program (see below). Assuming the sequence of the protein is present in an available database, the program compares the experimental peptide mass list to a theoretical digest of all database proteins, such that the number and accuracy of the peptide mass-matches to each protein is recorded. Usually, an entire sequence database belonging to the organism under investigation is searched in order to find the protein that best matches the data. A positive match should have a score significantly higher than all other proteins in the database (Fig. 2.1).

Using PMF, a protein is identified by having a significant number of peptide mass matches. However, each individual peptide by itself cannot usually be considered to be positively identified. In contrast, peptide fragmentation data (MS/MS) is used to identify specific peptides using both the mass of the whole peptide as well as the fragment ion masses. The sequence of the peptide can sometimes be elucidated directly from the MS/MS spectrum by advanced users (*de novo* sequencing) but, most frequently, an MS/MS ion search is conducted in much the same way as a PMF search is done, in order to match up the best peptide from a protein database. Usually, the identified peptide is unique to a particular protein and hence the protein can be positively identified. MS/MS is very powerful because a single peptide that fragments well can produce very high

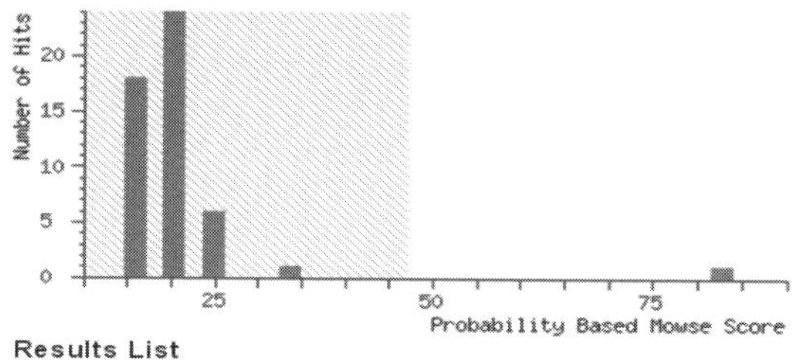

Results List

1. TF1331 **Mass:** 44534 **Score:** 83 **Expect:** 1.6e-005 **Queries matched:** 9
outer membrane protein

Observed	Mr(expt)	Mr(calc)	Error	Peptide	
848.41	847.40	847.39	0.01	K.NNNSFPR.A	
1013.61	1012.60	1012.59	0.01	R.IAPSGAISIGK.W	
1247.61	1246.60	1246.56	0.04	K.WYNPYMAFR.L	
1263.63	1262.62	1262.55	0.06	K.WYNPYMAFR.L	+ Oxidation (M)
1281.71	1280.70	1280.65	0.05	R.ENLQNVVYFR.L	
1319.65	1318.64	1318.58	0.06	K.RPEFCPECPK.C	
1636.86	1635.85	1635.79	0.06	K.HQEVSIFNTAEYAK.K	
1993.07	1992.07	1992.00	0.07	R.IDKHQEVSIFNTAEYAK.K	
2000.14	1999.13	1999.06	0.08	R.FIPFVGLGYALRPGYSDK.N	

2. TF0114 **Mass:** 49131 **Score:** 32 **Expect:** 2 **Queries matched:** 4
conserved hypothetical protein

Observed	Mr(expt)	Mr(calc)	Error	Peptide
1307.72	1306.72	1306.62	0.10	K.SPNLNSGESKMK.Y + Oxidation (M)
1614.82	1613.81	1613.79	0.02	K.AGFDNFAAWIDFLK.Q
1636.86	1635.85	1635.88	-0.03	R.TGHGHLLGISPYRTK.T
2000.14	1999.13	1998.99	0.14	R.YKPYNVQGFEVNETRR.I

Figure 2.1 Typical PMF search result for an identified protein. PMF search result of MALDI-TOF data using Mascot 2.1 (Matrix Science Ltd). The Mascot score shown for the top ranking protein is 83. This is considerably higher than the threshold of 47 (shaded), indicating a positive match. The second ranked protein has a much lower score of 32 which is below the threshold and therefore not significant.

scoring matches, even in large databases, and is particularly useful for identifying proteins that do not produce enough peptides to be routinely identified by PMF.

In a recent study of outer membrane preparations of *P. gingivalis* W50 cells grown under haem-limited conditions in continuous culture (see below), were subjected to 2D-PAGE. The pattern of spots obtained was reproducible, and included proteins represented by single spots and many proteins that were represented by a train of spots (Fig. 2.2). In addition, some proteins were represented as diffuse streaks suggesting large and potentially heterogeneous modifications, such as glycosylation and possibly lipidation (marked with a box in Fig. 2.2). Mass spectrometric analysis of these proteins resulted in the identification of four major *P. gingivalis* proteins, RgpB, P59, $RgpA_{A4}$ and P27. Confident identification of any of these proteins by N-terminal sequencing was difficult, as P59 and P27 are N-terminally blocked with pyroglutamate, and RgpB and $RgpA_{A4}$ share N-terminal sequence identity with related proteins. The predicted mass of each of these proteins was substantially less than that observed on the gels, consistent with extensive modification. Furthermore, the MS-identified peptides were all in the N-terminal portions of the proteins suggesting that the C-terminal 80 amino acid residues played some role in modification (Veith *et al.*, 2002). The C-terminal 80 amino acid residues had significant similarity to each other and as a result of this work, the conserved C-terminal domain (CTD) of these *P. gingivalis* proteins was identified and is the subject of continuing research to investigate its role in the secretion and attachment of the novel CTD family proteins (Nguyen *et al.*, 2007; Seers *et al.*, 2006) (see above). Further to the discovery of the CTD, spots identified as $RgpA_{A1}$ and Kgp_{A1} were found not only at their expected MW but also at elevated MW, suggesting that these proteins may also be modified (Fig. 2.2), albeit in a manner different to the CTD family proteins. The nature and location of these modifications remains unclear.

MS analysis of outer membrane preparations of *P. gingivalis* separated by 2D-PAGE was used to determine the proteolytic processing sites of the gingipains. MS analyses of protein spots

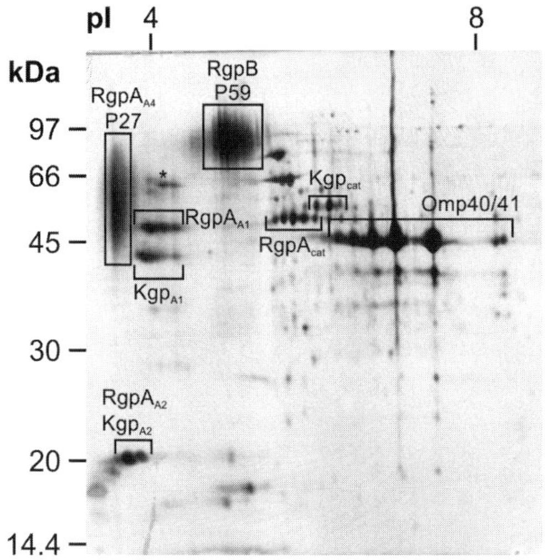

Figure 2.2 2D-PAGE of *P. gingivalis* outer membrane. 2D-PAGE was performed using 17 cm pH 3–10 IPG strips and a 20 cm 2nd dimension (Veith *et al.*, 2002) and silver stained. Labelled spots were identified by PMF (Veith *et al.*, 2002). The asterisk (*) indicates the location of higher MW forms of $RgpA_{A1}$ and Kgp_{A1}.

labelled $RgpA_{cat}$, $RgpA_{A1}$, $RgpA_{A2}$, $RgpA_{A3}$, Kgp_{cat}, Kgp_{A1}, Kgp_{A3} revealed that each one had been C-terminally processed N-terminal to a Lys residue (Fig. 2.2). The enzymes responsible for this processing were shown to be the Lys-specific cysteine proteinase Kgp, and CPG70 which is a surface located Lys/Arg-specific carboxypeptidase (Chen *et al.*, 2002b). This was achieved by identifying differently processed C-terminal peptides in individual isogenic mutants that lacked the respective enzyme (Veith *et al.*, 2004; Veith *et al.*, 2002).

In 2D gels of the *P. gingivalis* outer membrane preparations many of the protein spots were in horizontal trains, which may indicate the presence of stable protein modifications, such as phosphorylation and glycosylation. However, they may also be artefacts, as was shown for the Omp40/41 train (Fig. 2.2) (Veith *et al.*, 2001). In this study, single 2D gel spots corresponding to Omp40 or Omp41 were extracted and subjected to a second round of 2D-PAGE. It was found that each single spot was able to reproduce the whole train of spots and so it was concluded that the different spots represented different, but interchangeable, conformations of the protein rather than the result of chemical modification (Veith *et al.*, 2001).

In addition to its ability to provide detailed information about specific proteins, 2D-PAGE with MS identification always provides a useful reference for subsequent studies; especially for comparing isogenic mutants to each other and the wild-type and allowing the confirmation of hypothetical proteins. In the *P. gingivalis* study described above, the majority of outer membrane proteins identified had not been previously observed.

For large numbers of gel-separated proteins, automated tools become very helpful. Available tools include robotics for spot excision, tryptic digestion, and preparation of MALDI-TOF target plates, and automated MS analysis including database searching. Laboratory information management systems that manage and integrate the data are also available. These allow, for example, a spot on a gel to be linked to its associated mass spectra and database search results. For instance, outer membranes of *T. forsythia* were separated on a mini-2D gel system and stained with colloidal Coomassie Blue G250. 192 protein spots were excised with a robot and placed into two 96-well plates. The gel pieces were manually digested with trypsin and the resultant peptide solution was manually spotted onto the target plate of a MALDI-TOF/TOF instrument. PMF was automatically performed on each sample and this was followed by MS/MS analyses on up to five peptides per sample. In this way 120 gel spots have been confidently identified (unpublished data). Robots are also available for in-gel digestion and sample preparation on to MALDI-MS target plates. However, however caution must be exercised in the use of robots, especially for in-gel digestion as their performance is frequently lower than can be routinely achieved manually, and their optimization can be very tedious.

A.a. immunoproteomics

The identification of the major immunoreactive proteins from a bacterium, that is, 'immunoproteomics' can be achieved using reactive host sera to probe transblots of 2D gels that have been used to separate outer membrane protein preparations in order to identify proteins that are antigenic. Using this technique, a strongly immunoreactive outer membrane peptidoglycan-associated lipoprotein (*A.a.*PAL) from *A.a.* has been identified (Paul-Satyaseela *et al.*, 2006).

Functional identification of a *P. intermedia* outer membrane protein

A proteomic approach has recently been used for the identification of a *P. intermedia* extracellular matrix binding protein (Yu *et al.*, 2006). Prior to the production of an outer membrane protein preparation, *P. intermedia* cells were labelled with the succinimidyl ester of the fluorescent cyanine dye Cy5. This procedure labelled only surface-exposed proteins, thereby acting as an internal control for the quality of the outer membrane protein preparation. Proteins were then separated by conventional 2D gel electrophoresis and immediately transblotted to nitrocellulose membranes of the same size as the 2D gels. These membranes were then incubated with a fibronectin solution, washed and the location of any bound fibronectin determined by Western Blot analysis. Eight potential fibronectin-binding protein spots were detected, although five were considered to have been the result of non-specific binding. A single spot with a M_r of 29 kDa and a pI of 6.5 was identified by in-gel tryptic digestion of a corresponding spot on a 2D gel, followed by peptide mass fingerprinting. PMF analysis identified the protein as PIN A1012 (TIGR annotation), a protein predicted to be composed of 381 amino acids with a mass of 42 kDa. Only those peptides from the C-terminal two thirds of the predicted protein were identified, starting at residue 134. This indicates that the protein may have been N-terminally truncated, which is supported by the mass of 29 kDa calculated from the migration of the protein in the second dimension of 2D GE which is considerably smaller than the predicted mass. Although the protein was predicted to have an outer membrane location it does not contain an identifiable leader sequence which is usually found in exported proteins. It is still unclear as to whether the N-terminal domain of this protein is removed by post-translational processing or by artifactual cleavage during protein preparation.

A recombinant protein composed of the C-terminal 263 amino acids of the predicted PIN A1012 protein was then expressed in *E. coli*, purified and used to confirm fibronectin binding (Yu *et al.*, 2006). This protein was designated AdpB (**Ad**hesin from **P**revotella B). In addition to binding fibronectin this recombinant AdpB also bound laminin and fibrinogen and, given its ability to bind to a range of extracellular matrix proteins, it may play a role in binding to, and invasion of, epithelial cells. Analysis of mutants lacking AdpB will provide further insights into the role of this protein in the binding to, and invasion of, epithelial cells by *P. intermedia*.

This study demonstrates the way in which the proteomic approach can be used to uncover proteins having not only a specific subcellular location but also a specific function. However, the denaturing conditions of 2D PAGE render it unlikely that proteins will be able to refold to their native conformation when transblotted from the 2D gel to the nitrocellulose membrane, and so this technique has limited applicability. Properties and activities that require a native three dimensional structure of the protein are therefore unlikely to be detected. Nevertheless, the technique is useful in situations where binding to certain substrates only requires a linear sequence.

P. gingivalis proteomic analyses using multidimensional LC-MS

In a recent study, more than 1000 *P. gingivalis* proteins were successfully identified using a whole proteome in-solution enzymatic digest and three dimensional (3D) LC-MS separation and identification (Wang *et al.*, 2002). This method was not significantly biased against hydrophobic proteins, a phenomenon often reported for 2D-PAGE separation (Santoni *et al.*, 2000). In addition, the abundance of proteins in whole bacterial proteomes can have huge ($>10^5$:1) dynamic ranges. Multidimensional LC-MS has the ability to detect proteins across a large dynamic range thus enabling the identification of a larger number of proteins than is possible using 2D gel-based techniques (Washburn *et al.*, 2001). The same research group identified a comparable number of proteins using similar techniques as part of a proteomic approach to determine the effect of secreted epithelial cell components on the proteome of *P. gingivalis* (Zhang *et al.*, 2005). Each protein was identified by 2 or more unique peptides representing over 45% of the theoretical proteome of this species. This was the first time that the existence of 275 of these proteins had been demonstrated, as they were previously categorized as 'hypothetical' or 'conserved hypothetical'. As multidimensional LC-MS approaches are capable of identifying large numbers of proteins, they are frequently used for quantitative expression proteomics (see below).

Quantitative (differential) proteomics

A direct measurement of the changes in abundance of bacterial proteins, and interconversion of protein isoforms between two states in response to environmental stimuli, can provide useful data to further our understanding of oral microbial ecology and disease initiation and progression. The ability to determine quantitative protein changes on a global scale will allow the identification of groups of proteins that respond in concert to environmental stimuli (Andersen *et al.*, 2003; Dunkley *et al.*, 2004; Patwardhan *et al.*, 2006). In diseases that have a strain-based bacterial aetiology, such as has been proposed for chronic periodontitis, the ability to compare global protein abundances in pathogenic and non-pathogenic strains is of obvious benefit to understanding the pathogenicity of a bacterium. It may also be of use in the development of novel targeted therapeutics or the identification of molecular markers for the prediction or diagnosis of disease. The global responses of bacteria to environmental changes are the result of complex signal transduction pathways and available data examining bacterial physiology is currently helping to elucidate some of the environmental cues to which these pathways respond. Proteomic and transcriptomic studies are beginning to aid our understanding of the physiological data by providing a differential analysis between bacteria grown under different, but relevant, environmental conditions. This, in itself, provides a challenge for researchers as altering only one environmental parameter while keeping all others constant has lead to the resurgence of techniques such as continuous culture of bacteria in a chemostat (Hoskisson and Hobbs, 2005).

Quantitative, differential or comparative proteome analyses can be accomplished in a number of ways. These can be broadly divided into:

(1) Gel-based methods, usually involving 2D GE and comparison of individual spot staining intensity, followed by PMF protein identification.
(2) Label-free methodologies using the MS intensities of the tryptic peptides derived from a protein.
(3) Stable isotope labelling where one of a number of non-radioactive isotopes is incorporated into peptides either metabolically, *in vivo*, or during protein processing, *in vitro*.

Effect of growth pH on the *F. nucleatum* cytoplasmic proteome determined by 2D-PAGE-based differential proteomics

The wide diversity of roles that *F. nucleatum* plays in the development of plaque and its prevalence at periodontal sites suggest that this organism is a prime candidate for functional genomic investigations. Classified as a member of the Orange Complex (Socransky *et al.*, 1998), the organism contains a plethora of carbohydrate and protein adhesins which enable it to attach to both the antecedent and late colonizers of dental plaque, and so it has been recognized as a 'microbial bridge' during plaque development (Kolenbrander and London, 1993). Apart from its ability to coaggregate with other bacterial taxa, *F. nucleatum* has been shown to attach to human epithelial cells, collagen and gingival fibroblasts (Han *et al.*, 2000; Edwards, 2206; Xie, 1991). In addition to its role in attachment and coaggregation, it is also believed to play a role in oral diseases by initiating physico-chemical changes in the gingival sulcus which are conducive to the colonization and growth of Red Complex anaerobic periodontal pathogens (Socransky *et al.*, 1998). In order to fulfil this role, *F. nucleatum* must be able to grow and proliferate in the shallow periodontal pocket and initiate and respond to a number of environmental changes, including variations in oxygen levels, elevated temperature, nutrient sources and pH (Diaz *et al.*, 2002; Bickel, 1985; Haffajee, 1992).

An important change associated with transition from health to periodontal disease is the alkalinisation of the gingival sulcus (Bickel and Cimasoni, 1985). It has been proposed that the saccharolytic organisms *F. nucleatum* and

Prevotella intermedia initially colonise the gingival sulcus (Takahashi, 2005). During the early stages, the pH of the pocket may be acidic but the growth of these bacteria and the associated metabolism of host derived proteins found in gingival crevicular fluid leads to a rise in pH and creates a pH neutral environment. The initial rise in pH then allows acid-sensitive and more proteolytic species associated with the Red Complex to colonise and proliferate. The later and more severe stages of the development of periodontal disease (periodontitis) are associated with the destruction of host tissues and the increased efflux of gingival crevicular fluid. The continued fermentation of amino acids and subsequent generation of ammonia produces a rise in pH above neutrality which has been reported to be as high as 8.5 (Bickel and Cimasoni, 1985).

The role of *F. nucleatum* in periodontal disease has been equivocal due to its presence in both healthy and disease-associated dental plaque. The theory proposed by Takahashi (2005) may only partially explain this observation but it does highlight the ability of *F. nucleatum* to respond to, and successfully proliferate in, environments undergoing relatively large changes in external pH. Although the genome sequences of *F. nucleatum* subspecies *nucleatum* and *vincentii* are currently available (Kapatral *et al.*, 2002; Kapatral *et al.*, 2003), few applied genomic studies have been published.

In a recent study, Zilm *et al.* (2007) used a proteomic approach to investigate the response of *F. nucleatum* ATCC 10953 (subspecies *polymorphum*) to environmental pH changes which might be expected to occur in the gingival sulcus during the transition from health to disease. Cytoplasmic proteins prepared from the bacterium grown in continuous culture at pH 6.4, 7.4 and 7.8 were compared, by 2D-PAGE, with those derived from growth at pH 7.2. The continuous culture of *F. nucleatum* in a chemostat enabled the researchers to grow and harvest cells over several generations after the growth pH was altered, thereby providing an alternative to producing biological replicates normally required for conventional batch culture studies. *F. nucleatum* was cultured in a chemically defined medium and replicate samples were taken after steady state had been achieved (10 generations), following exposure to imposed changes in extracellular pH. In terms of the technical aspects of experimental design, the chemostat provides the means of harvesting relatively large culture volumes, a factor which must be considered if large (17–22 cm) IPG strips are to be used. 17 cm IPG strips were used to examine a subset of the proteome in which only proteins with a pI range of 4–7 were separated. The use of IPG strips with a wider pI range (e.g. 3–10) may be useful in maximizing proteome coverage but recent studies have highlighted the fact that poor resolution can sometimes occur as single spots have been found to contain 2–5 proteins (Campostrini *et al.*, 2005).

For each growth pH, replicate 2D gels ($n = 3$–6) were produced, stained with Coomassie blue R-250 and gel images acquired using a densitometer operated by the 2D-software program PD-Quest (version 7.2). While this program allows for automatic spot detection and matching, manual editing was found to be necessary. Gels were normalized by averaging the total density of all spots in all gels and replicate groups, representing only gels derived from each growth pH, were created. Differences in the means of protein quantity between replicate groups were analysed for statistical significance using Student's *t*-test.

Twenty two proteins with altered expression were excised and identified by μLC Tandem MS following tryptic digestion. The *de novo* sequence tags derived from the fragmentation data were used to interrogate the National Centre for Biotechnology Information (NCBI) non-redundant protein database for protein identification matches. The identification of proteins was correlated to the number of identified peptides from a particular protein but, due to the variability in protein sequences between the *nucleatum* and *vincentii* sub-species represented in the database, matching also relied upon manual *de novo* sequencing of peptides and BLAST searching of derived sequence tags to identify protein homology. Only those proteins having an isoelectric point (pI) between 4 and 7 and a molecular weight between 10–80 kDa were examined in this study. An *in silico* representation of the *F. nucleatum* (subsp. *nucleatum*) proteome predicted that, from the total of 2067 proteins, 890 should fall within this range.

When compared with protein expression at pH 7.2, eight proteins were found to have altered expression during steady state growth pH of 6.4 (Table 2.1). Upregulated proteins were almost exclusively linked to the generation of ATP via the catabolism of amino acids, such as glutamic acid and histidine. Physiological data showed that cell numbers were reduced following acidification, indicating that the generation of ATP may be needed for proton extrusion via an H^+-ATPase. The downregulation of peptidyl-prolyl *cis–trans* isomerase was consistent with the drop in cell numbers as this enzyme has been associated with the regulation of cell division in eukaryote cells. Of note was the upregulation of oxygen-insensitive NADPH nitroreductase dihydropteridine reductase, also reported to be upregulated in *Streptococcus oralis* following batch growth at pH 5.5 (Wilkins *et al.*, 2001).

Nine proteins were up- or downregulated following a shift in growth pH from 7.2 to 7.8 (Table 2.2). Proteins associated with protein translation were predominantly upregulated and included the 60 KDa chaperonin, a known stress-induced protein. The upregulation of formimino-tetrahyrofolate cyclodeaminase suggested that the resulting ATP and ammonia production may contribute to alkalinity. Enzymes associated with the Embden–Meyerhof pathway were also shown to be influenced by external growth pH. The apparent six fold upregulation of fructose-bisphosphate aldolase and downregulation of enolase and the allosterically controlled enzyme, pyruvate kinase, supported earlier work showing that intracellular polyglucose (IP) levels were influenced by the generation of ATP from glutamic acid metabolism and that increased IP was associated with growth at alkaline pH (Robrish *et al.*, 1987; Robrish and Thompson, 1990; Wilkins *et al.*, 2006; Zilm *et al.*, 2002). Fructose-bisphosphate aldolase has been proposed as a branch point in determining the fate of glucose (Robrish and Thompson, 1990) and the downregulation of pyruvate kinase and enolase may cause the build up of glycolytic intermediates and increase the synthesis of IP (Zilm *et al.*, 2007).

As discussed earlier, the use of global differential proteomic studies will continue to improve our understanding of bacterial physiology and ecology. One of the main advantages of this

Table 2.1 *F. nucleatum* proteins regulated by growth at pH 6.4

Protein name	NCBI accession no.*	Mean ratio of protein abundance pH 6.4 / 7.2[a]	Function
Glutamate forminotransferase	19714276	2	Histidine catabolism (ATP synthesis)
Anthranilate synthase	19713994	4.1	Tryptophan synthesis (ATP synthesis)
Acetoacetate butyrate CoA transferase	19713099	1.9	Butyrate synthesis (ATP synthesis)
Glutaconate CoA transferase subunit B	19713644	1.6	Butanoate metabolism (ATP synthesis)
O-insensitive NADPH nitroreductase dihydropteridine reductase	34764235	14	Co-factor biosynthesis
Peptidyl-prolyl *cis-trans* isomerase	19713714	0.2	Cell division
60 kDa chaperonin (GroEL)	19714198	0.6	Protein translation
Hypothetical protein	34762615	0.1	Unknown

*National Centre for Biotechnology Information (May 2005 release).

[a]Significantly up- or downregulated ($p<0.05$) as calculated using Student's *t*-test.

Table 2.2 *F. nucleatum* cytoplasmic proteins regulated by growth at pH 7.8

Protein name	NCBI accession no*.	Mean ratio of protein abundance pH 7.8/7.2[a]	Function
Formiminotetrahyrofolate cyclodeaminase	19714274	16.4	One carbon metabolism
Elongation factor Ts	60416365	2.1	Protein translation
70 kDa chaperonin (DnaK)	19713544	2.7	Protein translation
60 kDa chaperonin (GroEL)	19714198	1.6	Protein translation
FtsZ cell division protein	19715120	1.9	Cell division
Hypothetical cytosolic	34762615	2.7	Unknown
Fructose-bisphosphate aldolase	19713785	6.4	Glycolysis
Enolase	19713005	0.8	Glycolysis
Pyruvate kinase	19713006	0.5	Glycolysis

*National Centre for Biotechnology Information (May 2005 release).

[a]Significantly up- or downregulated ($p<0.05$) as calculated using Student's *t*-test.

approach is that the regulation of proteins not normally associated with certain environmental cues can be discovered. This will aid our understanding of the complex signal pathways used by bacteria as a coordinated response to environmental changes/stresses.

Effect of secreted epithelial cell components on the *P. gingivalis* proteome determined by label-free differential proteomics

Much work has been conducted on the ability of oral bacteria to invade epithelial cells (Lamont and Jenkinson, 1998), although the role of epithelial cell invasion by periodontopathogenic bacteria in disease initiation and progression is still debatable. It has been proposed that epithelial cell invasion during times of stress allows bacteria to survive *in situ* and can act as a reservoir for recolonization of a site. *P. gingivalis* 33277 has been shown to readily invade gingival epithelial cells whilst strain W50 has been shown to be internalized by epithelial cells at a much lower frequency (Belton *et al.*, 1999; Duncan *et al.*, 1993; Lamont *et al.*, 1995). Epithelial cell invasion and bacterial survival within the epithelial cell involves a variety of processes, including the interaction of fimbriae with integrin receptors on the epithelial cell surface; metabolic enzymes like haloacid dehalogenases that affect microtubules and subsequently cause remodelling of the actin and tubulin cytoskeleton of the epithelial cells and the utilization of bacterial endopeptidases, cation ATPases and specific ABC transporters for the intracellular lifestyle (Tribble *et al.*, 2006; Yilmaz *et al.*, 2002; Yilmaz *et al.*, 2003; Park, 2004). A label-free method has been used to perform a semi-quantitative determination of the changes in the *P. gingivalis* 33277 proteome when cells were exposed to secreted epithelial cell compounds (Zhang *et al.*, 2005). Relative quantification was carried out by summing all the MS intensities of peptides from a protein and dividing the total counts by the number of peptides identified. To produce a ratio of protein abundance between *P. gingivalis* cells incubated with epithelial cell components and those incubated without the average signal intensities of a particular protein from one condition, was divided by the average signal intensity of the same protein from the other condition. Using this method it was shown that the major fimbrial protein, FimA, was either of low abundance or not present when *P. gingivalis* cells were exposed to secreted epithelial cell compounds. This finding was supported using reverse transcriptase PCR to determine the level of mRNA transcript and a 2D-PAGE approach. The authors hypothesized that the suppression of FimA may be a way of avoiding the host immune response (Zhang *et al.*, 2005). An ATP-dependent Clp proteinase complex and a homologue of the *Listeria monocytogenes* internalin family of proteins (Lecuit *et al.*, 1997), thought to be involved in cellular invasion, were more abundant when *P. gingivalis* was exposed to secreted epithelial cell compounds. However, insertional inactivation of the gene encoding the putative internalin had little effect on the ability of the bacterium to invade gingival epithelial cells (Zhang *et al.*, 2005). It should be noted that quantification of proteins by measuring only the signal intensities of peptides (different peptides of the same protein) is difficult as the ionization intensities within the MS are largely dependent on the nature of the peptide, suppression effects and even the behaviour of the peptides during the chromatographic separation steps prior to MS (Cohen and Chait, 1996). Although internal standards can be used, accuracy is still affected if the chemical properties of these are different to the peptide of interest. A more accurate way of utilizing the label-free quantification approach is to measure the intensities of the same peptide in which the proteomes to be compared are treated and processed in exactly the same way (Ong and Mann, 2005). Advances in ultrahigh resolution HPLCs and high resolution mass spectrometers have recently made possible these types of analyses.

Stable isotope-labelled differential proteomics

Stable isotope labelling strategies, utilizing non-radioactive stable isotopes such as ^{13}C, ^{18}O, ^{2}H and ^{15}N, have recently been developed, thus enabling accurate relative quantification of protein abundance using MS. This labelling method is based on the stable isotope dilution theory which states that the relative signal intensity in a mass spectrometer of two analytes that are chemically identical but of different stable isotope composition (and thus distinguishable in a mass analyser) are a true representation of the relative abundance of the two analytes in the sample (de Leenheer and Thienpont, 1992). Stable isotopes can essentially be incorporated into the protein or peptide via two main methods; namely, metabolic labelling while the protein is being expressed (*in vivo*), and labelling (*in vitro*) through chemical derivatization of the protein/peptides.

Stable isotope labelling strategies can generally be classified as internal standard methods, since the proteins from one of the two conditions to be compared function as standards for determining the relative quantification. As a result of the use of internal standards, variables such as matrix effects, sample recovery and detection interferences are no longer detrimental to accurate quantification. These labelling strategies can be compared to DNA microarray methods except stable isotopes with isotopically distinguishable forms are used. Since isotopes have identical physico-chemical properties and the peptides labelled with the light isotope elute simultaneously with their heavy counterpart, they can be quantified using mass spectrometry. In addition to quantification the peptides can be identified by MS/MS.

In vivo quantification

The first differential proteomics for microorganisms using *in vivo* labelling was carried out by growing yeast cells in two separate media, one of which was ^{15}N enriched, and quantifying them through the isotopic distribution ratios (Oda *et al.*, 1999). In this methodology it is assumed that identical amounts of either ^{14}N or ^{15}N are incorporated into proteins as they are synthesized by the cell. Peptides from the different cultures can then be differentiated on mass differences. The *in vivo* labelling method is a relatively simple and straightforward method for the quantification of proteome changes but this technique has some challenging aspects, since the mass difference, introduced between the labelled and unlabelled peptide, depends on the amino acid sequence of the peptide in question. There is a need to use highly enriched ^{15}N to prevent partially labelled peptides and the partial resolution of ^{14}N and ^{15}N labelled peptides during chromatographic separation complicates the peptide identification process (Ong and Mann, 2005). A simpler way to quantify proteome changes in response to an environmental stimulus through *in vivo* labelling is to use the stable isotope labelling by amino acid in cell culture (SILAC) method (Fig. 2.3) (Ong

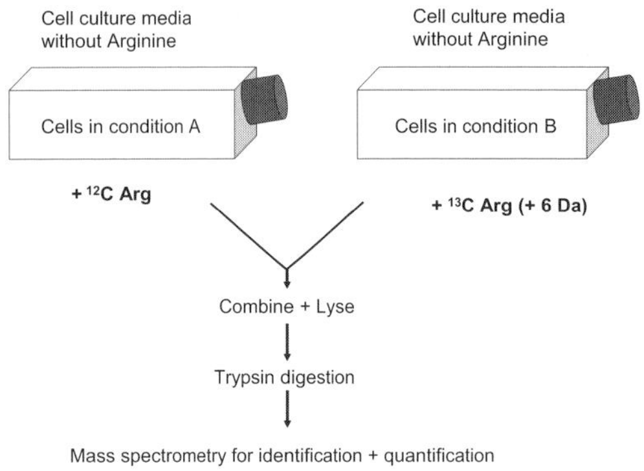

Figure 2.3 Flow diagram of the methodology for SILAC. Cells are cultured in the presence of either arginine or arginine uniformly labelled with ^{13}C, which is distinguishable by a 6 Da increase in mass of the [^{13}C]arginine due to the six carbon atoms in the arginine molecule.

et al., 2002). This technique involves growing two cultures of cells on media lacking an essential amino acid. One medium is supplemented with a stable heavy isotopic amino acid (e.g. Leu-d_3 or ^{13}C labelled Arg) whilst the other is supplemented with the same level of unlabelled amino acid. Different environmental conditions, such as an increase in oxidative stress, temperature or pH, can then be applied to the two culture. The heavy and normal amino acid will theoretically be incorporated, to the same degree, into all of the cellular proteins as they are synthesized. Quantitation then can be carried out through mass spectrometric methods, which measure the ratio of peak height of different leucine- or arginine-containing peptides, after identification through MS/MS. Theoretically, this method could be applied to bacterial cultures, although all published work has so far been done on eukaryotic cells (Ong and Mann, 2005). One of the major drawbacks of this technique is that to obtain reproducible results the bacterium should be grown in chemically defined media and asaccharolytic, fastidious, oral pathogens like *P. gingivalis*, *T. forsythia* and *T. denticola* require a complex medium for growth. In addition these species use amino acids as energy sources and the amino acids catabolized depend on growth conditions (Dashper *et al.*, 2001). These difficulties have hampered attempts to *in vivo* label *P. gingivalis* using ^{15}N ammonium sulphate and uniformly ^{15}N-labelled peptone nitrogen in minimal media (Lamont *et al.*, 2006).

Effect of haem-limitation on the proteome of P. gingivalis *as determined by* in vitro *stable isotope quantification*

We have recently utilized an isotope coded affinity tag (ICAT) labelling strategy to study the proteome changes of *P. gingivalis* W50 in response to growth under haem-limitation. For pathogenic or commensal bacteria to survive in the human host they must be able to acquire sufficient iron (Wooldridge and Williams, 1993). *P. gingivalis* is asaccharolytic and relies on the fermentation of amino acids usually obtained in the form of small peptides for metabolism (Dashper *et al.*, 2001; Dashper *et al* (2004); Shah *et al.*, 1993). Like most bacteria, *P. gingivalis* has an essential growth requirement for iron, preferentially utilizing haemoglobin-derived haem or its oxidation product haemin as its iron source (Shizukuishi *et al.*, 1995). This has been attributed to the ability of the gingipains, the major extracellular cysteine proteinase/adhesin complexes of *P. gingivalis*, to bind and hydrolyse haemoglobin, releasing haemin at the cell surface (Dashper *et al.*, 2004; DeCarlo *et al.*, 1999; Lewis *et al.*, 1999; Okamoto *et al.*, 1996). Unlike most bacteria, *P. gingivalis* cannot synthesize PPIX *de novo* (Roper *et al.*, 2000) and must acquire this from the environment, which may also explain the preferential utilisation of haem by this bacterium. Haem is stored on the *P. gingivalis* cell surface in the form of μ-oxo bishaem, which has inherent catalase activity helping to protect the cell from oxidative attack (Smalley *et al.*, 1998). This storage of haem may be advantageous for the bacterium to ensure its continued survival under haem-limited conditions, as seen in a healthy periodontal pocket. The concentration of iron in gingival crevicular fluid increases significantly in periodontally diseased sites and bleeding, as a result of gingival inflammation, elevates subgingival haem concentrations. Both phenomena are thought to encourage *P. gingivalis* proliferation (Lamont and Jenkinson, 1998). It is widely agreed that the level of haemin has an effect on the virulence of *P. gingivalis*, although the precise details remain unclear. Some studies have shown that *P. gingivalis* cells grown under haemin excess, as would pertain in diseased sites, were more virulent in the murine abscess model of disease than organisms grown under haemin limitation, representative of healthy sites (Kesavalu *et al.*, 2003; Marsh *et al.*, 1994; McKee *et al.*, 1986). However, Bramanti *et al.* (1993) showed that *P. gingivalis* cells grown under haemin-limited conditions were more virulent in the murine abscess model than cells grown in excess haemin. Genco *et al.* (1995) demonstrated that a mutant strain, defective in the transport of haemin and thus with an inferred low intracellular haemin level, increased the expression of several virulence factors and was more virulent than wild-type in a murine subcutaneous chamber model.

The use of stable isotopes for large scale *in vitro* labelling was initiated by Gygi and colleagues using the isotope coded affinity tag (ICAT) methodology (Gygi *et al.*, 1999b). Since then many groups have adopted the stable isotope labelling strategy, making modifications to further generate new methodologies for the labelling of different reactive functional groups within the peptide to suit different purposes. A brief summary of the many types of stable isotope labelling strategies is shown in Fig. 2.4. The ICAT approach has the capacity to reduce the complexity of proteins in a mixture and, at the same time, allow protein identification and quantification. The ICAT reagent has four moieties: a biotin affinity tag which has strong affinity for avidin; an acid-cleavable linker for subsequent removal of the affinity tag after digestion; an isotope-coded tag where the heavy tag includes 9 heavy or light isotopes and, lastly, a reactive group that covalently links the ICAT reagent to the target protein by alkylation of free cysteines (Fig. 2.5).

NH_2	
Labels	**Ref**
iTRAQ	A
ICPL	B
Formaldehydes (CH_2O/CD_2O)	C
IVICAT (N-terms peptide)	D

SH	
Labels	**Ref**
ICAT	H
HysTag	I
ALICE	J
Acrylamide	K
Vinyl pyridine	L

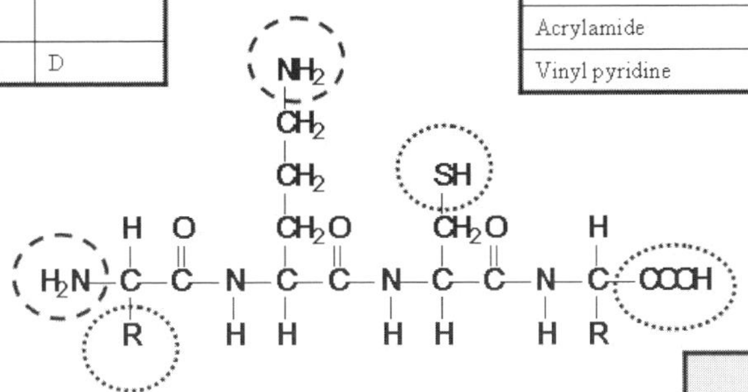

R	
Labels	**Ref**
2-nitrobenzenesulfenyl chloride (Tryptophan)	E
Dimethyl labeling after guanidination of Lys	F
MCAT (C-term lysine)	G

Others	
Labels	**Ref**
AQUA (For absolute quantification)	M
SISCAPA (For absolute quantification)	N
QCET (Cysteinyl enrichment and ^{18}O labeling	O
PhIAT (Phosphopeptides)	P

COOH	
Labels	**Ref**
^{18}O (C-terminal peptide)	Q
Sulfanilic acid (Carboxylic groups)	R
Methyl esterification of D, E residues and carboxyl terminal (d0/d3 methanol)	S

Figure 2.4 Overview of some of the available chemical-tagging approaches involving stable isotopes for comparative proteomics. This figure is representative of a few of the many different types of labelling. Many more, not mentioned here, are available. References: A (Ross *et al.*, 2004) B (Schmidt *et al.*, 2005) C (Hsu *et al.*, 2003) D (Simons *et al.*, 2006) E (Kuyama *et al.*, 2003) F (Ji *et al.*, 2005) G (Cagney and Emili, 2002) H (Gygi *et al.*, 1999a) I (Olsen *et al.*, 2004) J (Qiu *et al.*, 2002) K (Sechi, 2002) L (Sebastiano *et al.*, 2003) M (Gerber *et al.*, 2003) N (Anderson *et al.*, 2004) O (Liu *et al.*, 2004) P (Goshe *et al.*, 2001) Q (Yao *et al.*, 2001) R (Panchaud *et al.*, 2006) S (Goodlett *et al.*, 2001).

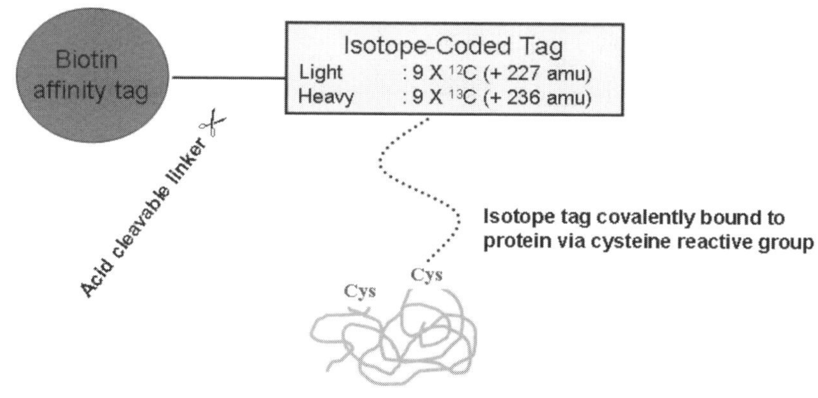

Figure 2.5 Structure of the isotope-coded affinity tag (ICAT) showing the nine carbon atoms that are either ^{12}C or ^{13}C, giving rise to a mass difference of 9 Da between the light and heavy tags.

Proteins in each of the two samples to be compared are reduced and labelled by reductive alkylation of cysteine using either the heavy or light ICAT reagent. The two isotopically distinct labelled protein samples are then combined ensuring that all following processes are identical, thereby reducing variation and enabling quantitative comparison. A prefractionation step (SDS-PAGE or HPLC) can be performed to further reduce the complexity of the proteome prior to proteolytic digestion (Fig. 2.6). Digestion of the protein mixture, usually with trypsin, results in a highly complex mixture of labelled and unlabelled peptides. The ICAT-labelled cysteine-containing peptides can be separated from the unlabelled peptides using an avidin column that binds the biotin-tagged, ICAT-labelled peptides. A heavy and light ICAT-labelled pair of peptides is chemically identical and will co-elute during the initial separation step using the online reversed phase HPLC which is coupled to a mass spectrometer. Detection of the heavy and light ICAT-labelled peptides is achieved by the 9 m/z (mass/charge) difference which is easily measurable by the mass spectrometer (Fig. 2.6). To identify and quantify the peptides, the mass spectrometer is operated alternately in MS and MS/MS mode. During the MS mode, precursor ions are selected for subsequent MS/MS and their intensities recorded. During the MS/MS mode, the selected precursor ions are fragmented by collision-induced dissociation (CID) and the peptide sequence identified by searching the CID spectrum against a database, using software such as Mascot (Perkins *et al.*, 1999). Calculation of the relative intensities of the peptides can be carried out using one of the following methods: (i) integration of the isotope envelope in a single MS spectrum or the average spectrum over the elution time of the peptide; (ii) integration of the monoisotopic peak in a single MS spectrum; or (iii) measurement of the monoisotopic peak intensity (signal intensity or peak area) in a single MS spectrum (Han *et al.*, 2001; Hansen *et al.*, 2003).

In our study, *P. gingivalis* was cultured in continuous culture as opposed to the more traditional methodology of batch culture. Batch culture introduces a large range and degree of variation into bacterial analyses due to interbatch variables such as the size and viability of the inoculum, exact time of harvesting, exact growth stage of the bacterium when harvested, levels of available nutrients remaining in the medium and redox potential of the medium amongst other factors. This is especially true when conditions of (micro)nutrient limitation are being investigated. In a bacterium, such as *P. gingivalis*, which has multiple accumulation and storage mechanisms for the essential micronutrients iron and proto-

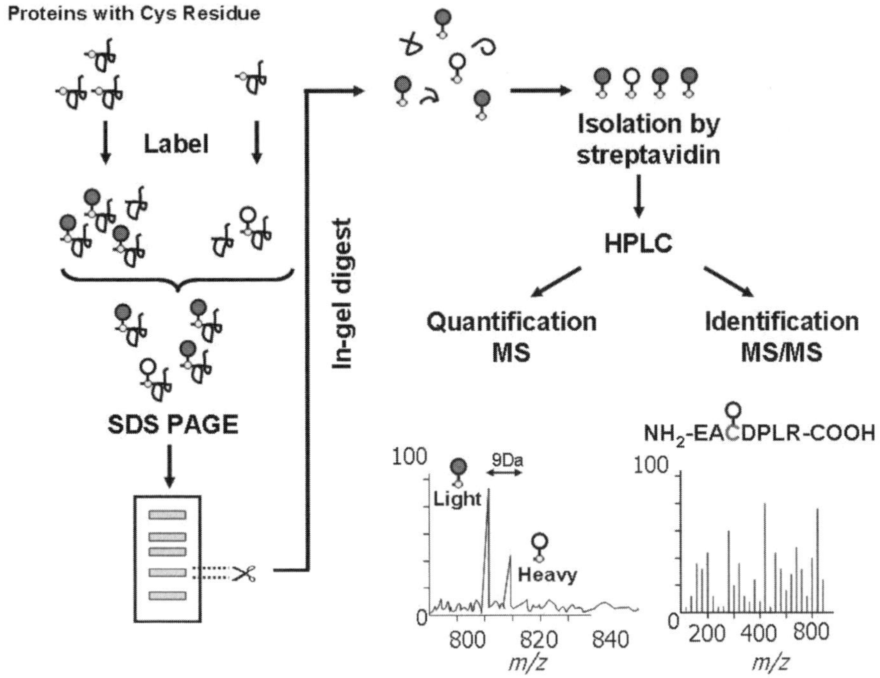

Figure 2.6 Schematic description of ICAT labelling of bacterial proteins derived from two growth conditions; followed by separation on a 1D gel, digestion using trypsin, isolation of the labelled peptides with streptavidin, followed by HPLC and quantification MS and identification MS/MS.

porphyrin IX, it is exceedingly difficult to achieve a consistent level of limitation of these nutrients in batch culture and avoid other variable stress responses. In continuous culture the bacterium is grown at a strictly controlled and reproducible growth rate, cell density, nutrient concentrations and physical conditions (temperature, pH, redox potential, agitation) for many generations. As has already been mentioned in the case of *F. nucleatum*, the data obtained are much more reliable than those obtained from batch culture studies. To study the effect of haem availability on the proteome, *P. gingivalis* W50 was cultured in a rich growth medium supplemented with either 5 mg/L or 0.1 mg/L haemin for haem excess and haem-limited growth, respectively. There was a significant decrease in cell density under the haem-limited conditions, indicating that lack of haem was, in fact, limiting the growth of the bacterium. Steady-state growth under each condition was considered to have been achieved after a minimum of ten bacterial generations. There are many examples of the high degree of reproducibility of transcriptomic and proteomic analyses of bacterial and fungal species when the organisms are grown in continuous culture (Hoskisson and Hobbs, 2005).

ICAT-labelling and analysis of *P. gingivalis* protein preparations derived from cells grown under haem-limitation and excess resulted in the identification of 142 proteins of which 53 increased in abundance by more than 2 fold and 17 decreased more than 2 fold (unpublished data). Interestingly, there was a coordinated increase in the abundance of enzymes of the glutamate catabolic pathway and a decrease in fumigate reductase indicating a significant shift in metabolism during haem-limited growth (Fig. 2.7). Many of these proteins are encoded by genes that are in operons, thus demonstrating the robust nature of the analysis. The shift in relative abundance of these enzymes is likely to be attributable in part to the fact that the two enzymes (FrdA and FrdB), catalysing the conversion of fumarate to succinate, require haem as a cofactor for activity. Growth in a haem-limited environment would not enable full expression of these enzymes. The co-ordinated increase in the

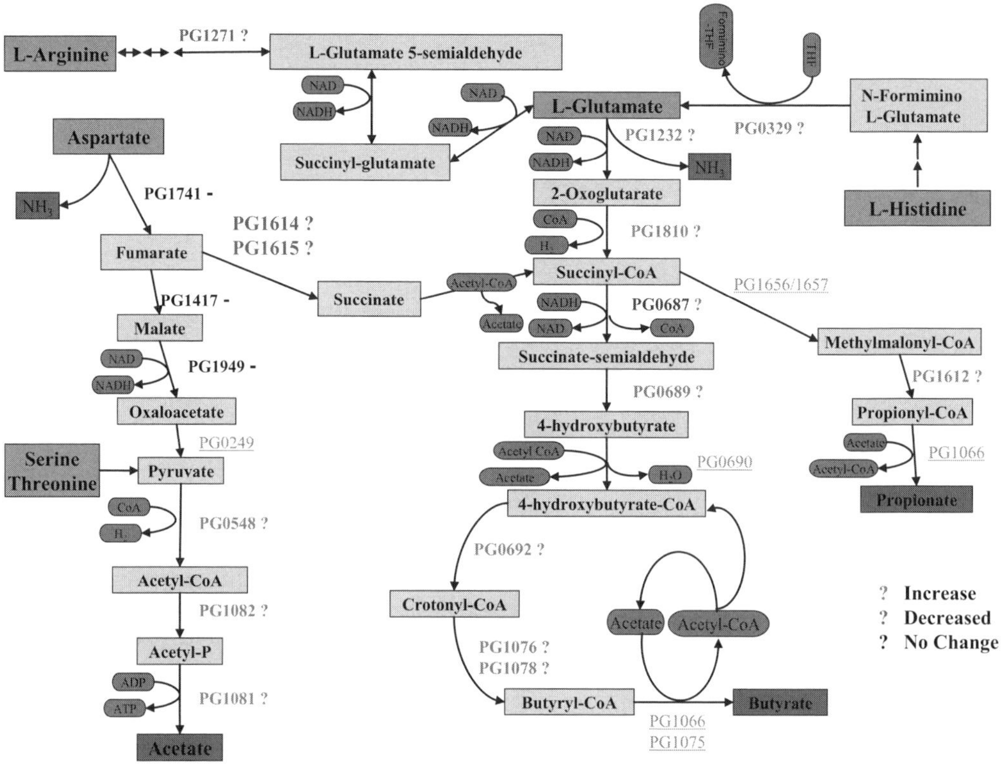

Figure 2.7 Summary of the effects of haem-limitation on the protein abundances of enzymes in the major glutamate and aspartate catabolic pathways of P. gingivalis. Bacteria were grown in continuous culture, with a mean generation time of 6.9 h, in brain heart infusion medium supplemented with 5 mg/mL filter-sterilized cysteine hydrochloride and either 5.0 µg/mL haemin (haem-excess) or 0.1 µg/mL haemin (haem-limitation). Substrates of the pathways are shown in green, intermediates are shown in light blue and end-products in red. P. gingivalis enzymes catalysing reactions are represented by their TIGR accession number. There was a co-ordinated increase in the abundance of enzymes in the glutamate pathway during haem-limited growth and a decrease in the level of the haem-containing fumarate reductase enzymes (PG1614-1615). (Unpublished data).A colour version of this figure is located in the plate section at the back of the book.

abundance of enzymes in the glutamate catabolic pathway may in part compensate for the lack of fumarate reductase activity.

The most obvious limitation of the ICAT method is that a proportion of proteins in a proteome will not contain cysteine and these will not be detected and proteins containing only a single cysteine will be difficult to detect (Rabilloud, 2002). For example, of the predicted 2000 proteins in the *P. gingivalis* proteome ~250 (12.5%) contain no cysteine and 380 (19%) contain only one cysteine. Another limitation is that post-translational modifications of the protein cannot be detected unless they occur on one of the peptides derived from hydrolytic digestion of the protein that contains cysteine. These limitations apply to all applications targeting single amino acids and this is the reason that more than one amino acid should be targeted by different labels in order to give a better representation of the entire proteome.

Effect of biofilm growth on the proteome of P. gingivalis *as determined by enzymatic incorporation of* ^{18}O

The stable isotope ^{18}O can be incorporated into peptides by the enzyme-catalysed (usually trypsin) incorporation of oxygen from $H_2{}^{18}O$ water in the surrounding solution into the C-terminal carboxylic acid during hydrolysis (Fig. 2.8) (Bender and Kemp, 1957). ^{18}O labelling in proteomics has been used for the identification of protein C-terminus, identification of N-linked glycosylation after enzymatic removal of the gly-

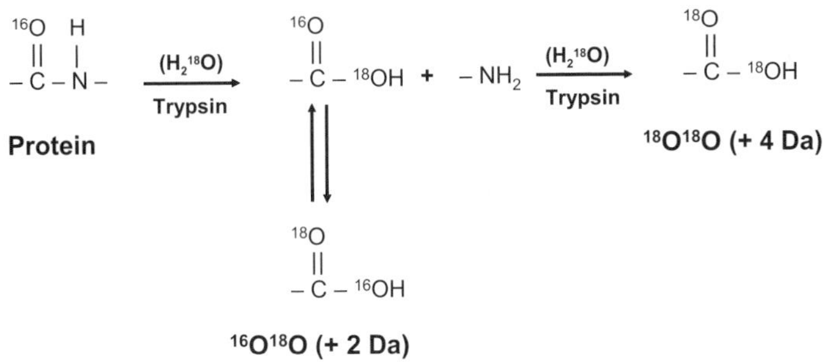

Figure 2.8 Trypsin-catalysed incorporation of ^{18}O into the C-terminal lysine or arginine of peptides. In this scheme, labelling takes place during hydrolysis in $H^{18}O$, a single digestion method. Incorporation of ^{18}O can also take place after hydrolysis.

can and the simplification of MS/MS data interpretation (Kuster and Mann, 1999; Shevchenko *et al.*, 1997; Takao *et al.*, 1991).

To date, there are no published reports on the application of $^{16}O/^{18}O$ labelling to oral bacteria, although we have recently utilized $^{16}O/^{18}O$ labelling to determine the differences in the cell envelope proteome of *P. gingivalis* growing in planktonic and biofilm states (unpublished data). Bacteria such as *P. gingivalis*, *T. denticola* and other oral species exist *in vivo* as part of a biofilm accreted to the surface of the tooth. Bacterial biofilms are usually defined as matrix-enclosed bacterial populations adherent to each other and/or to surfaces or interfaces (Costerton *et al.*, 1995). These sessile bacterial cells adhering to and growing on a surface as a mature biofilm are able to survive in hostile environments which can include the presence of host antimicrobial agents, shear forces and nutrient deprivation. The surface of the tooth is a unique microbial habitat as it is the only hard, permanent, non-shedding surface in the human body, allowing the accretion of a substantial bacterial biofilm in which bacteria grow in very high cell densities. Bacterial cells within a biofilm (sessile cells) are significantly different metabolically to planktonic cells. The presence of an extracellular matrix, usually composed of polysaccharide, creates a further difference between planktonic and biofilm cells. Sessile bacterial cells, such as *P. gingivalis*, can release antigens, toxins and hydrolytic enzymes including the gingipains and haemagglutinins that can stimulate a host immune response. However, the host response is not very effective in killing bacteria within biofilms and may cause immune damage to surrounding tissues (Cochrane *et al.*, 1988). As such, biofilms often complicate treatment of chronic infections, such as periodontitis, by protecting bacteria from the immune system, decreasing antibiotic efficacy and allowing dispersal of planktonic cells to distant sites that can aid reinfection (Cvitkovitch *et al.*, 2003). Sophisticated gene expression regulatory systems that are able to determine cell density of the same species, total bacterial cell density and contact with both other bacterial and eukaryotic cells and surfaces have developed in bacteria (O'Toole *et al.*, 2000). Many biofilm bacteria, especially those that have pathogenic or symbiotic relationships with eukaryotic hosts, have been shown to utilize quorum sensing intercellular signalling systems (Fuqua *et al.*, 1996); this topic is discussed in detail in Chapter 4 of this book. Quorum sensing controls many cellular functions, including virulence factors, and is distributed amongst both Gram-negative and Gram-positive bacteria. In Gram-negative bacteria, signalling molecules (autoinducers or bacterial pheromones) are produced by the cell at a constant rate and are freely permeable to the cell membrane. At low cell densities, most of the pheromones diffuse into the environment and the intracellular concentration remains low. However, as the cell density increases an increase in intracellular pheromone concentration induces or represses the transcription of a number of genes. *P. gingivalis* and *F. nucleatum* have been shown to use LuxS-based interspecies signalling systems, non-specific systems enabling bacteria to determine the total cell density at a site (Frias *et al.*, 2001). A *P. gingivalis* *luxS* isogenic mutant was shown to have altered expression of at least three outer membrane proteins, including the major virulence factor RgpA (Chung *et al.*, 2001). The majority of studies conducted on bacteria rely on the culture of the organism in batch culture in the planktonic state. As already noted, this is a highly artificial model, especially for oral bacteria whose continued existence within the oral cavity depends upon their ability to form part of a biofilm attached to a surface of the host. Oral bacteria, like the majority of bacteria in nature have evolved to exist as part of a biofilm and the planktonic state is usually only transitory and unlikely to be a proliferative stage. In this context, the study of oral bacteria cultured as a biofilm has far more relevance to the *in vivo* situation and will enable us to identify proteins that are important for biofilm formation and maintenance and as well as those that mediate interactions with other species and the host. It will also enable the discovery of new therapeutic and vaccine targets that are expressed during biofilm growth.

Differential labelling in the presence of $H_2{}^{18}O$ involves the formation of an enzyme-peptide ester (acyl-enzyme intermediate) at the C-terminus of the newly formed peptide that is subsequently hydrolysed to form the free peptide carrying one ^{18}O. Once the free peptide is formed, further oxygen exchanges can take place when the enzyme-peptide ester complex is reformed and then subsequently hydrolysed. This process allows a second ^{18}O atom from the water to be incorporated into the carboxyl terminus of the peptide. When the kinetics for complex formation and hydrolysis are fast, compared to the digestion, a rapid equilibrium is established with exchange at either or both the C-terminal carboxyl oxygen atoms. During the multiple cycles of esterification and hydrolysis in $H_2{}^{18}O$ containing water, all natural occurring ^{16}O is eventually displaced from the C-terminal carboxyl group (Fig. 2.8). The ^{18}O labels in carboxylate groups are resistant to back-exchange in the absence of active enzyme under normal conditions of liquid chromatography, ESI and MALDI (Chen *et al.*, 2005a; Schnolzer *et al.*, 1996).

These unique properties of trypsin-catalysed ^{18}O incorporation have since been extended to quantify changes of protein abundance, using a relatively simple procedure (Schnolzer *et al.*, 1996). Because the ^{18}O atom is universally incorporated into all peptides during hydrolysis, a higher sequence coverage of peptides, potential quantification of post-translationally modified peptides and the ease in chemistry make it an ideal method. This quantification method has since been validated to be an effective differential isotope-coding technology for comparative proteomics, using mass spectrometry (Heller *et al.*, 2003; Yao *et al.*, 2001).

To determine the effect of the two different conditions to be tested on the proteome, both reactions are performed similarly, the only difference being that one sample is prepared using $H_2{}^{16}O$ while the other is prepared in $H_2{}^{18}O$. Essentially, the naturally abundant form of water, $H_2{}^{16}O$, is completely removed by lyophilization under vacuum. The protein mixture is then digested with an endoproteinase such as trypsin, (although the use of Glu-C and Lys-C have also been reported) in a solution made up of either $H_2{}^{16}O$ or $H_2{}^{18}O$ water. Post-digestion labelling can also be carried out by first digesting the protein in $H_2{}^{16}O$, followed by complete removal of the water by lyophilization and then using fresh enzyme to label the peptides in $H_2{}^{18}O$ (Yao *et al.*, 2001). The ^{16}O and ^{18}O labelled digests are then combined prior to analysis, using LC MS/MS, and peptides eluting from the column can be quantified by measuring the relative signal intensities of the peptide ion pairs in MS mode (Fig. 2.9). Discrimination of the isotope pairs can be carried out as trypsin will incorporate two atoms of ^{18}O into the C-terminus of digested peptides, resulting in a mass shift of +4 m/z. Identification of the proteins is achieved by searching the relevant database with the MS/MS spectra, using Mascot.

Due to the mixed incorporation of one or two ^{18}O atom into the peptide and the contribution of natural abundance of the ^{18}O isotope (Schnolzer *et al.*, 1996) the ratios of the peptides have to be mathematically corrected (Yao *et al.*, 2003; Yao *et al.*, 2001). In order for this method to be performed, the peptide sequence must be known and the theoretical relative intensities of the isotopic envelope of that peptide calculated using MS-Isotope (http://prospector.ucsf.edu).

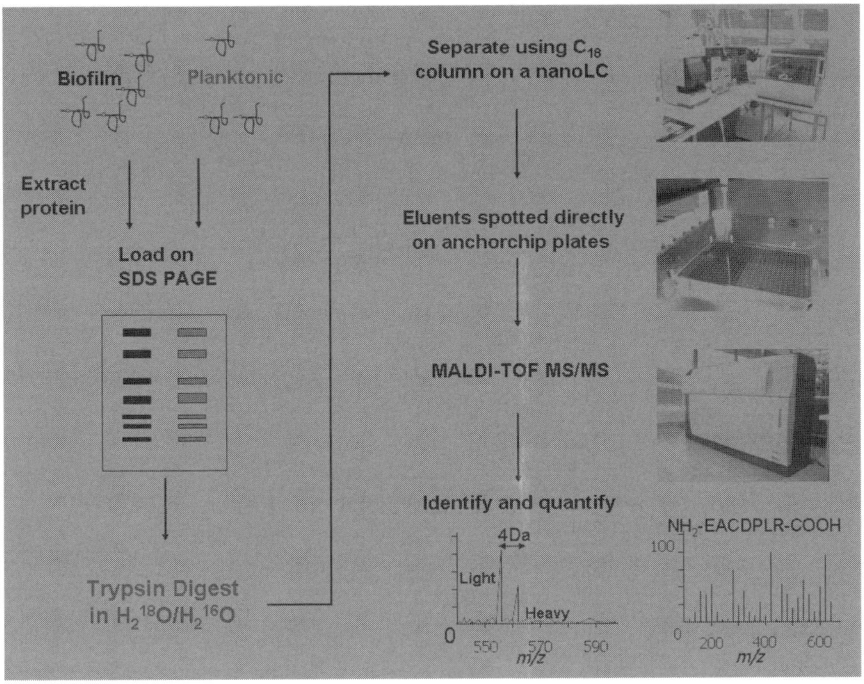

Figure 2.9 A typical scheme of performing $^{18}O^{16}O$ proteolytic labelling, coupled with mass spectrometry, for the study of bacterial proteome changes under two different growth conditions. A colour version of this figure is located in the plate section at the back of the book.

Using the intensities obtained from the first five peaks of the isotopic envelope: where I_0, I_2 and I_4 were the measured intensities of the monoisotopic peak for peptide without ^{18}O label, the peak with 2 Da higher than the monosiotopic peak, and the peak with 4 Da higher than the monoisotopic peak respectively (Fig. 2.9), the ratio is calculated using the equation:

$$\text{Ratio}_{\text{Heavy/Light}} = \{I_4 - (M_4/M_0)I_0 - (M_2/M_0)[I_2 - (M_2/M_0)I_0] + 0.5[I_2 - (M_2/M_0)I_0]\}/I_0$$

where I_0 primarily represents the light label, I_2 is a mixture of signals from the mix incorporation of one ^{18}O atom plus the natural occurring isotope of the light sample and I4 is a mixture of signals from the following three (different) sources: (1) The 2 ^{18}O incorporated peptide, (2) naturally occurring isotope of I_2 and (3) naturally occurring isotope from the light I_0 ($I_0 + 4$). M_0, M_2 and M_4 were the corresponding theoretical relative intensities of the isotopic envelope of the peptide.

In order to increase the data quality on samples, especially those with larger differences, the signal to noise (S/N) ratio has to be increased. However, with on-line LC-IT MS, this is often not possible since the rate of acquisition depends largely on the width of its chromatographic peak. Although modern ion traps have much higher cycle time for automated data-dependent MS and MS/MS than their predecessors, this often remains insufficient for complex samples (such as serum and bacterial proteomes). One alternative is through the coupling of LC to MALDI TOF-MS since the samples are static. Coupling of LC to MALDI TOF-MS not only reduces the complexity of the samples compared to direct MALDI-MS analysis but it also allows sample concentration from the entire chromatographic peak, increased time in sample acquisition and decreased ion suppression by contaminations, which lead to increased S/N. Using MALDI-MS/MS, the identity of the protein can be easily obtained.

There is great potential in the application of the ^{18}O labelling methods to quantify changes in the proteome under different conditions. Using a known protein, bovine serum albumin, with predetermined ratios as a proof of principal, we have shown that the ratios obtained using ^{18}O labelling methods were very similar to the expected ratios (unpublished data). Standard deviations were from 11% to 18% of the expected mean as the ratio of protein increased from 1:1 to 1:10. However there was little change in the overall accuracy of the analysis with an error of 5% at a ratio of 1:1 compared to 7% at a ratio of 1:10. An increase in the standard deviation for larger ratios as compared to the lower ratios has been previously observed (Johnson and Muddiman, 2004). It is probably due to the increasing difficulties in distinguishing the I_2 and I_4 peptides (mixed incorporation of $^{16}O/^{18}O$) and the low signal to noise ratio of the less abundant peak with respect to the higher abundance peak. Nevertheless, in reality, the detection of more than two-fold differences is often considered to be a significant change in a cellular system. As such, the higher standard deviations of larger ratios especially for 1:5 and 1:10 and should not be of major concern.

To achieve this, we have grown *P. gingivalis* in a chemostat as a biofilm and simultaneously as a planktonic culture in the same fermentation vessel. In this model system the major difference between the two cellular states will be the change from planktonic to biofilm. A total of 116 cell envelope proteins were identified with ~73% of proteins identified from two or more unique peptides, boosting the confidence of identification. Of these, 80 proteins were consistently identified from the two biological replicates, thus demonstrating the reproducibility of the bacterial culture system. A number of these proteins, involved in substrate and micronutrient transport, showed consistent changes in abundance as did a number of CTD family proteins (unpublished data). Many of the proteins identified were annotated as hypothetical and have not been previously described, strongly suggesting that there are many novel proteins required for the attachment process or the physiology of *P. gingivalis* during biofilm growth.

Future directions

The oral cavity is host to over 700 known taxa of bacteria, the majority of which exist as polymicrobial biofilms with complex ecologies on the tooth surface. The two most prevalent oral diseases, dental caries and chronic periodontitis, appear to result from shifts in the bacterial species composition of dental plaque (Marsh, 2003; Socransky *et al.*, 1998). The polymicrobial nature of chronic periodontal disease and the proposed ecological succession means that a better understanding of the microbial ecology of dental plaque will provide insights into the disease processes and aid in the identification of novel preventive and therapeutic strategies. Tools facilitating more detailed study of the bacterial ecology of the gingival crevice and the response of the host *in situ*, are now becoming available.

The ICAT methodology described above has recently been used to determine the effects of enteropathogenic *E. coli* (EPEC) secreted proteins on the proteome of intestinal epithelial cells (Hardwidge *et al.*, 2004). EPEC have been shown to use a type III secretion system (TTSS) to deliver bacterial effector proteins directly into intestinal epithelial cells. Intestinal epithelial cell monolayers were exposed to either wild-type EPEC or TTSS-deficient EPEC mutants to determine the effect of EPEC effector proteins on the epithelial cell proteome. Of the 2000 proteins identified in epithelial cells 13% were differentially expressed in the presence of EPEC effector proteins. This work has helped to define EPEC-host cell interactions and identify proteins that are potentially important in EPEC-induced diarrhoea. This methodology could be applied to study the effects of oral bacteria on the proteome of gingival epithelial cells.

A modification of the ICAT methodology has recently been used for the determination of protein cellular locations. This technique termed Localization of Organelle Proteins by Isotope Tagging (LOPIT) provides both qualitative and quantitative information on proteins within a cell (Dunkley *et al.*, 2004). In the Dunkley *et al.* (2004) study subcellular organelles of *Arabidopsis thaliana* cells were separated by density gradients. Protein distributions of known endoplasmic reticulum and Golgi apparatus were then determined through a series of pair-wise comparisons of gradient fractions with ICAT to enable relative quantitation of protein levels. The ability to accurately determine the cellular location of novel proteins could be extremely useful in the study of oral pathogens, such as *P. gingivalis* or *T. denticola*, that are able to invade epithelial cells as it could indicate the organelle with which the bacterium interacts when localized within the host cells.

An advanced application of the ^{18}O labelling technique was recently used to study and compare the proteomes of invasive ductal breast carcinoma cells with non-cancerous cells obtained from laser capture microdissection biopsies (Zang *et al.*, 2003). Relatively small amounts of protein (1–4 µg) obtained from these biopsies was sufficient to carry out $^{16}O/^{18}O$ labelling analysis to differentially quantify potential cancer biomarkers and proteins previously implicated with breast cancer. This method could potentially be applicable to the study of oral epithelial cells before and after invasion by oral microorganisms.

A new strategy for the complete labelling of peptides, but not yet been applied to the study of oral bacteria, is iTRAQ, a non-polymeric and isobaric reagent developed by Applied Biosystems (www.appliedbiosystems.com). This reagent labels peptides at the N-termini and lysine side chains for broad proteome coverage, detection of post-translational modifications and comparison of up to four samples in the same experiment (Ross *et al.*, 2004). The iTRAQ reagents are a set of four isobaric tagged reagents and each set consists of three main parts – a reporter group, a balance group and a peptide reactive group (Fig. 2.10). The peptide reactive group is amine-specific and should label all N-terminal amines and lysine side chains of a peptide. There are four balance groups, each specific for a reporter group and these balance groups ensure that the same peptide labelled with any of the iTRAQ reagents will be of the same mass. During the high-energy fragmentation in the mass spectrometer (MS/MS), the isobaric tag is cleaved (Fig. 2.10) and the generated reporter group ions appear as signal ion peaks in the low-mass region between 113 and 118 m/z. Signals found in this region are due to the reporter ions from the corresponding labelled sample digests and the relative amounts can be quantified by comparison of the respective peak areas. Since all peptidases are tagged, one major advantage of this method is its ability to provide information about post-translational modification. The complete tagging of peptides also increases the proteome coverage with multiple peptides per protein, thereby improving confidence in protein identification. The ability to include duplicates or triplicates in parallel into the experimental design, using two or three of the four reagents, is an added advantage. In addition, this technology could be applied to the temporal analysis of changes to the proteome of a bacterium in response to environmental change.

The iTRAQ method has been used to study the role of the *E. coli* RshA by comparing the effects of different expression level of this protein using an external inducer (Aggarwal *et al.*, 2005). Controlled expression of *rshA*, with an inducible promoter, was carried out using three different concentrations of IPTG. Including the control sample, this makes a comparison of four different biological conditions which would not be possible with other stable isotope-labelling strategies. 780 proteins were relatively quantified, including some of the low abundance proteins such as transcriptional factors, DnaB and DnaG. An added benefit of using this method as compared to the ICAT methodology is that more than 65% of the proteins were identified by 2 or more peptides, increasing the confidence of identification. Further comparison of these four physiological conditions of *E. coli* cells expressing RshA in parallel with 2D PAGE techniques showed that both techniques had coefficients of variance of 0.24 and 0.31, respectively, thereby showing the similarity and accuracy of both methods (Aggarwal. *et al.*, 2005). An examination of the *Bacillus subtilis* heat shock response has also been undertaken using the iTRAQ methodology (Wolff *et al.*, 2006). This analysis revealed a quantitative increase in proteins belonging to heat shock regulons (SigB, HrcA and CtsR) when the bacterium was exposed to elevated temperature, similar to that obtained from gel-based data; thus verifying the applicability of this methodology. The utility of this system in comparing peptide abundances, under four conditions, has obvious benefits for the study of oral bacteria as it will allow the design of time-course experiments to investigate aspects of bacterial biofilm formation, interspecies interactions and host-bacterial interactions.

Metagenomics (or 'community genomics'), the determination and sequencing of all the genes

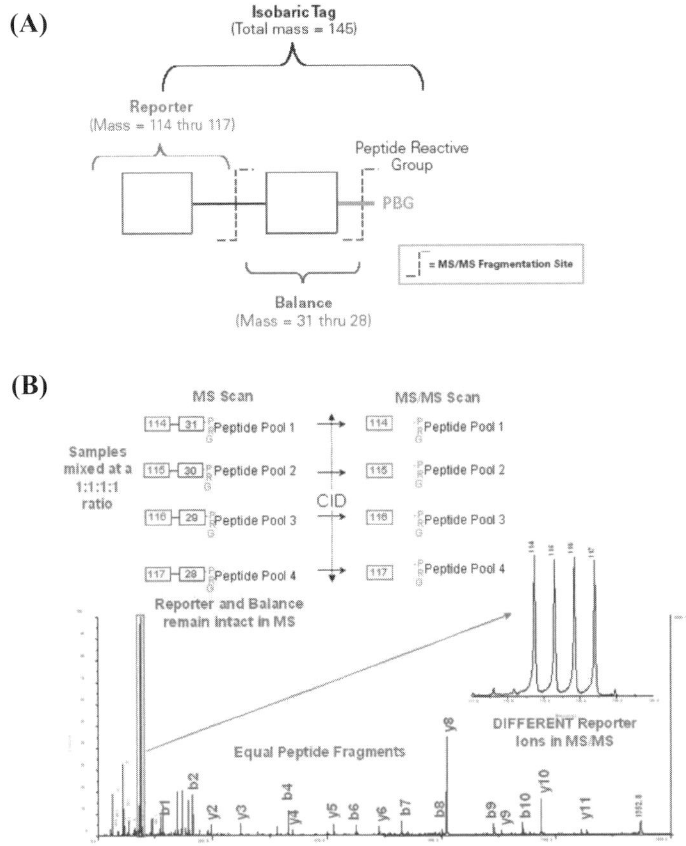

Figure 2.10 (A) The iTRAQ reagent consisting of the reporter, balance and peptide reactive groups. (B) Spectra showing the MS/MS spectra after CID and the different reporter ions (114–117 mz) at the 'quiet region' of the MS spectra which are used for quantitation. (http://docs.appliedbiosystems.com/pebiodocs/00113379.pdf). A colour version of this figure is located in the plate section at the back of the book.

that are present in an ecosystem has recently been carried out by a number of groups on various ecosystems. For example, genome shotgun-sequencing techniques, to determine the metagenome of microbial communities in the Sargasso Sea, have been applied; and significant numbers of hitherto-unknown genes, the protein products of which play major roles in the functioning of the ecosystem, have been discovered (Venter *et al.*, 2004). There is little doubt that these types of analyses will soon be applied to the study of oral microbial ecology. In itself, the application of these technologies will greatly extend our understanding of the genetic diversity of dental plaque and help to determine the role of uncultivable species in these complex polymicrobial systems. More importantly, the availability of this metagenomic sequence data, the ever decreasing amount of sample needed for proteomic analyses, the increasing ability to deconvolute complex protein samples and new technologies, such as iTRAQ, that enable the comparison of multiple samples will allow an unprecedented understanding of the molecular interactions at a community level in subgingival dental plaque; interactions that lead to the proliferative changes seen in plaque species composition, the development of chronic inflammation and the destruction of host tissue.

Websites

All of the genome sequences and annotations of the oral bacterial species referred to in this chapter can be found at the following websites, all of which allow free access. They also contain many tools for the analysis of these genomes and links to other sites for further analyses.

- The Institute for Genomic Research (TIGR) www.TIGR.gov/
- Los Alamos Oral Pathogens Sequence Databases www.oralgen.lanl.gov/
- Bacterial Resource for Oral Pathogens (BROP) http://www.brop.org/, which is hosted by the Forsyth Institute and supported by the NIDCR (Chen *et al.*, 2005b).

Acknowledgements

We thank Nada Slakeski, Christine Seers, Yu Yen Chen, Neil O'Brien-Simpson, Rishi Pathirana, Patricia Lissel, Dina Chen, Yan Tan and Rita Paolini for their contributions to this research. It was supported by the National Health and Medical Research Council (Australia), The National Institutes of Health (USA), the Co-operative Research Centre for Oral Health Science, Australian Dental Research Foundation and the Science Technology Innovation Initiative of the Victorian State Government.

References

Abe, N., Kadowaki, T., Okamoto, K., Nakayama, K., Ohishi, M., and Yamamoto, K. (1998) Biochemical and functional properties of lysine-specific cysteine proteinase (Lys-gingipain) as a virulence factor of *Porphyromonas gingivalis* in periodontal disease. J. Biochem. (Tokyo) *123*, 305–312.

Adamson, N., Riley, P.F., and Reynolds, E.C. (1993) The analysis of multiple phosphoseryl-containing casein peptides using capillary zone electrophoresis. J. Chromatog. *646*, 391–396.

Aggarwal, K., Choe, L.H., and Lee, K.H. (2005) Quantitative analysis of protein expression using amine-specific isobaric tags in *Escherichia coli* cells expressing rhsA elements. Proteomics 5, 2297–2308.

Andersen, J.S., Wilkinson, C.J., Mayor, T., Mortensen, P., Nigg, E.A., and Mann, M. (2003) Proteomic characterization of the human centrosome by protein correlation profiling. Nature *426*, 570–574.

Anderson, N.L., and Anderson, N.G. (1998) Proteome and proteomics: new technologies, new concepts, and new words. Electrophoresis *19*, 1853–1861.

Anderson, N.L., Anderson, N.G., Haines, L.R., Hardie, D.B., Olafson, R.W., and Pearson, T.W. (2004) Mass spectrometric quantitation of peptides and proteins using Stable Isotope Standards and Capture by Anti-Peptide Antibodies (SISCAPA). J. Proteome Res. 3, 235–244.

Baker, P.J., Dixon, M., Evans, R.T., and Roopenian, D.C. (2000) Heterogeneity of *Porphyromonas gingivalis* strains in the induction of alveolar bone loss in mice. Oral Microbiol. Immunol. *15*, 27–32.

Baker, P.J., Evans, R.T., and Roopenian, D.C. (1994) Oral infection with *Porphyromonas gingivalis* and induced alveolar bone loss in immunocompetent and severe combined immunodeficient mice. Arch Oral. Biol. *39*, 1035–1040.

Bartold, P.M., Gully, N.J., Zilm, P.S., and Rogers, A.H. (1991) Identification of components in *Fusobacterium nucleatum* chemostat-culture supernatants that are potent inhibitors of human gingival fibroblast proliferation. J. Periodont. Res. *26*, 314–322.

Belton, C.M., Izutsu, K.T., Goodwin, P.C., Park, Y., and Lamont, R.J. (1999) Fluorescence image analysis of the association between *Porphyromonas gingivalis* and gingival epithelial cells. Cell. Microbiol. *1*, 215–223.

Bender, M.L., and Kemp, K.C. (1957) The kinetics of the chymotrypsin-catalized oxygen exchange of carboxylic acids. J. Am. Chem. Soc. *79*, 116–120.

Bickel, M., and Cimasoni, G. (1985) The pH of human crevicular fluid measured by a new microanalytical technique. J. Periodont. Res. *20*, 35–40.

Bird, P.S., Shakibaie, F., Gemmell, E., Polak, B., and Seymour, G.J. (2001) Immune response to *Bacteroides forsythus* in a murine model. Oral Microbiol. Immunol. *16*, 311–315.

Bjellqvist, B., Ek, K., Righetti, P.G., Gianazza, E., Gorg, A., Westermeier, R., and Postel, W. (1982) Isoelectric focusing in immobilized pH gradients: principle, methodology and some applications. J. Biochem. Biophys. Meth. 6, 317–339.

Bolstad, A.I., Jensen, H.B., and Bakken, V. (1996) Taxonomy, biology, and periodontal aspects of *Fusobacterium nucleatum*. Clin. Microbiol. Rev. 9, 55–71.

Boyce, J.D., Wilkie, I., Harper, M., Paustian, M.L., Kapur, V., and Adler, B. (2002) Genomic scale analysis of *Pasteurella multocida* gene expression during growth within the natural chicken host. Infect. Immun. *70*, 6871–6879.

Bradshaw, D.J., and Marsh, P.D. (1998) Analysis of pH-driven disruption of oral microbial communities *in vitro*. Caries Res. 32, 456–462.

Bramanti, T.E., and Holt, S.C. (1993) Hemin uptake in *Porphyromonas gingivalis*: Omp26 is a hemin-binding surface protein. J. Bacteriol. *175*, 7413–7420.

Cagney, G., and Emili, A. (2002) *De novo* peptide sequencing and quantitative profiling of complex protein mixtures using mass-coded abundance tagging. Nat. Biotechnol. *20*, 163–170.

Campostrini, N., Areces, L.B., Rappsilber, J., Pietrogrande, M.C., Dondi, F., Pastorino, F., Ponzoni, M., and Righetti, P.G. (2005) Spot overlapping in two-dimensional maps: a serious problem ignored for much too long. Proteomics 5, 2385–2395.

Chan, E.C., Siboo, R., Keng, T., Psarra, N., Hurley, R., Cheng, S.L., and Iugovaz, I. (1993) *Treponema denticola* (ex Brumpt 1925) sp. nov., nom. rev., and identification of new spirochete isolates from periodontal pockets. Int. J. Syst. Bacteriol. 43, 196–203.

Chen, H.A., Weinberg, A., Darveau, R.P., Engel, D., and Page, R.C. (1995) Immunodominant antigens of *Porphyromonas gingivalis* in patients with rapidly

progressive periodontitis. Oral Microbiol. Immunol. *10*, 193–201.
Chen, J., Leblanc, D.J., and Galli, D.M. (2002a). DNA inversion on conjugative plasmid pVT745. J. Bacteriol. *184*, 5926–5934.
Chen, P.B., Davern, L.B., Katz, J., Eldridge, J.H., and Michalek, S.M. (1996) Host responses induced by co-infection with *Porphyromonas gingivalis* and *Actinobacillus actinomycetemcomitans* in a murine model. Oral Microbiol. Immunol. *11*, 274–281.
Chen, T., Abbey, K., Deng, W.J., and Cheng, M.C. (2005a). The bioinformatics resource for oral pathogens. Nucleic Acids Res. *33*, W734–740.
Chen, T., Hosogi, Y., Nishikawa, K., Abbey, K., Fleischmann, R.D., Walling, J., and Duncan, M.J. (2004) Comparative whole-genome analysis of virulent and avirulent strains of *Porphyromonas gingivalis*. J. Bacteriol. *186*, 5473–5479.
Chen, X., Cushman, S.W., Pannell, L.K., and Hess, S. (2005b). Quantitative proteomic analysis of the secretory proteins from rat adipose cells using a 2D liquid chromatography-MS/MS approach. J. Proteome Res. *4*, 570–577.
Chen, Y.Y., Cross, K.J., Paolini, R.A., Fielding, J.E., Slakeski, N., and Reynolds, E.C. (2002b). CPG70 is a novel basic metallocarboxypeptidase with C-terminal polycystic kidney disease domains from *Porphyromonas gingivalis*. J. Biol. Chem. *277*, 23433–23440.
Christersson, L., Zambon, J., Dunford, R., Grossi, S., and Genco, R. (1989) Specific subgingival bacteria and diagnosis of gingivitis and periodontitis. J. Dent. Res. *68*, 1633–1639.
Chung, W.O., Park, Y., Lamont, R.J., McNab, R., Barbieri, B., and Demuth, D.R. (2001) Signaling system in *Porphyromonas gingivalis* based on a LuxS protein. J. Bacteriol. *183*, 3903–3909.
Cochrane, D.M., Brown, M.R., Anwar, H., Weller, P.H., Lam, K., and Costerton, J.W. (1988) Antibody response to *Pseudomonas aeruginosa* surface protein antigens in a rat model of chronic lung infection. J. Med. Microbiol. *27*, 255–261.
Cohen, S.L., and Chait, B.T. (1996) Influence of matrix solution conditions on the MALDI-MS analysis of peptides and proteins. Anal. Chem. *68*, 31–37.
Costerton, J.W., Lewandowski, Z., Caldwell, D.E., Korber, D.R., and Lappin-Scott, H.M. (1995) Microbial Biofilms. Ann. Rev. Microbiol. *49*, 711–745.
Curtis, M.A., Kuramitsu, H.K., Lantz, M., Macrina, F.L., Nakayama, K., Potempa, J., Reynolds, E.C., and Aduse-Opoku, J. (1999) Molecular genetics and nomenclature of proteases of *Porphyromonas gingivalis*. J. Periodont. Res. *34*, 464–472.
Cvitkovitch, D.G., Li, Y.H., and Ellen, R.P. (2003) Quorum sensing and biofilm formation in Streptococcal infections. J. Clin. Invest. *112*, 1626–1632.
Dashper, S.G., Brownfield, L., Slakeski, N., Zilm, P.S., Rogers, A.H., and Reynolds, E.C. (2001) Sodium ion-driven serine/threonine transport in *Porphyromonas gingivalis*. J. Bacteriol. *183*, 4142–4148.
Dashper, S.G., Cross, K.J., Slakeski, N., Lissel, P., Aulakh, P., Moore, C., and Reynolds, E.C. (2004) Hemoglobin hydrolysis and heme acquisition by *Porphyromonas gingivalis*. Oral Microbiol. Immunol. *19*, 50–56.
Dashper, S.G., Hendtlass, A., Slakeski, N., Jackson, C., Cross, K.J., Brownfield, L., Hamilton, R., Barr, I., and Reynolds, E.C. (2000) Characterization of a novel outer membrane hemin-binding protein of *Porphyromonas gingivalis*. J. Bacteriol. *182*, 6456–6462.
de Leenheer, A.P., and Thienpont, L.M. (1992) Applications of isotope dilution-mass spectrometry in clinical chemistry, pharmacokinetics and toxicology. Mass Spectrom. Rev. *11*, 249–307.
DeCarlo, A.A., Paramaesvaran, M., Yun, P.L., Collyer, C., and Hunter, N. (1999) Porphyrin-mediated binding to hemoglobin by the HA2 domain of cysteine proteinases (gingipains) and hemagglutinins from the periodontal pathogen *Porphyromonas gingivalis*. J. Bacteriol. *181*, 3784–3791.
Diaz, P., Zilm, P., and Rogers, A. (2002) *Fusobacterium nucleatum* supports the growth of *Porphyromonas gingivalis* in oxygenated and carbon dioxide-depleted environments. Microbiology *148*, 467–472.
Duncan, M.J., Nakao, S., Skobe, Z., and Xie, H. (1993) Interactions of *Porphyromonas gingivalis* with epithelial cells. Infect. Immun. *61*, 2260–2265.
Dunkley, T.P., Watson, R., Griffin, J.L., Dupree, P., and Lilley, K.S. (2004) Localization of organelle proteins by isotope tagging (LOPIT). Mol. Cell Proteomics 3, 1128–1134.
Dzink, J.L., Sheenan, M.T., and Socransky, S.S. (1990) Proposal of three subspecies of *Fusobacterium nucleatum* Knorr 1922: *Fusobacterium nucleatum* subsp. *nucleatum* subsp. nov., comb. nov.; *Fusobacterium nucleatum* subsp. *polymorphum* subsp. nov., nom. rev., comb. nov.; and *Fusobacterium nucleatum* subsp. *vincentii* subsp. nov., nom. rev., comb. nov. Int. J. Syst. Bacteriol. *40*, 74–78.
Ebersole, J.L., Feuille, F., Kesavalu, L., and Holt, S.C. (1997) Host modulation of tissue destruction caused by periodontopathogens: effects on a mixed microbial infection composed of *Porphyromonas gingivalis* and *Fusobacterium nucleatum*. Microb. Pathog. *23*, 23–32.
Eke, P.I., Rotimi, V.O., and Laughon, B.E. (1989) Coaggregation of black-pigmented *Bacteroides* species with other oral bacteria. J. Med. Microbiol. *28*, 1–4.
Ellen, R.P., and Galimanas, V.B. (2005) Spirochetes at the forefront of periodontal infections. Periodontol. 2000 *38*, 13–32.
Evans, R.T., Klausen, B., Ramamurthy, N.S., Golub, L.M., Sfintescu, C., and Genco, R.J. (1992) Periodontopathic potential of two strains of *Porphyromonas gingivalis* in gnotobiotic rats. Arch. Oral Biol. *37*, 813–819.
Fleischmann, R.D., Adams, M.D., White, O., Clayton, R.A., Kirkness, E.F., Kerlavage, A.R., Bult, C.J., Tomb, J.F., Dougherty, B.A., Merrick, J.M., and *et al.* (1995) Whole-genome random sequencing and assembly of *Haemophilus influenzae* Rd. Science *269*, 496–512.
Frias, J., Olle, E., and Alsina, M. (2001) Periodontal pathogens produce quorum sensing signal molecules. Infect. Immun. 69, 3431–3434.
Fuqua, C., Winans, S.C., and Greenberg, E.P. (1996) Census and consensus in bacterial ecosystems: the

LuxR-LuxI family of quorum-sensing transcriptional regulators. Ann. Rev. Microbiol. *50*, 727–751.

Genco, C.A., Simpson, W., Forng, R.Y., Egal, M., and Odusanya, B.M. (1995) Characterization of a Tn4351-generated hemin uptake mutant of *Porphyromonas gingivalis*: evidence for the coordinate regulation of virulence factors by hemin. Infect. Immun. *63*, 2459–2466.

Gerber, S.A., Rush, J., Stemman, O., Kirschner, M.W., and Gygi, S.P. (2003) Absolute quantification of proteins and phosphoproteins from cell lysates by tandem MS. Proc. Natl. Acad. Sci. USA *100*, 6940–6945.

Gharbia, S.E., and Shah, H.N. (1992) *Fusobacterium nucleatum* subsp. *fusiforme* subsp. nov. and *Fusobacterium nucleatum* subsp. *animalis* subsp. nov. as additional subspecies within *Fusobacterium nucleatum*. Int. J. Syst. Bacteriol. *42*, 296–298.

Gmur, R., Strub, J.R., and Guggenheim, B. (1989) Prevalence of *Bacteroides forsythus* and *Bacteroides gingivalis* in subgingival plaque of prosthodontically treated patients on short recall. J. Periodont. Res. *24*, 113–120.

Goodlett, D.R., Keller, A., Watts, J.D., Newitt, R., Yi, E.C., Purvine, S., Eng, J.K., von Haller, P., Aebersold, R., and Kolker, E. (2001) Differential stable isotope labeling of peptides for quantitation and *de novo* sequence derivation. Rapid Commun. Mass Spectrom. *15*, 1214–1221.

Goshe, M.B., Conrads, T.P., Panisko, E.A., Angell, N.H., Veenstra, T.D., and Smith, R.D. (2001) Phosphoprotein isotope-coded affinity tag approach for isolating and quantitating phosphopeptides in proteome-wide analyses. Anal. Chem. *73*, 2578–2586.

Grenier, D. (1992) Nutritional interactions between two suspected periodontopathogens, *Treponema denticola* and *Porphyromonas gingivalis*. Infect. Immun. *60*, 5298–5301.

Griffen, A.L., Lyons, S.R., Becker, M.R., Moeschberger, M.L., and Leys, E.J. (1999) *Porphyromonas gingivalis* strain variability and periodontitis. J. Clin. Microbiol. *37*, 4028–4033.

Gygi, S.P., Rist, B., Gerber, S.A., Turecek, F., Gelb, M.H., and Aebersold, R. (1999a). Quantitative analysis of complex protein mixtures using isotope-coded affinity tags. Nat. Biotechnol. *17*, 994–999.

Gygi, S.P., Rochon, Y., Franza, B.R., and Aebersold, R. (1999b). Correlation between protein and mRNA abundance in yeast. Mol. Cell Biol. *19*, 1720–1730.

Haake, S.K., Yoder, S.C., Attarian, G., and Podkaminer, K. (2000) Native plasmids of *Fusobacterium nucleatum*: characterization and use in development of genetic systems. J. Bacteriol. *182*, 1176–1180.

Han, D.K., Eng, J., Zhou, H., and Aebersold, R. (2001) Quantitative profiling of differentiation-induced microsomal proteins using isotope-coded affinity tags and mass spectrometry. Nat. Biotechnol. *19*, 946–951.

Han, Y.W., Shi, W., Huang, G.T., Kinder Haake, S., Park, N.H., Kuramitsu, H., and Genco, R.J. (2000) Interactions between periodontal bacteria and human oral epithelial cells: *Fusobacterium nucleatum* adheres to and invades epithelial cells. Infect. Immun. *68*, 3140–3146.

Hansen, K.C., Schmitt-Ulms, G., Chalkley, R.J., Hirsch, J., Baldwin, M.A., and Burlingame, A.L. (2003) Mass spectrometric analysis of protein mixtures at low levels using cleavable 13C-isotope-coded affinity tag and miltidimensional chromatography. Mol. Cell Biol. *2*, 299–314.

Heller, M., Mattou, H., Menzel, C., and Yao, X. (2003) Trypsin catalyzed 16O-to-18O exchange for comparative proteomics: tandem mass spectrometry comparison using MALDI-TOF, ESI-QTOF, and ESI-ion trap mass spectrometers. J. Am. Soc. Mass Spectrom. *14*, 704–718.

Holt, S.C., Ebersole, J., Felton, J., Brunsvold, M., and Kornman, K.S. (1988) Implantation of *Bacteroides gingivalis* in nonhuman primates initiates progression of periodontitis. Science *239*, 55–57.

Hoskisson, P.A., and Hobbs, G. (2005) Continuous culture-making a comeback? Microbiology *151*, 3153–3159.

Hsu, J.L., Huang, S.Y., Chow, N.H., and Chen, S.H. (2003) Stable-isotope dimethyl labeling for quantitative proteomics. Anal. Chem. *75*, 6843–6852.

Ji, C., Guo, N., and Li, L. (2005) Differential dimethyl labeling of N-termini of peptides after guanidination for proteome analysis. J. Proteome Res. *4*, 2099–2108.

Johnson, K.L., and Muddiman, D.C. (2004) A method for calculating 16O/18O peptide ion ratios for the relative quantification of proteomes. J. Am. Soc. Mass Spectrom. *15*, 437–445.

Kapatral, V., Anderson, I., Ivanova, N., Reznik, G., Los, T., Lykidis, A., Bhattacharyya, A., Bartman, A., Gardner, W., Grechkin, G. (2002) Genome sequence and analysis of the oral bacterium *Fusobacterium nucleatum* strain ATCC 25586. J. Bacteriol. *184*, 2005–2018.

Kapatral, V., Ivanova, N., Anderson, I., Reznik, G., Bhattacharyya, A., Gardner, W.L., Mikhailova, N., Lapidus, A., Larsen, N., D'Souza, M., *et al.* (2003) Genome analysis of *F. nucleatum sub spp vincentii* and its comparison with the genome of *F. nucleatum* ATCC 25586. Genome Res. *13*, 1180–1189.

Katz, J., Ward, D.C., and Michalek, S.M. (1996) Effect of host responses on the pathogenicity of strains of *Porphyromonas gingivalis*. Oral Microbiol. Immunol. *11*, 309–318.

Kesavalu, L., Holt, S.C., and Ebersole, J.L. (1998) Virulence of a polymicrobic complex, *Treponema denticola* and *Porphyromonas gingivalis*, in a murine model. Oral Microbiol. Immunol. *13*, 373–377.

Kesavalu, L., Holt, S.C., and Ebersole, J.L. (2003) In vitro environmental regulation of *Porphyromonas gingivalis* growth and virulence. Oral Microbiol. Immunol. *18*, 226–233.

Kinder, S.A., and Holt, S.C. (1989) Characterization of coaggregation between *Bacteroides gingivalis* T22 and *Fusobacterium nucleatum* T18. Infect. Immun. *57*, 3425–3433.

Klose, J. (1975) Protein mapping by combined isoelectric focusing and electrophoresis of mouse tissues. A novel approach to testing for induced point mutations in mammals. Humangenetik *26*, 231–243.

Kolenbrander, P.E., and Andersen, R.N. (1989) Inhibition of coaggregation between *Fusobacterium nucleatum* and *Porphyromonas (Bacteroides) gingivalis* by lactose and related sugars. Infect. Immun. *57*, 3204–3209.

Kolenbrander, P.E., and London, J. (1993) Adhere today, here tomorrow: oral bacterial adherence. J. Bacteriol. *175*, 3247–3252.

Kolenbrander, P.E., Parrish, K.D., Andersen, R.N., and Greenberg, E.P. (1995) Intergeneric coaggregation of oral *Treponema spp.* with *Fusobacterium spp.* and intrageneric coaggregation among *Fusobacterium spp.* Infect. Immun. *63*, 4584–4588.

Kuster, B., and Mann, M. (1999) 18O-labeling of N-glycosylation sites to improve the identification of gel-separated glycoproteins using peptide mass mapping and database searching. Anal. Chem. *71*, 1431–1440.

Kuyama, H., Watanabe, M., Toda, C., Ando, E., Tanaka, K., and Nishimura, O. (2003) An approach to quantitative proteome analysis by labeling tryptophan residues. Rapid Commun. Mass Spectrom. *17*, 1642–1650.

Lamont, R.J., Chan, A., Belton, C.M., Izutsu, K.T., Vasel, D., and Weinberg, A. (1995) *Porphyromonas gingivalis* invasion of gingival epithelial cells. Infect. Immun. *63*, 3878–3885.

Lamont, R.J., and Jenkinson, H.F. (1998) Life below the gum line: pathogenic mechanisms of *Porphyromonas gingivalis*. Microbiol, Mol. Biol. Rev. *62*, 1244–1263.

Lamont, R.J., Meila, M., Xia, Q., and Hackett, M. (2006) Mass spectrometry-based proteomics and its application to studies of *Porphyromonas gingivalis* invasion and pathogenicity. Infect. Disord. Drug Targets *6*, 311–325.

Lecuit, M., Ohayon, H., Braun, L., Mengaud, J., and Cossart, P. (1997) Internalin of *Listeria monocytogenes* with an intact leucine-rich repeat region is sufficient to promote internalization. Infect. Immun. *65*, 5309–5319.

Lewis, J.P., Dawson, J.A., Hannis, J.C., Muddiman, D., and Macrina, F.L. (1999) Hemoglobinase activity of the lysine gingipain protease (Kgp) of *Porphyromonas gingivalis* W83. J. Bacteriol. *181*, 4905–4913.

Listgarten, M.A., and Hellden, L. (1978) Relative distribution of bacteria at clinically healthy and periodontally diseased sites in humans. J. Clin. Periodontol. *5*, 115–132.

Liu, T., Qian, W.J., Strittmatter, E.F., Camp, D.G., 2nd, Anderson, G.A., Thrall, B.D., and Smith, R.D. (2004) High-throughput comparative proteome analysis using a quantitative cysteinyl-peptide enrichment technology. Anal. Chem. *76*, 5345–5353.

Loesche, W.J. (1988) The role of spirochetes in periodontal disease. Adv. Dent. Res. *2*, 275–283.

Maiden, M.F., Cohee, P., and Tanner, A.C. (2003) Proposal to conserve the adjectival form of the specific epithet in the reclassification of *Bacteroides forsythus* Tanner *et al.* 1986 to the genus *Tannerella* Sakamoto *et al.* 2002 as *Tannerella forsythia* corrig., gen. nov., comb. nov. Request for an Opinion. Int. J. Syst. Evol. Microbiol. *53*, 2111–2112.

Marsh, P.D. (2003) Are dental diseases examples of ecological catastrophes? Microbiology *149*, 279–294.

Marsh, P.D., McDermid, A.S., McKee, A.S., and Baskerville, A. (1994) The effect of growth rate and haemin on the virulence and proteolytic activity of *Porphyromonas gingivalis* W50. Microbiology *140 (Pt 4)*, 861–865.

McKee, A.S., McDermid, A.S., Baskerville, A., Dowsett, A.B., Ellwood, D.C., and Marsh, P.D. (1986) Effect of hemin on the physiology and virulence of *Bacteroides gingivalis* W50. Infect. Immun. *52*, 349–355.

Mira, A., Pushker, R., Legault, B.A., Moreira, D., and Rodriguez-Valera, F. (2004) Evolutionary relationships of *Fusobacterium nucleatum* based on phylogenetic analysis and comparative genomics. BMC Evol. Biol. *4*, 50.

Moore, W.E., and Moore, L.V. (1994) The bacteria of periodontal diseases. Periodontol. 2000 *5*, 66–77.

Mortz, E., Krogh, T.N., Vorum, H., and Gorg, A. (2001) Improved silver staining protocols for high sensitivity protein identification using matrix-assisted laser desorption/ionization-time of flight analysis. Proteomics *1*, 1359–1363.

Neiders, M.E., Chen, P.B., Suido, H., Reynolds, H.S., Zambon, J.J., Shlossman, M., and Genco, R.J. (1989) Heterogeneity of virulence among strains of *Bacteroides gingivalis*. J Periodont. Res. *24*, 192–198.

Nelson, K.E., Fleischmann, R.D., DeBoy, R.T., Paulsen, I.T., Fouts, D.E., Eisen, J.A., Daugherty, S.C., Dodson, R.J., Durkin, A.S., Gwinn, M., *et al.* (2003) Complete genome sequence of the oral pathogenic bacterium *Porphyromonas gingivalis* strain W83. J. Bacteriol. *185*, 5591–5601.

Nguyen, K.A., Travis, J., and Potempa, J. (2007) Does the importance of the C-terminal residues in the maturation of RgpB from *Porphyromonas gingivalis* reveal a novel mechanism for protein export in a subgroup of Gram-Negative bacteria? J. Bacteriol. *189*, 833–843.

Norskov-Lauritsen, N., and Kilian, M. (2006) Reclassification of *Actinobacillus actinomycetemcomitans, Haemophilus aphrophilus, Haemophilus paraphrophilus* and *Haemophilus segnis* as *Aggregatibacter actinomycetemcomitans* gen. nov., comb. nov., *Aggregatibacter aphrophilus* comb. nov. and *Aggregatibacter segnis* comb. nov., and emended description of *Aggregatibacter aphrophilus* to include V factor-dependent and V factor-independent isolates. Int. J. Syst. Evol. Microbiol. *56*, 2135–2146.

O'Brien-Simpson, N.M., Paolini, R.A., Hoffmann, B., Slakeski, N., Dashper, S.G., and Reynolds, E.C. (2001) Role of RgpA, RgpB, and Kgp proteinases in virulence of *Porphyromonas gingivalis* W50 in a murine lesion model. Infect. Immun. *69*, 7527–7534.

O'Brien-Simpson, N.M., Veith, P.D., Dashper, S.G., and Reynolds, E.C. (2003) *Porphyromonas gingivalis* gingipains: the molecular teeth of a microbial vampire. Curr. Protein Pept. Sci. *4*, 409–426.

O'Farrell, P.H. (1975) High resolution two-dimensional electrophoresis of proteins. J. Biol. Chem. *250*, 4007–4021.

O'Toole, G., Kaplan, H.B., and Kolter, R. (2000) Biofilm formation as microbial development. Ann. Rev. Microbiol. *54*, 49–79.

Oda, Y., Huang, K., Cross, F.R., Cowburn, D., and Chait, B.T. (1999) Accurate quantitation of protein expression and site-specific phosphorylation. Proc. Natl. Acad. Sci. USA *96*, 6591–6596.

Okamoto, K., Tomoko, K., Nakayama, K., and Tamamoto, K. (1996) Cloning and sequencing of the gene encoding a nobel lysine-specific cysteine proteinase (Lys-Gingipain) in *Porphyromonas gingivalis*: Structural relationship with the Arginine-Specific Systeine Proteinase (Arg-Gingipain). J. Biochem. *120*, 398–406.

Olsen, I., and Shah, H. (2003) Minutes: International Committee on systematics of prokaryotes subcommittee on the taxonomy of Gram-negative anaerobic rods. Int. J. Syst. Evol. Microbiol. 53, 923–924.

Olsen, J.V., Andersen, J.R., Nielsen, P.A., Nielsen, M.L., Figeys, D., Mann, M., and Wisniewski, J.R. (2004) HysTag--a novel proteomic quantification tool applied to differential display analysis of membrane proteins from distinct areas of mouse brain. Mol. Cell. Proteomics 3, 82–92.

Onagawa, M., Ishihara, K., and Okuda, K. (1994) Coaggregation between *Porphyromonas gingivalis* and *Treponema denticola*. Bull. Tokyo Dent. Coll. 35, 171–181.

Ong, S.E., Blagoev, B., Kratchmarova, I., Kristensen, D.B., Steen, H., Pandey, A., and Mann, M. (2002) Stable isotope labeling by amino acids in cell culture, SILAC, as a simple and accurate approach to expression proteomics. Mol. Cell Proteomics *1*, 376–386.

Ong, S.E., and Mann, M. (2005) Mass spectrometry-based proteomics turns quantitative. Nat. Chem. Biol. *1*, 252–262.

Panchaud, A., Guillaume, E., Affolter, M., Robert, F., Moreillon, P., and Kussmann, M. (2006) Combining protein identification and quantification: C-terminal isotope-coded tagging using sulfanilic acid. Rapid Commun. Mass Spectrom. *20*, 1585–1594.

Pathirana, R.D., O'Brien-Simpson, N.M., Veith, P.D., Riley, P.F., and Reynolds, E.C. (2006) Characterization of proteinase-adhesin complexes of *Porphyromonas gingivalis*. Microbiology *152*, 2381–2394.

Patwardhan, A.J., Strittmatter, E.F., Camp, D.G., 2nd, Smith, R.D., and Pallavicini, M.G. (2006) Quantitative proteome analysis of breast cancer cell lines using 18O-labeling and an accurate mass and time tag strategy. Proteomics 6, 2903–2915.

Paul-Satyaseela, M., Karched, M., Bian, Z., Ihalin, R., Boren, T., Arnqvist, A., Chen, C., and Asikainen, S. (2006) Immunoproteomics of *Actinobacillus actinomycetemcomitans* outer-membrane proteins reveal a highly immunoreactive peptidoglycan-associated lipoprotein. J. Med. Microbiol. 55, 931–942.

Perkins, D.N., Pappin, D.J., Creasy, D.M., and Cottrell, J.S. (1999) Probability-based protein identification by searching sequence databases using mass spectrometry data. Electrophoresis *20*, 3551–3567.

Persson, G.R., Engel, D., Whitney, C., Darveau, R., Weinberg, A., Brunsvold, M., and Page, R.C. (1994) Immunization against *Porphyromonas gingivalis* inhibits progression of experimental periodontitis in nonhuman primates. Infect. Immun. *62*, 1026–1031.

Potempa, J., Pike, R., and Travis, J. (1995) The multiple forms of trypsin-like activity present in various strains of *Porphyromonas gingivalis* are due to the presence of either Arg-gingipain or Lys-gingipain. Infect. Immun. *63*, 1176–1182.

Qiu, Y., Sousa, E.A., Hewick, R.M., and Wang, J.H. (2002) Acid-labile isotope-coded extractants: a class of reagents for quantitative mass spectrometric analysis of complex protein mixtures. Anal. Chem. *74*, 4969–4979.

Rabilloud, T. (2002) Two-dimensional gel electrophoresis in proteomics: old fashioned, but it still climbs up the mountains. Proteomics *2*, 3–10.

Rabilloud, T., Valette, C., and Lawrence, J.J. (1994) Sample application by in-gel rehydration improves the resolution of two-dimensional electrophoresis with immobilized pH gradients in the first dimension. Electrophoresis *15*, 1552–1558.

Robrish, S.A., Oliver, C., and Thompson, J. (1987) Amino acid-dependent transport of sugars by *Fusobacterium nucleatum* ATCC 10953. J. Bacteriol. *169*, 3891–3897.

Robrish, S.A., and Thompson, J. (1990) Regulation of fructose metabolism and polymer synthesis by *Fusobacterium nucleatum* ATCC 10953. J Bacteriol *172*, 5714–5723.

Rogers, A.H., Zilm, P.S., Gully, N.J., Pfennig, A.L., and Marsh, P.D. (1991) Aspects of the growth and metabolism of *Fusobacterium nucleatum* ATCC 10953 in continuous culture. Oral Microbiol. Immunol. *6*, 250–255.

Roper, J.M., Raux, E., Brindley, A.A., Schubert, H.L., Gharbia, S.E., Shah, H.N., and Warren, M.J. (2000) The enigma of cobalamin (Vitamin B12) biosynthesis in *Porphyromonas gingivalis*. Identification and characterization of a functional corrin pathway. J. Biol. Chem. 275, 40316–40323.

Rose, J.E., Meyer, D.H., and Fives-Taylor, P.M. (2003) *Aae*, an autotransporter involved in adhesion of *Actinobacillus actinomycetemcomitans* to epithelial cells. Infect. Immun. *71*, 2384–2393.

Ross, B.C., Czajkowski, L., Hocking, D., Margetts, M., Webb, E., Rothel, L., Patterson, M., Agius, C., Camuglia, S., Reynolds, E., *et al.* (2001) Identification of vaccine candidate antigens from a genomic analysis of *Porphyromonas gingivalis*. Vaccine *19*, 4135–4142.

Ross, P.L., Huang, Y.N., Marchese, J.N., Williamson, B., Parker, K., Hattan, S., Khainovski, N., Pillai, S., Dey, S., Daniels, S., *et al.* (2004) Multiplexed protein quantitation in *Saccharomyces cerevisiae* using amine-reactive isobaric tagging reagents. Mol. Cell Proteomics *3*, 1154–1169.

Saito, T., Ishihara, K., Kato, T., and Okuda, K. (1997) Cloning, expression, and sequencing of a protease gene from *Bacteroides forsythus* ATCC 43037 in *Escherichia coli*. Infect. Immun. 65, 4888–4891.

Sakamoto, M., Suzuki, M., Umeda, M., Ishikawa, I., and Benno, Y. (2002) Reclassification of *Bacteroides forsythus* (Tanner *et al.* 1986) as *Tannerella forsythensis* corrig., gen. nov., comb. nov. Int. J. Syst. Evol. Microbiol. 52, 841–849.

Sanger, F., Nicklen, S., and Coulson, A.R. (1977) DNA sequencing with chain-terminating inhibitors. Proc. Natl. Acad. Sci. USA *74*, 5463–5467.

Santoni, V., Molloy, M., and Rabilloud, T. (2000) Membrane proteins and proteomics: un amour impossible? Electrophoresis *21*, 1054–1070.

Schmidt, A., Kellermann, J., and Lottspeich, F. (2005) A novel strategy for quantitative proteomics using isotope-coded protein labels. Proteomics 5, 4–15.

Schnolzer, M., Jedrzejewski, P., and Lehmann, W.D. (1996) Protease-catalyzed incorporation of 18O into peptide fragments and its application for protein sequencing by electrospray and matrix-assisted laser desorption/ionization mass spectrometry. Electrophoresis *17*, 945–953.

Schultz, C.P., Wolf, V., Lange, R., Mertens, E., Wecke, J., Naumann, D., and Zahringer, U. (1998) Evidence for a new type of outer membrane lipid in oral spirochete *Treponema denticola*. Functioning permeation barrier without lipopolysaccharides. J. Biol. Chem. *273*, 15661–15666.

Sebaihia, M., Bentley, S., Thompson, N., and Parkhill, J. (2002) Tales of the unexpected. Trends Microbiol. *10*, 261–262.

Sebastiano, R., Citterio, A., Lapadula, M., and Righetti, P.G. (2003) A new deuterated alkylating agent for quantitative proteomics. Rapid Commun. Mass Spestrom. *17*, 2380–2386.

Sechi, S. (2002) A method to identify and simultaneously determine the relative quantities of proteins isolated by gel electrophoresis. Rapid Commun. Mass Spectrom. *16*, 1416–1424.

Seers, C.A., Slakeski, N., Veith, P.D., Nikolof, T., Chen, Y.Y., Dashper, S.G., and Reynolds, E.C. (2006) The RgpB C-terminal domain has a role in attachment of RgpB to the outer membrane and belongs to a novel C-terminal-domain family found in *Porphyromonas gingivalis*. J. Bacteriol. *188*, 6376–6386.

Seshadri, R., Myers, G.S., Tettelin, H., Eisen, J.A., Heidelberg, J.F., Dodson, R.J., Davidsen, T.M., DeBoy, R.T., Fouts, D.E., Haft, D.H., *et al.* (2004) Comparison of the genome of the oral pathogen *Treponema denticola* with other spirochete genomes. Proc. Natl. Acad. Sci. USA *101*, 5646–5651.

Shah, H.N., Mayrand, D., and Genco, R.J. (1993) Biology of the species *Porphyromonas gingivalis* (London: CRC Press).

Sharma, A., Inagaki, S., Sigurdson, W., and Kuramitsu, H.K. (2005) Synergy between *Tannerella forsythia* and *Fusobacterium nucleatum* in biofilm formation. Oral Microbiol. Immunol. *20*, 39–42.

Shevchenko, A., Chernushevich, I., Ens, W., Standing, K.G., Thomson, B., Wilm, M., and Mann, M. (1997) Rapid 'de novo' peptide sequencing by a combination of nanoelectrospray, isotopic labeling and a quadrupole/time-of-flight mass spectrometer. Rapid Commun. Mass Spectrom. *11*, 1015–1024.

Shevchenko, A., Wilm, M., Vorm, O., and Mann, M. (1996) Mass spectrometric sequencing of proteins silver-stained polyacrylamide gels. Anal. Chem. *68*, 850–858.

Shizukuishi, S., Tazaki, K., Inoshita, E., Kataoka, K., Hanioka, T., and Amano, A. (1995) Effect of concentration of compounds containing iron on the growth of *Porphyromonas gingivalis*. FEMS Microbiol. Lett. *131*, 313–317.

Simons, B.L., Wang, G., Shen, R.F., and Knepper, M.A. (2006) In vacuo isotope coded alkylation technique (IVICAT); an N-terminal stable isotopic label for quantitative liquid chromatography/mass spectrometry proteomics. Rapid Commun. Mass Spectrom. *20*, 2463–2477.

Simonson, L.G., McMahon, K.T., Childers, D.W., and Morton, H.E. (1992) Bacterial synergy of *Treponema denticola* and *Porphyromonas gingivalis* in a multinational population. Oral Microbiol. Immunol. *7*, 111–112.

Slakeski, N., Cleal, S.M., Bhogal, P.S., and Reynolds, E.C. (1999) Characterization of a *Porphyromonas gingivalis* gene prtK that encodes a lysine-specific cysteine proteinase and three sequence-related adhesins. Oral Microbiol. Immunol. *14*, 92–97.

Smalley, J.W., Silver, J., Marsh, P.J., and Birss, A.J. (1998) The periodontopathogen *Porphyromonas gingivalis* binds iron protoporphyrin IX in the mu-oxo dimeric form: an oxidative buffer and possible pathogenic mechanism. Biochem. J. *331 (Pt 3)*, 681–685.

Socransky, S.S., Haffajee, A.D., Cugini, M.A., Smith, C., and Kent, R.L., Jr. (1998) Microbial complexes in subgingival plaque. J. Clin. Periodont. *25*, 134–144.

Stintzi, A., Marlow, D., Palyada, K., Naikare, H., Panciera, R., Whitworth, L., and Clarke, C. (2005) Use of genome-wide expression profiling and mutagenesis to study the intestinal lifestyle of *Campylobacter jejuni*. Infect. Immun. *73*, 1797–1810.

Takahashi, N. (2005) Microbial ecosystem in the oral cavity: Metabolic diversity in an ecological niche and its relationship with oral diseases. Int. Cong. Series *1284*, 103–112.

Takao, T., Hori, H., Okamoto, K., Harada, A., Kamachi, M., and Shimonishi, Y. (1991) Facile assignment of sequence ions of a peptide labelled with 18O at the carboxyl terminus. Rapid Commun. Mass Spectrom. *5*, 312–315.

Tanner, A., Listergarten, M., and Ebersole, J. (1986) Bacteroides forsythus sp. nov., a slow-growing, fusiform *Bacteroides sp.* from the human oral cavity. Internat. J. System. Bacteriol. *36*, 213–221.

Tribble, G.D., Mao, S., James, C.E., and Lamont, R.J. (2006) A *Porphyromonas gingivalis* haloacid dehalogenase family phosphatase interacts with human phosphoproteins and is important for invasion. Proc. Natl. Acad. Sci. USA *103*, 11027–11032.

van Steenbergen, T.J., Kastelein, P., Touw, J.J., and de Graaff, J. (1982) Virulence of black-pigmented *Bacteroides* strains from periodontal pockets and other sites in experimentally induced skin lesions in mice. J. Periodont. Res. *17*, 41–49.

Veith, P.D., Chen, Y.Y., and Reynolds, E.C. (2004) *Porphyromonas gingivalis* RgpA and Kgp proteinases and adhesins are C terminally processed by the carboxypeptidase CPG70. Infect. Immun. *72*, 3655–3657.

Veith, P.D., Talbo, G.H., Slakeski, N., Dashper, S.G., Moore, C., Paolini, R.A., and Reynolds, E.C. (2002) Major outer membrane proteins and proteolytic processing of RgpA and Kgp of *Porphyromonas gingivalis* W50. Biochem. J. *363*, 105–115.

Veith, P.D., Talbo, G.H., Slakeski, N., and Reynolds, E.C. (2001) Identification of a novel heterodimeric outer membrane protein of *Porphyromonas gingivalis* by two-dimensional gel electrophoresis and peptide mass fingerprinting. Eur. J. Biochem. *268*, 4748–4757.

Venter, J.C., Remington, K., Heidelberg, J.F., Halpern, A.L., Rusch, D., Eisen, J.A., Wu, D., Paulsen, I., Nelson, K.E., Nelson, W. (2004) Environmental genome shotgun sequencing of the Sargasso Sea. Science *304*, 66–74.

Vesey, P.M., and Kuramitsu, H.K. (2004) Genetic analysis of *Treponema denticola* ATCC 35405 biofilm formation. Microbiology *150*, 2401–2407.

Wilkins, J.C., Homer, K.A., and Beighton, D. (2001) Altered protein expression of *Streptococcus oralis* cultured at low pH revealed by two-dimensional gel electrophoresis. Appl. Environ. Microbiol. *67*, 3396–3405.

Wilkins, M.R., Appel, R.D., Van Eyk, J.E., Chung, M.C., Gorg, A., Hecker, M., Huber, L.A., Langen, H., Link, A.J., Paik, Y.K., *et al.* (2006) Guidelines for the next 10 years of proteomics. Proteomics 6, 4–8.

Wolf, V., Lange, R., and Wecke, J. (1993) Development of quasi-multicellular bodies of *Treponema denticola*. Arch. Microbiol. *160*, 206–213.

Wolff, S., Otto, A., Albrecht, D., Zeng, J.S., Buttner, K., Gluckmann, M., Hecker, M., and Becher, D. (2006) Gel-free and gel-based proteomics in *Bacillus subtilis*: a comparative study. Mol. Cell Proteomics 5, 1183–1192.

Wooldridge, K.G., and Williams, P.H. (1993) Siderophores: structure and function of microbial iron transport compounds. FEMS Microbiol. Rev. *12*, 325–348.

Ximenez-Fyvie, L.A., Haffajee, A.D., and Socransky, S.S. (2000) Comparison of the microbiota of supra- and subgingival plaque in health and periodontitis. J. Clin. Periodont. *27*, 648–657.

Yang, H.W., Huang, Y.F., and Chou, M.Y. (2004) Occurrence of *Porphyromonas gingivalis* and *Tannerella forsythensis* in periodontally diseased and healthy subjects. J. Periodont. *75*, 1077–1083.

Yao, X., Afonso, C., and Fenselau, C. (2003) Dissection of proteolytic 18O labeling: endoprotease-catalyzed 16O-to-18O exchange of truncated peptide substrates. J. Proteome Res. *2*, 147–152.

Yao, X., Freas, A., Ramirez, J., Demirev, P.A., and Fenselau, C. (2001) Proteolytic 18O labeling for comparative proteomics: model studies with two serotypes of adenovirus. Anal. Chem. *73*, 2836–2842.

Yilmaz, O., Watanabe, K., and Lamont, R.J. (2002) Involvement of integrins in fimbriae-mediated binding and invasion by *Porphyromonas gingivalis*. Cell. Microbiol. *4*, 305–314.

Yilmaz, O., Young, P.A., Lamont, R.J., and Kenny, G.E. (2003) Gingival epithelial cell signalling and cytoskeletal responses to *Porphyromonas gingivalis* invasion. Microbiology *149*, 2417–2426.

Yoneda, M., Hirofuji, T., Anan, H., Matsumoto, A., Hamachi, T., Nakayama, K., and Maeda, K. (2001) Mixed infection of *Porphyromonas gingivalis* and *Bacteroides forsythus* in a murine abscess model: involvement of gingipains in a synergistic effect. J. Periodont. Res. *36*, 237–243.

Yu, F., Iyer, D., Anaya, C., and Lewis, J.P. (2006) Identification and characterization of a cell surface protein of *Prevotella intermedia* 17 with broad-spectrum binding activity for extracellular matrix proteins. Proteomics *6*, 6023–6032.

Zhang, Y., Wang, T., Chen, W., Yilmaz, O., Park, Y., Jung, I., and Hackett, M. (2005) Differential protein expression by *Porphyromonas gingivalis* in response to secreted epithelial cell components. Proteomics 5, 198–211.

Zilm, P., Gully, N., and Rogers, A.H. (2002) Growth pH and transient increases in amino acid availability influence polyglucose synthesis by *Fusobacterium nucleatum*. FEMS Microbiol. Lett. *21*, 5203–208.

Zilm, P.S., Bagley, C.J., Rogers, A.H., Milne, I.R., and Gully, N.J. (2007) The proteomic profile of *Fusobacterium nucleatum* is regulated by growth pH. Microbiology *153*, 148–159.

Horizontal Gene Transfer in Oral Bacteria

3

Alex Mira

Abstract

Analyses of gene phylogenies and DNA-composition reveal that an important source of genetic innovation in prokaryotes is the transfer of sequences from other bacteria. The advent of genomic data now allows determination of the importance of horizontally transferred genes on a genomic scale. By comparative analysis of fully sequenced oral bacteria and between them and non-oral relatives, it is clear that horizontal gene transfer (HGT) is extremely frequent in oral pathogens and that this process probably also takes place between bacteria and oral fungi. It is likely that HGT is facilitated, in oral biofilms such as dental plaque, by the close physical contact between phylogenetically distant bacteria. In some cases, certain sets of genes are shared between more than two species co-inhabiting the mouth ecosystem, supporting the idea that there is an environment-associated gene pool formed by sequences of special adaptive importance. This *habitat gene reservoir* is also present in other natural ecosystems. A survey of HGT in oral bacteria is presented, as well as the evolutionary and biological consequences of this process. From the applied viewpoint, some of these genes could be potential vaccine candidates, targeting more than one potentially pathogenic species.

Introduction

The generation of new functional capabilities is a vital component of bacterial adaptability and the genomic era is shedding light on the source of new genes. A large proportion of novel genes arises by DNA duplication, followed by sequence divergence (Ohno, 1970). In prokaryotes, the proportion of these paralogous genes can be very important, reaching up to fifty percent of the DNA in versatile species with large genomes (Pushker *et al.*, 2004). New gene functions can also be generated by fusion of pre-existing domains, although the overall importance of this process appears to be limited in prokaryotes (Snel, Bork and Huynen, 2002). Moreover, there is strong evidence for a third mechanism, which has an overwhelming impact on bacterial evolution. Namely, lateral transfer of DNA sequences between bacterial cells, which has been proved experimentally and by sequence analysis, to occur frequently between different species (see, for example, Ochman *et al.*, 2000), and even between bacterial and archaeal species (Nelson *et al.*, 1999).

The existence of horizontal gene transfer (HGT) in prokaryotes was, of course, known many years before the completion of bacterial genomic sequences. For example, the transmission of antibiotic resistance among clinical isolates has long been identified as the outcome of plasmid transmission (Spratt and Maiden, 1999); and many bacterial species are naturally competent and can incorporate DNA from the environment (Ogunseitan, 1995). The contribution of bacterial genomics to our understanding of these phenomena has mainly been the quantification of their importance. Whole genomic sequences can now be screened for unusual compositional features or phylogenetic incongruence and different methods point to a major role played by exogenous genes in the lifestyle of bacteria. Some estimates put the level of HGT in model organisms, such as *Escherichia coli*, at 15–20% of its genome (Lawrence and Ochman, 1998); and trans-domain transfers between Eubacteria and *Archaea* have been estimated to represent one fourth of the genes in *Thermotoga maritima* (Nelson *et al.*, 1999). When the DNA sequences of the uropathogenic and enterohaemorragic strains of *E. coli* were completed and compared to the saprophytic laboratory strain K12, it was found that only 40% of the overall gene pool is shared by the three genomes (Welch *et al.*, 2002). A similar situation has been found in other cases where multiple strains of the same species have been sequenced, giving rise to the pan-genome concept (Tettelin *et al.*, 2005), a term reflecting the fact that the total amount of DNA material in a given species is far greater than that of a sequenced representative. As expected, a large portion of the unique, strain-specific genes appears to have a xenologous origin (e.g. Marri *et al.*, 2006). Therefore, the interspecific transfer of DNA material has important consequences for the way in which evolution takes place in bacteria; and a prokaryotic evolutionary model must necessarily include HGT as a crucial factor in understanding phylogenies, population genetics or the species concept (Gogarten *et al.*, 2002).

Horizontal gene transfer is also present in the oral cavity, where at least 500 bacterial species may co-inhabit in various niches such as the tongue, the teeth or gingivae (Paster *et al.*, 2001), presenting the possibility for exchange among them. To these, we have to add the common presence of fungal species, both in healthy and diseased individuals (Qi *et al.*, 2005; Lyon *et al.*, 2006). Full genome sequences of bacteria inhabiting the oral niche have recently become available, and high levels of HGT have been proposed for these environments (Mira *et al.*, 2004), probably facilitated by the close physical contact such as occurs in dental plaque. An excellent review on this subject has recently been published (Roberts and Mullany, 2006). The present chapter will focus on the impact of genomic data on our understanding of the frequency of this phenomenon and its relevance for the ecology of oral bacteria. Firstly, the different mechanisms responsible for gene transfer, with special emphasis on the cases identified in oral microbes, will be described. The methods currently used to identify these processes at a genomic scale will then be reviewed and, based on them, estimates of the sources and impact of HGT on the biology of microorganisms inhabiting the mouth will be provided. Finally, the presence of niche-associated genes among oral bacteria will be discussed, as will their evolutionary consequences and practical applications.

Mechanisms of gene transfer in oral bacteria

All the main mechanisms responsible for HGT have been identified in oral bacteria; namely, conjugation, transformation and transduction (Fig. 3.1). It has to be noted, however, that these three processes are not totally independent since a modular structure has been identified in many

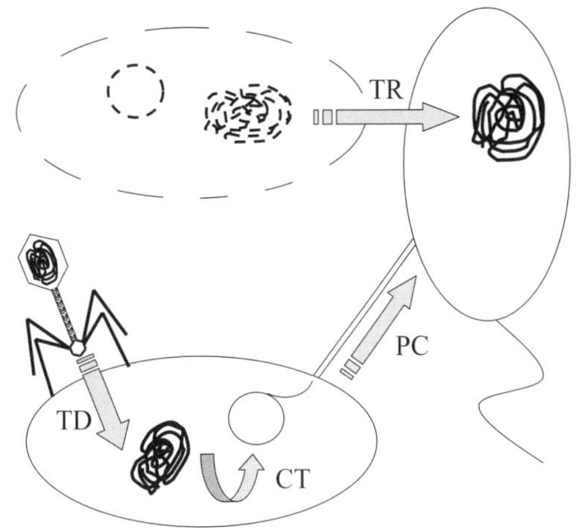

Figure 3.1 Mechanisms of HGT in oral species. When a cell undergoes lysis (top left) its DNA can be incorporated by competent cells, such as oral streptococci, in a process called Transformation (TR in the figure). Intercellular contact may transmit DNA horizontally by conjugation, transferring plasmids (PC) or conjugative transposons (CT), as has frequently been documented for antibiotic resistance genes in different oral species. Finally, bacteriophages have been repeatedly isolated from human saliva and can transmit the DNA packaged in their capsids by transduction (TD) and exchange DNA with its bacterial hosts.

mobile elements (Osborn and Boltner, 2002). This means that different mobile units can recombine and exchange DNA fragments, in such a way that genes from plasmids or bacteriophages can form part of conjugative transposons (Roberts and Mullany, 2006) and *vice versa*.

Conjugative plasmids and transposons

Bacterial conjugation is a genetic transfer mechanism requiring cell-to-cell contact, which has been frequently identified in oral species. Plasmids are common among oral bacteria and some of them are responsible for broad host-range transfers, including 'trans-Gram' transfers and transfers between bacteria and yeast (Adamczyk and Jagura-Burdzy, 2003). Multiple examples of HGT between oral bacteria, mediated by conjugation, have been identified (Roberts and Mullany, 2006) and the process can occur at significantly high rates. The transfer rate of a derivative of plasmid RK2 between *E coli* and an oral *Actinobacillus* was found to be as high as 1 every 2.5 donor cells (Goncharoff *et al.*, 1993). Large plasmids encoding antibiotic resistance have been identified in oral streptococci and have been shown to be highly promiscuous, transferring not only themselves but also other smaller plasmids between Gram-positive bacteria and between species such as *Enterococcus faecalis* and *E. coli* (Kurenbach *et al.*, 2003). Apart from antibiotic resistance, plasmids encoding cell wall antigens have been identified in oral *Actinomyces* (Hammond *et al.*, 1976).

In addition, conjugative transposons are able not only to transpose intragenomically but can also be transferred from the genome of a donor into a recipient (Mullany *et al.*, 2002); and do not undergo the constraints imposed by incompatibility (Norgren and Scott 1991). These elements appear to be excised from the chromosome and join their ends to form a circular molecule, which can then be transferred by conjugation to a different cell (see, for example, Salyers *et al.*, 1995). These elements are very common in prokaryotes and have been extensively studied for their implication in the spread of antibiotic resistance (Rice, 1998). In the mouth, the presence of conjugative transposons is well established (Roberts *et al.*, 2003), including oral streptococci (Hartley *et al.*, 1984) and *Fusobacterium* species (McKay *et al.*, 1995). Using saliva as an inoculum to create a dental microcosm (Peters *et al.*, 1991), it was demonstrated that tetracycline resistance, coded by the conjugative transposon Tn5397 of non-oral *Bacillus subtilis*, could be transferred and preserved in oral bacteria after a short six-hour exposure (Roberts *et al.*, 1999). Filter-mating experiments demonstrated that oral streptococci directly isolated from the saliva were able to transmit Tn916-like elements to a recipient *Enterococcus faecalis*. Conjugative transposition between oral bacteria was also demonstrated between a tetracycline resistant strain of *Streptococcus salivarius* and *S. parasanguinis* growing for 150 hours in an artificial dental biofilm (Roberts *et al.*, 2001). It is important to note that antibiotic resistance is not the only property coded for in conjugative transposons. For example, the element Tn5276 contains a sucrose metabolic pathway (Rauch and de Vos, 1992) and Tn5252 codes for the production of the bacteriocin Nisin, to which it also confers immunity (Horn *et al.*, 1991).

Transduction

Bacterial DNA can be packaged into phage capsids and then transferred by the virus when it infects a new host cell (see, for example Ochman *et al.*, 2000). This transduction process, although widespread in prokaryotes, is not well documented in oral bacteria. There could be a mechanistic reason for this, in the sense that heterologous microbial cells and extracellular structures, such as polymers present in the oral biofilm, may impede viral access to the bacterial cell surface (Sutherland *et al.*, 2004). However, many, although not all, phages can produce polysaccharases or polysaccharide lyases and are known to infect bacterial species forming part of biofilms (Sutherland *et al.*, 2004 and references therein). Phages are even used as a control strategy against biofilms (Sillankorva *et al.*, 2004). Phages therefore have an enormous potential for HGT in the oral environment. In fact, the phenomenon of genetic exchange through bacteriophages is considered a major vehicle for HGT, especially the transfer of virulence factors and invasion-related functions (Boyd and Brussow, 2002) and it is fundamental for gene innovation in the microbial world (Daubin and Ochman, 2004a; 2004b).

The presence of bacteriophage has been demonstrated in saliva, being found in oral *Actinomyces* as well as in *E. faecalis* (Bachrach *et al.*, 2003) and other transient inhabitants of the mouth (Hitch *et al.*, 2004). Bacteriophage particles have also been recovered from subgingival plaque samples in healthy and periodontitis patients (Sandmeier *et al.*, 1995; Willi *et al.*, 1997). They also appear to survive well in saliva samples held at different temperatures (Bachrach *et al.*, 2003). In addition, the availability of genomic data from oral inhabitants shows a variable number of genes of putative viral origin (Seshadri *et al.*, 2004; Nelson *et al.*, 2003). They are clearly distinguishable by encoding typical phage proteins, such as integrases, or by appearing as 'sister' branches to phage genes in phylogenetic trees (Mira *et al.*, 2004). Finally, phage transduction has been demonstrated in chloramphenicol-resistance experiments *in vitro* (Willi *et al.*, 1997b).

Transformation

Bacteria can also take up DNA directly from the environment (Ogunseitan 1995). This process, called transformation, requires that the cells are physiologically competent and is also influenced by physico-chemical parameters such as pH, salt, temperature, ultraviolet light and concentration of DNA. Oral bacteria from the genera *Streptococcus, Neisseria* and *Actinobacillus* are naturally competent (Cvitkovitch, 2001; Roberts and Mullany, 2006 and references therein). Transformation is well demonstrated in *S. mutans*, a species in which six erythromycin-sensitive strains were transformed by DNA from dead cells that were resistant to the antibiotic (Li *et al.*, 2001). It is interesting that the transformation frequency increased up to 600 times when the cells were grown in biofilms but not planktonically, suggesting that competence is probably facilitated in dental plaque. In another experiment, *S. gordonii* was shown to acquire a plasmid when naked DNA from *Treponema denticola* was exposed to the recipient grown both in liquid culture and in biofilm cultures (Wang *et al.*, 2002).

The competence of oral bacteria is not sufficient to allow the uptake of genes, as the DNA has to be able to persist in the mouth environment, where low pH, temperature, DNAses, etc., can degrade the molecule. Thus, the longevity of DNA in saliva samples and its ability to transform different bacterial species has been tested experimentally on a few occasions. Estimates of persistence times vary in different experiments; from 24 hours in ovine saliva (Duggan *et al.*, 2000) to under 4 minutes in human saliva at neutral pH and 1 minute at acidic pH (Mercer *et al.*, 1999). Even after extensive DNA degradation, specific regions could be amplified by PCR, and *Lactococcus lactis* and *Streptococcus gordonii* cells were successfully transformed, becoming antibiotic-resistant after a short incubation of several minutes (Mercer *et al.*, 1999). Nevertheless, transformation efficiency always decreased after exposure to saliva (Mercer *et al.*, 2001). It is important that DNA, freely available in human saliva, derives not only from other oral bacteria, but has also been shown to originate from other species that do not form part of the normal oral flora and that enter the mouth via food items (Roberts and Mullany, 2006). This should also be extended to fungal species that are contaminants or part of human food.

Barriers to gene transfer

In the conjugation process, an important constraint is imposed by plasmids' incompatibility groups, which make incompatible plasmids unable to coexist in the same cell. Thus, only those conjugation transfers involving plasmids from different Inc groups will be successful in the oral environment. It is also known that certain elements in oral bacteria require the transfer machinery of conjugative elements to be mobilized. For example, the transfer of antibiotic-resistance genes in *E. faecium*, where a chromosomally located tetracycline resistance gene could only be transferred to *E. faecalis* and *B. subtilis* when the streptomycin resistance plasmid was also present (Roberts *et al.*, 2003). Moreover, the constraints on HGT do not end with the intercellular gene transmission. Once the new sequence has been inserted, it has to be expressed in the new environment and it needs to confer an advantage on its new host. If these two conditions are not met, the xenologous DNA is quickly eliminated and turnover rates of foreign genetic material have been estimated to be extremely high (Berg and Kurland, 2002). It is clear that these two requirements are not always met and the genomes of oral bacteria contain remnants of genes that appear to be unsuccessful horizontal transfers. In *F. nucleatum*, for example, there are short sequences with some sequence similarity to other genes scattered along the intergenic spacers (Mira *et al.*, 2004). Most of them are contiguous to gene clusters transferred from *Proteobacteria*, and contain eroded genes with diverse functions, including outer membrane proteins, periplasmic-like proteins, permeases or hemolysin activators.

For a gene to be active in its new environment, it must be adapted to the intracellular physicochemical characteristics of the recipient cell. This is probably constraining transfers between species thriving at different environmental conditions and is reflected in the high incidence of HGT events between unrelated thermophiles or other extremophiles inhabiting similar ecological niches (see, for example, Nelson *et al.*, 1999; Tyson *et al.*, 2004). In the mouth, physical conditions are not entirely homogenous, with differences in pH or oxygen availability depending on the location of the bacteria and exposure to food, but other parameters, such as temperature, are stable. Once a DNA segment is transferred, its functionality will be influenced by the new genomic context. In *E. coli*, for example, it was found that the emergence of extended-spectrum α-lactamases was influenced by the genetic background of the strains, as a consequence of complex interactions between virulence factors and other genes, such as fluoroquinolone resistance (Branger *et al.*, 2005). In addition, transcription and translation must be especially problematic in HGTs across distantly related species. This is very common in oral species, where transfers have been detected between taxonomic groups as distantly related as *Proteobacteria*, *Fusobacteria*, *Spirochaetes* and *Bacteroidetes*. Because regulatory pathways, promoter regions and ribosome-binding sites are very different among these bacterial taxa, it is unlikely that the foreign ORFs can be expressed using the new transcription and translation machinery. This problem, homologous to the one faced by geneticists when rare proteins are cloned and expressed into model organisms like *E coli* (Daniel, 2005) has not been studied in depth. In oral bacteria, the constraints extend to genes that appear to have been transferred from yeast species (see the section 'Fungal donors'). We have identified several instances in which likely horizontal transfers in oral bacteria are located contiguously to, and orientated in the same direction as, native genes. Because the intergenic region between these gene pairs was found to be always shorter than the average spacer length between co-orientated genes, it is likely that foreign genes 'hitchhike' upon resident genes to be co-transcribed with them. For example, in *F. nucleatum*, a proline iminopeptidase (gene FN0752) which appears to be transferred from *Proteobacteria*, is co-oriented with the native ORF FN0753 (a Glutamyl-tRNA amidotransferase) and separated by 44 nucleotides. Also of *proteobacterial* origin, a gene coding for a large-conductance mechanosensitive channel (FN0766) is separated by 30 bp from its native predecessor (a tRNA methyltransferase). A systematic study of this phenomenon should be further studied both *in-silico* and experimentally and may reveal interesting insights into the ways in which genetic transfers circumvent expression constraints in the recipient cell.

HGT detection meets genomics

Experimental methods

How can HGT among oral bacteria be detected? Because transferred regions are normally associated to mobile elements, the search for specific sequences, such as transposon and IS-element associated regions, phage integrases or repeat structures, have commonly been used (Dobrindt *et al.*, 2004). In addition, the screening of common regions utilized for mobile-element insertion, such as specific tRNAs, can also reveal transfers (Alsmark *et al.*, 2004); and studying the stability of a DNA region, using counterselectable markers, may indicate whether a genomic island is expandable and not part of the core genome (Dobrindt *et al.*, 2004). Whole genome approximations can be made by suppressive subtractive hybridization, which will identify regions specific for a given strain (see, for example, Zhang *et al.*, 2005), as well as by the use of microarrays to detect absent genes in bacteria-related isolates (e.g. Lindroos et. al. 2006). This technique is known as 'comparative genome hybridization'. A sequenced strain is normally used to amplify fragments of each gene, spotting them into an array where whole DNA from related strains is hybridized using a fluorescent dye. However, these methods rely on the use of a bacterial strain as a reference and do not, therefore, detect novel genes absent in that strain. The accessory regions of genomes are normally more prone to HGT. In the oral pathogen *Streptococcus mutans*, for example, Waterhouse and collaborators have used whole genome hybridization to compare the genetic composition of 9 strains against a microarray containing 1948 ORFs of the reference strain UA159. About twenty percent of the genome was found to be dispensable in the tested strains and at least half of these genes had signs of HGT (Waterhouse *et al.*, 2007). By PCR amplification of specific regions suspected to contain horizontally transferred genes, Waterhouse and Russell (2006) had previously shown important heterogeneity among 39 *S. mutans* strains of different geographic origin. This experimental approach demonstrated the clear existence of a 'core' and a 'dispensable' genome in *S. mutans*, paving the way for relating physiological and virulence characteristics of a given strain to each genomic composition.

Experimental detection of a specific transfer is usually linked to the identification of the trait. For that reason, antibiotic resistance has been the character most widely studied in oral bacteria (Roberts *et al.*, 2003). However, the transmission of gene functions among oral bacteria surely extends to numerous metabolic routes, in the same way as HGT has been demonstrated not only in pathogenic species but also in most environmental bacteria studied (Doolittle *et al.*, 2003). For this reason, the advent of genomics has allowed researchers to extend their studies to the HGT of multiple traits, as well as to quantify the phenomenon on a genomic scale. Systematic analysis of phylogenies, compositional features, genomic context and sequence similarity comparisons can now be undertaken in sequenced oral microbes simply by analysing their sequences 'in silico'.

DNA composition

DNA compositional features can now be measured at a genomic level, characterizing bacterial genomes by specific signatures. One of them is the GC content that, contrary to eukaryotes, is fairly constant throughout the bacterial chromosome, typically showing discrete peaks at islands of HGT (Dobrindt *et al.*, 2004); especially when first or third codon positions are considered (Lawrence and Ochman, 1997). The G+C content alone is, nevertheless a crude indicator of external origin and more complex features have been used to detect potential lateral transfers. Among these, codon usage and oligonucleotide frequencies have received special attention. Due to the genetic code's degeneracy, there is more than one codon for the same amino acid, and bacteria appear to be quite specific in the preferred codons (Nakamura *et al.*, 1997). Analyses of the use of preferred codons in bacteria suggests that a high proportion of the genome of many species may be of external origin (Garcia-Vallve *et al.*, 2000), with values for model organisms such as *E. coli* reaching around 17% of its genes (Lawrence and Ochman 1998). It must be borne in mind that the use of GC content and codon bias, to detect HGT, can provide many false positives (Wang *et al.*, 2001) and can also be incomplete, since many genes with similar compositional features show other evidence of lateral transfer (Koski *et al.*, 2001). When a combination of GC content and codon usage is used in a conservative test, a list of at least 365, 198, 49 and 106 likely transfers are found in the oral species *Treponema denticola, Porphyromonas gingivalis, Fusobacterium nucleatum* and *Streptococcus mutans*, respectively (see the HGT Database by Garcia-Vallvé *et al.*, 2003). These are probably underestimates of the impact of HGT in those species, because only recent transfer events are identified by compositional features. The reason for this is that unusual compositional features tend to disappear with time, as the donor's genomic signature becomes homogenized with that of the recipient, in an ongoing amelioration process (Lawrence and Ochman, 1997). Finally, the frequency of different oligonucleotides, (typically 2–7 nucleotides in length) across whole bacterial genomes, show distinctive patterns for each species; and this 'genomic signature' can be used to find regions of unusual oligonucleotide usage (Dufraigne *et al.*, 2005). When the frequency of each oligonucleotide is computed in windows of around 10 000 bp along a bacterial chromosome and compared with the average oligo-use of the rest of the genome, a 'distance' value can be obtained. If this oligo-use distance is plotted along the genomes of oral bacteria, distinctive peaks are observed (Fig. 3.2), representing regions of unusual oligonucleotide frequencies of probable HGT origin.

Sequence similarity

An obvious instrument to detect HGT in a given genome is the use of DNA or protein-sequence similarity searches against public databases such as GeneBank. The most frequently used program is BLAST, a fast and versatile algorithm producing a local alignment against the most similar sequences available (Altschul *et al.*, 1990). However, the best BLAST hit is not always indicative

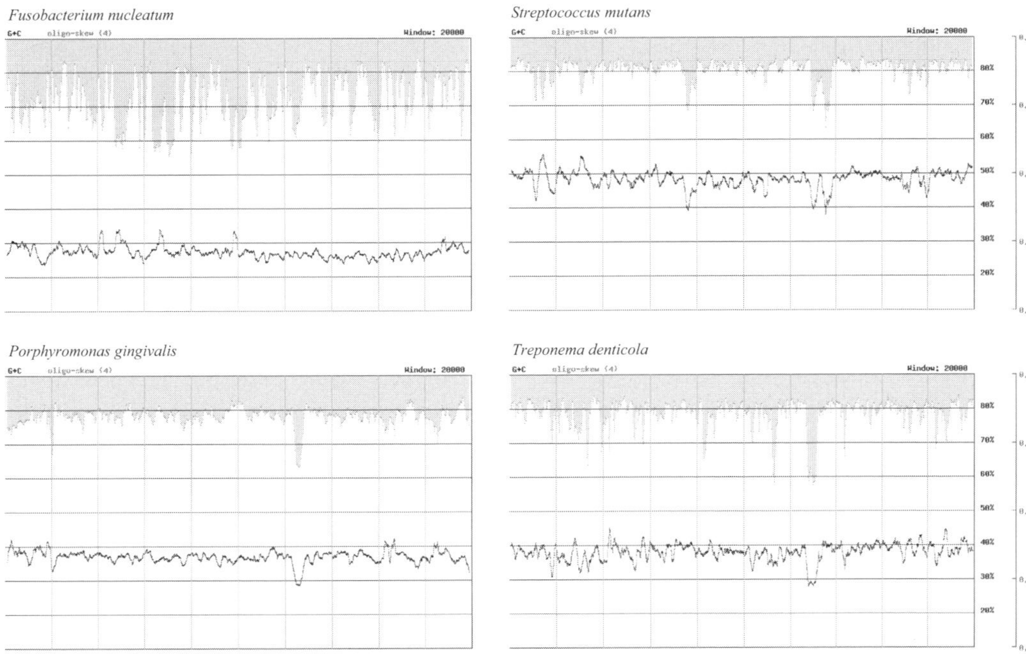

Figure 3.2 Oligonucleotide use (grey) and GC content (black) across the genomes of four fully sequenced oral bacteria. The frequency of all possible tetranucleotides was computed at windows of 20 000 nucleotides and compared against the oligo-use of the rest of the genome, calculating a genomic distance usage to draw the 'oligoskew' plot. Peaks deviating from the normal tetranucleotide use are indicative of genomic islands of putative external origin. These peaks tend to coincide with regions of unusual GC content, supporting the view that they are recent horizontal gene transfers. Oligo-skews for users sequences can be calculated at the oligoweb (http://insilico.ehu.es/oligoweb).

of true relatedness (Koski and Golding, 2001). If a gene is vertically inherited, it will most likely be present in phylogenetically related species; if on the other hand that gene is absent from related bacteria but highly similar to a taxonomically distant taxon, it is likely that a transfer event is responsible for the sequence similarity. Basic' best-hit' searches by BLAST have been improved by several authors by analysing the pattern of BLAST hits against a structured database separated by species. Unusual patterns of gene distribution are indicative of HGT (Clarke *et al.*, 2002; Ragan and Charlebois, 2002). When repeated BLAST searches are done systematically for all genes in sequenced oral bacteria, between 5 and 35% of the best sequence-similarities hit unrelated microorganisms. This method is less reliable for oral species belonging to a taxon with few sequenced representatives, such as the Fusobacteria, due to the weak phylogenetic signal remaining and to the high chance of finding best-hits to distant bacteria, because sequences of related bacteria are unavailable (Ragan, 2001).

An interesting case of HGT in oral bacteria, easily identified by sequence similarity, is that of CRISPR associated proteins. CRISPR elements are repeats evenly spaced in many bacterial chromosomes and showing remnants of mobile elements, such as phage sequences (Mojica *et al.*, 2005). When CRISPR-associated proteins are compared across prokaryotic species, those of *P. gingivalis*, *T. denticola*, *S. mutans* and *F. nucleatum* are most highly related to one another (Bolotin *et al.*, 2005), despite their distant phylogenetic relatedness. To this has to be added other oral species, whose genomes are incomplete, where CRISPR-associated proteins are also found to be highly related to the oral variants of these puzzling genes.

Tree incongruence

A direct and intuitive method to detect HGT is the use of phylogenetic trees (Nesbø *et al.*, 2001; Ochman *et al.*, 2000). The phylogeny of genes under study can be compared to that of the species, and incongruence in the tree, if its resolution is robust, is a strong indication of HGT (Brown and Doolitle, 1997); suggesting, in addition, the potential source of the transfer. Naturally, this method relies on the accuracy of the 'true' tree used as reference, traditionally based on rRNA genes (Woese, 2000) and also on concatenated multiple sequences (Lerat *et al.*, 2003) or 'genome trees' (Snel *et al.*, 1999). Several factors mitigate against a purely phylogenetic approach. These include the difficulty of inferring high-quality trees, the models used and the possibility that HGT has been so omnipresent that vertical signal has been lost (Brocchieri, 2001; Ragan, 2001). In the dental-plaque forming *F. nucleatum*, 1236 phylogenetic trees could be constructed from a total of 2128 genes in the sequenced strain, from which 38% were unresolved. Using several rules to determine the potential source of horizontal transfers, between 8–34% of the resolved trees were constant with the standard phylogeny (estimated by a 16S-23S gene and the concatenated sequences of 40 ribosomal proteins). In addition, some 20–45% were consistent with potential transfers from *Archaea, Spirochaetes, Firmicutes, Bacteroidetes* and *Proteobacteria* (Mira *et al.*, 2005). An example of an incongruent tree is shown in Fig. 3.3, where a gene from *F. nucleatum* is found to be related to proteobacterial species and locates in a cluster where other oral species are also present.

Genomic context

In order to detect lateral gene transfers, Koski and collaborators (2001) chose to study chromosomal positions of homologous genes between related bacteria. If the location of a gene was conserved between species or genera, vertical inheritance is the most parsimonious explanation. By determining this set of conserved genes, regions of multiple horizontal transfers could be identified. When the *E. coli* genome was compared with that of *Salmonella typhimurium*, for example, 70% of the genes were positionally conserved (Koski *et al.*, 2001). A similar approach was undertaken in *Fusobacterium nucleatum*, where a program was designed to find clusters of genes whose relative order is conserved across taxa (Mira *et al.*, 2005). Interestingly, some of these clusters are absent in taxonomically related bacteria but keep an identical, or highly similar gene order, in other oral bacteria (see Table 3.2). In general, it can be concluded that evidence for HGT is more robust if the genomic context of the putatively trans-

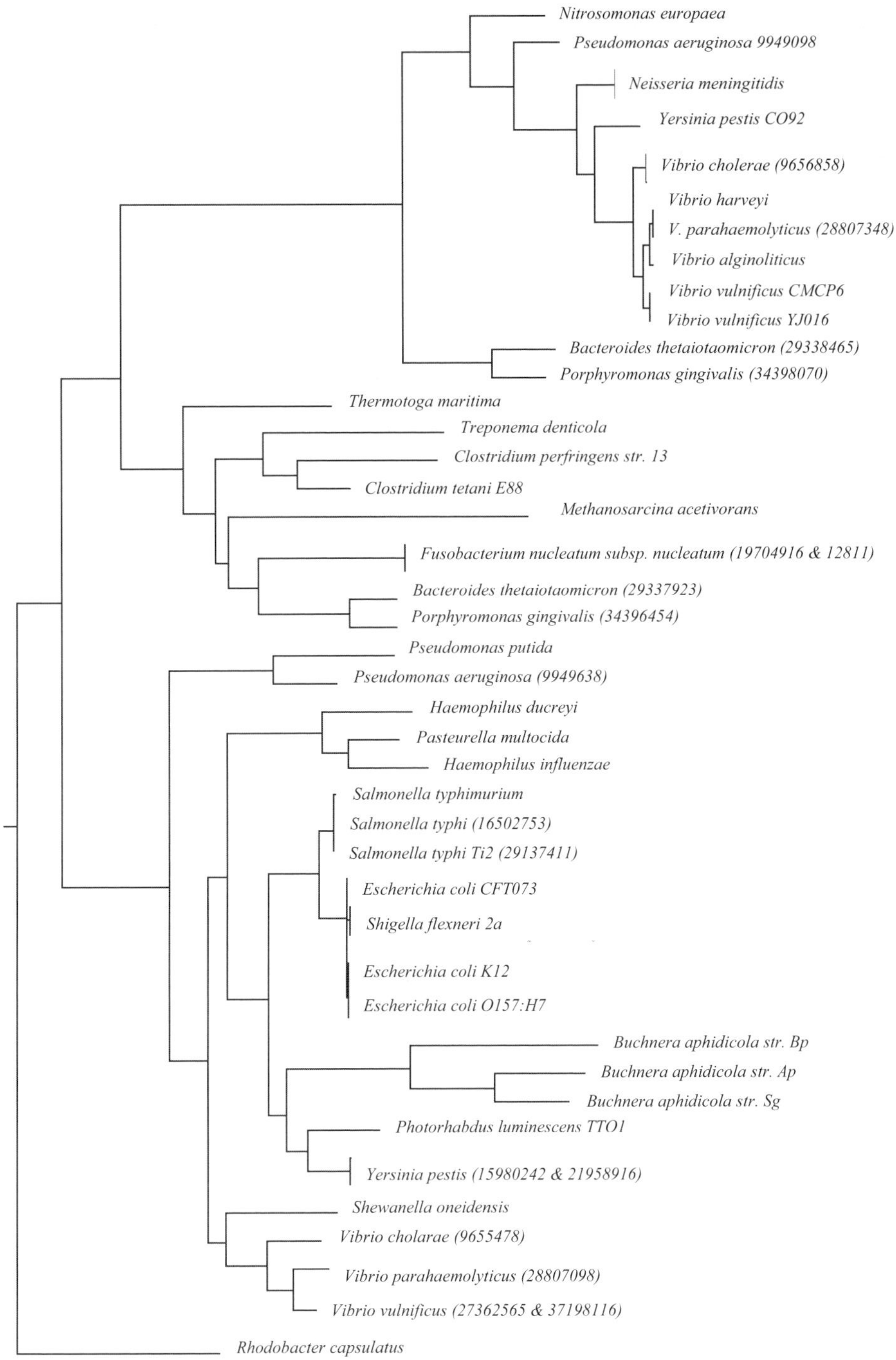

Figure 3.3 Sequence similarity searches and phylogenetic incongruence as indicators of HGT. A BLAST analysis was performed with the amino acid sequence of Na(+)-translocating NADH-quinone reductase (gene FN1595) of the oral microbe *Fusobacterium nucleatum*. The top 50 best-hits were selected and aligned to construct this maximum-likelihood tree. Most BLAST hits are within the gamma-proteobacteria, suggesting that the origin of the gene is in that taxonomic group. The *F. nucleatum* sequence appears highly related to other phylogenetically distant species that share the oral niche; for example, *P. gingivalis* and *T. denticola*, which belong to the unrelated *Bacteroidetes* and *Spirochaetes*, respectively. This gene forms part of a nitrogen-fixation cluster, where all other genes (FN1591-FN1596) give similar tree topologies. Numbers between brackets indicate gene PID number.

ferred gene is also supportive of being non-native; and the weaknesses of individual methods can be largely minimized by combining evidence from different methods. Now that some other major players in the oral environment are in the process of being sequenced (Duncan, 2003), a detailed study of HGT in the mouth will be highly desirable. Moreover, it will be possible to confirm whether or not the massive levels of transferred genes found in *Fusobacterium* are due to the peculiarity of this bacterium, as a prominent member of the dental plaque community, or a general phenomenon of the oral ecosystem. In addition, proteomic approaches will enable researchers to identify those transferred genes which are actually expressed in its new host genome and which ones are not functional.

The impact of HGT in the mouth

What is the importance of HGT for the biology of oral species? If we look at the extent of the phenomenon in sequenced representatives, it can be seen that putatively transferred genes account for between 5 and 45% of the genes in different species. The values obtained depend, as expected, on the method used. Oligonucleotide frequencies, for example, suggest several islands of extraneous composition across the different genomes (see Fig. 3.2); and other DNA compositional features suggest that 5–20% of the genomes from different oral bacteria have undergone HGT. Because these methods mainly indicate recent transfers (before the genomic signature of the incoming genes is homogenized with that of the recipient genome), they probably underestimate the magnitude of the process. Comparative genome hybridization to a *S. mutans* reference DNA microchip indicates that 20% of the genes in this species are variable among strains (Waterhouse *et al.*, 2007). The use of phylogenetic trees and conservation of gene-order increase the values up to 45% in other species. Among the few oral species fully sequenced, *F. nucleatum* would appear to be the most promiscuous species and *S. mutans* the one with more native genes. When more fully sequenced genomes from oral bacteria are available, it may be possible to determine whether or not the degree of HGT is a reflection of the species niche. Streptococci, for instance, might have fewer chances for interspecific interactions due to their unique acidogenic and acidophilic lifestyle, whereas Fusobacteria are key components of dental plaque and aggregate with many other bacteria, thereby facilitating the chance for interspecific gene transfer.

However, not only the quantity but mainly the quality of the transfers have to be considered in order to evaluate the impact of HGT in oral pathogens. In *F. nucleatum*, for example, a large portion of the transfers has cell-wall related functions (Mira *et al.*, 2005). This could be vital for the bacterium because while Fusobacteria appear to be highly related to Gram-positives, the transfer of many outer membrane genes (from *Bacteroidetes* and *Proteobacteria*) confers upon *F. nucleatum* a complex Gram-negative cell wall. In the oral environment, an outer membrane could be vital for adhesion and evasion of the immune system. Clear examples of the consequences of horizontally transferred sequences in the oral environment are given by the transmission of antibiotic-resistance genes. For example, resistance to erythromycin or tetracycline, has been shown to be transmitted by conjugative transposons across a wide range of oral bacteria, thus affecting the overall resistance of the biofilm to these (and other) antibiotics (Roberts *et al.*, 2001; 2003). Analysis of whole-genome sequences reveals that a whole operon involved in glutaconyl-coA metabolism appears to have been transferred from *Acidaminococcus fermentans* into *F. nucleatum*, which has also received different hemolysin activators, presumably of *proteobacterial* origin (Mira *et al.*, 2004). Other putative transfers in *F. nucleatum* include gene clusters involved in nitrogen fixation or peptide transport, strongly indicating that HGT is instrumental in shaping the current ecology of oral species. Other oral pathogens, such as *T. denticola*, show multiple transfers of iron and manganese binding and transport proteins (A. Mira, *unpublished data*). Such genes could be vital for exploiting nutrient resources from the gingival crevicular fluid (Marsh, 2003). A study of gene-gain and -loss in different streptococci suggests that a considerable portion of HGTs play an adaptive role since many of the transferred genes are under positive selection (Marri *et al.*, 2006). In *S. mutans*, these authors identified a long list of potential transfers, with functions related to carbohydrate transport and metabolism. This is consistent with HGT being driven by niche adaptation. In non-oral species, HGT events have been shown to determine a change in lifestyle for the recipient genome. A classic case is that of pathogenicity islands, which can transform commensal into pathogenic lifestyles (Hacker and Carniel, 2001). Dramatic examples include the transmission of the cholera toxin through a phage in *Vibrio cholerae* (Karaolis *et al.*, 1999) or the acquisition of genes that allow transmission, through a rodent vector, of the causative agent of plague (Parkhill *et al.*, 2001). There is no doubt that HGT of virulence-related factors among oral pathogens has directly and indirectly influenced their adhesion and biofilm-forming capabilities, as well as their chances of success in different environmental conditions within the oral cavity. The completion of genomic sequences of new oral bacteria, together with the shot-gun sequencing of the dental plaque metagenome by The Institute of Genomic Research (TIGR) is likely to provide numerous fascinating examples.

Sources of HGT

What is the origin of horizontally transferred genes detected in oral bacteria? By the use of sequence similarity searches, phylogenetic trees or DNA composition best-matches, potential donors can be recognized. Nevertheless, identification of the direction of the transfer is a common problem; that is, whether a gene of extraneous composition has been received by an oral bacterium or transferred from it into another species. In some cases, looking at the distribution of that gene in taxonomically related organisms can help to identify the donor and the recipient. For example, if a potentially transferred gene found in *S. mutans* is absent in related streptococci but present in species related to the potential donor, *S. mutans* would likely be the recipient. If all streptococci contain the gene, it is more likely that the transfer has been produced in the opposite direction. Analysis of the regions of xenologous origin indicates that the main donors and recipients of transferred genes in oral bacteria are other inhabitants of the oral niche and the gut, as well as transient bacteria accidentally ingested and swallowed.

Intra-plaque transfers

Physical interactions between bacteria inhabiting the oral environment are intense, probably facilitating transfer mechanisms. Excellent work by Kolenbrander and collaborators has demonstrated that many of the associations between oral bacteria are specific (Kolenbrander *et al.*, 2006), through the use of a complex system of receptors and adhesins that create an intricate biofilm (Marsh, 2006). Now that detailed maps of specific associations are emerging, it will be of interest to determine whether HGT events take place more frequently between species that form natural aggregations. In this sense, some species, such as *F. nucleatum*, are expected to show high levels of genome promiscuity. This species is considered a 'bridge' organism between the early and late plaque colonizers (Kolenbrander *et al.*, 2002) and forms numerous aggregations with other oral microorganisms (Weiss *et al.*, 2000; Kolenbrander *et al.*, 2006). In agreement with this, between 20 and 45% of the genes in this species show signs of HGT (Mira *et al.*, 2005).

As explained in the section 'Mechanisms of gene transfer in oral bacteria', *in vitro* experiments have shown that HGT can, indeed, occur between oral species (Roberts and Mullany, 2006). However, more recent work with *S. mutans* does not point to oral species as the main source of gene transfers in this bacterium (Waterhouse *et al.*, 2007). This finding could, nevertheless, be a consequence of the under-represented set of sequences from oral microbes in public databases, making it highly desirable to approach the study of the oral microbiota by a metagenomics approach. This could provide a good sampling of the genetic repertoire in the oral cavity, including those species that cannot be cultured using standard procedures (Rodriguez-Valera, 2004).

Fungal donors

The possibility for HGT between oral bacteria and fungi has not previously been considered. However, when sequence similarity searches are performed for all genes of oral bacteria, a reduced but significant number of top matches hits different yeast species (Table 3.1). At the beginning of the genomics era, significant hits to eukaryotic species were considered likely artefacts due to the limited size of prokaryotic databases (Ragan

Table 3.1 HGT in sequenced oral bacteria

(a) Impact of HGT in different species

Species	Method to detect HGT	Genome of potential HGT (%)	Reference
S. mutans	Codon usage + GC content	5.4	Garcia-Vallvé *et al.*, 2003
F. nucleatum	Codon usage + GC content	2.4	Garcia-Vallvé *et al.*, 2003
P. gingivalis	Codon usage + GC content	10.4	Garcia-Vallvé *et al.*, 2003
T. denticola	Codon usage + GC content	13.2	Garcia-Vallvé *et al.*, 2003
F. nucleatum	Sequence similarity	25.4	Mira *et al.*, 2004
F. nucleatum	Phylogenetic trees	27.1	Mira *et al.*, 2004
F. nucleatum	Gene order	42.0	Mira *et al.*, 2004
S. mutans	Oligonucleotide use	14.8	Bikandi and Mira, unpubl.
T. denticola	Oligonucleotide use	17.4	Bikandi and Mira, unpubl.
P. gingivalis	Oligonucleotide use	13.2	Bikandi and Mira, unpubl.
S. mutans	Whole genome hybridization	10.0	Waterhouse *et al.*, 2007

(b) Potential HGT from fungal donors

Gene function	Oral bacterial species	Putative fungal donor	Sequence similarity[1]
Manganese co-factored superoxide dismutase	*S. mutans* *P. gingivalis*	*Candida albicans*	38–34% $6e^{-33}$
Glutamine cyclotransferase-related protein	*P. gingivalis*	*Candida albicans*	27% 0.011
SUA5 protein	*F nucleatum* *S. mutans* *T. denticola*	*Candida albicans*	33% $7e^{-27}$
TPR domain protein	*T. denticola* *F. nucleatum* *P. gingivalis*	*Candida albicans*	24% $8e^{-7}$
Amino acid permease	*S. mutans* *F. nucleatum* *T. denticola*	*Candida albicans*	28% $1e^{-28}$
Unknown function; potential Cdc28p substrate (putative)	*Prev. intermedia* *P. gingivalis* *T. denticola*	*Saccharomyces cerevisiae*	26% $1e^{-14}$
Gastric mucin	*Prev. intermedia*	*Candida glabrata*	24% $1e^{-5}$
Hypothetical protein SPAPB1E7.04c	*P. gingivalis*	*Schizosaccharomyces pombe*	23% $2e^{-5}$
Putative calcium-transporting ATPase	*F nucleatum* *S. mutans* *P. gingivalis*	*Candida albicans*	37–34% $7e^{-144}$
Cysteine desulfhydrase	*F. nucleatum* *Camp. concisus*	*Candida albicans*	43–38% $5e^{-77}$

[1]Numbers indicate percentage amino acid identity and BLAST E-value.

2001). Today, however, the likelihood of these hits reflecting true relationships is fairly high. Fungal species appear to be normal inhabitants of the mouth flora (Lyon *et al.*, 2006; Qi *et al.*, 2005) and their presence can be significantly elevated if the oral ecosystem homeostasis is disturbed due to antibiotic treatment, immune depression, diabetes or other health conditions (Vozza *et al.*, 2005; Manfredi *et al.*, 2006). Apart from the ecological settlement of these species as part of the oral flora, contact with yeast species (or more likely with their naked DNA) can also come from exposure to food, where yeast can be a contaminant or a baking/fermenting agent. The fungal species most frequently found in the sequence similarity hits are *Candida*, *Saccharomyzes* and *Schizosaccharomyzes*. The gene types involved in potential transfers are normally of unknown or hypothetical functions but some similarity to metabolic genes can also be observed. Although limited, the contribution of fungal genes to oral bacteria cannot, therefore, be neglected and should be further studied. Important mechanistic issues also arise; including such as the functioning of eukaryotic genes in a prokaryotic genomic context, where regulatory regions, transcription factors and splicing mechanisms are expected to be radically different. It has to be remembered that yeast species are also common inhabitants of the gut (Bernhardt and Knoke, 1997), opening the possibility that fungal-bacterial HGT events can also take place in the intestinal tract, after which the bacteria could return to the oral cavity via the faecal–oral route.

Transient microbiome

One has to look at the mouth not only as a particular microbial niche but also as an obligatory route to the digestive tract and, therefore, a site for the constant transit of microorganisms. It is, for example, a known natural reservoir of *Helicobacter pylori* (Song *et al.*, 2000). As noted above, the oral flora can dramatically change under stress, antibiotic treatment, depression of the immune system (Brady *et al.*, 1996) or dental health status (Marshal *et al.*, 1992), increasing the probability of genetic exchange with microorganisms that are not usual members of the oral microbiota. In addition, oral residents will constantly be in contact with bacteria carried in food. These include lactic acid bacteria from yoghurt or cheese, food fermenters of all types, bacteria surviving in pickles and drinks, etc. To this, we have to add viruses that can be numerous in raw vegetables, as has recently been shown in metagenomic studies of the human gut (Zhang *et al.*, 2006). Finally, bacterial contaminants in food or on fingers, as well as the contact with other individuals' saliva, creates more opportunities for genetic contact between resident and transient oral bacteria. It has been demonstrated that several hours of contact between a saliva sample and a tetracycline-resistant *Bacillus subtilis* strain was sufficient to allow transfer of antibiotic-resistance to the resident, normally sensitive microbiota (Roberts *et al.*, 1999).

The fact that oral bacteria are continuously being swallowed and, therefore, exposed to genetic interaction with the gut flora must also be taken into account. Indeed, several instances of horizontal gene transfer between gut and oral bacteria have been documented. For instance, the transfer of tetracycline-resistance genes from Gram-positive bacteria inhabiting the mammalian colon to species of *Prevotella*, *Porphyromonas* or *Bacteroides*; or the potential transfer of tetQ from *Prevotella ruminicola* (commonly found in the digestive tract of cattle and pigs) to the oral anaerobe *P. gingivalis* (reviewed in Salyers *et al.*, 2004). As explained above for fungal-bacterial genetic exchange, once HGT has occurred in the gut environment, the oral microbe could, as already mentioned, re-enter the mouth ecosystem by the faecal–oral transmission route.

Habitat gene reservoir (HGR) in the oral environment

When the genomes of oral bacteria are screened for potential gene transfer events, it is frequently found that certain gene clusters are shared by phylogenetically distant species, while these genes are absent in related bacteria that do not inhabit the oral ecosystem (Mira *et al.*, 2005). For instance, present in *Treponema denticola* is a cluster of nitrogen fixation genes that is absent from the genome of its sequenced, non-oral relative *T. pallidum* (Table 3.2). What makes this and other gene clusters unique is their presence, with a high

Table 3.2 Examples of gene clusters (a) and individual genes (b) forming part of the habitat gene reservoir in the oral cavity

(a)

Genes function	Size[b]	Spp with identical or similar gene order
Branched-chain amino acid transport permeases	5	*F. nucleatum, T. denticola, S. mutans*
Nitrogen fixation, NADH-quinone reductase	6	*F. nucleatum, P. gingivalis, T. denticola*
Permease proteins, ABC transporters	5	*T. denticola, Bifidobacterium longum*[a], *F. nucleatum, Campylobacter jejuni*[a]
dTDP-4-dehydrorhamnose reductase and epimerase	2	*P. gingivalis, S. typhimurium*[2], *F. nucleatum, Enterococcus faecalis*[2]
Thiazole and thiamine biosynthesis	5	*P. gingivalis, F. nucleatum, C. jejuni*[a]
glutaconyl-COA decarboxylase	2	*P. gingivalis, S. mutans, F. nucleatum, E. faecalis*[a]
Export ABC transporter	2	*F. nucleatum, T. denticola, S. mutans, P. gingivalis*
High-affinity zinc uptake system	4	*T. denticola, F. nucleatum, E coli*[a]
Cobalamine synthesis	3	*T. denticola, F. nucleatum, Actinobacteria*
ABC transporter permease, integral membrane protein	5	*T. denticola, F. nucleatum, B. longum*[a], *C. jejuni*[a]

[a]Transient oral inhabitants or gut bacteria

[b] Number of genes in the cluster

(b)

Gene function	Species	Oral species within Top-5 BLAST hits
Urocanate hydratase	*F nucleatum*	*P. gingivalis, T. denticola, S. pyogenes*[a]
Glutamate formiminotransferase	*F nucleatum*	*T. denticola, B. thetaiotaomicron*[a], *P. gingivalis*
Flavoprotein	*F. nucleatum*	*B. thetaiotaomicron, P. intermedia, P. gingivalis*
Tryptophanase	*F nucleatum*	*P. gingivalis, T. denticola, P. intermedia*
Deoxyuridine nucleotidohydrolase	*F. nucleatum*	*P. intermedia, P. gingivalis, T. forsythensis, B. thetaiotaomicron*[a]
Transcription regulator tetR family	*T. denticola*	*S. mutans, P. intermedia, P. gingivalis, B. thetaiotaomicron*[a]
Hypothetical protein	*T. denticola*	*P. gingivalis, P. intermedia, F. nucleatum*
Hypothetical protein	*S. mutans*	*T. denticola, F. nucleatum, A. naeslundii*
Transcription regulator	*S. mutans*	*S. gordonii, T. denticola, T. forsythensis*
Putative hydrolase or acyltransferase	*S. mutans*	*T. denticola, S. gordonii, A. naeslundii*
ATP-dependent helicase	*Bacteroides thetaiotaomicron*	*P. gingivalis, P. intermedia, F. nucleatum*
Transcription regulator	*Bacteroides thetaiotaomicron*	*P. gingivalis, P. intermedia, F. nucleatum*

[a]Inhabitant of human gut.

degree of sequence similarity, in other sequenced oral bacteria, such as *P. gingivalis* and *F. nucleatum*, and their general absence in related, non-oral *Bacteroidetes* and *Firmicutes*. It is, therefore, most likely that these genes are transferred among the species inhabiting the oral environment, forming a 'niche-associated gene pool' that could be formed by gene functions that are advantageous in the environment and are thus under a high selective pressure to be retained after HGT events. The term Habitat Gene Reservoir (HGR) has been coined for this shared gene pool in natural environments (Legault *et al.*, 2005). Antibiotic resistance genes are obviously highly selected in western societies where the use of antibiotics is common; and some of these genes appear to be shared among oral species, as evidenced by the high incidence of ampicillin-resistance in oral bacteria (Ready *et al.*, 2003) and tetracycline-resistance genes in streptococci and *Veillonella* species (Ready *et al.*, 2006). The functions coded for by these niche-associated genes are, however, diverse. Table 3.2 shows individual genes in members of dental plaque whose BLAST best-hits are oral inhabitants. Given the low number of oral bacteria sequenced, it is remarkable that those genes do not show hits on non-oral bacteria and it is to be expected that the phenomenon is widespread. Unfinished analyses of oral bacteria, available at the BROP resource (Chen *et al.*, 2005), confirm the presence of oral-niche-associated gene pools (Mira *et al.*, unpublished data). Table 3.2 also shows examples of gene clusters and operons that are part of this Habitat Gene Reservoir. The gene functions forming part of the HGR include metabolic functions such as nitrogen fixation, peptide transport, iron binding and cell wall-related proteins. Metagenomes can be determined by shot-gun of DNA samples extracted directly from the environment, thereby obviating the need for bacterial culture. Once this has been accomplished for dental plaque by TIGR (http://genomesonline.org/GOLD_CARDS/Gm00029.html), it should be possible to confirm the pervasiveness of HGR in the oral cavity.

HGT versus HGR

The presence of environmental gene pools is not unique to the mouth. Probably the first reported case came from a metagenomic study of an acid mine drainage (Tyson *et al.*, 2004), where over 75 Mbp of DNA sequences were retrieved by a shot-gun approach. Because of the extreme conditions, a limited biodiversity was found in the sample, which was dominated by a bacterial and archaeal representation. Interestingly, a pool of shared genes was found in the genomes of these species, demonstrating trans-domain transfer of genes that were probably vital for survival in this environment. Hypersaturated brines also display a wide repertoire of niche-associated genes, with fascinating examples of single genes and whole operons shared between bacterial and archaeal halophiles (Mongodin *et al.*, 2005). The suggestion is therefore that a Habitat Gene Repertoire is fairly common in complex natural environments, offering a large degree of flexibility at a genomic and ecosystem level. This means, for example, that oral bacteria are, in a sense, the result of gene accumulation by duplication and horizontal transfer in a constant 'tinkering' process; and that the oral microenvironment is a flexible and dynamic ecosystem characterized, from a holistic point of view, not only by the addition of independent bacterial partners but also by a particular HGR. Thus, the study of HGT must go further than the acceptance of this phenomenon in the microbial evolutionary framework (Lawrence and Hendrickson, 2003) and must begin to analyse new features, such as the gene pool forming part of the HGR, to unravel the characteristics of a given habitat, and even to use this information in applied issues; for example, vaccine development.

HGR and reverse vaccinology

It has been proposed that niche-associated genes shared by oral pathogens be used as targets for vaccines against oral diseases (Mira *et al.*, 2005). The goal is to combat the complex aetiology of oral diseases, especially periodontitis. Contrary to most bacterial infections, in which a single microbe can be found responsible for the disease, oral disorders are normally the result of a combination of pathogenic organisms (Tatakis and Kumar, 2005). For example, *T. denticola*, *P. gingivalis* and *Tannerella forsythia* form the so-called 'red complex' of periodontal disease (Socransky *et al.*, 1998). Such a combination of bacteria is over-represented in diseased sites (Kumar *et al.*, 2005). Thus, vaccine targets directed to a single species would probably be ineffective. By focusing on antigens shared by different species involved in the disease, a single target may be more successful. A relevant example would be outer membrane genes shared not only by suspected pathogens but also by vital contributors to dental plaque formation, such as *Fusobacterium* species. In addition, niche-associated genes are probably vital for survival in that particular environment, making more likely their expression and representation in the population. In the past, selection of suitable antigens required bacterial culturing. This is not always possible in oral bacteria, using current culturing techniques; in addition, not all genes are expressed under *in vitro* conditions, thus reducing the number of potential vaccine targets. The inverse process or 'reverse vaccinology' has its origins in the genomic data from which to select appropriate targets (see, for example, Rappuoli and Covacci, 2003). These typically include excreted proteins, genes expressed on the cell wall and those matching virulence factors (Serruto and Rappuoli, 2006). Use of bioinformatics, taking HGT into account, should be made in selecting effective vaccine targets. This would also apply not only to oral diseases but also to other illnesses for which more than one causative agent can be found.

Conclusions

What is the nature of gene innovation in the mouth ecosystem? Although gene duplication is considered one of the main sources of novel gene functions in bacteria (Jordan *et al.*, 2001; Pushker *et al.*, 2004), the appearance of new genes in the oral ecosystem is greatly influenced by HGT. Transfer is undoubtedly favoured by the close physical interactions in dental biofilms, facilitating cell to cell contacts (conjugation). It also promotes transformation, a mechanism found to occur more frequently in biofilms of mutans streptococci than in free-living cells of this species (Li *et al.*, 2001). The importance of HGT in the oral niche is due not only to the widespread occurrence of the process but also by the relevance of transferred traits and the speed with which the new features can be incorporated into the new host. In addition, it will be of interest to determine whether the numerous 'orphan' genes identified in oral bacteria (short genes without detectable sequence similarity to available genomic data) are non-functional pseudogene remnants (Mira *et al.*, 2002) or functional genes of unknown function and likely viral origin, as has been demonstrated for some free-living species (Daubin and Ochman, 2004a). In this sense, the contribution of bacteriophage to gene innovation in oral bacteria and their infection capabilities within the biofilm remains to be elucidated.

The impact of HGT on the biology of oral bacteria and of the oral ecosystem as a whole is enormous. Oral microorganisms can acquire, from other species, genes that could confer survival advantage in the mouth. Apart from the well-studied antibiotic resistance genes, oral bacteria have been shown to transfer bacteriocinogeny (Horn *et al.*, 1991), mercury resistance (Roberts *et al.*, 2003) and outer membrane genes (Mira *et al.*, 2004). It is interesting to note that a strikingly large number of Gram-positive bacteria (classified as such by molecular phylogenies) but strangely exhibiting a typical Gram-negative cell wall, inhabit the mouth. These include many members of the *Sporomusa* branch (*Acidaminococcaceae* family). Although this group is phylogenetically placed at the base of the Clostridiales, they present a typical Gram-negative morphology in electron micrographs; and confirmed by the unpublished genome sequence of *Dialister pneumosintes*. To this we have to add the Gram-negative outer membrane of *Fusobacteria*, typical inhabitants of oral niches, which are phylogenetically placed at the base of *Firmicutes* (Mira *et al.*, 2004). The maintenance of this outer membrane structure, together with the frequent transfer of proteobacterial cell wall genes, suggests that the combination of a Gram-positive metabolic repertoire with a Gram-negative adherence and immune-evasive membrane is fairly successful for bacteria in the oral environment.

From a theoretical viewpoint, the comparative genomics of oral bacteria corroborate the existence of niche-associated genes and the presence of a habitat-associated gene reservoir that would provide an adaptive gene pool for many species inhabiting the mouth. From an applied perspective, the search for antigenic niche-associated genes could be a promising strategy for vaccine development. In addition, gene transfer appears as one of the main mechanisms involved in the transfer of antibiotic resistance in oral bacteria and, therefore, also affects control strategies that could be adopted in general dental practice (Sweeney *et al.*, 2004), given the not-inconsiderable consumption of antibiotics in oral infections and disease prevention.

It has to be remembered that the mouth is the entry-point for microbes finding their way to the digestive tract. Given the high levels of HGT and the relevance of the traits made available to the oral bacterial community, it is not unlikely that new gene combinations are 'tested' in the mouth ecosystem before the modified organisms colonize other host-associated niches. The oral cavity could act as a 'genomic playground' in which a tinkering evolutionary process adds, removes and shuffles genes and their corresponding functions from resident and transient organisms to create new genomic variants. These variants could be pre-adapted to thrive not only in the oral niche itself but also in other environments. In fact, it has been suggested that HGT is involved in the formation of new bacterial species (de la Cruz and Davies, 2000). This putative genomic playground could actually work both ways, since the faecal–oral route could reintroduce gut bacteria into the oral ecosystem and the digestive tract could itself be a testing arena for bacteria entering the mouth (Roberts *et al.*, 2003). Potential transfers from gut bacteria to the genomes of oral species, such as *Fusobacterium nucleatum* (Mira *et al.*, 2004), *Treponema denticola* (Seshadri *et al.*, 2004) and other unfinished genomes (see, for example the BLAST resource at www.brop.org), are commonly found. In this sense, it is interesting that the gut environment appears to act as an antibiotic-resistance reservoir (Salyers *et al.*, 2004). There is increasing evidence that gut bacteria can act as traffickers of antibiotic-resistance genes (Shoemaker *et al.*, 1992; Stinear *et al.*, 2001; Salyers *et al.*, 2004) and, by extension, they could also transfer genes related to virulence, adherence, novel metabolic functions and so on. Thus, microorganisms associated with the digestive system of animals and humans, from the mouth to the colon, may form a complex intertwined ecosystem where genetic exchange offers endless possibilities for bacterial adaptation.

The genomics era will presumably revolutionize the study of the oral ecosystem and the diagnostic and treatment of oral diseases. The emerging picture of the oral niche ecology (Marsh and Percival, 2006) will be enhanced by new genomic and metagenomic data from mouth bacterial species, in the same way as these data are already clarifying many aspects of gut flora biology (Gill *et al.*, 2006). The oral niche is extremely diverse in resident species (Paster *et al.*, 2001) and the limited number of sequenced oral bacteria has only begun to unravel unexpected features of the ecosystem. Not only is there a low presence of oral microbes among sequenced bacteria but there are even some well-represented bacterial groups in the mouth, such as the TM7 taxon (previously believed to be exclusive inhabitants of sediment) in which not a single member has been sequenced. Future years will provide a wealth of genomic data on oral genomes (Duncan, 2003) and promise to provide exciting information about the genetic exchange mechanisms of oral microbes as well as the basis for their oral adaptation and interactions.

References

Adamczyk M., and Jagura-Burdzy G. (2003) Spread and survival of promiscuous IncP-1 plasmids. Acta Biochim. Pol. *50*, 425–453.

Alsmark C.M., Frank, A.C., Karlberg, E.O, Legault, B.A., Ardell, D.H., Canback, B., Eriksson, A.S., Naslund, A.K., Handley, S.A., Huvet, M., La Scola, B., Holmberg, M., and Andersson, S.G.E. (2003) The louse-borne human pathogen Bartonella quintana is a genomic derivative of the zoonotic agent *Bartonella henselae*. Proc. Natl. Acad. Sci. USA *101*, 9716–9721.

Altschul, S.F., Gish, W., Miller, W., Myers, E.W., and Lipman, D.J. (1990) Basic local alignment search tool. J. Mol. Biol. *215*, 403–410.

Berg, O.G., and Kurland, C.G. (2002) Evolution of microbial genomes: sequence acquisition and loss. Mol. Biol. Evol. *19*, 2265–2276.

Bachrach, G., Leizerovici-Zigmond M., Zlotkin A., Naor R., and Steinberg D. (2003) Bacteriophage isolation from human saliva. Lett. Appl. Microbiol. *36*, 50–53.

Bernhardt, H., and Knoke, M. (1997) Mycological aspects of gastrointestinal microflora. Scand. J. Gastroenterol. Suppl. *222*, 102–106.

Bolotin, A., Quinquis, B., Sorokin, A., and Ehrlich, S.D. (2005) Clustered regularly interspaced short palindrome repeats (CRISPRs) have spacers of extrachromosomal origin. Microbiology *151*, 2551–2561.

Boyd, E.F., and Brussow, H. (2002) Common themes among bacteriophage-encoded virulence factors and diversity among the bacteriophages involved. Trends Microbiol. *10*, 521–529.

Brady, L.J., Walker, C., Oxford, G.E., Stewart, C., Magnusson, I., and McArthur, W. (1996) Oral diseases, mycology and periodontal microbiology of HIV-1-infected women. Oral Microbiol. Immunol. *11*, 371–380.

Branger, C., Zamfir, O., Geoffroy, S., Laurans, G., Arlet, G., Thien, H.V., Gouriou, S., Picard, B., and Denamur, E. (2005) Genetic background of Escherichia coli and extended-spectrum beta-lactamase type. Emerg. Infect. Dis. *11*, 54–61.

Brocchieri, L. (2001) Phylogenetic inferences from molecular sequences: review and critique. Theor. Popul. Biol. *59*, 27–40.

Brown, J.R, and Doolittle, W.F. (1997) Archaea and the prokaryote-to-eukaryote transition. Microbiol Mol. Biol. Rev. *61*, 456–502.

Chen, T., Abbey, K., Deng, W.J., and Cheng, M.C. (2005) The bioinformatics resource for oral pathogens. Nucleic Acids Res. *33* (Web Server issue): W734–740.

Clarke, G.D., Beiko, R.G., Ragan, M.A., and Charlebois, R.L. (2002) Inferring genome trees by using a filter to eliminate phylogenetically discordant sequences and a distance matrix based on mean normalized BLASTP scores. J. Bacteriol. *184*, 2072–2080.

Cvitkovitch D.G. (2001) Genetic competence and transformation in oral streptococci. Crit. Rev. Oral Biol. Med. *12*, 217–243.

Daniel, R. (2005) The metagenomics of soil. Nat. Rev. Microbiol. 3, 470–478.

Daubin, V., and Ochman, H. (2004) Start-up entities in the origin of new genes. Curr Opin Genet Dev. *14*, 616–619.

Daubin, V., and Ochman, H. (2004b). Bacterial genomes as new gene homes: the genealogy of ORFans in *E. coli*. Genome Res. *14*, 1036–1042.

de la Cruz, F. and Davies, J. (2000) Horizontal gene transfer and the origin of the species: lessons from bacteria. Trends Microbiol. *8*, 128–133.

Dobrindt, U., Hochnut, B., Hentschel, U., and Hacker, J. (2004) Genomic islands in pathogenic and environmental microorganisms. Nature Rev. Microb. *2*, 414–424.

Doolitle, W.F., Boucher, Y., Nesbo, C.L., Douady, C.J., Andersson, J.O., and Roger, A.J. (2003) How big is the iceberg of which organellar genes in nuclear genomes are but the tip? Philos. Trans. R. Soc. London Ser. B *358*, 39–57.

Dufraigne, C., Fertil, B., Lespinats, S., Giron, A., and Deschavanne, P. (2005) Detection and characterization of horizontal transfers in prokaryotes using genomic signature. Nucleic Acids Res. *33*, e6.

Duggan, P.S., Chambers, P.A., Heritage, J., and Forbes, J.M. (2000) Survival of free DNA encoding antibiotic resistance from transgenic maize and the transformation activity of DNA in ovine saliva, ovine rumen fluid and silage effluent. FEMS Microbiol Lett. *191*, 71–77.

Duncan, M.J. (2003) Genomics of oral bacteria. Crit. Rev. Oral Biol. Med. *14*, 175–187.

Garcia-Vallve, S., Romeu.A., and Palau, J. (2000) Horizontal gene transfer in bacterial and archaeal complete genomes. Genome Res. *10*, 1719–1725.

Garcia-Vallve, S., Guzman, E., Montero, M.A., and Romeu, A. (2003) HGT-DB: a database of putative horizontally transferred genes in prokaryotic complete genomes. Nucleic Acids Res. *31*, 187–189.

Gill, S.R., Pop, M., Deboy, R.T., Eckburg, P.B., Turnbaugh, P.J., Samuel, B.S., Gordon, J.I., Relman, D.A., Fraser-Liggett, C.M., and Nelson, K.E. (2006) Metagenomic analysis of the human distal gut microbiome. Science *312*, 1355–1359.

Gogarten, J.P., Doolittle, W.F., and Lawrence, J.G. (2002) Prokaryotic evolution in light of gene transfer. Mol. Biol. Evol. *19*, 2226–2238.

Goncharoff, P., Yip, J.K., Wang, H., Schreiner, H.C., Pai, J.A., Furgang, D., Stevens, R.H., Figurski, D.H., and Fine, D.H. (1993) Conjugal transfer of broad-host-range incompatibility group P and Q plasmids from *Escherichia coli* to *Actinobacillus actinomycetemcomitans*. Infect. Immun. *61*, 3544–3547.

Hacker, J., and Carniel, E. (2001) Ecological fitness, genomic islands and bacterial pathogenicity. A Darwinian view of the evolution of microbes. EMBO Rep. *2*, 376–381.

Hammond, B.F., Steel, C.F., and Peindl, K.S. (1976) Antigens and surface components associated with virulence of *Actinomyces viscosus*. J. Dent. Res. 55, A19–25.

Hartley, D.L., Jones, K.R., Tobian, J.A., LeBlanc, D.J., and Macrina, F.L. (1984) Disseminated tetracycline resistance in oral streptococci: implication of a conjugative transposon. Infect. Immun. *45*, 13–17.

Hitch, G., Pratten, J., and Taylor, P.W. (2004) Isolation of bacteriophages from the oral cavity. Lett. Appl. Microbiol. *39*, 215–219.

Horn, N., Swindell, S., Dodd, H., and Gasson, M. (1991) Nisin biosynthesis genes are encoded by a novel conjugative transposon. Mol. Gen. Genet. 228, 129–135.

Jordan, I.K., Makarova, K.S., Spouge, J.L., Wolf, Y.I., and Koonin, E.V. (2001) Lineage-specific gene expansions in bacterial and archaeal genomes. Genome Res. *11*, 555–565.

Karaolis, D.K., Somara, S., Maneval, D. R. Jr,, Johnson, J.A., and Kaper, J.B. (1999) A bacteriophage encoding a pathogenicity island, a type-IV pilus and a phage receptor in cholera bacteria. Nature *399*, 375–379.

Kolenbrander, P.E., Andersen, R.N., Blehert, D.S., Egland, P.G., Foster, J.S., and Palmer, R. J. Jr. (2002) Communication among oral bacteria. Microbiol. Mol. Biol. Rev. *66*, 486–505

Kolenbrander, P.E., Palmer, R. J. Jr., Rickard, A.H., Jakubovics, N.S., Chalmers, N.I., and Diaz, P.I. (2006) Bacterial interactions and successions during plaque development. Periodontol. 2000 *42*, 47–79.

Koski, L.B., Morton, R.A., and Golding, G.B. (2001) Codon bias and base composition are poor indicators of horizontally transferred genes. Mol. Biol. Evol. *18*, 404–412.

Koski, L.B., and Golding, G.B. (2001) The closest BLAST hit is often not the nearest neighbor. J. Mol. Evol. 52, 540–542.

Kumar, P.S., Griffen, A.L., Moeschberger, M.L., and Leys, E.J. (2005) Identification of candidate periodontal pathogens and beneficial species by quantitative 16S clonal analysis. J. Clin. Microbiol. *43*, 3944–3955.

Kurenbach, B., Bohn, C., Prabhu, J., Abudukerim, M., Szewzyk, U., and Grohmann, E. (2003) Intergeneric transfer of the *Enterococcus faecalis* plasmid pIP501 to *Escherichia coli* and *Streptomyces lividans* and sequence analysis of its tra region. Plasmid *50*, 86–93.

Lawrence, J.G., and Hendrickson, H. (2003) Lateral gene transfer: when will adolescence end? Mol. Microbiol. *50*, 739–749.

Lawrence, J.G., and Ochman, H. (1997) Amelioration of bacterial genomes: rates of change and exchange. J. Mol. Evol. *44*: 383–397.

Lawrence, J.G., and Ochman, H. (1998) Molecular archaeology of the *Escherichia coli* genome. Proc. Natl. Acad. Sci. USA 95: 9413–9417.

Legault, B.A., Lopez-Lopez, A., Alba-Casado, J.C., Doolittle, W.F., Bolhuis, H., Rodriguez-Valera, F., and Papke, R. T. Environmental genomics of 'Haloquadratum walsbyi' in a saltern crystallizer indicates a large pool of accessory genes in an otherwise coherent species. BMC Genomics 7171.

Lerat, E., Daubin, V., and Moran, N.A. (2003) From gene trees to organismal phylogeny in prokaryotes: the case of the gamma-Proteobacteria. PloS. Biol. *1*, E19

Li, Y.H., Lau, P.C., Lee, J.H., Ellen, R.P., and Cvitkovitch, D.G. (2001) Natural genetic transformation of *Streptococcus mutans* growing in biofilms. J. Bacteriol. *183*, 897–908.

Lindroos, H.L., Vinnere, O., Mira, A., Repsilber, D., Boulouis, A., Näslund, K., and Andersson, S. G.E. (2006) Genome Rearrangements, Deletions and Amplifications in the Natural Population of *Bartonella henselae*. J. Bacteriol. *188*, 7426–7439.

Lyon, J.P., da Costa, S.C., Totti, V.M., Muñoz, M.F., and de Resende, M.A. (2006) Predisposing conditions for *Candida* spp. carriage in the oral cavity of denture wearers and individuals with natural teeth. Can. J. Microbiol. 52, 462–467.

Manfredi, M., McCullough, M.J., Al-Karaawi, Z.M., Vescovi, P., and Porter, S.R. (2006) Analysis of the strain relatedness of oral *Candida albicans* in patients with diabetes mellitus using polymerase chain reaction-fingerprinting. Oral Microbiol. Immunol. *21*, 353–359.

Marri, P.R., Hao, W., and Golding, G.B. (2006) Gene gain and gene loss in streptococci: Is it driven by habitat? Mol. Biol. Evol. 23, 2379–2391.

Marsh, P.D. (2003) Are dental diseases examples of ecological catastrophes? Microbiology *149*, 279–294.

Marsh, P.D. (2006) Dental plaque as a biofilm and a microbial community – implications for health and disease. BMC Oral Health 6 Suppl 1:S14.

Marsh, P.D., and Percival, R.S. The oral microflora–friend or foe? Can we decide? Int. Dent. J. 2006 Aug;56(4 Suppl 1):233–239.

Marshall, J.R., Graham, S., and Haughey, B.P. (1992) Smoking, alcohol, dentition and diet in the epidemiology of oral cancer. Eur. J. Cancer B. Oral Oncol. *28B*: 9–15.

McKay, T.L., Ko, J., Bilalis, Y., and DiRienzo, J.M. (1995) Mobile genetic elements of *Fusobacterium nucleatum*. Plasmid 33, 15–25.

Mercer, D.K., Scott, K.P., Bruce-Johnson, W.A., Glover, L.A., and Flint, H.J. (1999) Fate of free DNA and transformation of the oral bacterium *Streptococcus gordonii* DL1 by plasmid DNA in human saliva. Appl. Environ. Microbiol. 65, 6–10.

Mercer, D.K., Scott, K.P., Melville, C.M., Glover, L.A., and Flint, H.J. (2001) Transformation of an oral bacterium via chromosomal integration of free DNA in the presence of human saliva. FEMS Microbiol. Lett. *200*, 163–167.

Mira, A., Klasson, L., and Andersson, S.G. (2002) Microbial genome evolution: sources of variability. Curr. Opin. Microbiol. 5, 506–512.

Mira, A., Pushker, R., Legault, B.A., Moreira, D., and Rodriguez-Valera, F. (2004) Evolutionary relationships of *Fusobacterium nucleatum* based on phylogenetic analysis and comparative genomics. BMC Evol. Biol. *4*, 50.

Mojica, F.J., Diez-Villasenor, C., Garcia-Martinez, J., and Soria, E. (2005) Intervening sequences of regularly spaced prokaryotic repeats derive from foreign genetic elements. J. Mol. Evol. *60*, 174–182.

Mongodin, E.F., Nelson, K.E., Daugherty, S., Deboy, R.T., Wister, J., Khouri, H., Weidman, J., Walsh, D.A., Papke, R.T., Sánchez-Perez G., Sharma, A.K., Nesbo, C.L., MacLeod, D., Bapteste, E., Doolittle, W.F., Charlebois, R.L., Legault, B., and Rodriguez-Valera, F. (2005) The genome of *Salinibacter ruber*: convergence and gene exchange among hyperhalophilic bacteria and archaea. Proc. Natl. Acad. Sci. USA *102*, 18147–18152.

Mullany, P., Roberts, A.P., and Wang, H. (2002) Mechanism of integration and excision in conjugative transposons. Cell. Mol. Life Sci. 59, 2017–2022.

Nakamura, Y., Gojobori, T., and Ikemura, T. (1997) Codon usage tabulated from the international DNA sequence databases. Nucleic Acids Res. 25, 244–245.

Nelson, K.E., Clayton, R.A., Gill, S.R., Gwinn, M.L., Dodson, R.J., Haft, D.H., Hickey, E.K., Peterson, J.D., Nelson, W.C., Ketchum, K.A., McDonald, L., Utterback, T.R., Malek, J.A., Linher, K.D., Garrett, M.M., Stewart, A.M., Cotton, M.D., Pratt, M.S., Phillips, C.A., Richardson, D., Heidelberg, J., Sutton, G.G., Fleischmann, R.D., Eisen, J.A., White, O., Salzberg, S.L., Smith, H.O., Venter, J.C., and Fraser, C.M. (1999) Evidence for lateral gene transfer between Archaea and bacteria from genome sequence of *Thermotoga maritima*. *Nature* 399, 323–329.

Nelson, K.E., Fleischmann, R.D., DeBoy, R.T., Paulsen, I.T., Fouts, D.E., Eisen, J.A., Daugherty, S.C., Dodson, R.J., Durkin, A.S., Gwinn, M., Haft, D.H., Kolonay, J.F., Nelson, W.C., Mason, T., Tallon, L., Gray, J., Granger, D., Tettelin, H., Dong, H., Galvin, J.L., Duncan, M.J., Dewhirst, F.E., and Fraser, C.M. (2003) Complete genome sequence of the oral pathogenic Bacterium *Porphyromonas gingivalis* strain W83. J. Bacteriol. *185*, 5591–5601.

Nesbø, C.L., L'Haridon, S., Stetter, K.O., and Doolittle, W.F. (2001) Phylogenetic analyses of two 'archaeal' genes in *Thermotoga maritima* reveal multiple transfers between archaea and bacteria. Mol. Biol. Evol. *18*, 362–375.

Norgren, M., and Scott, J.R. (1991) The presence of conjugative transposon Tn916 in the recipient strain does not impede transfer of a second copy of the element. J. Bacteriol. 173, 319–324.

Ochman, H., Lawrence, J.G., and Groisman, E.A. (2000) Lateral gene transfer and the nature of bacterial innovation. Nature *405*, 299–304.

Ogunseitan, O.A. (1995) Bacterial genetic exchange in nature. Sci. Prog. *78*, 183–204.

Ohno, S. (1970) Evolution by Gene Duplication. (New York: Springer).

Osborn, A.M., and Boltner, D. (2002) When phage, plasmids, and transposons collide: genomic islands, and conjugative- and mobilizable-transposons as a mosaic continuum. Plasmid *48*, 202–212.

Parkhill, J., Wren, B.W., Thomson, N.R., Titball, R.W., Holden, M.T., Prentice, M.B., Sebaihia, M., James, K.D., Churcher, C., Mungall, K.L., Baker, S., Basham, D., Bentley, S.D., Brooks, K., Cerdeno-Tarraga, A.M., Chillingworth, T., Cronin, A., Davies, R.M., Davis, P., Dougan, G., Feltwell, T., Hamlin, N., Holroyd, S., Jagels, K., Karlyshev, A.V., Leather, S., Moule, S., Oyston, P.C., Quail, M., Rutherford, K., Simmonds, M., Skelton, J., Stevens, K., Whitehead, S., and Barrell, B.G. (2001) Genome sequence of *Yersinia pestis*, the causative agent of plague. Nature *413*, 523–527.

Paster, B.J., Boches, S.K., Galvin, J.L., Ericson, R.E., Lau, C.N., Levanos, V.A., Sahasrabudhe, A., AND Dewhirst, F.E. (2001) Bacterial diversity in human subgingival plaque. J. Bacteriol. *183*, 3770–3783.

Peters, A.C., Thomas, L., and Wimpenny, J.W. (1991) Effects of salt concentration on bacterial growth on plates with gradients of pH and temperature. FEMS Microbiol. Lett. *61*, 309–314.

Pushker, R., Mira, A., and Rodriguez-Valera, F. (2004) Comparative genomics of gene-family size in closely related bacteria. Genome Biol. 5, R27.

Qi, Q.G., Hu, T., and Zhou, X.D. (2005) Frequency, species and molecular characterization of oral *Candida* in hosts of different age in China. J. Oral Pathol. Med. *34*, 352–356.

Ragan, M.A., and Charlebois, R.L. (2002) Distributional profiles of homologous open reading frames among bacterial phyla: implications for vertical and lateral transmission. Int. J. Syst. Evol. Microbiol. *52*, 777–787.

Ragan, M.A. (2001) Detection of lateral gene transfer among microbial genomes. Curr. Opin. Genet. Dev. *11*, 620–626.

Rappuoli, R., and Covacci, A. (2003) Reverse vaccinology and genomics. Science *302*, 602.

Rauch, P.J., and De Vos, W.M. (1992) Characterization of the novel nisin-sucrose conjugative transposon Tn5276 and its insertion in *Lactococcus lactis*. J. Bacteriol. 174, 1280–1287.

Ready, D., Bedi, R., Spratt, D.A., Mullany, P., and Wilson, M. (2003) Prevalence, proportions, and identities of antibiotic-resistant bacteria in the oral microflora of healthy children. Microb. Drug Resist. *9*, 367–372.

Ready, D., Pratten, J., Roberts, A.P., Bedi, R., Mullany, P., and Wilson, M. (2006) Potential role of *Veillonella* spp. as a reservoir of transferable tetracycline resistance in the oral cavity. Antimicrob. Agents Chemother. *50*, 2866–2868.

Rice, L.B. (1998) Tn916 family conjugative transposons and dissemination of antimicrobial resistance determinants. Antimicrob. Agents Chemother. *42*, 1871–1877.

Roberts, A.P., and Mullany, P. (2006) Genetic basis of horizontal gene transfer among oral bacteria. Periodontol. 2000 *42*, 36–46.

Roberts, A.P., Pratten, J., Wilson, M., and Mullany, P. (1999) Transfer of a conjugative transposon, Tn5397 in a model oral biofilm. FEMS Microbiol. Lett. *177*, 63–66.

Roberts, A.P., Cheah, G., Ready, D., Pratten, J., Wilson, M., and Mullany, P. (2001) Transfer of TN916-like elements in microcosm dental plaques. Antimicrob. Agents Chemother. *45*, 2943–2946.

Roberts, A.P., Davis, I.J., and Mullany, P. (2003) Antibiotic resistance gene transfer in the oral microflora. Res. Adv. in Antimicrob. Agents Chemother. *3*, 1–6.

Rodriguez-Valera, F. (2004) Environmental genomics, the big picture? FEMS Microbiol. Lett. *231*, 153–158.

Salyers, A.A., Shoemaker, N.B., Stevens, A.M., and Li, L.Y. (1995) Conjugative transposons: an unusual and diverse set of integrated gene transfer elements. Microbiol. Rev. *59*, 579–590.

Salyers, A.A., Gupta, A., and Wang, Y. (2004) Human intestinal bacteria as reservoirs for antibiotic resistance genes. Trends Microb. *12*, 412–416.

Sandmeier, H., van Winkelhoff, A.J., Bar, K., Ankli, E., Maeder, M., and Meyer, J. (1995) Temperate bacteriophages are common among *Actinobacillus actinomycetemcomitans* isolates from periodontal pockets. J. Periodontal Res. *30*, 418–425.

Serruto, D., and Rappuoli, R. (2006) Post-genomic vaccine development. FEBS Lett. *580*, 2985–2992.

Seshadri, R., Myers, G.S., Tettelin, H., Eisen, J.A., Heidelberg, J.F., Dodson, R.J., Davidsen, T.M., DeBoy, R.T., Fouts, D.E., Haft, D.H., Selengut, J., Ren, Q., Brinkac, L.M., Madupu, R., Kolonay, J., Durkin, S.A., Daugherty, S.C., Shetty, J., Shvartsbeyn, A., Gebregeorgis, E., Geer, K., Tsegaye, G., Malek, J., Ayodeji, B., Shatsman, S., McLeod, M.P., Smajs, D., Howell, J.K., Pal, S., Amin, A., Vashisth, P., McNeill, T.Z., Xiang, Q., Sodergren, E., Baca, E., Weinstock, G.M., Norris, S.J., Fraser, C.M., and Paulsen, I.T. (2004) Comparison of the genome of the oral pathogen *Treponema denticola* with other spirochete genomes. Proc. Natl. Acad. Sci. USA 101, 5646–5561.

Shoemaker, N.B., Wang, G.R., and Salyers, A.A. (1992) Evidence for natural transfer of a tetracycline resistance gene between bacteria from the human colon and bacteria from the bovine rumen. Appl. Environ. Microbiol. *58*, 1313–1320.

Sillankorva, S., Oliveira, R., Vieira, M.J., Sutherland, I.W., and Azeredo, J. (2004) Bacteriophage Phi S1 infection of *Pseudomonas fluorescens* planktonic cells versus biofilms. Biofouling. *20*, 133–138.

Snel, B., Bork, P., and Huynen, M.A. (1999) Genome phylogeny based on gene content. Nat. Genet. 21, 108–110.

Snel, B., Bork, P., and Huynen, M.A. (2002) Genomes in flux: the evolution of archaeal and proteobacterial gene content. Genome Res. *12*, 17–25.

Socransky, S.S., Haffajee, A.D., Cugini, M.A., Smith, C., and Kent, R. L. Jr. (1998) Microbial complexes in subgingival plaque. J. Clin. Periodontol. *25*, 134–44.

Song, Q., Lange, T., Spahr, A., Adler, G., Bode, G. (2000) Characteristic distribution pattern of *Helicobacter pylori* in dental plaque and saliva detected with nested PCR. J. Med. Microbiol. 49, 349–353.

Spratt, B.G., and Maiden, M.C. (1999) Bacterial population genetics, evolution and epidemiology. Philos. Trans. R. Soc. Lond. B Biol. Sci. *354*, 701–710.

Stinear, T.P., Olden, D.C., Johnson, P.D., Davies, J.K., and Grayson, M.L. (2001) Enterococcal vanB resistance locus in anaerobic bacteria in human faeces. Lancet *357*, 855–856.

Sutherland, I.W., Hughes, K.A., Skillman, L.C., and Tait, K. (2004) The interaction of phage and biofilms. FEMS Microbiol. Lett. *232*, 1–6.

Sweeney, L.C., Dave, J., Chambers, P.A., and Heritage, J. (2004) Antibiotic resistance in general dental practice--a cause for concern? J. Antimicrob. Chemother. *53*, 567–576.

Tatakis, D.N., and Kumar, P.S. (2005) Etiology and pathogenesis of periodontal diseases. Dent. Clin. North Am. *49*, 491–516.

Tettelin, H., Masignani, V., Cieslewicz, M.J., Donati, C., Medini, D., Ward, N.L., Angiuoli, S.V., Crabtree, J., Jones, A.L., Durkin, A.S., Deboy, R.T., Davidsen, T.M., Mora, M., Scarselli, M., Margarit, y Ros, I., Peterson, J.D., Hauser, C.R., Sundaram, J.P., Nelson, W.C., Madupu.R., Brinkac, L.M., Dodson, R.J., Rosovitz, M.J., Sullivan, S.A., Daugherty, S.C., Haft, D.H., Selengut, J., Gwinn, M.L., Zhou, L., Zafar, N., Khouri, H., Radune, D., Dimitrov, G., Watkins, K., O'Connor, K.J., Smith, S., Utterback, T.R., White, O., Rubens, C.E., Grandi, G., Madoff, L.C., Kasper, D.L., Telford, J.L., Wessels, M.R., Rappuoli, R., and Fraser, C.M. (2005) Genome analysis of multiple pathogenic isolates of *Streptococcus agalactiae*: implications for the microbial 'pan-genome'. Proc. Natl. Acad. Sci. USA *102*, 13950–13955.

Tyson, G.W., Chapman, J., Hugenholtz, P., Allen, E.E., Ram, R.J., Richardson, P.M., Solovyev, V.V., Rubin, E.M., Rokhsar, D.S., and Banfield, J.F. (2004) Community structure and metabolism through reconstruction of microbial genomes from the environment. Nature *428*, 37–43.

Vozza, I., Zino, G., Puddu, P., and Quaranta, M. (2005) Study on the frequency of protozoa and mycetes in the oral cavity. Minerva Stomatol. *54*, 575–581.

Wang, H.C., Badger, J., Kearney, P., and Li, M. (2001) Analysis of codon usage patterns of bacterial genomes using the self-organizing map. Mol. Biol. Evol. *18*, 792–800.

Wang, B.Y., Chi, B., and Kuramitsu, H.K. (2002) Genetic exchange between *Treponema denticola* and *Streptococcus gordonii* in biofilms. Oral Microbiol. Immunol. *17*, 108–112.

Waterhouse, J.C., and Russell, R.R. (2006) Dispensable genes and foreign DNA in *Streptococcus mutans*. Microbiology *152*, 1777–1788.

Waterhouse, J.C., Swan, D.C., and Russell, R.R.B. (2007) Comparative genome hybridization of *Streptococcus mutans* strains. Oral Microbiol. Immunol. *22*, 103–110.

Weiss, E.I., Shaniztki, B., Dotan, M., Ganeshkumar, N., Kolenbrander, P.E., and Metzger, Z. (2000) Attachment of *Fusobacterium nucleatum* PK1594 to mammalian cells and its coaggregation with periodontopathogenic bacteria are mediated by the same galactose-binding adhesin. Oral Microbiol. Immunol. *15*, 371–377.

Welch, R.A., Burland, V., Plunkett, G. 3rd, Redford, P., Roesch, P., Rasko, D., Buckles, E.L., Liou, S.R., Boutin, A., Hackett, J., Stroud, D., Mayhew, G.F., Rose, D.J., Zhou, S., Schwartz, D.C., Perna, N.T., Mobley, H.L., Donnenberg, M.S., and Blattner, F.R. (2002) Extensive mosaic structure revealed by the complete genome sequence of uropathogenic *Escherichia coli*. Proc. Natl. Acad. Sci. USA 99, 17020–17024.

Willi, K., Sandmeier, H., Asikainen, S., Saarela, M., and Meyer, J. (1997) Occurrence of temperate bacteriophages in different *Actinobacillus actinomycetemcomitans* serotypes isolated from periodontally healthy individuals. Oral Microbiol. Immunol. *12*, 40–46.

Willi, K., Sandmeier, H., Kulik, E.M., and Meyer, J. (1997b). Transduction of antibiotic resistance markers among *Actinobacillus actinomycetemcomitans* strains by temperate bacteriophages *Aa* phi 23. Cell. Mol. Life Sci. 53, 904–910.

Woese, C.R. (2000) Interpreting the universal phylogenetic tree. Proc. Natl. Acad. Sci. USA *97*, 8392–8396.

Zhang, Q., Melcher, U., Zhou, L., Najar, F.Z., Roe, B.A., and Fletcher, J. (2005) Genomic comparison of plant pathogenic and nonpathogenic *Serratia marcescens* strains by suppressive subtractive hybridization. Appl Environ Microbiol. *71*, 7716–7723.

Zhang, T., Breitbart, M., Lee, W.H., Run, J.Q., Wei, C.L., Soh, S.W., Hibberd, M.L., Liu, E.T., Rohwer, F., and Ruan, Y. (2006) RNA viral community in human feces: prevalence of plant pathogenic viruses. PLoS. Biol. *4*, e3.

Cell–Cell Communication in Oral Microbial Communities

4

Alex H. Rickard, Gilad Bachrach and David G. Davies

Abstract

Most of the bacterial species found in the mouth belong to surface-attached microbial communities called biofilms. A common feature of biofilms is interbacterial communication which can be mediated by two distinct phenomena. The first is through direct cell–cell contact, which is mediated by specific protein 'adhesins' and often, as in the case of interspecies aggregation, by complementary polysaccharide receptors. Intra (auto-) and interspecies (co-) aggregation both promote ordered successional integration of species into the biofilm. The second method of communication involves cell–cell signalling molecules, which are of two classes; those used for intraspecies and those used for interspecies signalling. An example of the former is 'quorum sensing', a process in which acyl-homoserine lactones (AHLs) induce members of the same population to produce and release specific enzymes or to initiate biofilm formation. As yet, no oral bacteria have been shown to produce AHLs. However, they can produce small peptides, such as 'competence stimulating peptides', which have been shown to mediate intraspecies signalling and to promote single-species biofilm formation. A common form of interspecies signalling is mediated by 4,5-dihydroxy-2,3-pentanedione (DPD). This spontaneously forms a family of inter-convertible compounds, collectively called autoinducer-2 (Al-2). With respect to oral biofilm communities, the present review will focus on the molecular basis of communication and the effects of cell–cell contact and signal molecules on gene expression. A model relating interspecies cell–cell communication and biofilm development is proposed.

Introduction

Bacteria are not solitary units of life. The vast majority exist as part of surface-associated communities called 'biofilms' (Costerton *et al.*, 1994). They are typically composed of many species of bacteria and are found in nearly every environment, including the oral cavity (Kolenbrander *et al.*, 2006; Rickard *et al.*, 2003a; Stoodley *et al.*, 2002). Within the oral cavity, surfaces that are colonized by bacteria include the tongue (Danser *et al.*, 2003), buccal epithelia (Van der Velden *et al.*, 1986), gingivae (Muller *et al.*, 1997) and teeth (Socransky and Haffajee, 2005). Biofilms associated with teeth can form below the gingival margin and are called sub-gingival plaque. Those on exposed tooth surfaces are described as supra-gingival dental plaque. The properties and microbial composition of dental plaque, as a whole, has been extensively studied for nearly 60 years and is the subject of over 10 000 research and review articles. Through direct culturing and, recently, culture-independent methods, estimates of the bacterial species diversity in the oral cavity indicate the presence of at least 500 species in supragingival plaque and approximately 20% less in subgingival plaque (Diaz *et al.*, 2006; Nyvad and Kilian, 1987; Paster *et al.*, 2001). Many of the species found in subgingival plaque are found in supragingival plaque. Cultivable species account for between 60–80% of the total identified by molecular methods (Kolenbrander *et al.*, 2002). Indeed, research to define the species composition of dental-plaque biofilms has been important as the findings indicate that many of the species within dental plaque biofilms can be studied by routine culture methods. Such a high proportion of cultivable biofilm species is uncommon and most environmental biofilms contain <1% of cultivable species (Davey and O'Toole G, 2000; Wuertz *et al.*, 2004). Therefore, it is clear that dental plaque is one of the best-described mixed-species bacterial communities containing a large number of culturable species. Thus, dental plaque represents a suitable biofilm from which to isolate indigenous oral bacteria and study interspecies interactions within multispecies biofilms. This is an enticing proposition and accounts, in part, for the increasing number of studies of communication between oral bacteria (Kolenbrander *et al.*, 2006).

Before considering communication mechanisms between species in dental plaque and its various roles in biofilms, it is important to understand how biofilm communities develop. In this regard, a general overview of biofilm development will be presented, followed by a break-down of the various mechanisms of communication that contribute to the development of multispecies biofilms. The different forms of communication will be addressed and an insight into the molecular mechanism for expression, reciprocal effects on other species, and the contribution to community development will be discussed.

Dental plaque biofilm formation

There are a variety of models that attempt to describe the development of biofilms and some of them address aspects of biofilm biology that are beyond the scope of this review (Wimpenny, 2000). A model that is gaining increasing attention describes biofilm development as a succession of adhesion/integration and multiplication events (Kolenbrander *et al.*, 2006; Lyautey *et al.*, 2005; Marsh and Bradshaw, 1995; Rickard *et al.*, 2003a). This model will be discussed here, in the context of dental plaque biofilm development, with supporting reference to studies conducted outside the oral cavity.

In the 'successional integration' model, the first organisms to attach to a surface are called primary or early colonizers. Adhesion can be through non-specific physico-chemical interactions with the surface (Kolenbrander and London, 1993; van Loosdrecht *et al.*, 1990) or the conditioning-film coating the surface or through specific protein-saccharide interactions with the conditioning film (Fig. 4.1). In the oral cavity, a professionally cleaned tooth surface is bathed, post-cleaning, in saliva. As a consequence, salivary proteins, glycoproteins and polysaccharides (such as mucins, salivary agglutinins, amylases and statherin) as well as bacterial constituents rapidly coat the surface. These polymers are targets for the adherence of primary colonizers, such as the streptococci and actinomyces (Kolenbrander and London, 1993; Scannapieco, 1994). *Streptococcus gordonii*, for example, has been shown to exhibit improved adhesion to surfaces coated with proline-rich proteins and alpha-amylases than to non-coated surfaces (Scannapieco *et al.*, 1995). Molecular studies have shown that SspA and SspB (antigen I/II family proteins) aid in binding *S. gordonii* to other oral bacteria and also to salivary agglutinin glycoprotein; another constituent of the salivary conditioning film that coats the tooth (Zhang *et al.*, 2005). Binding to other bacteria can occur through non-specific physico-chemical interactions, with members of the same species or those from different species, or through specific adhesion–receptor interactions with other species. The specific interactions are described as coaggregation interactions (Cisar *et al.*, 1979; Kolenbrander, 1997) and aid in the adhesion of other species to the surfaced-attached primary colonizers. Coaggregation interactions are not restricted to oral bacteria and increasing evidence indicates that coaggregation in aquatic environments also contributes to the integration of species into biofilms (Ivanov *et al.*, 2006; Jiang *et al.*, 2006; Rickard *et al.*, 2003b). Coaggregation is one form of interspecies communication and will be further described later in this review. The incoming species that attach to the primary colonizers are called secondary colonizers. An example of a secondary colonizer is *Veillonella atypica*, which coaggregates with *Streptococcus* species including *S. oralis* and *S. gordonii*, which are among the first to colonize freshly cleaned tooth enamel (Diaz *et al.*, 2006; Hughes *et al.*, 1992). As more secondary-colonizing species adhere to the developing biofilm community it is suggested that more species can integrate through coaggregation interactions (and additionally, non-specific interactions). The numbers of species increases through continued integrations and cell-divisions. As more secondary species integrate into the biofilm the chance/opportunity/conditions for pathogens to integrate also increases. This

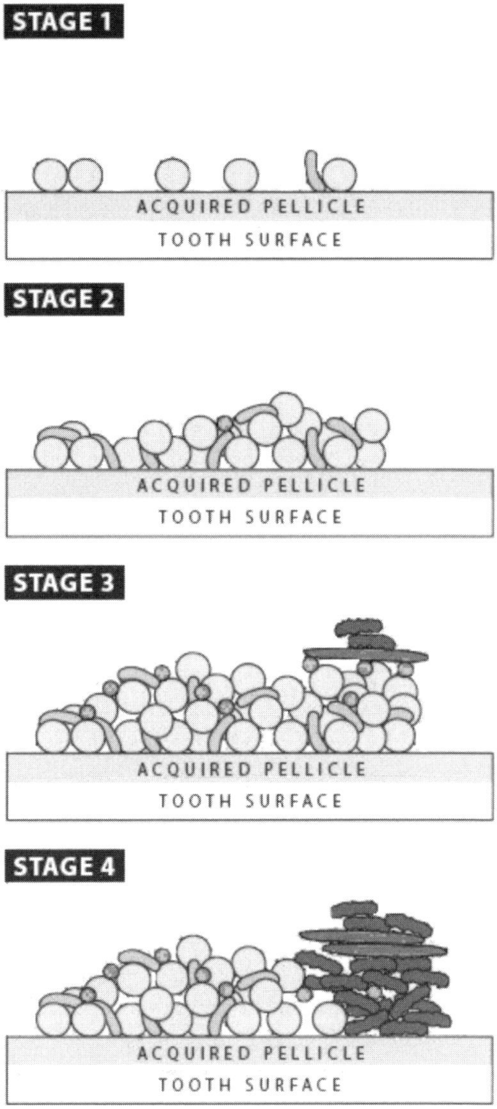

Figure 4.1 A diagrammatic representation showing the successional development of multispecies biofilms on an enamel surface under flowing conditions. Light grey cell shapes are typically associated with healthy plaque, while those that are coloured dark grey are associated with periodontal disease. Stage 1. Primary colonizing cells-types such as *Streptococcal* and *Actinomyces* species are the first to colonize a cleaned tooth surface. Interactions leading to attachment can be with the tooth surface but is more likely with acquired pellicle. Stage 2. Primary colonizing cells grow on the tooth surface and interactions between surface-attached cells and between other surface-attached cells and planktonic cells promote further growth and integrations of other species such as *Veillonella* species (shown as the integration of a small coccus-shaped cell). The species that integrate into the biofilm containing primary colonizers are called secondary colonizers. Stage 3. The biofilm containing primary and secondary colonizers grow and develop and this constitutes a biofilm of early colonizers (Figure 4.2). *Fusobacterium nucleatum* (dark grey long rod-like cells) is a late colonizer and integrates into the biofilm through interactions typically with secondary colonizers. The presence of *F. nucleatum* is associated with the integration and presence or opportunistic or disease-causing species such as *Porphyromonas gingivalis* (dark grey small rod-like cells). *F. nucleatum* promotes integration of oral pathogens by acting as a bridging organism to the healthy primary and secondary colonizing species. Stage 4. Through as yet undefined processes, the pathogenic organisms can start to rapidly divide and the numbers of cells increases at the expense of the number of primary and secondary colonizers. The processes involves interspecies communication. This stage represents the onset of periodontal disease.

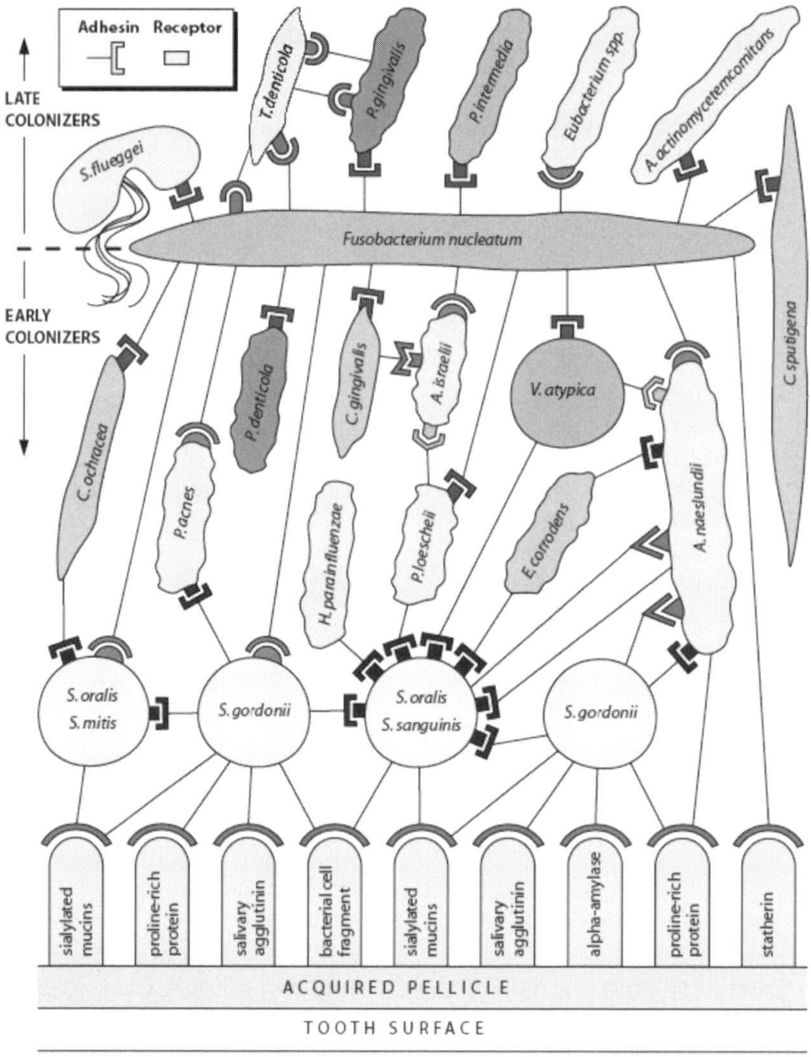

Figure 4.2 A diagrammatic representation showing the coaggregation interactions between species of oral bacteria. Many species coaggregate and this image shows some of the coaggregation interactions between species that are common to dental plaque. Stick-like symbols indicate coaggregation adhesins and block-like symbols indicate coaggregation receptors. Different shapes and shades of grey denote functionally distinct coaggregation adhesin-receptor reactions. Adhesins are typically proteinacious and receptors are composed of polysaccharides. Cell shapes indicate groups of different genera that are associated with oral health and disease. Sizes of cells are not to scale. Image modified from Kolenbrander *at al.* (2002).

is thought to be primarily due to the increasing numerical presence of *Fusobacterium nucleatum*, which is found in mature oral biofilms and is considered a 'bridging organism' to bacteria that would otherwise not integrate/coaggregate into the biofilm (Bolstad *et al.*, 1996; Bradshaw *et al.*, 1998; Kolenbrander *et al.*, 2002). With the integration of pathogens, the biofilm is subject to significant changes; including increased biofilm cell numbers and decreased species diversity (Lovegrove, 2004). It is the successional integration of primary, secondary and opportunistic species that is thought to be key to dental plaque species structure and composition (Kolenbrander and London, 1993; Kolenbrander *et al.*, 2006). Moreover, the spatio-temporal changes, role of specific species as primary or secondary colonizers, and subsequent increasing species diversity is highly reproducible between human individuals (Diaz *et al.*, 2006).

Growth in a biofilm is distinct from planktonic growth and, whilst little molecular work has been performed to compare biofilm versus planktonic growth in oral biofilms, there is much information on the differences in other ecosystems. For example, elegant work by Waite *et al.* (2006), using the opportunistic pathogen *Pseudomonas aeruginosa*, demonstrated a genome-wide change in gene expression when cells were grown in two planktonic phases (logarithmic and stationary) and multiple biofilm time points (early to mature/climax biofilm). In another study, a comparison of the effects of biofilm and planktonic growth was performed on *Staphylococcus aureus* and it was shown that many genes encoding binding proteins, proteins involved in the glucosaminoglycan polysaccharide intercellular adhesin synthesis and murein synthesis, and other enzymes involved in cell envelope synthesis and function were significantly upregulated within biofilm cells (Resch *et al.*, 2005). Thus, a comparison of oral planktonic cells to dental-plaque biofilm cells, within single and multispecies biofilms, needs to be performed. Indeed, an analysis of communities in multispecies dental plaque biofilms could be realized with the use of community microarrays; a new technology that allows for the simultaneous analysis of gene expression of different species in a biofilm (Hall, 2007).

Communication between bacteria in dental plaque

Inferred by studies of successional development of multispecies oral biofilms, communication between and within oral species contributes to the formation of stable communities (Kolenbrander *et al.*, 2006). Inter- and intraspecies communication can occur via two routes. The first is through physical cell–cell adhesion events that are mediated by specific surface-attached protein adhesins and polysaccharide receptors on the complementary cell types. The second is through the production of metabolites or specific signal molecules by the cells. Whilst physical cell–cell communication has received much research attention over the past 35 years, understanding of chemical cell–cell signalling is still in its infancy and is the subject of much debate and active research (Marsh, 2005; Sbordone and Bortolaia, 2003).

Intra-species communication

From an ecological perspective, intraspecies communication can represent a 'selfish' mechanism to form biofilm. In nature, biofilms are multispecies and often contain a single, numerically dominant, species (or genus). Granted, numerical dominance in a given environment is often a reflection of the environmental conditions, but may also be enhanced by the ability of the given species to maintain its presence by intraspecies communication. For example, streptococci, constitute between 60–80% of the species in healthy dental-plaque biofilms (Diaz *et al.*, 2006; Paster *et al.*, 2006; Socransky and Haffajee, 2005). Many streptococcal species are able to auto-aggregate and hyper-adhere to surfaces. This promotes their retention and contributes to their aggressive ability to remain the dominant species in healthy and diseased plaque. Other forms of intraspecies communication include the production of a class of cell–cell signal molecules called 'autoinducer-1'. Also, to be considered is genetic communication, through conjugation, and this is more likely between members of the same species in a biofilm, than between members of different species (Roberts and Mullany, 2006). Finally, the uptake of genetic material can be further enhanced by the production of intraspecies, extra-cellular signal molecules that act as a cue for the active uptake of cell-free DNA by that sole species (Miller and Bassler, 2001; Suntharalingam and Cvitkovitch, 2005). Such signal molecules are known to be produced by oral streptococci and to contribute to their ability to take up foreign DNA. Different forms of intraspecies communication, including the production of intraspecies signal molecules, are discussed below.

Communication by auto-aggregation

Auto-aggregation, or the propensity for cells of the same bacterial species to aggregate, is a characteristic that is common to many primary colonizers and some pathogens. Specifically, *Streptococcus*, *Actinomyces*, *Porphyromonas* and *Actinobacillus* species are known to auto-aggregate (Koop *et al.*, 1989; Narimatsu *et al.*, 2004; Shen *et al.*, 2005). A key question is, 'Why would these primary colonizers possess a predilection for auto-aggregation?' It is possible, that autoaggregation conveys a selective advantage over non-autoaggregating species by enhancing attachment within a developing biofilm (Elliott *et al.*, 2006; Rickard *et al.*, 2004). However, this form of communication has received less attention than interspecies communication (discussed below) and only a few autoaggregating oral species have been studied in-depth (Koop *et al.*, 1989; Narimatsu *et al.*, 2004; Soukka *et al.*, 1993). Insights into the mechanism for autoaggregation have only recently been revealed by molecular investigations. Studies of the surface polymers that mediate autoaggregation between *Porphyromonas gingivalis* cells by Narimatsu and colleagues (2004) resulted in the identification of a putative glycosyltransferase gene, designated *gtfA*, which had a consensus domain for glycosyltransferase in its N terminus. Microscopy demonstrated that *gtfA*-deficient mutants of *P. gingivalis* lacked mature fimbriae. Also, *gtfA*-deficient *P. gingivalis* also showed a very low ability for autoaggregation, and its ability to attach to surfaces was severely impaired. In several other species of bacteria, sugar-containing metabolites or glycoconjugates can act as bacterial adhesins, and glycosyltransferases participate in the synthesis of these products as well as in the adherence of bacteria to other surfaces. For example, in the 'laboratory-grade' Gram negative species, *Escherichia coli* K-12, *rfaG* also encodes for a glycosyltransferase and this, too, is also required for adhesion to surfaces as well as for lipopolysaccharide synthesis (Genevaux *et al.*, 1999). Thus, sugar-recognizing glycosyltransferases might have important roles in bacterial adhesion in various species, but little is known about their expression and functionality in *P. gingivalis*. However, these findings indicate that gtfA is required for auto-aggregation of *P. gingivalis* and potentially in their integration into biofilms. Such a finding is exciting, since an understanding of the mechanisms by which *P. gingivalis* (and, of course, other oral species) integrate into biofilms could contribute to novel treatment strategies of periodontal disease. However, no such research focus is, as yet, evident from the literature.

Communication by autoinducer-1

Autoinducer-1 (AI-1) was first discovered as the chemical signalling agent responsible for the generation of bioluminescence in the marine bacterium, *Vibrio fischeri*. This organism was observed to produce light only when present at high cell densities, such as are found within the light organs of host fish and squid species. In 1970, Nealson and colleagues (Nealson *et al.*, 1970) reported that an extracellular factor was responsible for inducing *V. fischeri* bioluminescence in a cell density-dependent manner. When grown in liquid culture at low cell densities, the inducer was found to be ineffective in stimulating bioluminescence, while at a population density of around 1×10^{10} bacteria per ml, as is found in the light organ of the bacterial host, the inducer was able to stimulate production of the enzyme luciferase, resulting in the emission of light. An extracellular factor, subsequently called 'autoinducer-1', regulating the transcription of the light-producing enzyme luciferase, was found in culture supernatants (Nealson and Hastings, 1979). Synthesized by the protein LuxI, this autoinducer was identified as the small extracellular messenger, N-3-oxohexanoyl-L-homoserine lactone, (Eberhard *et al.*, 1981). The autoinducer-dependent transcriptional activator was determined to be a protein LuxR, which when bound intracelluarly to the auto-inducer, is able to promote transcription of the *lux* operon genes (Engebrecht *et al.*, 1983; Engebrecht and Silverman, 1984; Engebrecht and Silverman, 1987). Autoinduction of the *lux* operon is therefore dependent upon the concentration of a specific acyl homoserine lactone (AHL) which increases with the culture growth of *V. fischeri*. When the autoinducer concentration reaches the micromolar range, its efflux from the cells becomes balanced by an influx, so that it can interact with LuxR. Because a minimum number of cells is required to activate transcription mediated by AHL autoinduction, this type of intercellular communication has been named 'quorum sensing'.

For several years, it was assumed that quorum sensing in *V. fischeri*, and a similar system in *V. harveyi*, was a unique method of communication restricted to the production of light by the luciferin/luciferase pathway. However, in the 1990s, several other members of the *Proteobacteria* group of Gram-negative bacteria were shown to use quorum sensing to regulate the production of proteins not related to bioluminescence. These bacteria were all initially shown to be pathogens of plants or humans. They included *Erwinia carotovora* (Bainton *et al.*, 1992; Jones *et al.*, 1993) and *Agrobacterium tumefaciens* (Fuqua and Winans, 1994; Piper *et al.*, 1993; Zhang *et al.*, 1993) in the case of the former and *Pseudomonas aeruginosa* (Gambello and Iglewski, 1991; Pearson *et al.*, 1994) in the case of the latter. These findings led to the belief that quorum sensing was a regulatory mechanism restricted to pathogenesis or symbiosis in selected groups of *Proteobacteria*. However, continued focus on quorum sensing revealed that a wide range of members from the group are capable of using specific AHL molecules to modulate the expression of a number of unrelated functions ranging from antibiotic production (Pearson *et al.*, 1994) to biofilm formation (Davies *et al.*, 1998) to pigment production (Chernin *et al.*, 1998). Each of these functions appears to play a role in the development or maintenance of populations within the biofilm context. AI-1 has also been shown to be involved in limited interspecies communication. For example, cell-free extracts of wild-type *P. aeruginosa* have been shown induce the formation of quorum sensing-regulated exoproducts in *Burkholderia cepacia* (McKenney *et al.*, 1995); and the use of reporters for the detection of autoinducer-1 by non-specific organisms is widely recognized.

The manner in which quorum sensing operates is relatively uncomplicated. Participating cells produce basal levels of AHL synthase, which is structurally homologous to the LuxI synthase of *V. fischeri*. Following production, AHL molecules rapidly diffuse from the cell along a concentration gradient and disperse into the surrounding medium. All AHL molecules share a common homoserine lactone core possessing an acyl side chain, the length and substitutions of which provide specificity to the molecule and influence its hydrophobicity. At sufficiently high cell densities, quorum sensing inducer molecules accumulate to a threshold concentration in the extra-cellular environment and drive an increase in the intracellular AHL concentration. When present at sufficient levels within the cell, the AHL signal binds to a transcription factor homologous to the LuxR protein of *V. fischeri*, and together these orchestrate the activation of quorum-sensing genes by direct interaction at target sites on the DNA. Typically, the LuxI homologue is one of the genes induced in response to LuxR/autoinducer binding and resulting in an amplification of the quorum response (for a review, see Fuqua *et al.*, 2001).
As more studies are performed, it is becoming increasingly evident that AI-1-based quorum sensing is a generalized phenomenon among members of the *Proteobacteria*. In 2001, a screen was performed by Jorge Frias and colleagues (2001) to test for the presence of autoinducer-1 produced by cultures of oral bacteria. Using a reporter strain of *V. harveyi* inducible for bioluminesce in response to extra-cellular autoinducer-1, they failed to detect the presence of this chemical messenger among the Gram-negative oral bacteria tested. This finding has led to the assumption that autoinducer-1 is not used as a communication molecule amongst oral bacteria. However, allowing for the breadth of species diversity in the oral cavity, and that a substantial part of dental plaque consists of species belonging to the *Proteobacteria* (Paster *et al.*, 2001), a more in-depth study needs to be performed before a definitive role (or, indeed, absence of a role) for AI-1 can truly be assigned.

Communication by competence-stimulating peptides

Streptococci are the numerically dominant species in dental plaque biofilms (Aas *et al.*, 2005; Diaz *et al.*, 2006; Li *et al.*, 2004; Nyvad and Kilian, 1987). One reason for the vast array of streptococcal species is their ability to consume foreign DNA from their local environment and incorporate it into their own genome (Martin *et al.*, 2006). As such, their genomes possess a plasticity that is distinct from other oral organisms. Natural transformation probably provides a selective advantage by allowing competent cells to acquire new traits, such as antibiotic resistance, by incorporation of DNA released from other closely related cell-types.

Recent studies have demonstrated that the uptake of DNA and substantial phenotypic changes leading to enhanced biofilm retention is due to the production of cell–cell signal peptides called 'competence-stimulating peptides' (CSP) by streptococci (Li *et al.*, 2002; Loo *et al.*, 2000). The presence of CSP, at very specific concentrations, initiates a phentotypic change in streptococci and is believed to play an important role in cell–cell signalling in the development of dental-plaque biofilms (Petersen *et al.*, 2005). Elegant work by Peterson *et al.* (2006) demonstrated that addition of micro-molar concentrations of synthetic CSP to developing wild-type biofilms of *S. mutans*, increased biofilm mass by 85% over the wild-type *S. mutans* that were not supplemented with synthetic CSP. Moreover, the addition of 1μM of synthetic CSP to wild-type cells enhanced the uptake of plasmid DNA. Changes in this optimum concentration of synthetic CSP resulted in reduced biofilm formation and transformation ability (Petersen *et al.*, 2005).

CSP is also known to be produced by members of the *mitis* and *anginosus*, groups of streptococci (Havarstein *et al.*, 1997; Petersen *et al.*, 2004). In the *anginosus* group, distinct species commonly encode and respond to identical CSPs. In other groups of streptococci, the CSPs are usually species-specific and, in several instances, are extremely specific to certain groups of strains (Havarstein *et al.*, 1997; Li *et al.*, 2001). The ecological reason for the species-specific nature of CSPs is unknown but, from an ecological perspective, is a likely mechanism by which individual species maximize their presence in dental-plaque biofilms (Ichihara *et al.*, 2006).

When considering the genes involved in CSP-based communication, a well-characterized system for the regulation of competence has been described in *Streptococcus pneumoniae* (Claverys *et al.*, 2000; Morrison and Lee, 2000) and knowledge of the mechanism of CSP-mediated communication is similar to that found in oral streptococci. In this system, the comCDE locus which encodes CSP, a histidine kinase receptor and a cognate response regulator, is transcribed at low levels during growth. The 41 amino acid product of comC is secreted from cells, as an active mature heptadecapeptide, by a dedicated ATP-binding-cassette (ABC) transporter encoded by the unlinked comAB locus. Once a threshold level of CSP is detected by cells expressing a surface-associated ComD histidine kinase, the response regulator ComE is phosphorylated. Activated, ComE switches on the genes required for DNA uptake and processing. Communication by CSP is also known to regulate a number of physiological activities, including sporulation in *Bacillus subtilis* (Grossman, 1995), antibiotic biosynthesis in *Lactococcus lactis* (Gutowski-Eckel *et al.*, 1994), and induction of virulence factors in *Staphylococcus aureus* (Ji *et al.*, 1995). Thus, CSP is regarded as a cell–cell communication peptide in many Gram-positive organisms and is very likely to play an important role in the colonization and retention of oral streptococci in dental-plaque biofilms.

Communication by conjugation

Gene transfer by conjugation can be regarded primarily as a uni-directional form of communication. Whilst not considered by some to be a form of communication, it meets the criteria of such a description. Communication could be defined as the act of conveying or imparting information and the transfer of genes from one cell to another therefore constitutes a type of bacterial communication. Conjugation is typically facilitated by the physical interaction of two cells and successful communication of genetic material occurs mainly by members of the same or very closely related species. Gene transfer may accelerate physiological and catabolic performance in multispecies communities such as dental-plaque biofilms (Molin and Tolker-Nielsen, 2003). Plasmids or conjugative transposons are accessory units of genetic information that can carry genes conveying a selective advantage to the harbouring species (Salyers *et al.*, 1995; Sorensen *et al.*, 2005). As such, the dissemination of genetic material through conjugal transfer will assist in the adaptation of the biofilm to changes in the localized environment. Data on the physiological and molecular mechanisms of conjugation between oral bacteria has accumulated and has been extensively reviewed (Roberts and Mullany, 2006). There are some excellent papers demonstrating the breadth of conjugative interactions in oral dental plaque. Indeed, very recently it has been shown that the macrolide resistance genes *mef*(E) and *erm*(B) can be transferred by interspecies conjugation from commensal *S. mitis* to the pathogenic *Streptococcus pyogenes* (Jonsson and Swedberg, 2006). Intergeneric conjugation has also been shown to occur between *B. subtilis* and oral streptococci in a multispecies model oral biofilm system (Roberts *et al.*, 1999). Another study has indicated that tetracycline resistance, linked to a Tn916 transposon on a conjugative plasmid, is rapidly transferred between oral streptococci in patients receiving tetracycline (Hartley *et al.*, 1984). Thus, in a natural dental-plaque biofilm, conjugation offers the potential for species to disseminate information and thus promote their retention/expansion in the community.

In general, genetic transfer between and within species requires intimate contact between the donor and recipient cells. As such, communication by auto-aggregation and co-aggregation (discussed below) will enhance such an interaction. A synergy between specific cell–cell aggregative interactions, mating pheromones and conjugal transfer systems will undoubtedly promote communication via genetic exchange.

Interspecies communication

Dental plaque represents an interactive community that develops through a complex web of communication mechanisms. Not only does communication occur between cells of the same species but also between evolutionarily distantly related species. Similar to intraspecies communication, interspecies interactions can occur through specific cell–cell aggregations or through cell–cell signalling. Examples of these forms of interspecies communication are discussed below.

Co-aggregation

Retention on a solid support is crucial for bacteria life in many flowing environments. In the oral cavity, non-sessile bacteria are washed away by saliva and swallowed. During the first four hours after professional teeth-cleaning, the earliest colonizers attach to the acquired pellicle (salivary proteins and glycoproteins, together with some bacterial molecules) coating on the tooth surface. These early/primary colonizers (or pioneers), consist primarily of oxygen-tolerant, Gram-positive bacteria (Nyvad and Kilian 1987; Kolenbrander *et al.*, 2005); they are mainly streptococci, equipped for pellicle binding (Jenkinson, 1994; Heddle *et al.*, 2003).

If left undisturbed, an increasingly complex population of bacteria develops, shifting from a primarily Gram-positive to a mainly Gram-negative anaerobic flora. This shift is also often accompanied by the advent of gingivitis or more severe periodontal diseases (Socransky *et al.*, 1998). Unlike early colonizers that adhere directly to the enquired pellicle, late colonizers rely mostly on co-aggregation and co-adherence for their continuing existence in the developing dental plaque.

Co-aggregations are non-random interactions between genetically distinct microorganisms. Co-aggregation differs from auto-aggregation, which is the cell–cell interaction between genetically identical cells, and from agglutination, which results from cell interaction with soluble molecules, such as host antibodies. Co-aggregations are mediated by a cell surface component on one cell type recognizing and binding to a complementary cell surface component on the partner cell. In many cases a heat- and protease-sensitive protein adhesin on one bacterium recognizes a heat- and protease-resistant receptor on the associated cell (Kolenbrander and London, 1993). All tested human oral bacteria co-aggregate *in vitro* with at least one partner cell type; and most tested oral bacteria exhibit co-aggregation with numerous bacterial species (Kolenbrander *et al.*, 2002). Bacterial isolates from a particular site are often co-aggregating species, suggesting that co-aggregation is involved in, if not vital to, colonization on oral surfaces (Kolenbrander *et al.*, 2002). Co-aggregation of planktonic/suspended cells with those already sessile is called 'co-adhesion'.

The organized and highly specific nature of co-aggregation is evident in the non-random bacterial succession in the oral biofilm (Socransky *et al.*, 1998). Evidence for the involvement of co-aggregation mediators (adhesins and receptors) in the spatio-temporal development of early dental plaque was recently provided in biofilms developed *in vivo* in human volunteers wearing a retrievable enamel chip (Palmer *et al.*, 2003). Scanning of the biofilms developed on the retrievable enamel chip, using laser confocal microscopy, revealed that fluorescent-labelled antibodies generated against *Actinomyces* type 2 fimbriae adhesins were juxtaposed with antibodies specific to a streptococcal complementary receptor-polysaccharide. The confocal microscopy images also demonstrated that the biofilms developed *in vivo* were generated by an interdigitated growth of mixed species rather than by mono-species communities growing side-by-side.

It seems reasonable to suggest that co-aggregation might act as a trigger to signal transition from a planktonic to a sessile life-style in a multispecies community-biofilm. Such transition involves alteration in gene response and gene expression. Although modification in gene expression, following transition to biofilm growth, has been studied (Kolter and Greenberg, 2006; Schembri *et al.*, 2003; Waite *et al.*, 2006; Waite *et al.*, 2005), alteration in gene response following co-aggregation signalling has not yet been reported.

Among the oral isolates tested for co-aggregation, *Fusobacterium nucleatum* is unique in its versatile co-aggregative capabilities. *F. nucleatum* is a long, Gram-negative, non motile, non spore-forming, anaerobic, rod. It can be found in dental plaque 12 hours after teeth-cleaning, following colonization of early colonizers and preceding colonization by late colonizers species associated with periodontal disease. These include *P. gingivalis, Actinobacillus actinomycetemcomitans (A.a), Treponema denticola* and *Tannerella forsythensis* (Socransky *et al.*, 1998). *In vitro, F. nucleatum* can mediate co-aggregation between all the species of early and late colonizers (Kolenbrander and London, 1993). Late colonizers such as *Eubacterium* species (George and Falkler, 1992), and *Helicobacter pylori* (Andersen *et al.*, 1998), co-aggregate only with fusobacteria. These observations led to the suggestion that *F. nucleatum* functions as bridging organism, that facilitating the attachment of late colonizers to the developing dental biofilm by co-adhering them to already surface-attached early colonizers (Kolenbrander and London, 1993) (see Fig. 4.2). It is interesting to note that these late colonizers are either directly or indirectly associated with disease. Thus, it seems appropriate to describe these later colonizers as either 'opportunists' or pathogens.

It is tempting to speculate that bridging by *F. nucleatum* also plays an important role in facilitating and enabling communication between bacteria that do not co-aggregate with one another but do so with *F. nucleatum*. In this case, *F. nucleatum* can bridge and proximate bacteria that would otherwise have been too distant for efficient communication, bringing them to the micrometer proximity required for efficient diffusion-dependent signalling under the flowing oral environmental conditions. Evidence for the mediation of *F. nucleatum*-bridging in metabolic communication between bacteria that do not co-aggregate will be presented below.

Identifying proteins relevant in co-aggregation

The realization of the potential key role that co-aggregation plays in the development of the dental plaque stimulated the search for genes associated with co-aggregation. The streptococcal antigen I/II surface proteins are among the most investigated adhesins (Jenkinson and Demuth, 1997). Members of this protein family share seven conserved structural domains, (i) signal peptide, (ii) N-terminal region, (iii) alanine rich repeat blocks, (iv) divergent central region, (v) proline-rich repeats, (vi) C-terminal region, and (vii) cell wall-anchoring sequences. Earlier studies revealed that members of this antigen I/II polypeptide family are involved in streptococcal binding to salivary agglutinin glycoproteins and can thus enhance streptococcal adherence to the acquired pellicle-coated oral surfaces (Demuth *et al.*, 1988). The finding that transcription of the genes coding for the two antigen I/II proteins of *S. gordonii* (SspA and SspB) is upregulated in the presence of saliva (Du and Kolenbrander, 2000), highlighted the involvement of these antigens in streptococcal colonization of oral surfaces. Local oral passive immunization of human volunteers with antibodies generated against antigen I/II of *S. mutans* prevented recolonization by indigenous *S. mutans* and conveyed long-term protection against caries (Ma *et al.*, 1990).

Later studies, in which biotinylated surface molecules of *S. gordonii* were incubated with *P. gingivalis* cells, revealed the capacity of the *S. gordonii* SspB protein to bind *P. gingivalis* (Lamont *et al.*, 1994). This biding activity was confirmed by expressing *SspB* in *Enterococcus faecalis* which consequently gained the ability to bind *P. gingivalis* (Lamont *et al.*, 1994). Generation of truncated recombinant SspB polypeptides enabled the mapping of the *P. gingivalis* binding-domain of the *S. gordonii* SspA and SspB (Brooks *et al.*, 1997). It was suggested that colonization of at least some strains of *P. gingivalis* (a late colonizer) in dental plaque is facilitated by *S. gordonii* via SspA and SspB (Lamont *et al.*, 1994). Inactivation of the genes coding for the *S. gordonii* sspA, sspB and sspAB adhesins enabled detailed characterization of their function in co-aggregation with different *A. naeslundii* strains. SspA was found to participate in lactose-inhibitable and non-inhibitable interactions with *A. naeslundii*, while SspB mediated only lactose-non-inhibitable co-aggregations with the *A. naeslundii* partner cells. The sspAB double-mutant revealed a third co-aggregation interaction with one of the tested *A. naeslundii* strains (Egland *et al.*, 2001).

Random insertional inactivation mutagenesis using transposons, followed by selection for loss of co-aggregation function, has proven useful for identifying co-aggregation-associated genes. Random insertion of the Tn916 transposon into *S. gordonii* DL1 identified a gene coding for a 100-kDa surface protein involved in auto-coaggregation with streptococcal partners of DL1 (Whittaker *et al.*, 1996).

Another adhesin found to mediate co-aggregation is the Glyceraldehyde-3-phosphate dehydrogenase (GAPDH) of *S. oralis*. GAPDH was found to function as a co-adhesin for the *P. gingivalis* major fimbriae. Interestingly, the gene coding for the *S. oralis* GAPDH shares a high degree of homology with the sequences of GAPDHs of other bacteria, such as *F. nucleatum*, known to coaggregate with *P. gingivalis* (Maeda *et al.*, 2004).

Genetic tools, which are currently being developed for random gene inactivation in *F. nucleaum*, are expected to enable the discovery of the multiple adhesins involved in the versatile coaggregation capability of these bacteria, and to enable thorough exploration of the role of fusobacterial bridging in the development of the dental biofilm.

Metabolic communication

Co-aggregation creates spatial proximity among distinct cells types. In the oral environment, which is under constant salivary flow, this proximity has been shown to contribute and, in some cases, to be essential for growth of certain bacterial species (Vesey and Kuramitsu, 2004). Under salivary-flow conditions, proximity can enable and enhance metabolic communication or metabolic synergism; that is, sequential cooperation of multiple species to completely metabolize a complex nutrient (Bradshaw *et al.*, 1994; Kolenbrander *et al.*, 2002). A co-aggregation-dependent mutualistic relationship, in which simultaneous growth of species can occur where neither could survive alone, was demonstrated in glass flowcell biofilm experiments, using saliva as the sole carbon and nitrogen source. In these experiments, *A. naeslundii* was unable to grow in either the planktonic or biofilm phases. *S. oralis* grew planktonically, but did not grow well as a biofilm. However, when mixed and inoculated together, both *S. oralis* and *A. naeslundii* showed luxuriant, interdigitated growth (Palmer *et al.*, 2001).

Co-aggregation has also been shown to enable the signalling involved in the metabolic synergism between *V. atypica* and *S. gordonii*. For growth, *V. atypica* utilizes lactic acid which is the major end-product of glucose fermentation by *S. gordonii*. Agar-plate co-culture of these organisms revealed that a signalling event, resulting in increased expression of the *S. gordonii* alpha-amylase-encoding gene *amyB*, takes place. Using an *S. gordonii* plasmid carrying the green fluorescent reporter-gene fused to, and expressed by the *amyB* promoter, it has been shown, in flowcells, that only those streptococci in mixed-species micro-colonies containing *V. atypica* expressed GFP. The *V. atypica*–*S. gordonii* signalling resulting in optimal PamyB-′gfp expression occurred over short distances, of the order of 1 μm (Egland *et al.*, 2004).

Metabolic communication might also explain the survival of obligate anaerobes in the aerated oral cavity. Experiments using a two-stage chemostat model system, in which the second-stage chemostat vessel was aerated (Bradshaw *et al.*, 1994), showed that in the tested conditions, anaerobes were able to grow in both the planktonic and biofilm phases only in the presence of facultative species (Bradshaw *et al.*, 1996). Another study showed that the presence of *F. nucleatum* enables growth of *P. gingivalis* in oxygenated- and carbon-dioxide-depleted environments (Diaz *et al.*, 2002). These studies suggested that intimate interactions between oxygen-tolerant and anaerobic species, mediate the latter's survival under aerated conditions. Thus, such survival may depend on oxygen consumption by aerotolerant species, generating local redox-potential gradients within planktonic multispecies co-aggregates and plaque biofilms (Bradshaw *et al.*, 1996)

F. nucleatum has been used to bridge between otherwise non-co-aggregating obligately anaerobic and oxygen-tolerant species. One such study demonstrated that the survival of anaerobes was reduced in both in planktonic and biofilm communities when *F. nucleatum* was omitted (Bradshaw *et al.*, 1998). It further showed not only that co-aggregation provides the close physical juxtaposition of partner cells necessary for the anaerobes growth in the presence of oxygen, but also that fusobacterial bridging can mediate this kind of metabolic communication.

Autoinducer-2

Over the last five years a great deal of research has focused on the possibility that the vast majority of oral bacteria in dental plaque biofilms produce and detect a family of interconvertible signal molecules. These molecules, that change their structure as a consequence of the environmental conditions, are collectively called 'autoinducer-2' (AI-2). Before focusing on the effects of AI-2 mediated communication and the ramifications of such a universal language in dental-plaque biofilms, it is important to understand how, why and when cells produce AI-2.

Recently, it was discovered that two cell–cell concentration-dependent signal transduction cascades are present in marine *Vibrio* species; they converge to induce bioluminescence in *V. harveyi* (Mok *et al.*, 2003) and virulence gene transcription in *V. cholerae* (Miller *et al.*, 2002; Zhu *et al.*, 2002). The two chemically unique signals that are involved in cell–cell signalling are described as auto-inducers. The first is autoinducer-1 (AI-1), which is an hydroxybutanoyl-l-homoserine lactone (often described as a member of the AHL family of autoinducers, described earlier), and this acts as an intraspecies cell–cell signalling molecule. With the exception of some work by Fias *et al.* (2001), studies of AI-1 production and detection by oral Gram-negative bacteria has generally not been studied but might be of significance in mature dental-plaque biofilms harbouring pathogens such as *Porphromonas* and *Fusobacterium* species. The second cell–cell signalling system actually consists of a family of interconvertible molecules, derived from a progenitor molecule called 4,5-dihydroxy-2,3-pentanedione (DPD) (Duerre and Miller, 1966; Semmelhack *et al.*, 2005). DPD is a product of bacterial LuxS enzyme in the catabolism of S-ribosylhomocysteine in the activated methyl cycle (Fig. 4.3). The activated methyl cycle is essential for bacterial growth (Winzer *et al.*, 2003) and through this cycle the toxic molecule S-adenosylhomocysteine (SAH) has to be converted to a non-toxic homocysteine for re-use into the cycle (Fig. 4.3). The conversion can be through a one- or two-step process. The former requires the activity of the enzyme S-adenosylhomocysteine hydrolase and might generate a signal molecule that is described as autoinducer-3 (AI-3) (Walters *et al.*, 2006). Little is known of this process and the role of AI-3 will not be discussed here. The latter requires the activity of two enzymes; Pfs and LuxS (Winzer *et al.*, 2002). The activity of SAH, Pfs and LuxS has a role in the central metabolism of bacteria and inactivation of these enzymes can significantly alter cell metabolism and physiology (Doherty *et al.*, 2006; Lee *et al.*, 2006; Reading and Sperandio, 2006; Winzer *et al.*, 2003). Pfs converts SAH to S-ribosylhomocysteine and this is, in turn, acted upon by luxS to generate homocysteine. A by-product of the activity of LuxS is DPD. Upon production, DPD spontaneously forms a family of interconvertible molecules (Fig. 4.3) collectively called AI-2 (Semmelhack *et al.*, 2005). Depending upon the environmental conditions, such as the availability of boron and pH, the different forms of AI-2 reach equilibrium. Excitingly, evidence suggests that these different forms of AI-2 can be identified by different genera of bacteria. To date, two structurally similar signalling molecules have been co-crystallized, with their respective receptors, from the human pathogenic bacterium *Salmonella typhimurium* and the bioluminescent marine bacterium *Vibrio harveyi* (Chen *et al.*, 2002; Miller *et al.*, 2004). *V. harveyi* receptors are specific for a boronated form of AI-2 and *S. typhimurium* receptors bind specifically to a non-boronated form of AI-2. It is likely that one of these forms or another, as yet uncharacterized form, or even a blend of forms may be required for communication between oral dental-plaque bacteria. Thus, AI-2 possesses

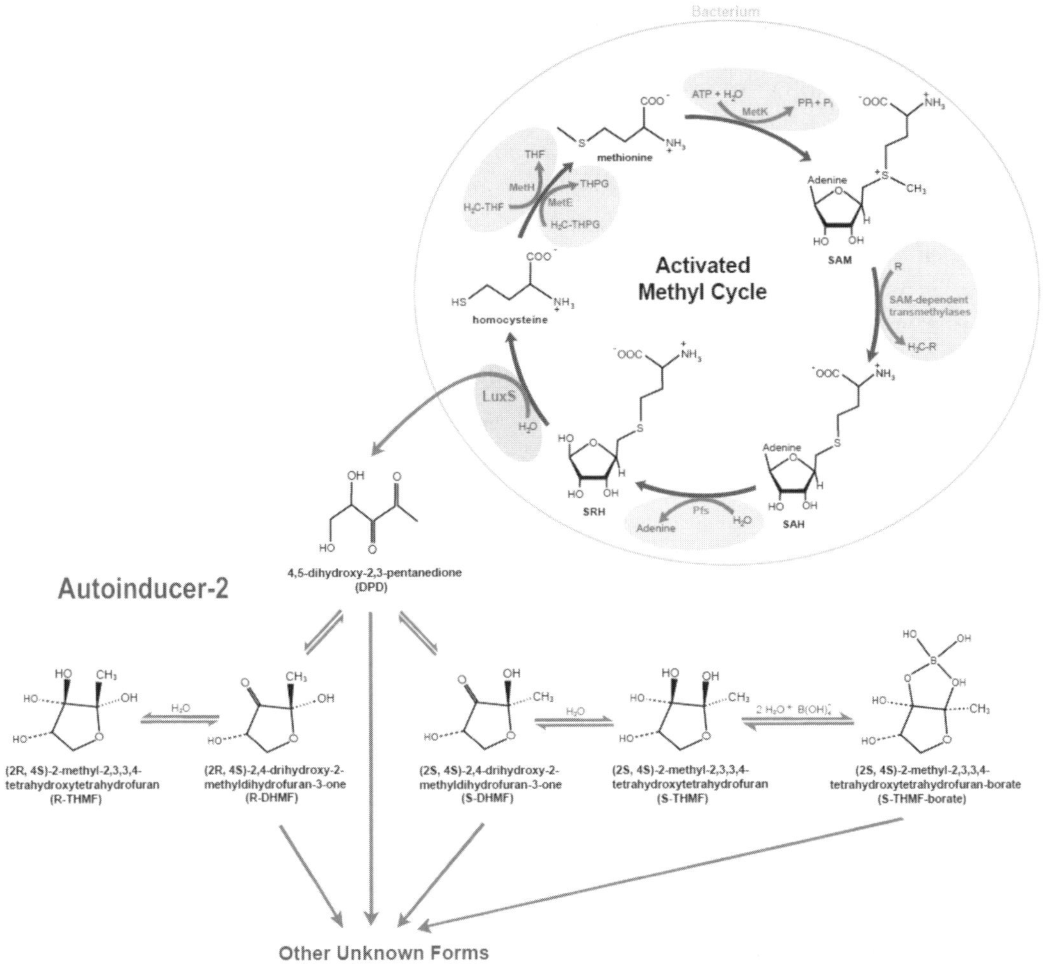

Figure 4.3 Diagram showing the activated methyl cycle and the family of interconvertible forms of AI-2 that is produced as a consequence of the activity of LuxS. LuxS is required in the two-step conversion of S-adenosyl-homocysteine to homocysteine in the activated methyl-cycle. The product, 4,5-dihydroxy-2,3-pentandione (DPD), is produced and spontaneously generates a family of interconvertible molecules that is collectively called AI-2. It is possible that AI-2 can exist in more forms than shown here and two forms of AI-2 are currently known to have biological cell–cell signalling activity. A furanosyl borate diester is used by the marine bacterium *Vibrio harveyi* for cell–cell signalling. The second is a non-boronated form that is known to be used by *Salmonella* species for cell–cell signalling. Depending upon environmental conditions (for example, pH and ionic composition), equilibrium of different forms of AI-2 will form with one form being potentially favoured. For example, in the presence of boron, AI-2 complexes with born to form (2S, 4S)-2-methyl-2,3,3,4-tetrahydroxytetrahydofuran-borate. This form is recognized by receptors on the cell-surface of *V. harveyii*. Other forms may be involved in signalling between oral bacteria.

a signalling activity that is dependent upon the environment into which it is excreted and the type(s) of receptor expressed by the detecting species.

Different species of bacteria respond to AI-2 in different ways. With respect to bacterial behaviour, there has been a recent increase in studies of environmental, pathogenic and oral bacteria to determine the effect of (i) *luxS* inactivation (Azakami *et al.*, 2006; Chung *et al.*, 2001; James *et al.*, 2006b; Merritt *et al.*, 2003); (ii) the addition of synthetic (*in-vitro* produced) AI-2 (Auger *et al.*, 2006; Babb *et al.*, 2005; Gonzalez Barrios *et al.*, 2006); and (iii) the addition of chemically synthesized AI-2 (De Keersmaecker *et al.*, 2005; Rickard *et al.*, 2006; Semmelhack *et al.*, 2005; Xavier and Bassler, 2005a; Xavier and Bassler, 2005b). The ability to form biofilms and communicate by AI-2 has been linked in many of these studies.

Interestingly, as with the ability to coaggregate, AI-2 production by oral dental-plaque forming bacteria seems to be very common (Frias *et al.*, 2001).

Linking co-aggregation and succession to the formation of dental-plaque biofilms, the role of AI-2-based communication will be reviewed; starting with the primary colonizers. When considering AI-2, oral streptococci have been the subject of a number of studies. One of the first insights into the possible importance of AI-2 in communication between oral dental-plaque bacteria was given by McNab *et al.* (McNab *et al.*, 2003) who demonstrated that *luxS* contributes to the ability of *S. gordonii* and *P. gingivalis* to form dual-species biofilms in an in-vitro model biofilm system. Inactivation of *luxS* (and consequently AI-2 production) in either of these two organisms resulted in retardation in dual-species biofilm forming ability of these two organisms. Also, the findings of these workers suggest that AI-2 regulates aspects of carbohydrate metabolism in *S. gordonii*. Differential display PCR demonstrated that the inactivation of *S. gordonii luxS* downregulated the expression of *fruA*, encoding extracellular exo-beta-D-fructosidase; *gtfG*, encoding glucosyltransferase; and *lacD*, encoding tagatose 1,6-diphosphate aldolase. Interestingly, *S. gordonii* cell-surface expression of SspA and SspB proteins that, as mentioned above, are implicated in mediating coaggregation between *S. gordonii* and *P. gingivalis*, was not affected by inactivation of *luxS*. The link to carbohydrate metabolism is an especially exciting finding as work by Egland *et al.*, (2004) demonstrated that carbohydrate metabolism is important in forming biofilm associations with *V. atypica*. AI-2 expression may therefore be linked, or act in synergy, with metabolic communication (described above). Further, *Veilonella* species can produce AI-2 and studies aimed at better understanding the importance of AI-2, are currently under way (Rickard, *unpublished data*). *S. oralis* is another primary colonizing species that is also found in early biofilms, in association with *S. gordonii* and other species. It has also been shown to produce AI-2 and this property has been shown to mediate mutualism with *Actinomyces naeslundii* in saliva (Rickard *et al.*, 2006). Unlike many other studies on the role of AI-2-based communication, chemically synthesized AI-2 was used in this study. Neither wild-type *S. oralis* nor *A. naeslundii* forms single-species biofilms in saliva (Palmer *et al.*, 2001). However, in dual-species biofilms, they both grow luxuriantly to form interdigitated biofilms, which is characteristic of mutualistic interactions. Disruption of the *luxS* gene in *S. oralis* results in the retardation of mutualism with *A. naeslundii*. Supplementing saliva with chemically synthesized AI-2 restores mutualism between *S. oralis* and *A. naeslundii*. Critically, however, restoration of mutualism only occurs if picomolar concentrations of chemically synthesized AI-2 are added to saliva. If the concentration is reduced or increased beyond the optimal picomolar concentration, mutualism is abolished. Such a finding is intriguing as it is known that many oral bacteria associated with healthy dental plaque produce very low concentrations of Al, whereas the levels are 10–100 times greater in those associated with disease (Frias *et al.*, 2001). This suggests that AI-2 levels may contribute to shifts in the species composition of dental plaque. Other notable studies on the production of low concentrations of AI-2 by oral streptococci include the work of Anne Scheie's oral microbiology group, who have also shown that the *luxS* gene is important in the biofilm-forming ability of *S. anginosus* (Petersen *et al.*, 2006). The *luxS* gene in this species was disrupted and while the direct effects of AI-2 on gene expression remain unknown, a significant loss in biofilm-forming ability was observed.

Studies of AI-2 based communication between secondary colonizers of dental-plaque biofilms are currently less extensive. One interesting study (Azakami *et al.*, 2006) focused on the secondary colonizer *Eikenella corrodens*, a facultive Gram-negative anaerobe which is found in both healthy and diseased dental plaque (Fig. 4.2) and which readily co-aggregates with primary colonizing streptococci. An orthologue of *luxS* was found in the *E. corrodens* genome and insertional inactivation of the gene resulted in the abolition of AI-2 production. Biofilm studies, using a microtitre plate assay, indicated that loss of the *luxS* gene, and associated AI-2 signalling, abolished the biofilm-forming ability of the species. Whether or not this ability could be recovered by the addition of exogenous AI-2 was not determined.

The role of AI-2 in communication between dental-plaque biofilm pathogens is receiving much attention from a number of laboratories. In particular, much work has been done on the role of AI-2 on biofilm formation by *Streptococcus mutans*. For example, Wen and Burne (2004) demonstrated that a mutant of this organism, having a disrupted *luxS* gene, formed significantly less biofilm on hydroxylapatite disks than the wild-type strain. In addition, the mutant phenotypes were complemented by exposure to Al-2-containing supernatants from wild-type cultures of *S. mutans*. This step unequivocally demonstrated that AI-2 is essential for *S. mutans* to form biofilms. These findings were confirmed by Yoshida and colleagues (2005) who took the concept of using AI-2-containing culture supernatants one step further. In an elegant series of experiments, involving a novel two-compartment system to monitor AI-2 production, detection and response by *S. mutans* to other oral bacteria was followed. Using this system, the biofilm-forming ability of the *S. mutans luxS* mutant was re-established by strains of *S. gordonii*, *S. anginosus*, *S. sobrinus*, *P. gingivalis* and *A.a.* but not by a *P. gingivalis luxS* mutant. Interestingly, biofilm-forming ability was not re-established by *S. oralis*, *S. salivarius*, or *S. sanguinis*. Such a findings indicate that, either AI-2 needs to be in a particular form or blend of forms, or indeed that the required concentration for AI-2 is critical for mediating the ability of the *S. mutans luxS* mutant to form a biofilm.

The concept that AI-2 signal strength mediates interspecies communication extends to *A.a.* For instance, Fong and co-workers (2001) demonstrated that the organism possesses *luxS* and produces AI-2. A *luxS* mutant that was incubated with spent growth medium from *A.a.* or from an AI-2-producing *E. coli*, resulted in a three-fold induction of leukotoxic activity and an associated increase in leukotoxin polypeptide. In contrast, no such increase in leukotoxin expression occurred when cells were exposed to sterile medium or to spent medium from the *E. coli* strain in which *luxS* was insertionally inactivated. Following these findings, an analysis of the complete genome sequence of *A. actinomycetemcomitans* was performed and it was discovered that RbsB protein (a ribose-binding protein) was related to LuxP, the AI-2 receptor of *V harveyi* (James *et al.*, 2006b). The same authors then determined whether or not RbsB interacts with AI-2. This was done by mixing RbsB with AI-2 from *V. harveyi* or from *A.a.* and examining the effect of mixtures on the bioluminescence of *V. harveyi* BB170 (which detects and bioluminesces only in response to AI-2). RbsB efficiently inhibited *V. harveyi* bioluminescence that would ordinarily be induced, in a dose-dependent manner, by AI-2 from both *A.a* and *V. harveyi*. This suggested that *A.a.* RbsB competed with *V. harveyi* LuxP for AI-2. Fifty percent inhibition occurred with a 0.3 nanomolar (nM) concentration of RbsB in the case of *A.a.* AI-2 and 15 nM RbsB for *V. harveyi* AI-2. In addition, RbsB-mediated inhibition of *V. harveyi* bioluminescence was reversed by the addition of 50 mM ribose, suggesting that *A.a.* AI-2 and ribose bind to the same site of RbsB. This thorough study was concluded by construction of an isogenic *A.a.* mutant that was unable to express rbsB. It was unable to deplete *A.a.* AI-2 from solution, thus supporting the notion that RbsB acts as a receptor for AI-2. In relation to other studies on AI-2-based communication, this study confirmed that *A.a.* responds to AI-2 at nanomolar concentrations. These concentrations are approximately 5–10 times higher than those used by primary colonizers.

In concert with studies of AI-2-based interspecies communication of the pathogen *A.a.*, *P. gingivalis* has also received attention. Initial work by Chung and colleagues (2001) demonstrated that the *luxS* gene in *P. gingivalis* may function to control the expression of genes involved in the acquisition of haemin. Following this study, microarray analysis of wild-type and *luxS* mutants of *P. gingivalis* was performed (Yuan *et al.*, 2005). Their results suggested that *luxS* might be involved in promoting survival of *P. gingivalis* in subgingival dental plaque biofilms by regulating the response to stresses, such as exposure to hydrogen peroxide, temperature and strongly alkaline conditions. This was further confirmed by James and co-workers (James *et al.*, 2006a). Using quantitative real-time reverse transcription-PCR, the expression of genes involved in hemin and iron uptake was determined in parent and *luxS* mutant strains of *P. gingivalis*. Compared to the parental strain, the *luxS* mutant demonstrated reduced expression of genes coding for Tlr, a TonB-linked hemin binding protein, and also Kgp, a lysine-specific protease which can degrade host haem-containing proteins. Addition of partially purified AI-2 from the culture supernatant of wild-type *P. gingivalis* re-established expression of Tlr and Kgp. Thus, as with the oral pathogens *A.a.* and *S. mutans*, AI-2 is required for intra and interspecies communication. It would be interesting to know whether these pathogens can communicate with one-another, as such a finding would suggest an AI-2-based synergy. Available evidence would suggest that such a finding is possible; and ecologically this notion has support, since where one species is present, other pathogens are also found (Lovegrove, 2004; Papapanou *et al.*, 1993; Saarela *et al.*, 1993).

The ecological significance of communication between oral bacteria

Throughout this review the concept that oral bacteria do not exist in dental-plaque biofilms as an unorganized mass of species, has been presented. There is an ordered, reproducible, succession that leads to a defined structure. We propose that inter- and intraspecies communication promotes the formation of biofilm echelons. This could be achieved through interspecies communication-mediated mutualism, as is the case between co-aggregating *S. gordonii* and *V. atypica* and also mutualism between *S. oralis* and *A. naeslundii*. In addition, this could occur through interspecies competition. This, in turn, could be mediated through the manipulation and utilization, by pathogens, of the propensity for healthy bacteria to co-aggregate and then produce different concentrations (and, possibly, types) of AI-2 that would disrupt communication between mutualistic species. Such a process would account for shifts in community species composition in 'healthy' plaque during the onset of disease (Ramberg *et al.*, 2003; Socransky and Haffajee, 1997; Socransky and Haffajee, 2005; Socransky *et al.*, 1998). A model for this process is presented in Fig. 4.4. In order for healthy bacteria to combat such invasions, the ability to communicate at the intraspecies level is just as likely to play a role as interspecies communication. Through intraspecies communication, cells can auto-aggregate to maximize their presence in dental-plaque biofilms and communicate with members of their own or closely related species to exchange genetic material. This could account for the great diversity of streptococci found in dental plaque, their ability to sustain substantial changes in the environment and continued survival in the biofilm.

It is appreciated that species in a multispecies biofilm possess a greater combined metabolic potential than if they were in mono-species biofilms; and this allows them to survive substantial natural or artificial environmental changes (Gilbert *et al.*, 2002). Inter- and intracommunication may enhance such an ability to form adaptable and stable communities. Indeed, it has been suggested that biofilms promote 'altruism' (Kreft, 2004; Kreft and Bonhoeffer, 2005) and this partially explains biofilm characteristics, such as resilience and the structural organization into species mosaics (Gilbert *et al.*, 2002; Sutherland, 2001; Wimpenny, 1997); interdigitations (Palmer *et al.*, 2001); and the shedding of single cells and coaggregates (Stoodley *et al.*, 2001). These characteristics promote the formation and maintenance of the altruistic strategy. As such, via the many forms of inter- and intraspecies communication in dental–plaque biofilms, altruistic, mutualistic and competitive interactions are mediated through complex molecular interactions at the genus, species and strain level. Elucidating the complex web of communication interactions will aid our understanding of the development and evolution of dental plaque biofilms. This understanding will inevitably contribute to the development of strategies for controlling the onset of oral biofilm-related infections, such as periodontal disease.

Acknowledgements

The authors wish to thank to Mr Richard Palmer for drawing Figs 4.1, 4.2 and 4.5; and to Mr. Kyung Rok Min for drawing Fig. 4.3.

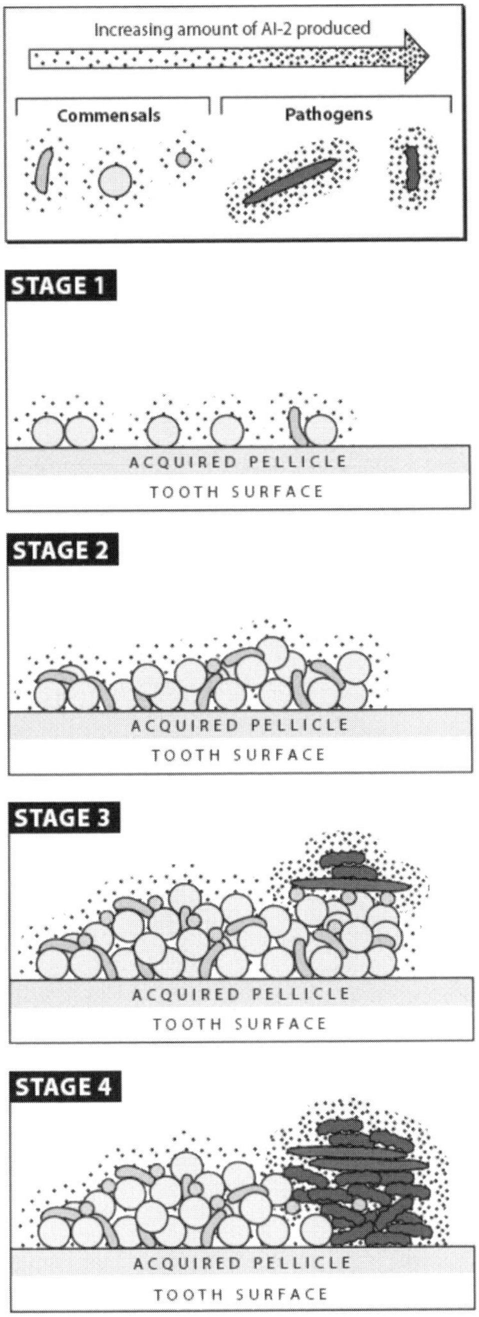

Figure 4.4 A model depicting the potential role of AI-2 in the development of dental-plaque biofilms under flowing conditions. Stages of development are described in Figure 4.1. Relative amounts of AI-2 produced by commensal and opportunistic/pathogenic species is shown by the intensity of dots around each cell-type. During biofilm development (stages 1 and 2) primary and secondary colonizers (commensal species) integrate into the biofilm and these produce low (picomolar) concentrations of AI-2. This concentration favours mutually beneficial interactions between commensal species. As the biofilm continues to develop, opportunistic and pathogenic species may integrate. Coaggregation may promote their integration. These species produce much higher concentrations of AI-2 and this elevated localized AI-2 concentration is not favourable for synergy between commensal species. The commensal species become less competitive and their ability to remain in the biofilm, under flowing conditions, is impaired. The pathogenic species expand in place of the commensal species (stage 4). Conceivable factors that contribute to the expansion of a pathogenic species in the commensal biofilm include salivary flow rate, numbers of colonizing pathogenic species and physiological status of the invading pathogenic species during stage 3. Image modified from Kolenbrander *at al.* (2006).

References

Aas, J.A., Paster, B.J., Stokes, L.N., Olsen, I., and Dewhirst, F.E. (2005) Defining the normal bacterial flora of the oral cavity. J. Clin. Microbiol. *43*, 5721–5732.

Andersen, R.N., Ganeshkumar, N., and Kolenbrander, P.E. (1998) *Helicobacter pylori* adheres selectively to *Fusobacterium* spp. Oral Microbiol. Immunol. *13*, 51–54.

Auger, S., Krin, E., Aymerich, S., and Gohar, M. (2006) Autoinducer 2 affects biofilm formation by *Bacillus cereus*. Appl. Environ. Microbiol. *72*, 937–941.

Azakami, H., Teramura, I., Matsunaga, T., Akimichi, H., Noiri, Y., Ebisu, S., and Kato, A. (2006) Characterization of autoinducer 2 signal in *Eikenella corrodens* and its role in biofilm formation. J. Biosci. Bioeng. *102*, 110–117.

Babb, K., von Lackum, K., Wattier, R.L., Riley, S.P., and Stevenson, B. (2005) Synthesis of autoinducer 2 by the lyme disease spirochete, *Borrelia burgdorferi*. J. Bacteriol. *187*, 3079–3087.

Bainton, N.J., Bycroft, B.W., Chhabra, S.R., Stead, P., Gledhill, L., Hill, P.J., Rees, C.E., Winson, M.K., Salmond, G.P., Stewart, G.S., and *et al.* (1992) A general role for the lux autoinducer in bacterial cell signalling: control of antibiotic biosynthesis in *Erwinia*. Gene *116*, 87–91.

Bolstad, A.I., Jensen, H.B., and Bakken, V. (1996) Taxonomy, biology, and periodontal aspects of *Fusobacterium nucleatum*. Clin. Microbiol. Rev. *9*, 55–71.

Bradshaw, D.J., Homer, K.A., Marsh, P.D., and Beighton, D. (1994) Metabolic cooperation in oral microbial communities during growth on mucin. Microbiology *140 (Pt 12)*, 3407–3412.

Bradshaw, D.J., Marsh, P.D., Allison, C., and Schilling, K.M. (1996) Effect of oxygen, inoculum composition and flow rate on development of mixed-culture oral biofilms. Microbiology *142 (Pt 3)*, 623–629.

Bradshaw, D.J., Marsh, P.D., Watson, G.K., and Allison, C. (1998) Role of *Fusobacterium nucleatum* and coaggregation in anaerobe survival in planktonic and biofilm oral microbial communities during aeration. Infect. Immun. *66*, 4729–4732.

Brooks, W., Demuth, D.R., Gil, S., and Lamont, R.J. (1997) Identification of a *Streptococcus gordonii* SspB domain that mediates adhesion to *Porphyromonas gingivalis*. Infect. Immun. *65*, 3753–3758.

Chen, X., Schauder, S., Potier, N., Van Dorsselaer, A., Pelczer, I., Bassler, B.L., and Hughson, F.M. (2002) Structural identification of a bacterial quorum-sensing signal containing boron. Nature *415*, 545–549.

Chernin, L.S., Winson, M.K., Thompson, J.M., Haran, S., Bycroft, B.W., Chet, I., Williams, P., and Stewart, G.S. (1998) Chitinolytic activity in *Chromobacterium violaceum*: substrate analysis and regulation by quorum sensing. J. Bacteriol. *180*, 4435–4441.

Chung, W.O., Park, Y., Lamont, R.J., McNab, R., Barbieri, B., and Demuth, D.R. (2001) Signaling system in *Porphyromonas gingivalis* based on a LuxS protein. J. Bacteriol. *183*, 3903–3909.

Cisar, J.O., Kolenbrander, P.E., and McIntire, F.C. (1979) Specificity of coaggregation reactions between human oral streptococci and strains of *Actinomyces viscosus* or *Actinomyces naeslundii*. Infect. Immun. *24*, 742–752.

Claverys, J.P., Prudhomme, M., Mortier-Barriere, I., and Martin, B. (2000) Adaptation to the environment: *Streptococcus pneumoniae*, a paradigm for recombination-mediated genetic plasticity? Mol. Microbiol. *35*, 251–259.

Costerton, J.W., Lewandowski, Z., DeBeer, D., Caldwell, D., Korber, D., and James, G. (1994) Biofilms, the customized microniche. J. Bacteriol. *176*, 2137–2142.

Danser, M.M., Gomez, S.M., and Van der Weijden, G.A. (2003) Tongue coating and tongue brushing: a literature review. Int. J. Dent. Hyg. *1*, 151–158.

Davey, M.E., and O'Toole G, A. (2000) Microbial biofilms: from ecology to molecular genetics. Microbiol. Mol. Biol. Rev. *64*, 847–867.

Davies, D.G., Parsek, M.R., Pearson, J.P., Iglewski, B.H., Costerton, J.W., and Greenberg, E.P. (1998) The involvement of cell-to-cell signals in the development of a bacterial biofilm. Science *280*, 295–298.

De Keersmaecker, S.C., Varszegi, C., van Boxel, N., Habel, L.W., Metzger, K., Daniels, R., Marchal, K., De Vos, D., and Vanderleyden, J. (2005) Chemical synthesis of (S)-4,5-dihydroxy-2,3-pentanedione, a bacterial signal molecule precursor, and validation of its activity in *Salmonella typhimurium*. J. Biol. Chem. *280*, 19563–19568.

Demuth, D.R., Davis, C.A., Corner, A.M., Lamont, R.J., Leboy, P.S., and Malamud, D. (1988) Cloning and expression of a *Streptococcus sanguis* surface antigen that interacts with a human salivary agglutinin. Infect. Immun. *56*, 2484–2490.

Diaz, P.I., Chalmers, N.I., Rickard, A.H., Kong, C., Milburn, C.L., Palmer, R.J., Jr., and Kolenbrander, P.E. (2006) Molecular characterization of subject-specific oral microflora during initial colonization of enamel. Appl. Environ. Microbiol. *72*, 2837–2848.

Diaz, P.I., Zilm, P.S., and Rogers, A.H. (2002) *Fusobacterium nucleatum* supports the growth of *Porphyromonas gingivalis* in oxygenated and carbon-dioxide-depleted environments. Microbiology *148*, 467–472.

Doherty, N., Holden, M.T., Qazi, S.N., Williams, P., and Winzer, K. (2006) Functional analysis of luxS in *Staphylococcus aureus* reveals a role in metabolism but not quorum sensing. J. Bacteriol. *188*, 2885–2897.

Du, L.D., and Kolenbrander, P.E. (2000) Identification of saliva-regulated genes of *Streptococcus gordonii* DL1 by differential display using random arbitrarily primed PCR. Infect. Immun. *68*, 4834–4837.

Duerre, J.A., and Miller, C.H. (1966) Cleavage of S-ribosyl-L-homocysteine by extracts from *Escherichia coli*. J. Bacteriol. *91*, 1210–1217.

Eberhard, A., Burlingame, A.L., Eberhard, C., Kenyon, G.L., Nealson, K.H., and Oppenheimer, N.J. (1981) Structural identification of autoinducer of *Photobacterium fischeri* luciferase. Biochemistry *20*, 2444–2449.

Egland, P.G., Du, L.D., and Kolenbrander, P.E. (2001) Identification of independent *Streptococcus gordonii* SspA and SspB functions in coaggregation with *Actinomyces naeslundii*. Infect. Immun. *69*, 7512–7516.

Egland, P.G., Palmer, R.J., Jr., and Kolenbrander, P.E. (2004) Interspecies communication in *Streptococcus gordonii-Veillonella atypica* biofilms: signaling in flow conditions requires juxtaposition. Proc. Natl. Acad. Sci. USA *101*, 16917–16922.

Elliott, D.R., Wilson, M., Buckley, C.M., and Spratt, D.A. (2006) Aggregative behavior of bacteria isolated from canine dental plaque. Appl. Environ. Microbiol. *72*, 5211–5217.

Engebrecht, J., Nealson, K., and Silverman, M. (1983) Bacterial bioluminescence: isolation and genetic analysis of functions from *Vibrio fischeri*. Cell *32*, 773–781.

Engebrecht, J., and Silverman, M. (1984) Identification of genes and gene products necessary for bacterial bioluminescence. Proc. Natl. Acad. Sci. USA *81*, 4154–4158.

Engebrecht, J., and Silverman, M. (1987) Nucleotide sequence of the regulatory locus controlling expression of bacterial genes for bioluminescence. Nucleic Acids Res. *15*, 10455–10467.

Fong, K.P., Chung, W.O., Lamont, R.J., and Demuth, D.R. (2001) Intra- and interspecies regulation of gene expression by *Actinobacillus actinomycetemcomitans* LuxS. Infect. Immun. *69*, 7625–7634.

Frias, J., Olle, E., and Alsina, M. (2001) Periodontal pathogens produce quorum sensing signal molecules. Infect. Immun. *69*, 3431–3434.

Fuqua, C., Parsek, M.R., and Greenberg, E.P. (2001) Regulation of gene expression by cell-to-cell communication: acyl-homoserine lactone quorum sensing. Ann. Rev. Genet. *35*, 439–468.

Fuqua, W.C., and Winans, S.C. (1994) A LuxR-LuxI type regulatory system activates *Agrobacterium* Ti plasmid conjugal transfer in the presence of a plant tumor metabolite. J. Bacteriol. *176*, 2796–2806.

Gambello, M.J., and Iglewski, B.H. (1991) Cloning and characterization of the *Pseudomonas aeruginosa* lasR gene, a transcriptional activator of elastase expression. J. Bacteriol. *173*, 3000–3009.

George, K.S., and Falkler, W.A., Jr. (1992) Coaggregation studies of the *Eubacterium* species. Oral Microbiol. Immunol. *7*, 285–290.

Gilbert, P., Maira-Litran, T., McBain, A.J., Rickard, A.H., and Whyte, F.W. (2002) The physiology and collective recalcitrance of microbial biofilm communities. Adv. Microb. Physiol. *46*, 202–256.

Gonzalez Barrios, A.F., Zuo, R., Hashimoto, Y., Yang, L., Bentley, W.E., and Wood, T.K. (2006) Autoinducer 2 controls biofilm formation in *Escherichia coli* through a novel motility quorum-sensing regulator (MqsR, B3022). J. Bacteriol. *188*, 305–316.

Grossman, A.D. (1995) Genetic networks controlling the initiation of sporulation and the development of genetic competence in *Bacillus subtilis*. Ann. Rev. Genet. *29*, 477–508.

Gutowski-Eckel, Z., Klein, C., Siegers, K., Bohm, K., Hammelmann, M., and Entian, K.D. (1994) Growth phase-dependent regulation and membrane localization of SpaB, a protein involved in biosynthesis of the lantibiotic subtilin. Appl. Environ. Microbiol. *60*, 1–11.

Hall, N.(2007) Advanced sequencing technologies and their wider impact in microbiology. J. Exp. Biol. *209*, 1518–1525

Hartley, D.L., Jones, K.R., Tobian, J.A., LeBlanc, D.J., and Macrina, F.L. (1984) Disseminated tetracycline resistance in oral streptococci: implication of a conjugative transposon. Infect. Immun. *45*, 13–17.

Havarstein, L.S., Hakenbeck, R., and Gaustad, P. (1997) Natural competence in the genus *Streptococcus*: evidence that streptococci can change pherotype by interspecies recombinational exchanges. J. Bacteriol. *179*, 6589–6594.

Hughes, C.V., Andersen, R.N., and Kolenbrander, P.E. (1992) Characterization of *Veillonella atypica* PK1910 adhesin-mediated coaggregation with oral *Streptococcus* spp. Infect. Immun. *60*, 1178–1186.

Ichihara, H., Kuma, K., and Toh, H. (2006) Positive selection in the ComC-ComD system of Streptococcal Species. J. Bacteriol. *188*, 6429–6434.

Ivanov, V., Wang, X.H., Tay, S.T., and Tay, J.H. (2006) Bioaugmentation and enhanced formation of microbial granules used in aerobic wastewater treatment. Appl. Microbio. l Biotechnol. *70*, 374–381.

James, C.E., Hasegawa, Y., Park, Y., Yeung, V., Tribble, G.D., Kuboniwa, M., Demuth, D.R., and Lamont, R.J. (2006a). LuxS involvement in the regulation of genes coding for hemin and iron acquisition systems in *Porphyromonas gingivalis*. Infect. Immun. *74*, 3834–3844.

James, D., Shao, H., Lamont, R.J., and Demuth, D.R. (2006b). The *Actinobacillus actinomycetemcomitans* ribose binding protein RbsB interacts with cognate and heterologous autoinducer 2 signals. Infect. Immun. *74*, 4021–4029.

Jenkinson, H.F., and Demuth, D.R. (1997) Structure, function and immunogenicity of streptococcal antigen I/II polypeptides. Mol. Microbiol. *23*, 183–190.

Ji, G., Beavis, R.C., and Novick, R.P. (1995) Cell density control of staphylococcal virulence mediated by an octapeptide pheromone. Proc. Natl. Acad. Sci. USA *92*, 12055–12059.

Jiang, H.L., Tay, J.H., Maszenan, A.M., and Tay, S.T. (2006) Enhanced phenol biodegradation and aerobic granulation by two coaggregating bacterial strains. Environ. Sci. Technol. *40*, 6137–6142.

Jones, S., Yu, B., Bainton, N.J., Birdsall, M., Bycroft, B.W., Chhabra, S.R., Cox, A.J., Golby, P., Reeves, P.J., Stephens, S., and *et al.* (1993) The lux autoinducer regulates the production of exoenzyme virulence determinants in *Erwinia carotovora* and *Pseudomonas aeruginosa*. Embo J. *12*, 2477–2482.

Jonsson, M., and Swedberg, G. (2006) Macrolide resistance can be transferred by conjugation from viridans streptococci to *Streptococcus pyogenes*. Int. J. Antimicrob. Agents *28*, 101–103.

Kolenbrander, P.E. (1997) Biofilm developmental biology. Trends Microbiol. 5, 475.

Kolenbrander, P.E., Andersen, R.N., Blehert, D.S., Egland, P.G., Foster, J.S., and Palmer, R.J., Jr. (2002) Communication among oral bacteria. Microbiol. Mol. Biol. Rev. *66*, 486–505.

Kolenbrander, P.E., and London, J. (1993) Adhere today, here tomorrow: oral bacterial adherence. J. Bacteriol. *175*, 3247–3252.

Kolenbrander, P.E., Palmer, R.J., Jr., Rickard, A.H., Jakubovics, N.S., Chalmers, N.I., and Diaz, P.I. (2006) Bacterial interactions and successions during plaque development. Periodontol. 2000 *42*, 47–79.

Kolter, R., and Greenberg, E.P. (2006) Microbial sciences: the superficial life of microbes. Nature *441*, 300–302.

Koop, H.M., Valentijn-Benz, M., Nieuw Amerongen, A.V., Roukema, P.A., and De Graaff, J. (1989) Aggregation of 27 oral bacteria by human whole saliva. Influence of culture medium, calcium, and bacterial cell concentration, and interference by autoaggregation. Antonie Van Leeuwenhoek 55, 277–290.

Kreft, J.U. (2004) Biofilms promote altruism. Microbiology *150*, 2751–2760.

Kreft, J.U., and Bonhoeffer, S. (2005) The evolution of groups of cooperating bacteria and the growth rate versus yield trade-off. Microbiology *151*, 637–641.

Lamont, R.J., Gil, S., Demuth, D.R., Malamud, D., and Rosan, B. (1994) Molecules of *Streptococcus gordonii* that bind to *Porphyromonas gingivalis*. Microbiology *140 (Pt 4)*, 867–872.

Lee, W.K., Ogura, K., Loh, J.T., Cover, T.L., and Berg, D.E. (2006) Quantitative effect of *luxS* gene inactivation on the fitness of *Helicobacter pylori*. Appl. Environ. Microbiol. *72*, 6615–6622.

Li, J., Helmerhorst, E.J., Leone, C.W., Troxler, R.F., Yaskell, T., Haffajee, A.D., Socransky, S.S., and Oppenheim, F.G. (2004) Identification of early microbial colonizers in human dental biofilm. J. Appl. Microbiol. *97*, 1311–1318.

Li, Y.H., Lau, P.C., Lee, J.H., Ellen, R.P., and Cvitkovitch, D.G. (2001) Natural genetic transformation of *Streptococcus mutans* growing in biofilms. J. Bacteriol. *183*, 897–908.

Li, Y.H., Tang, N., Aspiras, M.B., Lau, P.C., Lee, J.H., Ellen, R.P., and Cvitkovitch, D.G. (2002) A quorum-sensing signaling system essential for genetic competence in *Streptococcus mutans* is involved in biofilm formation. J. Bacteriol. *184*, 2699–2708.

Loo, C.Y., Corliss, D.A., and Ganeshkumar, N. (2000) *Streptococcus gordonii* biofilm formation: identification of genes that code for biofilm phenotypes. J. Bacteriol. *182*, 1374–1382.

Lovegrove, J.M. (2004) Dental plaque revisited: bacteria associated with periodontal disease. J. NZ Soc. Periodontol. *87*, 7–21.

Lyautey, E., Jackson, C.R., Cayrou, J., Rols, J.L., and Garabetian, F. (2005) Bacterial community succession in natural river biofilm assemblages. Microb. Ecol. *50*, 589–601.

Ma, J.K., Hunjan, M., Smith, R., Kelly, C., and Lehner, T. (1990) An investigation into the mechanism of protection by local passive immunization with monoclonal antibodies against *Streptococcus mutans*. Infect. Immun. *58*, 3407–3414.

Maeda, K., Nagata, H., Yamamoto, Y., Tanaka, M., Tanaka, J., Minamino, N., and Shizukuishi, S. (2004) Glyceraldehyde-3-phosphate dehydrogenase of *Streptococcus oralis* functions as a coadhesin for *Porphyromonas gingivalis* major fimbriae. Infect. Immun. *72*, 1341–1348.

Marsh, P.D. (2005) Dental plaque: biological significance of a biofilm and community life-style. J. Clin. Periodontol. *32 Suppl 6*, 7–15.

Marsh, P.D., and Bradshaw, D.J. (1995) Dental plaque as a biofilm. J. Ind. Microbiol. *15*, 169–175.

Martin, B., Quentin, Y., Fichant, G., and Claverys, J.P. (2006) Independent evolution of competence regulatory cascades in streptococci? Trends Microbiol. *14*, 339–345.

McKenney, D., Brown, K.E., and Allison, D.G. (1995) Influence of *Pseudomonas aeruginosa* exoproducts on virulence factor production in *Burkholderia cepacia*: evidence of interspecies communication. J. Bacteriol. *177*, 6989–6992.

McNab, R., Ford, S.K., El-Sabaeny, A., Barbieri, B., Cook, G.S., and Lamont, R.J. (2003) LuxS-based signaling in *Streptococcus gordonii*: autoinducer 2 controls carbohydrate metabolism and biofilm formation with *Porphyromonas gingivalis*. J. Bacteriol. *185*, 274–284.

Merritt, J., Qi, F., Goodman, S.D., Anderson, M.H., and Shi, W. (2003) Mutation of luxS affects biofilm formation in *Streptococcus mutans*. Infect. Immun. *71*, 1972–1979.

Miller, M.B., and Bassler, B.L. (2001) Quorum sensing in bacteria. Ann. Rev. Microbiol. *55*, 165–199.

Miller, M.B., Skorupski, K., Lenz, D.H., Taylor, R.K., and Bassler, B.L. (2002) Parallel quorum sensing systems converge to regulate virulence in *Vibrio cholerae*. Cell *110*, 303–314.

Miller, S.T., Xavier, K.B., Campagna, S.R., Taga, M.E., Semmelhack, M.F., Bassler, B.L., and Hughson, F.M. (2004) *Salmonella typhimurium* recognizes a chemically distinct form of the bacterial quorum-sensing signal AI-2. Mol. Cell *15*, 677–687.

Mok, K.C., Wingreen, N.S., and Bassler, B.L. (2003) *Vibrio harveyi* quorum sensing: a coincidence detector for two autoinducers controls gene expression. Embo J. *22*, 870–881.

Molin, S., and Tolker-Nielsen, T. (2003) Gene transfer occurs with enhanced efficiency in biofilms and induces enhanced stabilisation of the biofilm structure. Curr. Opin. Biotechnol. *14*, 255–261.

Morrison, D.A., and Lee, M.S. (2000) Regulation of competence for genetic transformation in *Streptococcus pneumoniae*: a link between quorum sensing and DNA processing genes. Res. Microbiol. *151*, 445–451.

Muller, H.P., Heinecke, A., Borneff, M., Knopf, A., Kiencke, C., and Pohl, S. (1997) Microbial ecology of *Actinobacillus actinomycetemcomitans*, *Eikenella corrodens* and *Capnocytophaga* spp. in adult periodontitis. J. Periodont. Res. *32*, 530–542.

Narimatsu, M., Noiri, Y., Itoh, S., Noguchi, N., Kawahara, T., and Ebisu, S. (2004) Essential role for the gtfA gene encoding a putative glycosyltransferase in the adherence of *Porphyromonas gingivalis*. Infect. Immun. *72*, 2698–2702.

Nealson, K.H., and Hastings, J.W. (1979) Bacterial bioluminescence: its control and ecological significance. Microbiol. Rev. *43*, 496–518.

Nealson, K.H., Platt, T., and Hastings, J.W. (1970) Cellular control of the synthesis and activity of the bacterial luminescent system. J. Bacteriol. *104*, 313–322.

Nyvad, B., and Kilian, M. (1987) Microbiology of the early colonization of human enamel and root surfaces *in vivo*. Scand. J. Dent. Res. *95*, 369–380.

Palmer, R.J., Jr., Gordon, S.M., Cisar, J.O., and Kolenbrander, P.E. (2003) Coaggregation-mediated interactions of streptococci and actinomyces detected in initial human dental plaque. J. Bacteriol. *185*, 3400–3409.

Palmer, R.J., Jr., Kazmerzak, K., Hansen, M.C., and Kolenbrander, P.E. (2001) Mutualism versus independence: strategies of mixed-species oral biofilms *in vitro* using saliva as the sole nutrient source. Infect. Immun. *69*, 5794–5804.

Papapanou, P.N., Sellen, A., Wennstrom, J.L., and Dahlen, G. (1993) An analysis of the subgingival microflora in randomly selected subjects. Oral Microbiol. Immunol. *8*, 24–29.

Paster, B.J., Boches, S.K., Galvin, J.L., Ericson, R.E., Lau, C.N., Levanos, V.A., Sahasrabudhe, A., and Dewhirst, F.E. (2001) Bacterial diversity in human subgingival plaque. J. Bacteriol. *183*, 3770–3783.

Paster, B.J., Olsen, I., Aas, J.A., and Dewhirst, F.E. (2006) The breadth of bacterial diversity in the human periodontal pocket and other oral sites. Periodontol. 2000 *42*, 80–87.

Pearson, J.P., Gray, K.M., Passador, L., Tucker, K.D., Eberhard, A., Iglewski, B.H., and Greenberg, E.P. (1994) Structure of the autoinducer required for expression of *Pseudomonas aeruginosa* virulence genes. Proc. Natl. Acad. Sci. USA *91*, 197–201.

Petersen, F.C., Ahmed, N.A., Naemi, A., and Scheie, A.A. (2006) LuxS-mediated signalling in *Streptococcus anginosus* and its role in biofilm formation. Antonie van Leeuwenhoek *90*, 109–121.

Petersen, F.C., Pecharki, D., and Scheie, A.A. (2004) Biofilm mode of growth of *Streptococcus intermedius* favored by a competence-stimulating signaling peptide. J. Bacteriol. *186*, 6327–6331.

Petersen, F.C., Tao, L., and Scheie, A.A. (2005) DNA binding-uptake system: a link between cell-to-cell communication and biofilm formation. J. Bacteriol. *187*, 4392–4400.

Piper, K.R., Beck von Bodman, S., and Farrand, S.K. (1993) Conjugation factor of *Agrobacterium tumefaciens* regulates Ti plasmid transfer by autoinduction. Nature *362*, 448–450.

Ramberg, P., Sekino, S., Uzel, N.G., Socransky, S., and Lindhe, J. (2003) Bacterial colonization during *de novo* plaque formation. J. Clin. Periodontol. *30*, 990–995.

Reading, N.C., and Sperandio, V. (2006) Quorum sensing: the many languages of bacteria. FEMS Microbiol. Lett. *254*, 1–11.

Resch, A., Rosenstein, R., Nerz, C., and Gotz, F. (2005) Differential gene expression profiling of *Staphylococcus aureus* cultivated under biofilm and planktonic conditions. Appl. Environ. Microbiol. *71*, 2663–2676.

Rickard, A.H., Gilbert, P., High, N.J., Kolenbrander, P.E., and Handley, P.S. (2003a). Bacterial coaggregation: an integral process in the development of multi-species biofilms. Trends Microbiol. *11*, 94–100.

Rickard, A.H., McBain, A.J., Ledder, R.G., Handley, P.S., and Gilbert, P. (2003b). Coaggregation between freshwater bacteria within biofilm and planktonic communities. FEMS Microbiol. Lett. *220*, 133–140.

Rickard, A.H., McBain, A.J., Stead, A.T., and Gilbert, P. (2004) Shear rate moderates community diversity in freshwater biofilms. Appl. Environ. Microbiol. *70*, 7426–7435.

Rickard, A.H., Palmer, R.J., Jr., Blehert, D.S., Campagna, S.R., Semmelhack, M.F., Egland, P.G., Bassler, B.L., and Kolenbrander, P.E. (2006) Autoinducer 2: a concentration-dependent signal for mutualistic bacterial biofilm growth. Mol. Microbiol. *60*, 1446–1456.

Roberts, A.P., and Mullany, P. (2006) Genetic basis of horizontal gene transfer among oral bacteria. Periodontol. 2000 *42*, 36–46.

Roberts, A.P., Pratten, J., Wilson, M., and Mullany, P. (1999) Transfer of a conjugative transposon, Tn5397 in a model oral biofilm. FEMS Microbiol. Lett. *177*, 63–66.

Saarela, M., von Troil-Linden, B., Torkko, H., Stucki, A.M., Alaluusua, S., Jousimies-Somer, H., and Asikainen, S. (1993) Transmission of oral bacterial species between spouses. Oral Microbiol. Immunol. *8*, 349–354.

Salyers, A.A., Shoemaker, N.B., Stevens, A.M., and Li, L.Y. (1995) Conjugative transposons: an unusual and diverse set of integrated gene transfer elements. Microbiol. Rev. *59*, 579–590.

Sbordone, L., and Bortolaia, C. (2003) Oral microbial biofilms and plaque-related diseases: microbial communities and their role in the shift from oral health to disease. Clin. Oral Investig. *7*, 181–188.

Scannapieco, F.A. (1994) Saliva-bacterium interactions in oral microbial ecology. Crit. Rev. Oral Biol. Med. *5*, 203–248.

Scannapieco, F.A., Torres, G.I., and Levine, M.J. (1995) Salivary amylase promotes adhesion of oral streptococci to hydroxyapatite. J. Dent. Res. *74*, 1360–1366.

Schembri, M.A., Kjaergaard, K., and Klemm, P. (2003) Global gene expression in *Escherichia coli* biofilms. Mol. Microbiol. *48*, 253–267.

Semmelhack, M.F., Campagna, S.R., Federle, M.J., and Bassler, B.L. (2005) An expeditious synthesis of DPD and boron binding studies. Org. Lett. *7*, 569–572.

Shen, S., Samaranayake, L.P., and Yip, H.K. (2005) Coaggregation profiles of the microflora from root surface caries lesions. Arch. Oral Biol. *50*, 23–32.

Socransky, S.S., and Haffajee, A.D. (1997) The nature of periodontal diseases. Ann. Periodontol. *2*, 3–10.

Socransky, S.S., and Haffajee, A.D. (2005) Periodontal microbial ecology. Periodontol. 2000 *38*, 135–187.

Socransky, S.S., Haffajee, A.D., Cugini, M.A., Smith, C., and Kent, R.L., Jr. (1998) Microbial complexes in subgingival plaque. J. Clin. Periodontol. *25*, 134–144.

Sorensen, S.J., Bailey, M., Hansen, L.H., Kroer, N., and Wuertz, S. (2005) Studying plasmid horizontal transfer in situ: a critical review. Nat. Rev. Microbiol. *3*, 700–710.

Soukka, T., Tenovuo, J., and Rundegren, J. (1993) Agglutination of Streptococcus mutans serotype C cells but inhibition of *Porphyromonas gingivalis* autoaggregation by human lactoferrin. Arch. Oral Biol. *38*, 227–232.

Stoodley, P., Sauer, K., Davies, D.G., and Costerton, J.W. (2002) Biofilms as complex differentiated communities. Ann. Rev. Microbiol. *56*, 187–209.

Stoodley, P., Wilson, S., Hall-Stoodley, L., Boyle, J.D., Lappin-Scott, H.M., and Costerton, J.W. (2001) Growth and detachment of cell clusters from mature mixed-species biofilms. Appl. Environ. Microbiol. *67*, 5608–5613.

Suntharalingam, P., and Cvitkovitch, D.G. (2005) Quorum sensing in streptococcal biofilm formation. Trends Microbiol. *13*, 3–6.

Sutherland, I.W. (2001) Exopolysaccharides in biofilms, flocs and related structures. Water Sci. Technol. *43*, 77–86.

Van der Velden, U., Van Winkelhoff, A.J., Abbas, F., and De Graaff, J. (1986) The habitat of periodontopathic micro-organisms. J. Clin. Periodontol. *13*, 243–248.

van Loosdrecht, M.C., Norde, W., and Zehnder, A.J. (1990) Physical chemical description of bacterial adhesion. J. Biomater. Appl. *5*, 91–106.

Vesey, P.M., and Kuramitsu, H.K. (2004) Genetic analysis of *Treponema denticola* ATCC 35405 biofilm formation. Microbiology *150*, 2401–2407.

Waite, R.D., Paccanaro, A., Papakonstantinopoulou, A., Hurst, J.M., Saqi, M., Littler, E., and Curtis, M.A. (2006) Clustering of *Pseudomonas aeruginosa* transcriptomes from planktonic cultures, developing and mature biofilms reveals distinct expression profiles. BMC Genomics *7*, 162.

Waite, R.D., Papakonstantinopoulou, A., Littler, E., and Curtis, M.A. (2005) Transcriptome analysis of *Pseudomonas aeruginosa* growth: comparison of gene expression in planktonic cultures and developing and mature biofilms. J. Bacteriol. *187*, 6571–6576.

Walters, M., Sircili, M.P., and Sperandio, V. (2006) AI-3 synthesis is not dependent on *luxS* in *Escherichia coli*. J. Bacteriol. *188*, 5668–5681.

Wen, Z.T., and Burne, R.A. (2004) LuxS-mediated signaling in *Streptococcus mutans* is involved in regulation of acid and oxidative stress tolerance and biofilm formation. J. Bacteriol. *186*, 2682–2691.

Whittaker, C.J., Clemans, D.L., and Kolenbrander, P.E. (1996) Insertional inactivation of an intrageneric coaggregation-relevant adhesin locus from *Streptococcus gordonii* DL1 (Challis). Infect. Immun. *64*, 4137–4142.

Wimpenny, J. W.T. (2000) An overview of biofilms as funtional communities. In Community Structure and Co-operation in Biofilms. D. G. Allison, ed. Symp. Soc. Gen. Microbiol. Vol. 59 (Cambridge UK: Cambridge University Press) pp1–24.

Wimpenny, J. W. T. and Colosanti, R.(1997) A unifying hypothesis for the structure of microbial biofilms based on cellular automaton models. FEMS Microbiol. Ecol. *22*, 1–16.

Winzer, K., Hardie, K.R., Burgess, N., Doherty, N., Kirke, D., Holden, M.T., Linforth, R., Cornell, K.A., Taylor, A.J., Hill, P.J., and Williams, P. (2002) LuxS: its role in central metabolism and the *in vitro* synthesis of 4-hydroxy-5-methyl-3(2H)-furanone. Microbiology *148*, 909–922.

Winzer, K., Hardie, K.R., and Williams, P. (2003) LuxS and autoinducer-2: their contribution to quorum sensing and metabolism in bacteria. Adv. Appl. Microbiol. *53*, 291–396.

Wuertz, S., Okabe, S., and Hausner, M. (2004) Microbial communities and their interactions in biofilm systems: an overview. Water Sci. Technol. *49*, 327–336.

Xavier, K.B., and Bassler, B.L. (2005a). Interference with AI-2-mediated bacterial cell–cell communication. Nature *437*, 750–753.

Xavier, K.B., and Bassler, B.L. (2005b). Regulation of uptake and processing of the quorum-sensing autoinducer AI-2 in *Escherichia coli*. J. Bacteriol. *187*, 238–248.

Yoshida, A., Ansai, T., Takehara, T., and Kuramitsu, H.K. (2005) LuxS-based signaling affects *Streptococcus mutans* biofilm formation. Appl. Environ. Microbiol. *71*, 2372–2380.

Yuan, L., Hillman, J.D., and Progulske-Fox, A. (2005) Microarray analysis of quorum-sensing-regulated genes in *Porphyromonas gingivalis*. Infect. Immun. *73*, 4146–4154.

Zhang, L., Murphy, P.J., Kerr, A., and Tate, M.E. (1993) *Agrobacterium* conjugation and gene regulation by N-acyl-L-homoserine lactones. Nature *362*, 446–448.

Zhang, Y., Lei, Y., Nobbs, A., Khammanivong, A., and Herzberg, M.C. (2005) Inactivation of *Streptococcus gordonii* SspAB alters expression of multiple adhesin genes. Infect. Immun. *73*, 3351–3357.

Zhu, J., Miller, M.B., Vance, R.E., Dziejman, M., Bassler, B.L., and Mekalanos, J.J. (2002) Quorum-sensing regulators control virulence gene expression in *Vibrio cholerae*. Proc. Natl. Acad. Sci. USA *99*, 3129–3134.

The Molecular Biology of Cariogenic Bacteria

5

Roy R.B. Russell

Abstract

Molecular biology techniques have made a major contribution to our understanding of dental plaque bacteria, helping to clarify the taxonomic relationships of the various species of oral bacteria and their association with dental caries. Methods developed for gene cloning and targeted gene knockout have allowed analysis of the function of individual genes and their contribution to aspects of cellular adhesion and physiology that may be important for cariogenicity. The availability of the *Streptococcus mutans* genome sequence has given access to new tools for discovering novel genes and exploring their regulation, as well means by which diversity within the species can be explored.

Introduction

It is always useful to consider the terms used in a title, and the extent to which they may determine the content of a chapter. Molecular biology can readily be defined as concerning itself with understanding the interactions between the various systems of a cell, including the interrelationship of DNA, RNA and protein synthesis and learning how these interactions are regulated. This brings us to the much more difficult task of defining what we mean by cariogenic bacteria. Caries lesions are formed as the end result of a complex process to which numerous different types of bacteria make a contribution. Certain species of bacteria are capable of inducing caries in gnotobiotic animals, when they are the only species present, and are thus demonstrably cariogenic under certain conditions. However, we need to remain conscious of the fact that the properties that make them cariogenic in the model system do not necessarily mean that they will be important in caries in humans. In human disease, the species that are most strongly associated with the development of caries are minor, late-arriving members of the dental plaque microflora; and in this sense, numerous other bacteria, unable to cause diseases on their own, play an essential part in creating conditions in which the pathogens can flourish and so contribute to the genesis of caries. From this it follows that a full understanding of caries demands that we focus our efforts on not just a limited number of pathogens but on all the players in the process. Thus, the task facing a dental researcher is a much greater one than that facing researchers in other areas of medical microbiology, where the classical model of each disease having its own specific aetiological agent holds sway (not least in persuading funding agencies of the necessity to study species not overtly pathogenic in themselves). It may, therefore, be useful to briefly survey our knowledge of caries microbiology and the logic behind selection of certain cariogenic bacteria to which molecular biology can be applied; at the same time considering the contribution of molecular biology techniques. A number of authors have presented more extensive arguments for the consideration of the caries process in terms of the complex interplay of the plaque microflora, reducing the focus on individual species (Beighton, 2005; Kleinberg, 2002; Marsh, 2003). Previous reviews have considered the impact of molecular biology on the field (Burne, 1998; Kuramitsu, 2003; Kuramitsu, 2006; Russell, 1994).

Caries-associated bacteria

The latter half of the nineteenth century saw rapid advances in the quest to identify the causes of infectious diseases. The fundamentals of our understanding of the pathogenesis of caries dates back to researchers such as Pasteur and Koch, who were developing the 'germ theory' of disease which was based on the search for specific aetiological agents. There were rapid and dramatic successes in identifying the bacteria responsible for diseases including anthrax, dysentery, cholera, plague and tuberculosis. The first report of the search for the causative agents of caries came in the same era, at the International Medical Congress in London in 1881, attended by leading figures including Pasteur, Koch and Lister. At the Congress, two London dentists, Underwood and Milles, presented a paper describing the microscopic observation of 'germs' in decaying teeth and experiments in which they looked for damage to the enamel of extracted teeth incubated in test tubes. They found that for enamel dissolution to occur in the teeth, not only live germs had to be present but also a source of carbohydrate; they used chewed bread. This concept of essential interaction between bacteria, carbohydrate and teeth was further developed by W.D. Miller, an American working in the same institute as Koch in Vienna. Miller's 'chemicoparasitic theory' of caries was published in 1890. Miller's book established two of the main themes that have dominated research in caries microbiology – what bacteria are involved and how do they use carbohydrate? A third major theme, that did not emerge until about 70 years later with the work of Ron Gibbons (Kuramitsu and Ellen, 2000), concerns the issue of how the bacteria attach to and colonize the tooth. As we shall see, molecular biology approaches have had a major influence in development of our understanding in these three areas.

Microbiology of caries

Mutans streptococci

Streptococci were noted in established caries lesions back in 1890 by Miller but not regarded as of any great significance. It was not until 1924 when Clarke, working at St. Mary's Hospital in London, examined the bacteria present in initial 'white-spot' enamel lesions. He identified an organism, that he named *Streptococcus mutans*, present in a high proportion of such early lesions and proposed it as the most important organism in the initiation of disease. Clarke's work went largely unnoticed until the late 1950s when new approaches to animal models of caries were developed, based on the selective use of antibiotics to modify the oral microflora and the introduction of isolation chambers where gnotobiotic animals could be raised and infected with specific bacteria. The great conceptual advances made during this period have been reviewed (Tanzer, 1995). The first organism implicated was actually an enterococcus but, subsequently, various streptococci could be ranked according to their ability to induce caries in rats or hamsters fed a sugar-rich diet and it came to be realized that the most cariogenic corresponded to Clarke's description of *S. mutans*. The earlier literature, up until the mid 1980s, is often difficult to interpret because what was at first known as *S. mutans* was gradually realized to consist of a number of distinct species. Biochemical tests and serotyping played an important part in resolving the species but it was only with the introduction of molecular methods, based on bacterial genotype, that the full diversity of streptococcal species came to be recognized; first with DNA–DNA hybridisation experiments (Coykendall, 1989) and later the sequencing of ribosomal RNA genes (Bentley *et al.*, 1991; Kawamura *et al.*, 1995). The organisms are now collectively referred to as the 'mutans group' or Mutans Streptococci (MS). It also became clear that some types that had been the object of study in animals (*S. cricetus* was first found in hamsters and *S. rattus* in rats) were rarely, if ever, found in humans. Other mammalian species also seem to harbour distinctive mutans streptococci. For example, *S. downei* and *S. macacae* were isolated from macaque monkeys, though there is one recent report of *S. downei* from a human (Yoo *et al.*, 2005a). *S. mutans* and *S. sobrinus* are the only two species consistently found in humans. The other groups of oral streptococci have also proved rather difficult to resolve taxonomically; a problem only partially resolved by ribosomal RNA gene sequencing. Phylogenetic analysis, based on the concatenated sequence of four housekeeping genes, has recently been used to clarify the relationship of the known species (Hoshino *et al.*, 2005) and Table 5.1 shows the current taxonomic status of the different groups of human oral streptococci.

The development of selective media for enumerating mutans streptococci led to the finding that the presence of high numbers of these organisms at a tooth site was associated with a high risk of subsequent caries development (Loesche, 1986). A large number of cross-sectional and longitudinal studies have confirmed a strong association between mutans streptococci and caries and a recent systematic review found 2730 papers on the topic (Tanzer *et al.*, 2001b). A limitation of most such studies is that the selective media used for counting the streptococci do not allow accurate identification to the species level; for example, *S. mutans* and *S. sobrinus* cannot reliably be distinguished just by colony morphology. *S. mutans* is by far the more common and is car-

Table 5.1 Streptococci commonly found in the human mouth

Group	Species	Occurrence
Mitis group	*S. australis* *S. gordonii** *S. infantis* *S. mitis** *S. oralis** *S. oligofermentans* *S. parasanguinis* *S. peroris* *S. sanguinis** *S. sinensis*	Pioneer species in plaque colonization. Some aciduric strains of species marked with an asterisk have been found in caries lesions. Most frequent cause of endocarditis
Mutans group	*S. mutans* *S. sobrinus*	Found mainly in established plaque, associated with dental caries
Salivarius group	*S. salivarius* *S. vestibularis*	Colonize principally mucosal surfaces
Anginosus group	*S. anginosus* *S. constellatus* *S. intermedius**	Found in gingival crevice, associated with abscesses. Some aciduric strains* have been found in caries lesions

ried by over 98% of adults and so its association with caries is clear. However, *S. sobrinus* has been reported to be found in only 5–40% of individuals in different countries. A number of studies have found that children with high caries activity are more likely to carry both *S. mutans* and *S. sobrinus*, though a role for *S. sobrinus* in the initiation stage of caries has not been established (de Soet *et al.*, 1993; Okada *et al.*, 2005; Rupf *et al.*, 2006; Tanzer *et al.*, 2001b). In terms of the range of characters generally considered to contribute to virulence of mutans streptococci, *S. sobrinus* strains are generally more cariogenic in rats, produce more acid, are more acid tolerant and show stronger sucrose-dependent adherence than strains of *S. mutans*. *S. sobrinus* strains, such as 6715, have thus proved a very attractive model organism for study of these properties, particularly in unravelling the complexities of polymer formation from sucrose and in work on experimental caries vaccines in rats. In interpreting the results from these studies, one must remain wary of assuming that all conclusions about virulence of a model organism in an artificial animal model can be directly extrapolated to *S. mutans*, though they can certainly point in the direction of aspects demanding exploration in the major human cariogenic bacterium.

So far as molecular biology goes, we as yet know very little about *S. sobrinus*. A number of its genes have been cloned and expressed in other bacteria, but it is not amenable to genetic manipulation since it is not naturally transformable (Westergren and Emilson, 1983); and no other method of introducing DNA has been described, apart from a single report using a conjugative plasmid (Buckley *et al.*, 1995). Knockout *gtf* mutants have however been constructed, by electroporation, in the closely related *S. downei* (Colby *et al.*, 1999). Genome sequencing of *S. sobrinus* is in progress, and should provide valuable insights into its properties.

Non-mutans streptococci

When he first described *S. mutans*, Clarke observed that caries sometimes develops in the absence of any detectable *S. mutans* and these early observations have been confirmed by others (Aamdal-Scheie *et al.*, 1996). This finding suggests that other species, or combinations of species, could produce similar amounts of acid to *S. mutans* and hence cause disease. The major organisms involved are now known to be the 'low-pH non-mutans streptococci' (van Houte *et al.*, 1996) and these have been identified as atypical strains of common plaque species such as *S. mitis*, *S. sanguinis* and *S. intermedius* (Beighton, 2005; de Soet *et al.*, 2000; Svensater *et al.*, 2003; van Ruyven *et al.*, 2000). Little is yet known about the genetic changes that may have led to the changed properties and nor is it yet clear whether they arise as a consequence of mutational change, recombination or acquisition of novel properties by horizontal gene transfer followed by selection of a sub-population (Alam *et al.*, 2000). While certain strains of non-mutans streptococci may contribute to caries, there is also extensive evidence (see below) that most members of these species are non-cariogenic and may, indeed, be beneficial. Such benefit may come from their ability to out-compete *S. mutans* or to have an inhibitory or lethal effect due to production of hydrogen peroxide (Kreth *et al.*, 2005; Tong *et al.*, 2006).

Other caries-associated organisms

A number of different molecular biology approaches have made a striking contribution to our understanding of the complexity of the microflora of dental caries, providing advantages over cultural approaches in the speed and precision of identification of the species present. Cultural methods, apart from being labour intensive, suffer from the disadvantage that not all plaque species are easily recovered and phenotypic tests are frequently inadequate for identification down to the species level. Furthermore, selective media, by definition, allow study of only a chosen sub-set of the population present and are rarely perfect in their selectivity; for example, one recent study, using 16s rDNA cloning and sequencing techniques for identification, found that of 21 colonies with the morphology typical of mutans streptococci on the commonly used Mitis Salivarius agar with Bacitracin were not, in fact, mutans streptococci but identified as *S. anginosus*, *S. sanguinis* and *Pantoea agglomerans* (Yoo *et al.*, 2005b). Molecular methods are based upon analysis of DNA extracted from a sample of plaque taken at the sampling site. While the results are not affected by the ease with which particular species can be grown in the laboratory or by the selectivity of the growth media, interpretation of quantitative data can be complicated by concerns about the relative ease with which DNA is extracted from different species.

An idea of the number of species present in a sample can be obtained by denaturing gradient gel electrophoresis (DGGE) and it has been estimated that about 100 different phylotypes of bacteria are present in supragingival plaque, though other experimental approaches have yielded somewhat lower numbers (Li *et al.*, 2007). The nature of the species present has been explored by several research groups, who amplified 16s RNA gene sequences from plaque DNA by PCR, then cloned and sequenced the products. Studies of plaque from healthy sites (Aas *et al.*, 2005; Diaz *et al.*, 2006) and active caries lesions (Becker *et al.*, 2002; Chhour *et al.*, 2005; Diaz *et al.*, 2006; Munson *et al.*, 2004) revealed the great variety of species present, including many that are uncultivable. The results can show the frequency with which particular organisms are found at different sites and, while confirming the strong association of *S. mutans* and lactobacilli with caries, also identify a range of others that are caries-associated. These include other streptococci and species of *Veillonella*, *Actinomyces*, *Bifidobacterium*, *Rothia* and *Propionibacterium*.

The PCR approach described above provides little information on the relative proportions of different bacteria in plaque but does identify those for which quantitative data are likely to be most fruitful. So far, the most informative methods have been variations on the checkerboard hybridisation technique, developed at the Forsyth Institute in Boston, which allows detection and rough quantitation of a panel of species in a single experiment. In one report examining 23 species, the strong relationship to caries for *S. mutans* was observed; *S. sobrinus* was not implicated though several other species were. A later study with 82 species in the hybridisation again found *S. mutans* and lactobacilli predominant, along with an *Actinomyces* sp. In contrast, *Abiotrophia defectiva*, *S. parasanguinis*, *S. mitis*, *S. oralis* and *S. sanguinis* exhibited an inverse relationship and could be regarded as beneficial species. The current list of species and their proposed association with caries is shown in Table 5.2.

Lactobacillus spp. in dental caries

Following the lead of Miller in identifying acid production as a key virulence determinant, early caries researchers regarded lactic acid-producing bacteria as the causative agents of caries; and for many years *Bacillus acidophilus odontolyticus*, a species name that no longer has any formal recognition, was the favoured culprit. The major groups of lactic acid bacteria found as commensals in humans belong to the genera *Lactobacillus* and *Streptococcus*. Lactobacilli were used in the first experimental attempts at inducing caries in rats and certain species have been repeatedly shown to be cariogenic in a rat model (Matsumoto *et al.*, 2005). Lactobacilli can also consistently be isolated from established caries lesions in humans but do not show an association with initiation of the disease. There have been numerous attempts over the years to identify which species of lactobacilli are found in human caries and until recently the

Table 5.2 Association of oral bacteria with dental caries

Initiation	Later invaders?	Health-associated
S. mutans	*Actinomyces* spp. *Bifidobacterium* spp. *Lactobacillus* spp. *S. sobrinus* *Veillonella* spp. Certain strains of *S. anginosus* *S. mitis* *S. oralis* *S. sanguinis*	*Abiotrophia defectiva* *Rothia dentocariosa* *Propionibacterium* spp. *S. gordonii* *S. mitis* *S. oligofermentans* *S. oralis* *S. parasanguinis* *S. sanguinis*

literature was extensive and thoroughly confusing. It now seems that the inconsistencies in reports were largely due to the inadequacies of biochemical schemes for species identification, exacerbated by geographical differences in the studies and hence variation in the exposure of the subjects to dietary lactobacilli. Our understanding of the taxonomy of lactobacilli was dramatically altered by the introduction of molecular biology techniques, particularly the construction of phylogenetic trees based on differences in sequence of 16s ribosomal RNA genes. The conclusions that can be drawn from a number of recent reports are firstly, that a wide range of *Lactobacillus* species can be isolated from carious lesions and second, that these are all found elsewhere in the body and/or in fermented foods (Byun *et al.*, 2004; Caufield *et al.*, 2007; Chhour *et al.*, 2005; Dal Bello and Hertel, 2006; Munson *et al.*, 2004). There is no evidence for a uniquely oral species of *Lactobacillus*; that is, no member of the genus has evolved to exploit the oral cavity as a habitat, in the way that some streptococci have done. Thus, lactobacilli are not significant members of the normal plaque microflora in the absence of caries but are opportunistic invaders taking advantage of the low-pH conditions established by other bacteria in the depth of caries lesions (van Houte, 1994). They may, therefore, contribute to further acid production in advancing caries lesions but, on their own, cannot be regarded as significantly cariogenic. The particular species found within the lesions will be largely dependent upon the environmental exposure (mainly diet) of the individual. Because lactobacilli are so prevalent in fermented foods, we are constantly exposed to infection by species including those present in probiotic foods such as yoghourt marketed as beneficial for intestinal health (Matsumoto *et al.*, 2005; Montalto *et al.*, 2004). Apart from the improvements in taxonomy and identification, molecular biology techniques have not been used to explore potential cariogenicity of *Lactobacillus* spp., though there are numerous examples of the fruitful two-way exchange of information derived from studies of cariogenic bacteria and other lactic acid bacteria. The physiological and evolutionary similarities of the genera *Streptococcus, Lactobacillus, Lactococcus, Leuconostoc* and *Pediococcus* means that considerable insights into the biology of oral organisms can be gained by comparative studies, including recent advances in genomics (Klaenhammer *et al.*, 2005; Makarova *et al.*, 2006).

Bifidobacteria

Up until the 1960s, *Bifidobacterium* species were collectively referred to as *Lactobacillus bifidus*. These anaerobic bacteria are high G+C bacteria taxonomically distant from the other lactic acid bacteria but can produce lactate and have very similar sugar metabolism (Parche *et al.*, 2007). Several recent studies, using molecular methods for bacterial identification, have found significant numbers of *Bifidobacterium* in caries lesions (Becker *et al.*, 2002; Modesto *et al.*, 2006); currently, however, virtually nothing is known about their origin or how they may contribute to the disease process. Bifidobacteria are normal inhabitants of the gut in man, have been associated with health effects and are often used as probiotics; that is, functional ingredients in food and pharmaceutical products. To match the observations on caries-association, there have also been reports of beneficial effects of bifidobacteria (Caglar *et al.*, 2005a; Caglar *et al.*, 2005b; Meurman, 2005) The current activity of researchers in the probiotic field is likely to generate, in the near future, detailed insight into the properties of these bacteria.

Other plaque bacteria

Current models of the formation of the dental plaque biofilm envisage a succession of stages, each of them involving a different range of bacterial species but all stages being characterized by a set of specific molecular interactions. In the initial stage, 'pioneer' species of bacteria colonize the tooth surface by binding to components of the salivary pellicle. Predominant among these initial colonizers are believed to be *S. sanguis, S. gordonii, S. oralis, S. mitis* and several species of *Actinomyces*. There is experimental evidence that at least certain strains of these organisms can be cariogenic in rats (Tanzer *et al.*, 2001a). While very little work has been done on this aspect, there is a wealth of information on other properties of central importance in plaque formation and metabolism; particularly mechanisms of adhesion and acid tolerance. A number of reviews provide a broad coverage of this topic (Beighton, 2005; Jenkinson and Vickerman, 2006; Yeung, 1999).

Properties affecting cariogenicity

The properties required for the growth of commensal bacteria found in plaque are precisely the same set of properties that cariogenic bacteria need if they too are to flourish as members of the complex plaque community. Indeed, the cariogenic organisms could be regarded as rather unsuccessful competitors under normal conditions since they comprise only a small proportion of the population and only increase under the stressed conditions occurring when acid is generated from dietary carbohydrate. Our understanding of properties common to all plaque bacteria thus helps to guide us in identifying the features which are important to cariogenic bacteria and investigation of which may help us gain insights into the disease process. These are:

- Sucrose-independent adhesive interactions with host and other plaque bacteria
- Metabolic interactions with other bacteria, both synergistic and antagonistic
- Sucrose-dependent adhesion
- Acid tolerance
- Carbohydrate metabolism and acid production

The remainder of this chapter will consider how molecular biology techniques have been applied to investigate these various aspects in *S. mutans*.

Working with *S. mutans*

S. mutans has proved itself amenable to most of the molecular biology techniques applied to other organisms but anyone embarking upon research with this organism needs to be aware of some of its quirks:

Safety concerns

S. mutans is a member of a normal human oral microflora that is only rarely a cause of systemic disease (such as infectious endocarditis) and can therefore be considered a much safer organism to work with than more obviously pathogenic streptococci such as *S. pyogenes* and *S. agalactiae*. Accordingly, in the early days of molecular biology, it was regarded as an attractive model *Streptococcus* for study. However, developments in the regulation of genetically manipulated organisms and handling of pathogenic organisms now mean that ALL streptococci are considered as Category 2 pathogens. Therefore, any work involving construction of novel genetic variants of *S. mutans* should be undertaken under appropriate containment conditions. This requires appropriate Biosafety 2 standards of laboratory procedure and registration of the work with the relevant bodies. Comparable rules apply in most countries and in the USA are implemented by the Recombinant Advisory Committee and in the United Kingdom by the Health and Safety Executive.

Antibiotic resistance

Targeted gene inactivation for construction of knockout mutants is a powerful technique discussed further below. A range of antibiotic-resistance markers have been successfully employed to allow positive selection of integrants but use of penicillin resistance is not permitted because such resistance does not occur naturally. This means that markers such as amp^r, widely used in *E. coli* vectors such as the pUC series, cannot be introduced into *S. mutans* (even as a non-selected part of a plasmid vector). There have been a number of incidents when workers, unaware of this rule and only discovering it when their manuscript was rejected by a journal, were required to destroy their mutant stocks and start again with another marker!

Getting DNA out: DNA extraction

The thick peptidoglycan layer and the presence of cellular polysaccharide can make preparation of stocks of pure chromosomal DNA from *S. mutans* problematical, particularly for fresh isolates. Commercial kits are rarely developed with streptococci in mind and vary in their effectiveness. For the most recalcitrant strains it may be necessary to grow the bacteria in an excess of threonine that weakens the cell wall by interfering with the formation of cross-links (Chassy and Giuffrida, 1980); this is followed by treatment with a mixture of lysozyme and mutanolysin, treatment with proteinase K and several phenol–chloroform extractions (Old *et al.*, 2006). If there is no requirement for chromosomal DNA, defined regions of the chromosome can be amplified from whole cells by PCR, simply ensuring that there is an initial denaturing step of 95° for 10 minutes, enabling the release of DNA from the cells (Rupf *et al.*, 1999).

Getting DNA in: the transformation scene

The ability to introduce genetically manipulated DNA into an organism is a vital step in any work involving construction of mutants and so exploration of the mechanism of DNA uptake has been important for progress. The majority of strains of *S. mutans* are not transformable by procedures that are effective for other oral streptococci (Westergren and Emilson, 1983) but careful control of conditions can improve the yield (Perry and Kuramitsu, 1981). Thus, considerable effort has gone into defining optimum conditions for 'natural transformation'. Competence can be increased by chemical mutagenesis (Tao *et al.*, 1993a), bearing in mind the risk of concomitant mutations. Electroporation has also been successfully used to introduce DNA (Dunny *et al.*, 1991; McLaughlin and Ferretti, 1995). A major advance came with the discovery that transformation rates are highest when growth is in a biofilm, where high cell density means that quorum-sensing molecules come into play (Li *et al.*, 2002). This is mediated by the competence-stimulating peptide (CSP) and addition of a synthetic 21-amino acid CSP can greatly increase the efficiency of transformation of most, if not all, strains (Lau *et al.*, 2002). The CSP undergoes post-translational processing by the removal of three C-terminal amino acids and the 18-CSP is even more potent (Petersen *et al.*, 2006). It should be noted that an excess of CSP is lethal, the discovery of which has led to the proposal that it may form the basis of a targeted antimicrobial (Eckert *et al.*, 2006; Qi *et al.*, 2005).

It is interesting to note that the cascade of regulatory molecules affecting competence in *S. mutans* differs from that in *S. pneumoniae* and probably other species in the mitis group (Martin *et al.*, 2006). The competence operon, including *comDE* that encodes CSP, is primarily concerned not with DNA uptake but in controlling the production of small peptide mutacins (Martin *et al.*, 2006; Petersen *et al.*, 2006; van der Ploeg, 2005). Over a dozen or more mutacins are thought to be expressed by various strains of *S. mutans* and appear to be highly significant in competition with other plaque bacteria (Kreth *et al.*, 2005). Indeed, it has been proposed that they are determinant factors leading to the dominance of mutacin-producing strains in plaque (Beighton *et al.*, 1982; Gronroos *et al.*, 1998). The competitive advantage conferred by production of an active mutacin has also been exploited to enhance the efficiency with which an introduced mutant can establish in dental plaque (Hillman *et al.*, 2000).

Cloning around with *S. mutans* genes

A large number of genes from *S. mutans* have now been successfully cloned in plasmids and expressed in heterologous hosts – most commonly *E. coli* but also in *S. typhimurium*, other streptococci, *Lactococcus lactis* and *Bacillus subtilis*. The majority of genes have not proved problematical but in some instances difficulties have been encountered. One source of difficulty is though to be the fact that the *S. mutans* DNA has a relatively greater AT content than does *E. coli*, and this may lead to the occurrence of 'false promoters', resulting in over-expression or initiation of transcription within an open reading frame. The solution to over-expression is to use a low copy number vector or one in which transcription is tightly regulated. False initiation can give rise to difficulties when trying to obtain a pure protein, but this can be overcome by fusion of N- or C-terminal tags such as His_6 that allow affinity purification of a just a single form (Joucla *et al.*, 2006; Monchois *et al.*, 1999a). The second class of troublesome genes are those encoding a product that is in some way toxic to *E. coli*. It includes the glucosyltransferases (GTF) and glucan binding proteins (GBP) discussed below, as well as some lipoproteins. Controlled expression is one possible solution but the first cloning of *gtf* and *gbp* genes was only achieved by the introduction of bacteriophage lambda as a vector (Gilpin *et al.*, 1985; Russell *et al.*, 1985). In this system, there is no requirement for long-term stability of the cloned DNA or its product in the *E. coli* cell and gene products can readily be detected in the plaques generated by lysed bacteria. Finally, it should be noted that the pH of an overnight

E. coli culture is around pH 8.0, well outside the normal growth range for *S. mutans* and hence likely to be sub-optimal for its enzymes. Simple buffering of the medium with 1M Tris has been found to improved detection of expression.

Going for the knockout

Targeted inactivation of specific genes is a powerful tool for exploring the function of particular genes, while random mutagenesis has been fruitfully employed to generate banks of mutants suitable for screening for loss of particular functions. The two main approaches to gene knockout are insertional inactivation and allele replacement, illustrated in Fig. 5.1.

A few things to remember:

- it is not permissible to introduce *amp* into streptococci
- any insert should be as close to the start of the gene as possible to ensure loss of function
- be aware of possible polar effects on downstream genes.

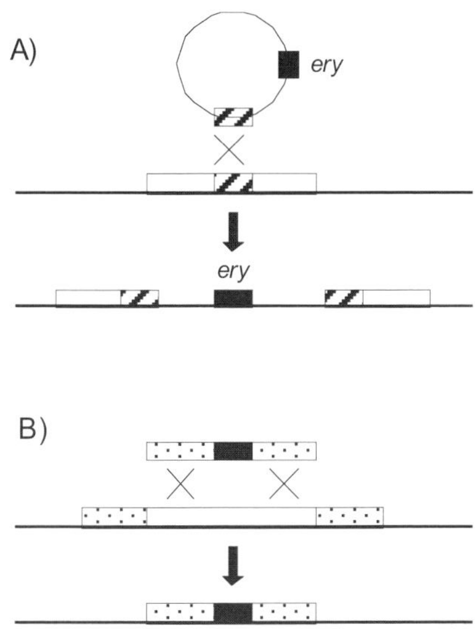

Figure 5.1 Targeted inactivation of a gene by (A) insertion-duplication mutagenesis with insertion of a plasmid carrying an internal fragment of the target gene (striped box) and an antibiotic resistance marker (black box) into the chromosome; (B) allele replacement mutagenesis by linking of a resistance marker to sequences flanking the target gene, followed by integration by double recombination.

Insertion-duplication mutagenesis

A short internal fragment near the start of the gene to be inactivated is cloned in *E. coli* on a suicide vector, such as pVA891 (Macrina *et al.*, 1983), that can replicate in *E. coli* but not in streptococci. The efficiency of the system is affected by the length of the cloned fragment (Lee *et al.*, 1998) but about 300 bp is adequate. The whole plasmid is inserted into the target gene on the chromosome by homologous Campbell-type recombination (single crossover) and recombinants selected by plating on medium containing erythromycin. A variety of suicide vectors, such as pVA8912, pSF151 or pSF143, has been constructed using different antibiotic resistance makers (erythromycin, kanamycin, spectinomycin) and including a multiple-cloning site to facilitate cloning (Faustoferri *et al.*, 1998; Malke *et al.*, 1994; Tao *et al.*, 1992; Wen and Burne, 2001). Examples of their use to knock out specific genes include those for *htrA* and the *msm* operon (Diaz-Torres and Russell, 2001; Russell *et al.*, 1992). It is also possible to generate banks of random mutants via insertional inaction, by cloning random small fragments of chromosomal DNA into the inactivation vector (Colby *et al.*, 1995; Sato *et al.*, 1992; Tsang *et al.*, 2005).

The advantage of this inactivation method is that there is efficient transformation with plasmid and efficient recombination because only a single crossover is required. The disadvantages are that there is uncertainty where insertion into the gene has occurred and it is difficult to introduce further mutations by the same method. The inserted plasmid provides a site for integration of any second suicide plasmid sharing sequence; and the introduction of multiple resistance markers has an adverse effect on growth rate. It must also be noted that the insertional mutagenesis method will fail if there are multiple copies of the gene (or even regions with a high degree of sequence homology), so the presence of a single insert should be checked by Southern blotting. It has been reported that insertion may occur at more than on site on the chromosome and result in chromosomal rearrangements (Sato *et al.*, 1998). A final problem is that regions of homology are generated by recombination and so the entire plasmid can, in theory, 'loop out' if antibiotic selection is not maintained. This occurs at a measurable frequency in *S. gordonii* and has been used to generate unmarked mutations, as a frameshift occurs as the plasmid is excised (Vickerman *et al.*, 1997a). However, there have been no reports of instability in *S. mutans*.

Allele-replacement mutagenesis

In this approach, an antibiotic marker such as the kanamycin-resistance element Omega-Km-2 (Perez-Casal *et al.*, 1991) is cloned into a site within the target gene or an internal fragment of the target gene is replaced with an antibiotic resistance marker (Simpson and Russell, 1998). The linear recombinant fragment is then integrated into the chromosome by a double recombination event (Fig. 5.1). Construction of the inactivated gene can be done by use of appropriate restriction sites if the gene has first been cloned into a plasmid (Gilmore *et al.*, 1993). However, there is a more modern variation of gene replacement/interruption where PCR is used to amplify regions of DNA flanking the gene to be inactivated and restrictions sites incorporated into the primers so that the PCR products can be ligated to an antibiotic resistance cassette that goes between the flanking sequences (Lau *et al.*, 2002). The composite molecule is used to transform *S. mutans*. An interesting variation has been reported in which inactivation is incomplete, resulting in construction of a 'knock-down' mutant with residual transcriptional activity (Lemos *et al.*, 2006). The advantages of the allele replacement method over the plasmid insertion-duplication method is that it is highly specific for its target site, mutants are stable, and careful choice of the primer sites means that any chance of residual activity can be avoided. The only disadvantages are that it is technically more demanding and there may be a lower recovery of mutants since two recombination events are required. However, high efficiency is not a necessity, since the recovery of only one mutant is required!

Both mutagenesis approaches described above result in an antibiotic marker being inserted into the chromosome. Apart from the possible effects that this may have on cell physiology, it makes it impossible to introduce further mutations using the same marker. A system for making unmarked mutations that overcomes these problems has been developed in *L. lactis* (Biswas *et al.*, 1993) and recently applied to *S. mutans* (Atlagic *et al.*, 2006). An improved version of the vector has been used for *S. pyogenes* (Fontaine *et al.*, 2003). It relies upon integration into the chromosome of a plasmid that carries left and right stretches of sequence flanking the gene to be removed (thus integrating at two points). This is followed by a temperature shift to prompt excision of the plasmid which can happen by recombination between the flanking sequences, and removal of the intervening target gene along with the plasmid DNA.

Transposon mutagenesis

Mention has already been made of a plasmid-based system for generating random mutations. An alternative approach is to use transposons and Tn*916* was first used for this purpose (Procino *et al.*, 1988). A useful vector called pTV1-OK based on Tn*917* has also been developed. In this vector, a temperature sensitive plasmid is used to deliver the transposon to *S. mutans*, thus yielding the advantage of efficient transformation with plasmid DNA. Following a shift to a temperature at which the plasmid cannot replicate (which may vary according to strain, in our experience), the transposon integrates into the chromosome at random locations to generate mutants at a high frequency (Cvitkovitch *et al.*, 2000; Cvitkovitch *et al.*, 1998). Appropriate screening methods are then used to detect the mutants of interest, and the DNA surrounding the site of insertion can be recovered to allow characterisation of the inactivated gene.

Plasmid shuttle vectors

Naturally occurring plasmids are rare in *S. mutans*, being found in only about 8% of isolates (Caufield *et al.*, 2006) but have been of value in allowing the construction of vectors that replicate in *E. coli* and are suitable for reintroducing cloned genes into *S. mutans*. This is particularly valuable for checking the proposed function of a gene that has been inactivated, by complementing it with an intact plasmid-borne gene (Zeng *et al.*, 2006). The first of these shuttle vectors were pSA3 and pVA838 (Dao and Ferretti, 1985; Macrina *et al.*, 1983) and an improved construct named pDL276 that is smaller, contains a multiple cloning site flanked by *E. coli* transcription termination sequences and a kanamycin-resistance determinant expressed in both Gram-positive and Gram-negative hosts has also been described (Dunny *et al.*, 1991).

Virulence determinants of *S. mutans*

One of the powerful drivers of research on *S. mutans* has been the search for particular cellular components that might prove suitable as candidates for the development of a dental caries vaccine or as targets for specific inhibitors. Vaccines against oral infections are considered in detail in Chapter 11 of this book.

Because vaccines work by inducing antibodies that react with accessible antigens, interest naturally focused on extracellular and surface components, particularly those involved in sucrose-dependent or sucrose-independent adhesion. It is clear that many aspects of intracellular physiology are of vital importance in determining the ability of *S. mutans* to grow, multiply and produce acid in plaque.

Areas such as intracellular regulatory circuits, acid tolerance, quorum sensing, interspecies signalling and biofilm formation are also areas of considerable research activity. However, the remainder of this chapter will focus upon extracellular and surface components that may contribute to cariogenicity.

Sucrose metabolism

S. mutans is capable of utilizing a wide range of carbohydrates and seems remarkably versatile in transporting and metabolizing a variety of substrates, which may go some way towards explaining its ability to flourish when exposed to a carbohydrate-rich diet. While any fermentable carbohydrate may be utilized by plaque bacteria to generate the acids which attack enamel substance, sucrose is recognized as being particularly important in the caries process because not only can it be fermented but it also serves as substrate for extra-cellular enzymes which synthesize sucrose-derived polymers. These polymers are of central importance in adhesive interactions in plaque, where they mediate attachment of bacteria to the tooth surface and to other bacteria. They thus stabilize the plaque biofilm; serve as energy stores aiding the survival of plaque bacteria (Burne, 1998); modulate the permeability of plaque and hence the level of acid at the enamel surface (Dibdin and Shellis, 1988; Thurnheer *et al.*, 2006; Van Houte *et al.*, 1989), and influence the local balance between mineralisation and demineralisation (Paes Leme *et al.*, 2006). Sucrose is thoroughly exploited by several different pathways (Fig. 5.2), with both glucose and fructose moieties of the molecule being metabolized by several different routes (Colby and Russell, 1997).

Intracellular sucrose metabolism

One of the first genes of *S. mutans* to be cloned was *scrB*, encoding sucrose-6-phosphate hydrolase that splits the disaccharide so that it can enter glycolysis (Hayakawa *et al.*, 1986). The existence of this intracellular enzyme implied the presence of a phosphotransferase system (PTS) that phosphorylates sucrose as it enters the cell and the EII component of a sucrose-specific transporter is encoded by the adjacent *scrA* gene (Sato *et al.*, 1989). Because of its structural similarity to trehalose, sucrose can also be transported and phosphorylated by the trehalose PTS (*treABC*) (Poy and Jacobson, 1990). It has long been known that there is a third, non-PTS system for sucrose uptake (Slee and Tanzer, 1982) and we now know that sucrose can enter by two ABC transporters; the ATP-driven multiple sugar metabolism *msm* and maltose *mal* transport systems (Kilic *et al.*, 2007; Tao *et al.*, 1993b) and subsequently enter glycolysis after catabolism by sucrose phosphorylase (*gtfA*) (Russell *et al.*, 1988a).

Extracellular sucrose metabolism

Fructans

S. mutans utilizes the fructose moiety of the sucrose molecule to synthesize inulin, a β-(2-1)-linked fructan, by fructosyltransferase encoded by a single *ftf* gene (Shiroza and Kuramitsu, 1988). Inactivation of *ftf* results in a strain less pathogenic in a rat-model, suggesting that FTF contributes to pathogenicity (Schroeder *et al.*, 1989). Fructans have a high turnover rate in plaque and are readily degraded and utilized by *S. mutans* and its neighbours in plague by the enzyme fructanase (*fruA*) that is subject to a complex regulatory circuit (Zeng *et al.*, 2006). Fructans are thus thought to act as an energy store, provid-

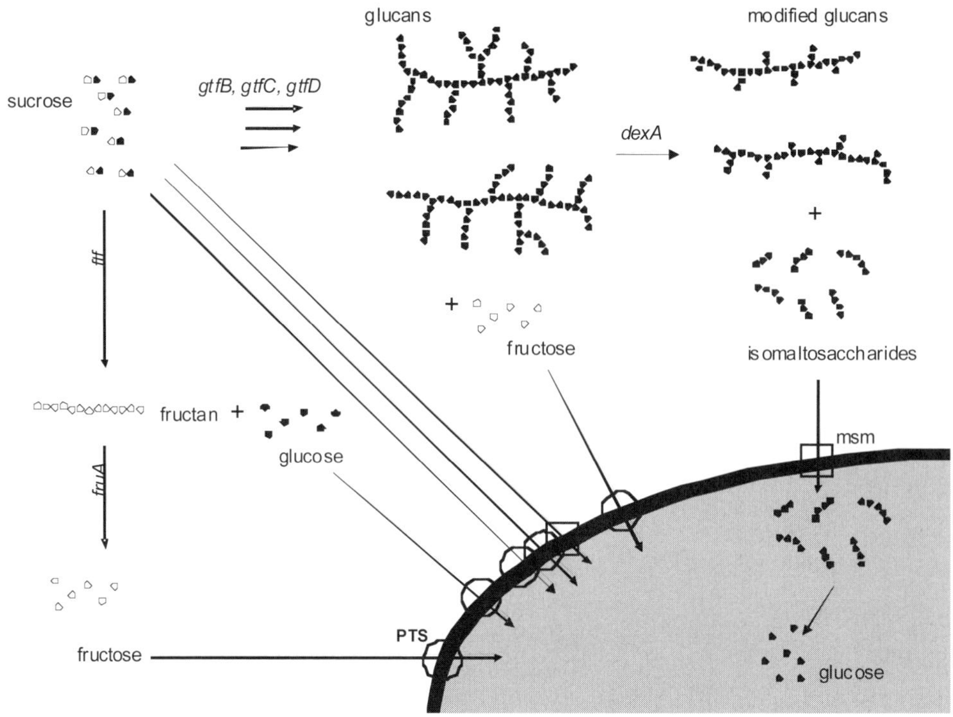

Figure 5.2 The pathways for metabolism of sucrose by *Streptococcus mutans*. ● glucose; ❍ fructose; *gtf*, glucosyltransferase; *dexA*, dextranase; *ftf*, fructosyltransferase; *fruA*, fructanase, *msm*, multiple sugar metabolism genes. Phosphotransferase (PTS) uptake systems are indicated by circles and ABC uptake systems by squares. Modified from Colby and Russell (1997) by permission of Blackwell Publishing.

ing *S. mutans* with an additional source of fermentable carbohydrate when exogenous sources are lacking (Burne, 1998). This would result in prolonging the exposure of the tooth surface to acids, therefore contributing to maintaining a low pH for dental caries to occur. FruA is therefore also considered a possible virulence factor (Burne *et al.*, 1996).

Glucans

The glucose moiety of sucrose is the substrate for glucosyltransferases (GTF). These are extracellular enzymes produced by several species of oral streptococci and they synthesize polymers consisting solely of glucose (glucans). They are also referred to as glucansucrases. Streptococcal GTF are capable of synthesizing either α1,6- or α1,3-linkages between the glucose units and the relative proportions of these two linkages and the degree of branching determines the ultimate properties of the glucan. A glucan which is essentially a linear α1,6-linked chain is referred to as a dextran and is soluble (the trivial name dextransucrase is often applied to GTF which produce dextran). In contrast, mutan is a water-insoluble glucan with a high proportion of α1,3-linkages. Both types of glucan are believed to be important in dental plaque, with dextran mediating bacterial aggregation and serving as a storage polymer while mutan has been shown to be the major contributor to adherence (Colby and Russell, 1997; Kuramitsu, 2003). In addition to this glucan-producing function, GTF at the surface of bacteria, or adsorbed to the tooth surface, act to bind bacteria and thus consolidate plaque (Rolla *et al.*, 1985).

Biochemical studies of GTF are greatly complicated by the fact the each species produces a number of different GTF but the cloning and expression of *gtf* genes in *E. coli* made it possible to study each GTF in isolation. Attempts to clone genes for sucrose-utilizing enzymes into plasmids led to discovery of the genes for sucrose-6-phosphate hydrolase (*scrB*) and sucrose phosphorylase (*gtfA*), but were not successful in identifying the polymer-forming GTF. The introduction of phage λ as cloning vector produced the desired results, as GTF activity could be detected by the formation of glucans by GTF released from cells lysed by recombinant phage carrying the *gtf* genes (Gilpin *et al.*, 1985). The first *gtf* cloned were from *S. downei* but the same screening method was later applied to clone *gtf* from *S. mutans*, *S. salivarius* and *S. gordonii*. Not all *gtf* can so easily be detected by expression of activity, so a variety of techniques of Southern blotting, reverse genetics and PCR using primers based on conserved sequences have been used to clone related enzymes.

It is now known that *S. mutans* produces three distinct GTF while *S. sobrinus* produces four. Each GTF is encoded by a separate gene and each enzyme has distinctive properties, varying in the proportion of α-1,6 and α1,3-linkages and hence the degree of branching it introduces into the glucan, and the total length of the glucan chain produced (Colby *et al.*, 1999; Colby and Russell, 1997).

The *S. mutans* GTF are encoded by *gtfB*, *gtfC* and *gtfD*. The enzyme encoded by *gtfB* synthesizes mainly α-(1-3)-linked insoluble glucans (GTF-I). The *gtfC* enzyme synthesizes a mixture of α-(1-3)-linked insoluble and α-(1-6)-linked soluble glucans (GTF-SI) and that of *gtfD* synthesizes α-(1-6)-linked soluble glucan (GTF-S). Inactivation of either *gtfB*, *gtfC* or *gtfD* or all three *gtf* genes has been shown to result in a reduction in smooth surface caries, induced by the relevant mutant, in an animal model (Fujiwara *et al.*, 1996; Munro *et al.*, 1991; Tsumori and Kuramitsu, 1997; Yamashita *et al.*, 1993). However, it is important to note that different results may be obtained with different animal diets (Yamashita *et al.*, 1993) and there may be differential effects at different tooth sites (Van Der Ploeg and Guggenheim, 2004).

It was initially thought that expression of *gtf* genes was constitutive but it is now clear that they are subject to regulation, though the detailed mechanism is still unclear. The gene regulation and hence the different proportions and relative activity of GTF determines the eventual properties of the glucan produced. Within dental plaque, the situation will be vastly more complex because the overall properties of the glucan represents the combined action of a dozen or more different GTF from different streptococci and also on their interactions, since one GTF may modify the product of another. Furthermore, glucans will be modified by the action of dextranase (*dexA*, Fig. 5.2), which is produced by *S. mutans* and various other plaque bacteria and degrades the α1,6-linked chains (Colby *et al.*, 1995).

As nucleotide sequences of GTF became available, it was possible to carry out multiple alignments of the deduced amino acid sequences in order to identify conserved and unique features which may explain which amino acid residues or domains are responsible for common actions of all GTF (such as sucrose hydrolysis) and which determine distinctive activities (such as type of linkage or chain termination). The basic organisation of GTF could be discerned as soon as more than one sequence became available (Russell *et al.*, 1988b) but the gene sequences from 9 species of streptococci, and several from *Leuconostoc mesenteroides* and *Lactobacillus reuterii* are now available. Over 50 sequences are listed on the CAZY database of Carbohydrate Active Enzymes (http://afmb.cnrs-mrs.fr/CAZY/) in Family 70 and this has confirmed the earlier model for a basic common organisation for all GTF (Russell, 1990). All GTF are of high-molecular weight and composed of distinct domains, illustrated in Fig. 5.2. Studies on glucansucrases from *Leuconostoc* and *Lactobacillus* spp. have shown a similar catalytic core, with variation in the N-terminal and C-terminal domains (Monchois *et al.*, 1999b; van Hijum *et al.*, 2006). The central catalytic core of all GTF is predicted to have alternating α-helices and β-sheets, in an arrangement the same as found in the $(\beta/\alpha)_8$ barrel found in amylase. However, in comparison to amylase, the order of the sheets and helices could be predicted to be circularly permuted so that helix 3 occurs first (Fig. 5.3) (MacGregor *et al.*, 1996). The similarity to amylases also allowed prediction of the amino acids contributing to the active site and these were confirmed by mutagenesis (Devulapalle *et al.*, 1997); further mutagenic alteration of individual amino acids confirmed the circular permutation (Monchois *et al.*, 2000). The reaction mechanism of glucansucrases has now been elucidated (Moulis *et al.*, 2006) though we do not yet have a full understanding of what determines the specificity of the linkage formed and a full 3-D structure has to date proved impossible to obtain, despite extensive efforts. High-level

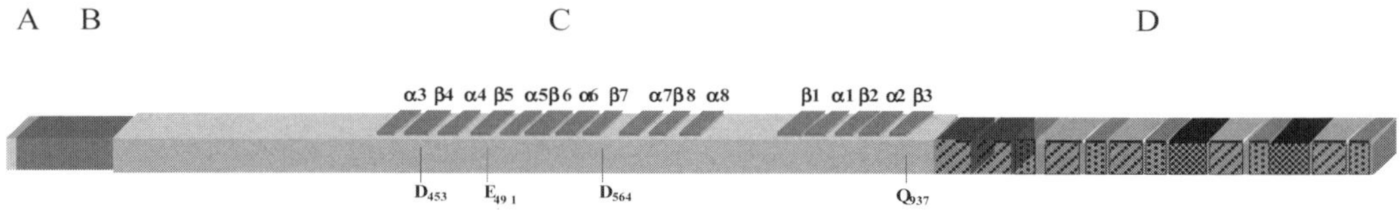

Figure 5.3 Schematic structure of *S. downei* GTF-I showing A, the N-terminal signal peptide; B, variable region; C, catalytic core; D, glucan-binding domain containing repeats. The predicted locations of the α helix and β strands corresponding to the components of the amylase family $(\beta/\alpha)_8$ barrel and residues contributing to the active site are indicated. Modified from Monchois *et al.* (2000) by permission of Blackwell Publishing.

expression of intact enzymes for purification and crystallisation has proved troublesome, though it has been possible to construct truncated proteins that retain catalytic activity (Joucla *et al.*, 2006; Monchois *et al.*, 1999a) and this approach is proving successful in obtaining a 3D structure of a glucansucrase from *L. reuterii* (S. Kralj, personal communication). Identification of the reaction mechanism and location of the active site is essential knowledge for the rational design of GTF inhibitors that mimic the sugar substrate and for targeting antibodies against the catalytic region. Much of the experimental work on a caries vaccine has utilized our knowledge of GTF structure and function to design a peptide representing the catalytic site and structural knowledge should improve the specificity of immunogen design. (See Chapter 11 of this book).

The terminal one-third of GTF consists of a series of related but non-identical amino acid repeats (Fig. 5.3). These characteristically have a pattern of repeated tyrosine (Y) and glycine (G) residues and it has been proposed that the basic building unit is a YG repeat of 19–21 amino acids (Giffard and Jacques, 1994; Kingston *et al.*, 2002). However, a variety of distinct repeats has been identified (Vickerman *et al.*, 1997b), the only one found in all GTF is 33 amino acids long, and named the A repeat (Shah *et al.*, 2004). This C-terminal region was first shown to be involved in binding glucans by truncating the *gtfI* enzyme and also by expressing the domain on its own (Ferretti *et al.*, 1987). Similar experiments have since been performed with other cloned GTF so that this region is widely referred to as a glucan binding domain (GBD), though it should be noted that such binding has not always been demonstrated experimentally. The precise function of the GBD in glucan synthesis is not yet clear and may not be the same in all GTF (Konishi *et al.*, 1999; Monchois *et al.*, 1999a; Shah *et al.*, 2004). In at least some GTF, the domain contributes to the synthesis of the glucan chain and it appears to be important in binding GTF to preformed or nascent glucan, thus mediating bacteria-bacteria or bacteria-surface adhesion. Because of these important functions that can contribute to the sucrose-dependent adhesion and virulence of *S. mutans*, a peptide based on the GBD is also a favoured target for developing caries vaccines (Chapter 11 of this book).

Glucan binding proteins

The sucrose-dependent adherence of *S. mutans* relies upon interaction between proteins produced by the bacterium and the polymers formed from sucrose. In addition, *S. mutans* can be aggregated by added soluble dextran. The proteins involved in these phenomena have been reviewed (Banas and Vickerman, 2003). The first glucan binding protein was discovered during experiments to exploit the binding properties of the GBD to purify GTF by affinity chromatography on columns containing dextran or other glucans (Russell, 1979). When this protein, encoded by *gbpA*, was sequenced it was found to consist of a series of repeats just like those found in GTF (Banas *et al.*, 1990). The consensus A repeat has more recently been used to search the *S. mutans* UA159 genome sequence for open reading frames predicted to contain this repeat; and this led to the discovery of *gpbD*, encoding a bifunctional extracellular protein with a GBD attached to a lipase (Shah, 2004). The binding and substrate specificity of GbpD suggested that it may attack the lipoteichoic acid of *S. sanguis*, though its possible effect on plaque ecology has yet to be determined. However, an important function of *gpbD* is indicated by the fact that its use as a vaccine results in reduced caries in rats (Zhao *et al.*, 2006). Knockout of either *gbpA* or *gbpD* results in altered *in vitro* biofilm architecture and so both are likely to be important in dental plaque structure (Banas *et al.*, 2007a; Lynch *et al.*, 2007). Rat experiments also show that knockout of *gbpA* alters conditions in plaque, leading to selection of mutants in which the tandem *gtfB*/*gtfC* genes recombine (Banas *et al.*, 2007b) and results in reduced cariogenicity (Matsumura *et al.*, 2003; Nakano *et al.*, 2002).

GbpB is an immunodominant antigen in children and shows modest glucan binding properties, sufficient to isolate it by affinity chromatography. It does not share the A repeats found in GbpA and GbpD but shows sequence homology with putative peptidoglycan hydrolases in other streptococci. Construction of knockout mutants of *gbpB* has posed difficulties, presumably associated with an essential cellular function, but are consistent with the protein's involvement in wall turnover and stress responses (Chia *et al.*, 2001; Mattos-Graner *et al.*, 2006). GbpB has also been used in successful experimental vaccines (Smith *et al.*, 2003).

GbpC again lacks the repeats characteristic of the GBD (Sato *et al.*, 1997) but does bind dextran tightly and is thought to be the major receptor involved in dextran-dependent aggregation, a function enhanced by the fact that it is anchored to the cell wall rather released into the supernatant (Matsumoto-Nakano *et al.*, 2007; Sato *et al.*, 2002b). As might be expected, inactivation of *gbpC* has the greatest influence on biofilm architecture (Lynch *et al.*, 2007; Matsumura *et al.*, 2003).

Wall-associated proteins

Many Gram-positive bacteria are known to have proteins covalently linked to their peptidoglycan. The proteins contain the LPXTG sequence motif near their C-terminal end and this is recognized by the sortase enzyme that performs the cross-linking (Marraffini *et al.*, 2006). The first of these wall-associated proteins, isolated from the culture supernatant, were identified independently by two research groups working on a caries vaccine in the mid-1970s. It is fortuitous that both laboratories were working with *S. mutans* strain Ingbritt, which was later found to have an inactive sortase, resulting in failure of the proteins to be completely attached to the cell (Igarashi, 2004). A consequence of this loss of anchoring is that surface hydrophobicity is altered and the ability to interact with saliva is lost (Lee and McGavin, 2004). The *spaP* gene of Ingbritt also has a frameshift mutation that leads to premature termination and secretion of an unanchored form (Jakubovics *et al.*, 2005).

Wall-associated protein WapA has been shown to protect moneys and other animals against caries when used as a pure vaccine (Hughes *et al.*, 1986; Russell *et al.*, 1982). It is identical to Antigen III (Russell *et al.*, 1995) and is released into the supernatant in a 29 kDa form, though its gene encodes a protein of predicted 45 000 kDa that is cleaved at the outside surface of the cell wall (Ferretti *et al.*, 1989). In addition, the intact molecule and the normally wall-linked section are released when a sortase knockout mutant is constructed (Lowes and Russell, unpublished). Knockout of *wapA* has an indirect effect on other surface components and on cell chain-length, surface ultrastructure and biofilm formation (Harrington and Russell, 1993; Zhu *et al.*, 2006). There are conflicting data on whether it is involved in both sucrose-independent and -dependent adhesion, though these may be explained by the fact that experiments were carried out with different strains. One of these, strain GS-5, is known to be atypical in a number of features (Qian and Dao, 1993; Zhu *et al.*, 2006).

Surface protein antigen A (SpaP) also known as Antigen B, Pac, SR and antigen I/II is a well-characterized adhesin of *S. mutans* and was discovered at the same time as WapA. The distinctive feature of the I/II family of proteins is their high molecular weight and, following the cloning and sequencing of a number of their genes, it has been possible to identify common features of their structure (Jenkinson and Demuth, 1997). All have the LPXTG motif characteristic of wall-anchored proteins in Gram-positive bacteria at the C-terminus and have a block of alanine-rich repeats and another of proline-rich repeats. Repeat regions often are associated with binding and these repeats have been implicated in binding of the salivary agglutinin glycoprotein gp340, of significance in saliva-mediated aggregation and adherence (Jakubovics *et al.*, 2005).

The importance of SpaP in adhesion was first demonstrated by showing that antibody raised against the purified protein could block attachment of *S. mutans* to saliva-coated hydroxyapatite (Douglas and Russell, 1982). Subsequently, gene knockout was shown to result in a decrease in hydrophobicity and a decrease in adherence to saliva-coated hydroxyapatite (Harrington and Russell, 1993; Lee *et al.*, 1989) as well as to decrease cariogenicity (though not colonisation ability) in a rat model (Crowley *et al.*, 1999). This evidence for involvement in caries provides the basis for the extensive use of SpaP (I/II) as the basis for experimental caries vaccines (see Chapter 11 of this book).

GbpC has already been mentioned as the major dextran receptor linked to the cell wall by its LPXTG anchorage. In the absence of an active sortase, GbpC is released from the cell and dextran-dependent agglutination is lost (Igarashi *et al.*, 2004b).

The availability of genome sequence information for *S. mutans* has made it possible to screen for the presence of ORFs containing the LPXTG motif, in the same manner as used to search for the A repeats of putative glucan binding proteins. This has resulted in the discovery of WapE, a 55 kDa protein (Ajdic *et al.*, 2002) that appears to alter cell surface and biofilm formation (Levesque *et al.*, 2005) and is only released when sortase in knocked out (Lowes and Russell, unpublished observations).

Besides SpaP, WapA, GbpC and WapE, all of which appear to contribute to cell surface properties and interaction with the environment, two enzymes contain LPXTG. These are the *dexA* dextranase that degrades the dextrans formed from sucrose. If the sortase gene is knocked out, dextranase is found in the supernatant instead of associated with the cell. The changed location has a dramatic effect on the appearance of colonies following growth on sucrose-containing media, resulting in a phenotype essentially the same as that caused by knockout of *dexA* completely (Colby *et al.*, 1995; Igarashi *et al.*, 2004a); and illustrating the importance of the dextranase activity being located close to the cell surface. A fructanase (FruA) also contains the LPXTG motif and the balance in its distribution between cell surface and supernatant has a corresponding effect on the degradation of dental plaque fructans (Burne and Penders, 1994).

The importance of the LPXTG proteins in a variety of functions and the crucial importance of sortase is shown by the extensive consequences of sortase knockout mutants examined by various laboratory assay methods (Igarashi *et al.*, 2004b; Levesque *et al.*, 2005). Nevertheless, the extent to which these functions are vital to the success of *S. mutans* in dental plaque or to its cariogenicity, is not so clear. Isolates of *S. mutans* vary widely in the extent to which they release surface proteins and hence display a spectrum of surface properties (Knox *et al.*, 1986). Two 'releasing' strains have been shown to have mutations in their sortase genes (Igarashi, 2004; Lee and McGavin, 2004) and it seems probable that these mutations may be widely distributed in the *S. mutans* population. Future studies on the association between characteristics such as wall anchorage and virulence in human caries will be needed to clarify this question. It is not unlikely that further LPXTG proteins will be discovered as information becomes available on a wider range of strains. At present, the only one known is a collagen-binding protein encoded by *cnm*, a gene that is absent from the genome-sequenced strain *S. mutans* UA159 but found in about 15% of other isolates (Sato *et al.*, 2004; Waterhouse and Russell, 2006). This brings us neatly to a consideration of the huge amount of information derived from the genome sequencing project, and to the question of variation within the species *S. mutans*.

The *S. mutans* genome sequence

The sequencing project was undertaken at the University of Oklahoma after discussion within the *S. mutans* research community led to the choice of strain UA159 for sequencing. This belongs to serotype c, the most common serotype and was known to be naturally transformable and to be cariogenic in rats. It was also checked in various laboratories and accepted as 'wild-type' for testable characters, such as presence of intact genes for *spaP*, *wapA*, *gtfB*, *gtfC*, *gtfD*, *ftf*, *dexA* and *gbpA*. A number of other widely used strains, including Ingbritt and GS-5, are known to have defects in one or more of these characters (Igarashi, 2004; Murakami *et al.*, 1997; Sato *et al.*, 2002a).

The paper reporting the genome sequence contains numerous insights into the genetic capabilities of *S. mutans* and the analysis of protein-coding sequences identified 1963 ORFs. At the time of publication, 63% were assigned a putative function, 21% had homologues in other species and 16% were unique to *S. mutans* (Ajdic *et al.*, 2002). These figures will have altered as others have availed themselves of the sequence for gene discovery – the searches that led to discovery of *gbpD* and *wapE* are just two examples. New genes can also be discovered by predictions of function based on the sequence information. This allows the formation of hypotheses about exactly what a gene might do, hypotheses that can tested by cloning or inactivating the specific gene. For example, the precise function of a some of the numerous predicted sugar transporter systems has now been experimentally demonstrated (Kilic *et al.*, 2007; Webb and Hosie, 2006) and the function of the 13 two-component regulators can be systematically explored by individual gene knockouts (Ahn *et al.*, 2006; Biswas and Biswas, 2006; Hussain *et al.*, 2006; Lee *et al.*, 2004). The UA159 sequence also lays the foundation for identification of peptide spots in 2D-gel electrophoresis proteomic studies of *S. mutans* (Len *et al.*, 2004; Luppens and ten Cate, 2005; Wilkins *et al.*, 2002) and in microarrays for monitoring changes in gene expression (Motegi *et al.*, 2006; Wen *et al.*, 2006).

The 2 030 921 bp genome sequence and its published annotation are available through GenBank, Accession No. AE014133. A total of 1960 putative protein-coding sequences were identified and these ORFs in the published annotation are named SMU.1–SMU.2165. Note the capital letters and the fact that some ORF numbers were discarded during the phase of completing the sequencing and the annotation. A convenient way of visualizing the GenBank data is through the genome viewer programme Artemis, a free genome viewer and annotation tool which can be downloaded from the Sanger Institute http://www.sanger/ac/uk. Artemis can also be used directly over the Internet through the Virulogenome site (http://www.vge.ac.uk), albeit without the full range of functions. Two other websites provide interfaces with the genome data and additional functions and updated annotations. These are located at the Los Alamos National Laboratory (http://www.oralgen.lanl.gov/) and the Forsyth Institute (http://www.brop/org). It is important to note that these sites use their own systems for numbering and naming ORFs. The original Oklahoma annotation has priority and should be regarded as being the 'official' system but it is essential, in order to avoid confusion in the literature, that authors make it very clear if they are using an alternative scheme and that they recognize the different annotation conventions (e.g. LANL employs Smu.xxx for its ORF designations). The NCBI Genome Project facility provides a further way of accessing the genome data (http://www.ncbi.nlm.nih.gov/).

Variation within *S. mutans*

It was acknowledged from the outset that no single strain could be expected to be entirely representative of its species, as it has long been known that *S. mutans* strains vary in a wide range of properties including sugar fermentation patterns, mutacin production, acidogenicity, cariogenicity and DNA fingerprint patterns (see Chapter 1 of this book). The arrival of the sequence for strain UA159 now provides a benchmark against which other strains can be tested, and provides valuable tools for exploring genetic variation with the sequence. An early example is provided by study of the multiple sugar metabolism (*msm*) operon, which had previously been found, by Southern blotting experiments, to be missing from certain strains of *S. mutans* unable to ferment melibiose (Ushiro *et al.*, 1991). The convenient provision of the sequences flanking *msm*, through the genome project, made it a simple matter to design primers that would amplify the intervening *msm* region and the resultant small amplicons, produced by PCR, showed the extent of the deletion (Fig. 5.4). Sequencing further showed that the deletion was associated with an insertion sequence (IS) element (Robinson *et al.*, 2003) of which there are numerous copies in UA159 (Ajdic *et al.*, 2002). The application of this approach to other suspected variable regions of the chromosome showed widespread occurrence of chromosomal

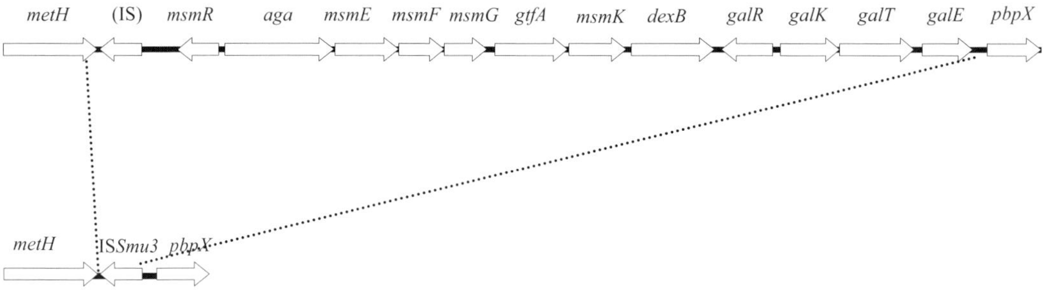

Figure 5.4 Gene arrangement in the *msm* region of the *S. mutans* chromosome. The upper line shows the arrangement in the genome sequenced strain UA159, the lower line that observed in melibiose-negative strains. Adapted from Robinson *et al*, (2003) by permission of S. Karger AG, Basel.

insertions and deletions relative to UA159 in strains isolated at different times and places (Waterhouse and Russell, 2006); and also the proposal that there exists a core genome of conserved genes and a dispensable genome of genes that are variable in occurrence but may be found in strains isolated from all over the world. This observation suggests that there has been widespread recombination and genetic assortment within the overall *S. mutans* population. Current investigations into the population structure of *S. mutans* using multilocus sequence typing (MLST), based on house-keeping genes, will throw further light on this topic and the possible co-evolution of *S. mutans* with its human host (Caufield *et al.*, 2006).

Additional evidence for genetic heterogeneity within *S. mutans* comes from subtractive DNA hybridisation studies (Guo *et al.*, 2006; Saxena *et al.*, 2005) and comparative genome hybridisation in which DNA from test strains is hybridized with a reference microarray based on strain UA159 ORFs (Waterhouse *et al.*, 2007). The latter study found that 20% of the UA159 ORFs were absent from one or more of a panel of 9 test strains. The ORFs that are present in UA159 but absent from the genomes of other strains, include both those that are considered to be normal components of the *S. mutans* core genome (such as *msm*, *gal*, *bgl*, *cel* and *gbpA* genes) and ORFs that are thought to represent foreign DNA acquired by horizontal transfer from other organisms (Waterhouse *et al.*, 2007). It appears that *S. mutans* has acquired foreign DNA from a wide range of sources, including other species of *Streptococcus*, *Bacillus* and the *Archaea*. Little is yet known about whether these acquired genes are expressed in *S. mutans* or confer a selective advantage (Marri *et al.*, 2006). It is somewhat surprising that none of the putatively foreign DNA has a recognisable source in other members of the oral microflora, though information on this aspect may well change as more genome sequences enter the public domain. The implications of horizontal gene transfer for the plaque microflora are discussed in Chapter 3 of this book.

Conclusions

The availability of the *S. mutans* genome sequence brings enormous benefits, not only in facilitating many of the established molecular biology approaches, such as gene cloning and gene knockout, but also in opening entirely new avenues for gene discovery, analysis of gene expression and genetic variation. Genome sequencing of a number of other oral streptococci is in progress or at the annotation stage but, while draft sequences are accessible on public databases, only limited interpretation of the data has yet appeared (Jenkinson and Vickerman, 2006). These genomes, along with those of additional strains of *S. mutans*, will provide exciting new opportunities helping us to explore distinctive features of the different plaque bacteria. Genome-based comparisons with other bacteria, such as non-oral species of *Streptococcus* and *Lactobacillus*, will also continue to give valuable insights and further our understanding of the molecular biology of cariogenic bacteria.

References

Aamdal-Scheie, A., Luan, W.M., Dahlen, G., and Fejerskov, O. (1996) Plaque pH and microflora of dental plaque on sound and carious root surfaces. J. Dent Res. *75*, 1901–1908.

Aas, J.A., Paster, B.J., Stokes, L.N., Olsen, I., and Dewhirst, F.E. (2005) Defining the normal bacterial flora of the oral cavity. J. Clin. Microbiol. *43*, 5721–5732.

Ahn, S.J., Wen, Z.T., and Burne, R.A. (2006) Multilevel control of competence development and stress tolerance in *Streptococcus mutans* UA159. Infect. Immun. *74*, 1631–1642.

Ajdic, D., McShan, W.M., McLaughlin, R.E., Savic, G., Chang, J., Carson, M.B., Primeaux, C., Tian, R., Kenton, S., Jia, H., *et al.* (2002) Genome sequence of *Streptococcus mutans* UA159, a cariogenic dental pathogen. Proc. Natl. Acad. Sci. USA *99*, 14434–14439.

Alam, S., Brailsford, S.R., Adams, S., Allison, C., Sheehy, E., Zoitopoulos, L., Kidd, E.A., and Beighton, D. (2000) Genotypic heterogeneity of *Streptococcus oralis* and distinct aciduric subpopulations in human dental plaque. Appl. Environ. Microbiol. *66*, 3330–3336.

Atlagic, D., Kilic, A.O., and Tao, L. (2006) Unmarked gene deletion mutagenesis of *gtfB* and *gtfC* in *Streptococcus mutans* using a targeted hit- and-run strategy with a thermosensitive plasmid. Oral Microbiol. Immunol. *21*, 132–135.

Banas, J.A., Fountain, T.L., Mazurkiewicz, J.E., Sun, K., and Margaret Vickerman, M. (2007a). *Streptococcus mutans* glucan binding protein A affects *Streptococcus gordonii* biofilm architecture. FEMS Microbiol. Lett. *267*, 80–88.

Banas, J.A., Miller, J.D., Fuschino, M.E., Hazlett, K.R., Toyofuku, W., Porter, K.A., Reutzel, S.B., Florczyk, M.A., McDonough, K.A., and Michalek, S.M. (2007b). Evidence that accumulation of mutants in a biofilm reflects natural selection rather than stress-induced adaptive mutation. Appl. Environ. Microbiol. *73*, 357–361.

Banas, J.A., Russell, R.R.B., and Ferretti, J.J. (1990) Sequence analysis of the gene for the glucan-binding protein of *Streptococcus mutans* Ingbritt. Infect. Immun. *58*, 667–673.

Banas, J.A., and Vickerman, M.M. (2003) Glucan-binding proteins of the oral streptococci. Crit. Rev. Oral Biol. Med. *14*, 89–99.

Becker, M.R., Paster, B.J., Leys, E.J., Moeschberger, M.L., Kenyon, S.G., Galvin, J.L., Boches, S.K., Dewhirst, F.E., and Griffen, A.L. (2002) Molecular analysis of bacterial species associated with childhood caries. J. Clin. Microbiol. *40*, 1001–1009.

Beighton, D. (2005) The complex oral microflora of high-risk individuals and groups and its role in the caries process. Comm. Dent. Oral Epidemiol. *33*, 248–255.

Beighton, D., Hayday, H., and Walker, J. (1982) The acquisition of *Streptococcus mutans* by infant monkeys (*Macaca fascicularis*) and its relationship to the initiation of dental caries. J. Gen. Microbiol. *128*, 1881–1892.

Bentley, R.W., Leigh, J.A., and Collins, M.D. (1991) Intrageneric structure of *Streptococcus* based on comparative analysis of small-subunit rRNA sequences. Int. J. Syst. Bacteriol. *41*, 487–494.

Biswas, I., Gruss, A., Ehrlich, S.D., and Maguin, E. (1993) High-efficiency gene inactivation and replacement system for gram-positive bacteria. J. Bacteriol. *175*, 3628–3635.

Biswas, S., and Biswas, I. (2006) Regulation of the glucosyltransferase (*gtfBC*) operon by CovR in *Streptococcus mutans*. J Bacteriol *188*, 988–998.

Buckley, N.D., Lee, L.N., and LeBlanc, D.J. (1995) Use of a novel mobilizable vector to inactivate the *scrA* gene of *Streptococcus sobrinus* by allelic replacement. J. Bacteriol. *177*, 5028–5034.

Burne, R.A. (1998) Oral streptococci.. products of their environment. J. Dent. Res. *77*, 445–452.

Burne, R.A., Chen, Y.Y., Wexler, D.L., Kuramitsu, H., and Bowen, W.H. (1996) Cariogenicity of *Streptococcus mutans* strains with defects in fructan metabolism assessed in a program-fed specific-pathogen-free rat model. J. Dent. Res. *75*, 1572–1577.

Burne, R.A., and Penders, J.E.C. (1994) Differential localization of the *Streptococcus mutans* GS-5 fructan-hydrolase enzyme, FruA. FEMS Microbiol. Lett. *121*, 243–249.

Byun, R., Nadkarni, M.A., Chhour, K.L., Martin, F.E., Jacques, N.A., and Hunter, N. (2004) Quantitative analysis of diverse *Lactobacillus* species present in advanced dental caries. J. Clin. Microbiol. *42*, 3128–3136.

Caglar, E., Kargul, B., and Tanboga, I. (2005a). Bacteriotherapy and probiotics' role on oral health. Oral Dis. *11*, 131–137.

Caglar, E., Sandalli, N., Twetman, S., Kavaloglu, S., Ergeneli, S., and Selvi, S. (2005b). Effect of yogurt with *Bifidobacterium* DN-173 010 on salivary mutans streptococci and lactobacilli in young adults. Acta Odontol. Scand. *63*, 317–320.

Caufield, P.W., Li, Y., Dasanayake, A., and Saxena, D. (2007) Diversity of lactobacilli in the oral cavities of young women with dental caries. Caries Res. *41*, 2–8.

Caufield, P.W., Saxena, D., Fitch, D., and Li, Y. (2007) Population structure of plasmid-containing strains of *Streptococcus mutans*, a member of the human indigenous biota. J. Bacteriol. *189*, 1238–1243.

Chassy, B.M., and Giuffrida, A. (1980) Method for the lysis of Gram-positive, asporogenous bacteria with lysozyme. Appl. Environ. Microbiol *39*, 153–158.

Chhour, K.L., Nadkarni, M.A., Byun, R., Martin, F.E., Jacques, N.A., and Hunter, N. (2005) Molecular analysis of microbial diversity in advanced caries. J. Clin. Microbiol. *43*, 843–849.

Chia, J.S., Chang, L.Y., Shun, C.T., Chang, Y.Y., Tsay, Y.G., and Chen, J.Y. (2001) A 60-kilodalton immunodominant glycoprotein is essential for cell wall integrity and the maintenance of cell shape in *Streptococcus mutans*. Infect. Immun. *69*, 6987–6998.

Colby, S.M., McLaughlin, R.E., Ferretti, J.J., and Russell, R.R.B. (1999) Effect of inactivation of *gtf* genes on adherence of *Streptococcus downei*. Oral Microbiol. Immunol. *14*, 27–32.

Colby, S.M., and Russell, R.R.B. (1997) Sugar metabolism by mutans streptococci. Soc. Appl. Bacteriol. Symp. Ser. *26*, 80S-88S.

Colby, S.M., Whiting, G.C., Tao, L., and Russell, R.R.B. (1995) Insertional inactivation of the *Streptococcus mutans dexA* (dextranase) gene results in altered adherence and dextran catabolism. Microbiology *141*, 2929–2936.

Coykendall, A.L. (1989) Classification and identification of the viridans streptococci. Clin. Microbiol. Rev. *2*, 315–328.

Crowley, P.J., Brady, L.J., Michalek, S.M., and Bleiweis, A.S. (1999) Virulence of a *spaP* mutant of *Streptococcus mutans* in a gnotobiotic rat model. Infect. Immun. *67*, 1201–1206.

Cvitkovitch, D.G., Gutierrez, J.A., Behari, J., Youngman, P.J., Wetz, J.E., Crowley, P.J., Hillman, J.D., Brady, L.J., and Bleiweis, A.S. (2000) Tn917-lac mutagenesis of *Streptococcus mutans* to identify environmentally regulated genes. FEMS Microbiol. Lett. *182*, 149–154.

Cvitkovitch, G.G., Gutierrez, J.A., Crowley, P.J., Wojciechowski, L., Hillman, J.D., and Bleiweis, A.S. (1998) Tn*917* transposon mutagenesis and marker rescue of interrupted genes of *Streptococcus mutans*. Methods Cell. Sci. *20*, 1–12.

Dal Bello, F., and Hertel, C. (2006) Oral cavity as natural reservoir for intestinal lactobacilli. Syst. Appl. Microbiol. *29*, 69–76.

Dao, M.L., and Ferretti, J.J. (1985) *Streptococcus-Escherichia coli* shuttle vector pSA3 and its use in the cloning of streptococcal genes. Appl. Environ. Microbiol. *49*, 115–119.

de Soet, J.J., Holbrook, W.P., Magnusdottir, M.O., and De Graaff, J. (1993) *Streptococcus sobrinus* and *Streptococcus mutans* in a longitudinal study of dental caries. Microb. Ecol. Health Dis. *6*, 237–243

de Soet, J.J., Nyvad, B., and Kilian, M. (2000) Strain-related acid production by oral streptococci. Caries Res. *34*, 486–490.

Devulapalle, K.S., Goodman, S.D., Gao, Q., Hemsley, A., and Mooser, G. (1997) Knowledge-based model of a glucosyltransferase from the oral bacterial group of mutans streptococci. Protein Sci. *6*, 2489–2493.

Diaz-Torres, M.L., and Russell, R. R.B. (2001) HtrA protease and processing of extracellular proteins of *Streptococcus mutans*. FEMS Microbiol. Lett. *204*, 23–28.

Diaz, P.I., Chalmers, N.I., Rickard, A.H., Kong, C., Milburn, C.L., Palmer, R.J., Jr., and Kolenbrander, P.E. (2006) Molecular characterization of subject-specific oral microflora during initial colonization of enamel. Appl. Environ. Microbiol. *72*, 2837–2848.

Dibdin, G.H., and Shellis, R.P. (1988) Physical and biochemical studies of *Streptococcus mutans* sediments suggest new factors linking the cariogenicity of plaque with its extracellular polysaccharide content. J. Dent. Res. *67*, 890–895.

Douglas, C.W., and Russell, R.R.B. (1982) Effect of specific antisera on adherence properties of the oral bacterium *Streptococcus mutans*. Arch. Oral Biol. *27*, 1039–1045.

Dunny, G.M., Lee, L.N., and LeBlanc, D.J. (1991) Improved electroporation and cloning vector system for gram-positive bacteria. Appl. Environ. Microbiol. 57, 1194–1201.

Eckert, R., He, J., Yarbrough, D.K., Qi, F., Anderson, M.H., and Shi, W. (2006) Targeted killing of *Streptococcus mutans* by a pheromone-guided 'smart' antimicrobial peptide. Antimicrob. Ag. Chemother. *50*, 3651–3657.

Faustoferri, R.C., Quivey, R.G., Smith, A.J., and Sanchez, R. (1998) A medium-copy-number plasmid for insertional mutagenesis of *Streptococcus mutans*. Plasmid *40*, 247–251.

Ferretti, J.J., Gilpin, M.L., and Russell, R.R.B. (1987) Nucleotide sequence of a glucosyltransferase gene from *Streptococcus sobrinus* MFe28. J. Bacteriol. *169*, 4271–4278.

Ferretti, J.J., Russell, R.R.B., and Dao, M.L. (1989) Sequence analysis of the wall-associated protein precursor of *Streptococcus mutans* antigen A. Mol. Microbiol. *3*, 469–478.

Fontaine, M.C., Lee, J.J., and Kehoe, M.A. (2003) Combined contributions of streptolysin O and streptolysin S to virulence of serotype M5 *Streptococcus pyogenes* strain Manfredo. Infect. Immun. *71*, 3857–3865.

Fujiwara, T., Tamesada, M., Bian, Z., Kawabata, S., Kimura, S., and Hamada, S. (1996) Deletion and reintroduction of glucosyltransferase genes of *Streptococcus mutans* and role of their gene products in sucrose dependent cellular adherence. Microb. Path. *20*, 225–233.

Giffard, P.M., and Jacques, N.A. (1994) Definition of a fundamental repeating unit in streptococcal glucosyltransferase glucan-binding regions and related sequences. J. Dent. Res. *73*, 1133–1141.

Gilmore, K.S., Russell, R.R.B., and Ferretti, J.J. (1993) Expression of Gtfs Is essential for normal Insoluble glucan synthesis by *Streptococcus downei*. Infect. Immun. *61*, 1246–1250.

Gilpin, M.L., Russell, R.R.B., and Morrissey, P. (1985) Cloning and expression of two *Streptococcus mutans* glucosyltransferases in *Escherichia coli* K-12. Infect. Immun. *49*, 414–416.

Gronroos, L., Saarela, M., Matto, J., Tanner-Salo, U., Vuorela, A., and Alaluusua, S. (1998) Mutacin production by *Streptococcus mutans* may promote transmission of bacteria from mother to child. Infect. Immun. 66, 2595–2600.

Guo, L.H., Shi, J.N., Zhang, Y., Liu, X.D., Duan, J., and Wei, S. (2006) Identification of genetic differences between two clinical isolates of *Streptococcus mutans* by subtractive hybridisation. Oral Microbiol. Immunol. *21*, 372–380.

Harrington, D.J., and Russell, R.R.B. (1993) Multiple changes in cell wall antigens of isogenic mutants of *Streptococcus mutans*. J. Bacteriol. *175*, 5925–5933.

Hayakawa, M., Aoki, H., and Kuramitsu, H.K. (1986) Isolation and characterization of the sucrose 6-phosphate hydrolase gene from *Streptococcus mutans*. Infect. Immun. 53, 582–586.

Hillman, J.D., Brooks, T.A., Michalek, S.M., Harmon, C.C., Snoep, J.L., and van Der Weijden, C.C. (2000) Construction and characterization of an effector strain of *Streptococcus mutans* for replacement therapy of dental caries. Infect. Immun. *68*, 543–549.

Hoshino, T., Fujiwara, T., and Kilian, M. (2005) Use of phylogenetic and phenotypic analyses to identify nonhemolytic streptococci isolated from bacteremic patients. J. Clin. Microbiol. *43*, 6073–6085.

Hughes, M., MacHardy, S., Sheppard, A., Langford, D., and Shepherd, W. (1986) Experiences in the Development of a Dental Caries Vaccine. In Molecular Microbiology and Immunobiology of *Streptococcus mutans*, S. Hamada, S. M. Michalek, H. Kiyono, L. Menaker, and J. R. McGhee, eds. (Amsterdam, Elsevier), pp. 349–357.

Hussain, H., Branny, P., and Allan, E. (2006) A eukaryotic-type serine/threonine protein kinase is required for biofilm formation, genetic competence, and acid resistance in *Streptococcus mutans*. J. Bacteriol. *188*, 1628–1632.

Igarashi, T. (2004) Deletion in sortase gene of *Streptococcus mutans* Ingbritt. Oral Microbiol. Immunol. *19*, 210–213.

Igarashi, T., Asaga, E., and Goto, N. (2004a). Roles of *Streptococcus mutans* dextranase anchored to the cell wall by sortase. Oral Microbiol. Immunol. *19*, 102–105.

Igarashi, T., Asaga, E., Sato, Y., and Goto, N. (2004b). Inactivation of *srtA* gene of *Streptococcus mutans* inhibits dextran-dependent aggregation by glucan-binding protein C. Oral Microbiol. Immunol. *19*, 57–60.

Jakubovics, N.S., Stromberg, N., van Dolleweerd, C.J., Kelly, C.G., and Jenkinson, H.F. (2005) Differential binding specificities of oral streptococcal antigen I/II family adhesins for human or bacterial ligands. Mol. Microbiol. *55*, 1591–1605.

Jenkinson, H.F., and Demuth, D.R. (1997) Structure, function and immunogenicity of streptococcal antigen I/II polypeptides. Mol. Microbiol. *23*, 183–190.

Jenkinson, H.F., and Vickerman, M.M. (2006) Genetics of *sanguinis* Group Streptococci. In Gram-Positive Pathogens, V. A. Fischetti, R. P. Novick, J. J. Ferretti, D. A. Portnoy, and J. I. Rood, eds. (Washington, D.C., ASM Press), pp. 347–355.

Joucla, G., Pizzut, S., Monsan, P., and Remaud-Simeon, M. (2006) Construction of a fully active truncated alternansucrase partially deleted of its carboxy-terminal domain. FEBS Lett. *580*, 763–768.

Kawamura, Y., Hou, X.G., Sultana, F., Miura, H., and Ezaki, T. (1995) Determination of 16s ribosomal-RNA sequences of *Streptococcus mitis* and *Streptococcus gordonii* and phylogenetic relationships among members of the genus *Streptococcus* Int. J. Syst. Bacteriol. *45*, 406–408 (erratum 445:407).

Kilic, A.O., Honeyman, A.L., and Tao, L. (2007) Overlapping substrate specificity for sucrose and maltose of two binding protein-dependent sugar uptake systems in *Streptococcus mutans*. FEMS Microbiol. Lett. *266*, 218–223.

Kingston, K.B., Allen, D.M., and Jacques, N.A. (2002) Role of the C-terminal YG repeats of the primer-dependent streptococcal glucosyltransferase, GtfJ, in binding to dextran and mutan. Microbiology *148*, 549–558.

Klaenhammer, T.R., Barrangou, R., Buck, B.L., Azcarate-Peril, M.A., and Altermann, E. (2005) Genomic features of lactic acid bacteria effecting bioprocessing and health. FEMS Microbiol. Rev. *29*, 393–409.

Kleinberg, I. (2002) A mixed-bacteria ecological approach to understanding the role of the oral bacteria in dental caries causation: An alternative to *Streptococcus mutans* and the specific-plaque hypothesis. Crit. Rev. Oral Biol. Med. *13*, 108–125.

Knox, K.W., Hardy, L.N., and Wicken, A.J. (1986) Comparative studies on the protein profiles and hydrophobicity of strains of *Streptococcus mutans* serotype c. J. Gen. Microbiol. *132*, 2541–2548.

Konishi, N., Torii, Y., Yamamoto, T., Miyagi, A., Ohta, H., Fukui, K., Hanamoto, S., Matsuno, H., Komatsu, H., Kodama, T., *et al.* (1999) Structure and enzymatic properties of genetically truncated forms of the water-insoluble glucan-synthesizing glucosyltransferase from *Streptococcus sobrinus*. J. Biochem. (Tokyo) *126*, 287–295.

Kreth, J., Merritt, J., Shi, W., and Qi, F. (2005) Competition and coexistence between *Streptococcus mutans* and *Streptococcus sanguinis* in the dental biofilm. J. Bacteriol. *187*, 7193–7203.

Kuramitsu, H.K. (2003) Molecular genetic analysis of the virulence of oral bacterial pathogens: an historical perspective. Crit. Rev. Oral Biol. Med. *14*, 331–344.

Kuramitsu, H.K. (2006) The virulence properties of *Streptococcus mutans*. In Gram-Positive Pathogens, V. A. Fischetti, R. P. Novick, J.J. Ferretti, D.A. Portnoy, and J.I. Rood, eds. (Washington, DC, ASM Press), pp. 340–346.

Kuramitsu, H.K., and Ellen, R.P., eds. (2000) Oral Bacterial Ecology: The Molecular Basis (Wymondam, England, Horizon Scientific Press).

Lau, P.C., Sung, C.K., Lee, J.H., Morrison, D.A., and Cvitkovitch, D.G. (2002) PCR ligation mutagenesis in transformable streptococci: application and efficiency. J. Microbiol. Meth. *49*, 193–205.

Lee, M.S., Seok, C., and Morrison, D.A. (1998) Insertion-duplication mutagenesis in *Streptococcus pneumoniae*: targeting fragment length is a critical parameter in use as a random insertion tool. Appl. Environ. Microbiol. *64*, 4796–4802.

Lee, S.F., Delaney, G.D., and Elkhateeb, M. (2004) A two-component *covRS* regulatory system regulates expression of fructosyltransferase and a novel extracellular carbohydrate in *Streptococcus mutans*. Infect. Immun. *72*, 3968–3973.

Lee, S.F., and McGavin, M.K. (2004) Identification of a point mutation resulting in loss of cell wall anchoring activity of SrtA of *Streptococcus mutans* NG5. Infect. Immun. *72*, 4314–4317.

Lee, S.F., Progulske-Fox, A., Erdos, G.W., Piacentini, D.A., Ayakawa, G.Y., Crowley, P.J., and Bleiweis, A.S. (1989) Construction and characterization of isogenic mutants of *Streptococcus mutans* deficient in major surface protein antigen P1 (I/II). Infect. Immun. *57*, 3306–3313.

Lemos, J.A., Luzardo, Y., and Burne, R.A. (2007) Physiologic effects of forced down-regulation of *dnaK* and *groEL* expression in *Streptococcus mutans*. J. Bacteriol. *189*, 1582–1588.

Len, A.C., Harty, D.W., and Jacques, N.A. (2004) Proteome analysis of *Streptococcus mutans* metabolic phenotype during acid tolerance. Microbiology *150*, 1353–1366.

Levesque, C.M., Voronejskaia, E., Huang, Y.C., Mair, R.W., Ellen, R.P., and Cvitkovitch, D.G. (2005) Involvement of sortase anchoring of cell wall proteins in biofilm formation by *Streptococcus mutans*. Infect. Immun. *73*, 3773–3777.

Li, Y., Ge, Y., Saxena, D., and Caufield, P.W. (2007) Genetic profiling of the oral microbiota associated with severe early-childhood caries. J. Clin. Microbiol. *45*, 81–87.

Li, Y.H., Tang, N., Aspiras, M.B., Lau, P.C., Lee, J.H., Ellen, R.P., and Cvitkovitch, D.G. (2002) A quorum-sensing signaling system essential for genetic competence in *Streptococcus mutans* is involved in biofilm formation. J. Bacteriol. *184*, 2699–2708.

Loesche, W.J. (1986) Role of *Streptococcus mutans* in human dental decay. Microbiol. Rev. *50*, 353–380.

Luppens, S.B., and ten Cate, J.M. (2005) Effect of biofilm model, mode of growth, and strain on *Streptococcus mutans* protein expression as determined by two-dimensional difference gel electrophoresis. J. Proteome Res. *4*, 232–237.

Lynch, D.J., Fountain, T.L., Mazurkiewicz, J.E., and Banas, J.A. (2007) Glucan-binding proteins are essential for shaping *Streptococcus mutans* biofilm architecture. FEMS Microbiol. Lett. *268*, 158–165.

MacGregor, E.A., Jespersen, H.M., and Svensson, B. (1996) A circularly permuted alpha-amylase-type alpha/beta-barrel structure in glucan-synthesizing glucosyltransferases. FEBS Lett. *378*, 263–266.

Macrina, F.L., Evans, R.P., Tobian, J.A., Hartley, D.L., Clewell, D.B., and Jones, K.R. (1983) Novel shuttle plasmid vehicles for *Escherichia-Streptococcus* transgeneric cloning. Gene *25*, 145–150.

Makarova, K., Slesarev, A., Wolf, Y., Sorokin, A., Mirkin, B., Koonin, E., Pavlov, A., Pavlova, N., Karamychev, V., Polouchine, N., *et al.* (2006) Comparative genomics of the lactic acid bacteria. Proc. Natl. Acad. Sci. USA *103*, 15611–15616.

Malke, H., Mechold, U., Gase, K., and Gerlach, D. (1994) Inactivation of the streptokinase gene prevents *Streptococcus equisimilis* H46A from acquiring cell-associated plasmin activity in the presence of plasminogen. FEMS Microbiol. Lett. *116*, 107–112.

Marraffini, L.A., Dedent, A.C., and Schneewind, O. (2006) Sortases and the art of anchoring proteins to the envelopes of gram-positive bacteria. Microbiol. Mol. Biol. Rev. *70*, 192–221.

Marri, P.R., Hao, W., and Golding, G.B. (2006) Gene gain and gene loss in streptococcus: is it driven by habitat? Mol. Biol. Evol. *23*, 2379–2391.

Marsh, P.D. (2003) Are dental diseases examples of ecological catastrophes? Microbiology *149*, 279–294.

Martin, B., Quentin, Y., Fichant, G., and Claverys, J.P. (2006) Independent evolution of competence regulatory cascades in streptococci? Trends Microbiol. *14*, 339–345.

Matsumoto-Nakano, M., Fujita, K., and Ooshima, T. (2007) Comparison of glucan-binding proteins in cariogenicity of *Streptococcus mutans*. Oral Microbiol. Immunol. 22, 30–35.

Matsumoto, M., Tsuji, M., Sasaki, H., Fujita, K., Nomura, R., Nakano, K., Shintani, S., and Ooshima, T. (2005) Cariogenicity of the probiotic bacterium *Lactobacillus salivarius* in rats. Caries Res. *39*, 479–483.

Matsumura, M., Izumi, T., Matsumoto, M., Tsuji, M., Fujiwara, T., and Ooshima, T. (2003) The role of glucan-binding proteins in the cariogenicity of *Streptococcus mutans*. Microbiol. Immunol. *47*, 213–215.

Mattos-Graner, R.O., Porter, K.A., Smith, D.J., Hosogi, Y., and Duncan, M.J. (2006) Functional analysis of glucan binding protein B from *Streptococcus mutans*. J. Bacteriol. *188*, 3813–3825.

McLaughlin, R.E., and Ferretti, J.J. (1995) Electrotransformation of Streptococci. Meth. Mol. Biol. *47*, 185–193.

Meurman, J.H. (2005) Probiotics: do they have a role in oral medicine and dentistry? Eur. J. Oral Sci. *113*, 188–196.

Modesto, M., Biavati, B., and Mattarelli, P. (2006) Occurrence of the family bifidobacteriaceae in human dental caries and plaque. Caries Res. *40*, 271–276.

Monchois, V., Arguello-Morales, M., and Russell, R.R.B. (1999a). Isolation of an active catalytic core of *Streptococcus downei* MFe28 GTF-I glucosyltransferase. J. Bacteriol. *181*, 2290–2292.

Monchois, V., Vignon, M., Escalier, P.C., Svensson, B., and Russell, R.R.B. (2000) Involvement of Gln937 of *Streptococcus downei* GTF-I glucansucrase in transition-state stabilization. Eur. J. Biochem. *267*, 4127–4136.

Monchois, V., Willemot, R.M., and Monsan, P. (1999b). Glucansucrases: mechanism of action and structure-function relationships. FEMS Microbiol. Rev. *23*, 131–151.

Montalto, M., Vastola, M., Marigo, L., Covino, M., Graziosetto, R., Curigliano, V., Santoro, L., Cuoco, L., Manna, R., and Gasbarrini, G. (2004) Probiotic treatment increases salivary counts of lactobacilli: a double-blind, randomized, controlled study. Digestion *69*, 53–56.

Motegi, M., Takagi, Y., Yonezawa, H., Hanada, N., Terajima, J., Watanabe, H., and Senpuku, H. (2006) Assessment of genes associated with *Streptococcus mutans* biofilm morphology. Appl. Environ. Microbiol. *72*, 6277–6287.

Moulis, C., Joucla, G., Harrison, D., Fabre, E., Potocki-Veronese, G., Monsan, P., and Remaud-Simeon, M. (2006) Understanding the polymerization mechanism of glycoside-hydrolase family 70 glucansucrases. J. Biol. Chem. *281*, 31254–31267.

Munro, C., Michalek, S.M., and Macrina, F.L. (1991) Cariogenicity of *Streptococcus mutans* V403 glucosyltransferase and fructosyltransferase mutants constructed by allelic exchange. Infect. Immun. *59*, 2316–2323.

Munson, M.A., Banerjee, A., Watson, T.F., and Wade, W.G. (2004) Molecular analysis of the microflora associated with dental caries. J. Clin. Microbiol. *42*, 3023–3029.

Murakami, Y., Nakano, Y., Yamashita, Y., and Koga, T. (1997) Identification of a frameshift mutation resulting in premature termination and loss of cell wall anchoring of the PAc antigen of *Streptococcus mutans* GS-5. Infect. Immun. *65*, 794–797.

Nakano, K., Matsumura, M., Kawaguchi, M., Fujiwara, T., Sobue, S., Nakagawa, I., Hamada, S., and Ooshima, T. (2002) Attenuation of glucan-binding protein C reduces the cariogenicity of *Streptococcus mutans*: Analysis of strains isolated from human blood. J. Dent. Res. *81*, 376–379.

Okada, M., Soda, Y., Hayashi, F., Doi, T., Suzuki, J., Miura, K., and Kozai, K. (2005) Longitudinal study of dental caries incidence associated with *Streptococcus mutans* and *Streptococcus sobrinus* in pre-school children. J. Med. Microbiol. *54*, 661–665.

Old, L.A., Lowes, S., and Russell, R.R.B. (2006) Genomic variation in *Streptococcus mutans*: deletions affecting the multiple pathways of β-glucoside metabolism. Oral Microbiol. Immunol. *21*, 21–27.

Paes Leme, A.F., Koo, H., Bellato, C.M., Bedi, G., and Cury, J.A. (2006) The role of sucrose in cariogenic dental biofilm formation – new insight. J. Dent. Res. *85*, 878–887.

Parche, S., Amon, J., Jankovic, I., Rezzonico, E., Beleut, M., Barutcu, H., Schendel, I., Eddy, M.P., Burkovski, A., Arigoni, F., *et al.* (2007) Sugar transport systems of *Bifidobacterium longum* NCC2705. J. Mol. Microbiol. Biotechnol. *12*, 9–19.

Perez-Casal, J., Caparon, M.G., and Scott, J.R. (1991) Mry, a trans-acting positive regulator of the M protein gene of *Streptococcus pyogenes* with similarity to the receptor proteins of two-component regulatory systems. J. Bacteriol. *173*, 2617–2624.

Perry, D., and Kuramitsu, H.K. (1981) Genetic transformation of *Streptococcus mutans*. Infect. Immun. *32*, 1295–1297.

Petersen, F.C., Fimland, G., and Scheie, A.A. (2006) Purification and functional studies of a potent modified quorum-sensing peptide and a two-peptide bacteriocin in *Streptococcus mutans*. Mol. Microbiol. *61*, 1322–1334.

Poy, F., and Jacobson, G.R. (1990) Evidence that a low-affinity sucrose phosphotransferase activity in *Streptococcus mutans* GS-5 is a high-affinity trehalose uptake system. Infect. Immun. *58*, 1479–1480.

Procino, J.K., Marri, L., Shockman, G.D., and Daneo-Moore, L. (1988) Tn916 insertional inactivation of multiple genes on the chromosome of *Streptococcus mutans* GS-5. Infect. Immun. *56*, 2866–2870.

Qi, F., Kreth, J., Levesque, C.M., Kay, O., Mair, R.W., Shi, W., Cvitkovitch, D.G., and Goodman, S.D. (2005) Peptide pheromone induced cell death of *Streptococcus mutans*. FEMS Microbiol. Lett. *251*, 321–326.

Qian, H., and Dao, M.L. (1993) Inactivation of the *Streptococcus mutans* wall-associated protein A gene (WapA) results in a decrease in sucrose-dependent adherence and aggregation. Infect. Immun. *61*, 5021–5028.

Robinson, W.G., Old, L.A., Shah, D. S.H., and Russell, R.R.B. (2003) Chromosomal insertions and deletions in *Streptococcus mutans*. Caries Res. *37*, 148–156.

Rolla, G., Scheie, A.A., and Ciardi, J.E. (1985) Role of sucrose in plaque formation. Scand. J. Dent. Res. *93*, 105–111.

Rupf, S., Kneist, S., Merte, K., and Eschrich, K. (1999) Quantitative determination of *Streptococcus mutans* by using competitive polymerase chain reaction. Eur. J. Oral Sci. *107*, 75–81.

Rupf, S., Merte, K., Eschrich, K., and Kneist, S. (2006) *Streptococcus sobrinus* in children and its influence on caries activity. Eur. Arch. Paediatr. Dent. *7*, 17–22.

Russell, M.W., Harrington, D.J., and Russell, R.R.B. (1995) Identity of *Streptococcus mutans* surface protein antigen III and wall-associated protein antigen A. Infect. Immun. *63*, 733–735.

Russell, R.R.B. (1979) Glucan-binding proteins of *Streptococcus mutans* serotype c. J. Gen. Microbiol. *112*, 197–201.

Russell, R.R.B. (1990) Molecular genetics of glucan metabolism in oral streptococci. Arch. Oral Biol. *35*, S53-S58.

Russell, R.R.B. (1994) The application of molecular genetics to the microbiology of dental caries. Caries Res. *28*, 69–82.

Russell, R.R.B., Aduse-Opoku, J., Sutcliffe, I.C., Tao, L., and Ferretti, J.J. (1992) A binding protein-dependent transport system in *Streptococcus mutans* responsible for multiple sugar metabolism. J. Biol. Chem. *267*, 4631–4637.

Russell, R.R.B., Beighton, D., and Cohen, B. (1982) Immunisation of monkeys (*Macaca fascicularis*) with antigens purified from *Streptococcus mutans*. Br. Dent. J. *152*, 81–84.

Russell, R.R.B., Coleman, D., and Dougan, G. (1985) Expression of a gene for glucan-binding protein from *Streptococcus mutans* in *Escherichia coli*. J. Gen. Microbiol. *131* 295–299.

Russell, R.R.B., Mukasa, H., Shimamura, A., and Ferretti, J.J. (1988a). *Streptococcus mutans gtfA* gene specifies sucrose phosphorylase. Infect. Immun. *56*, 2763–2765.

Russell, R.R.B., Shiroza, T., Kuramitsu, H.K., and Ferretti, J.J. (1988b). Homology of glucosyltransferase gene and protein sequences from *Streptococcus sobrinus* and *Streptococcus mutans*. J. Dent. Res. *67*, 543–547.

Sato, S., Yamamoto, Y., and Kizaki, H. (1998) Generation of duplication or deletion mutants in *Streptococcus mutans* following homologous recombination-mediated random mutagenesis by integrational vector pVA891. Jap. J. Oral Biol. *40*, 506–514.

Sato, Y., Okamoto, K., Kagami, A., Yamamoto, Y., Igarashi, T., and Kizaki, H. (2004) *Streptococcus mutans* strains harboring collagen-binding adhesin. J. Dent. Res. *83*, 534–539.

Sato, Y., Okamoto, K., and Kizaki, H. (2002a). *gbpC* and *pac* gene mutations detected in *Streptococcus mutans* strain GS-5. Oral Microbiol. Immunol. *17*, 263–266.

Sato, Y., Poy, F., Jacobson, G.R., and Kuramitsu, H.K. (1989) Characterization and sequence analysis of the *scrA* gene encoding enzyme IIScr of the *Streptococcus mutans* phosphoenolpyruvate-dependent sucrose phosphotransferase system. J. Bacteriol. *171*, 263–271.

Sato, Y., Senpuku, H., Okamoto, K., Hanada, N., and Kizaki, H. (2002b). *Streptococcus mutans* binding to solid phase dextran mediated by the glucan-binding protein C. Oral Microbiol. Immunol. *17*, 252–256.

Sato, Y., Yamamoto, Y., and Kizaki, H. (1997) Cloning and sequence analysis of the *gbpC* gene encoding a novel glucan-binding protein of *Streptococcus mutans*. Infect. Immun. *65*, 668–675.

Sato, Y., Yamamoto, Y., Kizaki, H., and Kuramitsu, H.K. (1992) Cloning of the phospho-beta-galactosidase gene in *Escherichia coli* from lactose-negative mutants of *Streptococcus mutans* isolated following random mutagenesis with plasmid pVA891 clone banks. FEMS Microbiol. Lett. *70*, 219–224.

Saxena, D., Li, Y., and Caufield, P.W. (2005) Identification of unique bacterial gene segments from *Streptococcus mutans* with potential relevance to dental caries by subtraction DNA hybridization. J. Clin. Microbiol. *43*, 3508–3511.

Schroeder, V.A., Michalek, S.M., and Macrina, F.L. (1989) Biochemical characterization and evaluation of virulence of a fructosyltransferase-deficient mutant of *Streptococcus mutans* V403. Infect. Immun. *57*, 3560–3569.

Shah, D.S., Joucla, G., Remaud-Simeon, M., and Russell, R.R.B. (2004) Conserved repeat motifs and glucan binding by glucansucrases of oral streptococci and *Leuconostoc mesenteroides*. J. Bacteriol. *186*, 8301–8308.

Shah, D.S., Russell, R.R.B. (2004) A novel glucan-binding protein with lipase activity from the oral pathogen, *Streptococcus mutans*. Microbiology.

Shiroza, T., and Kuramitsu, H.K. (1988) Sequence analysis of the *Streptococcus mutans* fructosyltransferase gene and flanking regions. J. Bacteriol. *170*, 810–816.

Simpson, C.L., and Russell, R.R.B. (1998) Intracellular alpha-amylase of *Streptococcus mutans*. J. Bacteriol. *180*, 4711–4717.

Slee, A.M., and Tanzer, J.M. (1982) Sucrose transport by *Streptococcus mutans*. Evidence for multiple transport systems. Biochim. Biophys. Acta *692*, 415–424.

Smith, D.J., King, W.F., Barnes, L.A., Peacock, Z., and Taubman, M.A. (2003) Immunogenicity and protective immunity induced by synthetic peptides associated with putative immunodominant regions of *Streptococcus mutans* glucan-binding protein B. Infect. Immun. *71*, 1179–1184.

Svensater, G., Borgstrom, M., Bowden, G.H., and Edwardsson, S. (2003) The acid-tolerant microbiota associated with plaque from initial caries and healthy tooth surfaces. Caries Res. *37*, 395–403.

Tanzer, J.M. (1995) Dental caries is a transmissible infectious disease: the Keyes and Fitzgerald revolution. J. Dent. Res. *74*, 1536–1542.

Tanzer, J.M., Baranowski, L.K., Rogers, J.D., Haase, E.M., and Scannapieco, F.A. (2001a). Oral colonization and cariogenicity of *Streptococcus gordonii* in specific pathogen-free TAN:SPFOM(OM)BR rats consuming starch or sucrose diets. Arch. Oral Biol. *46*, 323–333.

Tanzer, J.M., Livingston, J., and Thompson, A.M. (2001b). The microbiology of primary dental caries in humans. J. Dent. Educ. *65*, 1028–1037.

Tao, L., LeBlanc, D.J., and Ferretti, J.J. (1992) Novel streptococcal-integration shuttle vectors for gene cloning and inactivation. Gene *120*, 105–110.

Tao, L., MacAlister, T.J., and Tanzer, J.M. (1993a). Transformation efficiency of EMS-induced mutants of *Streptococcus mutans* of altered cell shape. J. Dent. Res. *72*, 1032–1039.

Tao, L., Sutcliffe, I.C., Russell, R. R.B., and Ferretti, J.J. (1993b). Transport of sugars, including sucrose, by the *msm* transport system of *Streptococcus mutans*. J. Dent. Res. *72*, 1386–1390.

Thurnheer, T., van der Ploeg, J.R., Giertsen, E., and Guggenheim, B. (2006) Effects of *Streptococcus mutans gtfC* deficiency on mixed oral biofilms *in vitro*. Caries Res. *40*, 163–171.

Tong, H., Zhu, B., Chen, W., Qi, F., Shi, W., and Dong, X. (2006) Establishing a genetic system for ecological studies of *Streptococcus oligofermentans*. FEMS Microbiol. Lett. *264*, 213–219.

Tsang, P., Merritt, J., Nguyen, T., Shi, W., and Qi, F. (2005) Identification of genes associated with mutacin I production in *Streptococcus mutans* using random insertional mutagenesis. Microbiol*ogy151*, 3947–3955.

Tsumori, H., and Kuramitsu, H. (1997) The role of the *Streptococcus mutans* glucosyltransferases in the sucrose-dependent attachment to smooth surfaces: essential role of the GtfC enzyme. Oral Microbiol. Immunol. *12*, 274–280.

Ushiro, I., Lumb, S.M., Aduse-Opoku, J., Ferretti, J.J., and Russell, R.R.B. (1991) Chromosomal deletions in melibiose-negative isolates of *Streptococcus mutans*. J. Dent. Res. *70*, 1422–1426.

van der Ploeg, J.R. (2005) Regulation of bacteriocin production in *Streptococcus mutans* by the quorum-sensing system required for development of genetic competence. J. Bacteriol. *187*, 3980–3989.

Van Der Ploeg, J.R., and Guggenheim, B. (2004) Deletion of *gtfC* of *Streptococcus mutans* has no influence on the composition of a mixed-species *in vitro* biofilm model of supragingival plaque. Eur. J. Oral Sci. *112*, 433–438.

van Hijum, S.A., Kralj, S., Ozimek, L.K., Dijkhuizen, L., and van Geel-Schutten, I.G. (2006) Structure-function relationships of glucansucrase and fructansucrase enzymes from lactic acid bacteria. Microbiol. Mol. Biol. Rev. *70*, 157–176.

van Houte, J. (1994) Role of micro-organisms in caries etiology. J. Dent. Res. *73*, 672–681.

van Houte, J., Lopman, J., and Kent, R. (1996) The final pH of bacteria comprising the predominant flora on sound and carious human root and enamel surfaces. J. Dent. Res. *75*, 1008–1014.

Van Houte, J., Russo, J., and Prostak, K.S. (1989) Increased pH-lowering ability of *Streptococcus mutans* cell masses associated with extracellular glucan-rich matrix material and the mechanisms involved. J. Dent. Res. *68*, 451–459.

van Ruyven, F.O., Lingstrom, P., van Houte, J., and Kent, R. (2000) Relationship among mutans streptococci, 'low-pH' bacteria, and iodophilic polysaccharide-producing bacteria in dental plaque and early enamel caries in humans. J. Dent. Res. *79*, 778–784.

Vickerman, M.M., Jones, G.W., and Clewell, D.B. (1997a). Molecular analysis of representative *Streptococcus gordonii* Spp phase variants reveals no differences in the glucosyltransferase structural gene, *gtfG*. Oral Microbiol. Immunol. *12*, 82–90.

Vickerman, M.M., Sulavik, M.C., Nowak, J.D., Gardner, N.M., Jones, G.W., and Clewell, D.B. (1997b). Nucleotide sequence analysis of the *Streptococcus gordonii* glucosyltransferase gene, *gtfG*. DNA Seq. *7*, 83–95.

Waterhouse, J.C., and Russell, R.R.B. (2006) Dispensable genes and foreign DNA in *Streptococcus mutans*. Microbiology *152*, 1777–1788.

Waterhouse, J.C., Swan, D.C., and Russell, R.R.B. (2007) Comparative genome hybridisation of *Streptococcus mutans* strains. Oral Microbiol. Immunol. *in press*.

Webb, A.J., and Hosie, A.H. (2006) A member of the second carbohydrate uptake subfamily of ATP-binding cassette transporters is responsible for ribonucleoside uptake in *Streptococcus mutans*. J. Bacteriol. *188*, 8005–8012.

Wen, Z.T., Baker, H.V., and Burne, R.A. (2006) Influence of BrpA on critical virulence attributes of *Streptococcus mutans*. J Bacteriol *188*, 2983–2992.

Wen, Z.T., and Burne, R.A. (2001) Construction of a new integration vector for use in *Streptococcus mutans*. Plasmid *45*, 31–36.

Westergren, G., and Emilson, C.G. (1983) Prevalence of transformable *Streptococcus mutans* in human dental plaque. Infect. Immun. *41*, 1386–1388.

Wilkins, J.C., Homer, K.A., and Beighton, D. (2002) Analysis of *Streptococcus mutans* proteins modulated by culture under acidic conditions. Appl. Environ. Microbiol. *68*, 2382–2390.

Yamashita, Y., Bowen, W.H., Burne, R.A., and Kuramitsu, H.K. (1993) Role of the *Streptococcus mutans gtf* genes in caries induction in the specific-pathogen-free rat model. Infect. Immun. *61*, 3811–3817.

Yeung, M.K. (1999) Molecular and genetic analyses of *Actinomyces* spp. Crit. Rev. Oral Biol. Med. *10*, 120–138.

Yoo, S.Y., Kim, K.J., Lim, S.H., Kim, K.W., Hwang, H.K., Min, B.M., Choe, S.J., and Kook, J.K. (2005a). First isolation of *Streptococcus downei* from human dental plaques. FEMS Microbiol. Lett. *249*, 323–326.

Yoo, S.Y., Kim, P.S., Hwang, H.K., Lim, S.H., Kim, K.W., Choe, S.J., Min, B.M., and Kook, J.K. (2005b). Identification of non-mutans streptococci organisms in dental plaques recovering on mitis-salivarius bacitracin agar medium. J. Microbiol. *43*, 204–208.

Zeng, L., Wen, Z.T., and Burne, R.A. (2006) A novel signal transduction system and feedback loop regulate fructan hydrolase gene expression in *Streptococcus mutans*. Mol. Microbiol. *62*, 187–200.

Zhao, H., Wu, B., Wu, H., Su, L., Pang, J., Yang, T., and Liu, Y. (2006) Protective immunity in rats by intranasal immunization with *Streptococcus mutans* glucan-binding protein D encapsulated into chitosan-coated poly(lactic-co-glycolic acid) microspheres. Biotechnol. Lett. *28*, 1299–1304.

Zhu, L., Kreth, J., Cross, S.E., Gimzewski, J.K., Shi, W., and Qi, F. (2006) Functional characterization of cell wall associated protein WapA in *Streptococcus mutans*. Microbiology *152*, 2395–2404.

Molecular Windows into the Pathogenic Properties of *Aggregatibacter actinomycetemcomitans*: a status report with a view to the future

6

Daniel H. Fine, David Figurski, Scott Kachlany and Jeffrey Kaplan

Abstract

Three lines of evidence implicate *A.a.* as an oral pathogen: (1) virulence factors established at the genetic level that affect; attachment to soft and hard tissues, and interference with both innate and acquired arms of host defence, (2) a powerful association of *A.a.* with localized aggressive periodontitis in young adolescents, and (3) prospective evidence that *A.a.* initiates bone loss in an animal model. The chapter will review molecular proofs that have provided evidence of *A.a.* genes that are critical for its pathogenic nature. In addition the chapter will describe new strategies designed to reflect complex bacterial-host interactions. It is suggested that these new approaches may help to develop insights into the role of *A.a.* in disease initiation and progression as well as its role in adaptation to its host. The chapter concludes with a recommendation that our focus on eliminating bacteria might profit from a shift toward exploration of bacterial strategies for adaptation and survival in the environment of the human host. The lessons learned from this paradigm shift might serve to allow us to gain a better understanding of the beneficial outcomes derived from the lives we share with bacteria.

Discovery of *A.a.*: its history and relationship to disease in the oral cavity

A. actinomycetemcomitans (*A.a.*) is a Gram-negative, capnophilic, coccobacillary microorganism, first identified in 1912 by Klinger as a co-inhabiting partner in Actinomycotic lesions identified in man (Zambon, 1985). Newman and colleagues (1976) and Slots (1976) first implicated *A.a.* as a potential oral pathogen in a form of periodontal disease found in young adolescents now called Localized Aggressive Periodontitis (LAP). Further evidence showed that a close correlation existed between LAP and the presence of *A.a.* at the site of infection (Mandell, 1984). It was also shown that a strong immune response could be induced at locally infected sites, as well as systemically in subjects with a well-documented *A.a.* infection (Zambon, 1985). Moreover, *in vitro* experiments suggested that *A.a.* possessed several virulence factors that produced effects resembling pathogenic events; such as the bone loss seen as the disease progressed (Fine *et al.*, 2005).

A.a. has been identified in several systemic infections, including subacute bacterial endocarditis, meningitis, lung, brain and urinary tract infections (van Winkelhoff and Slots, 1999). In all cases in which *A.a.* has been implicated in systemic disease its reservoir was thought to be the oral cavity (van Winkelhoff and Slots, 1999). Clinical microbiologists have placed *A.a.* in the HACEK group of microorganisms; that is, *Haemophilus, Actinobacillus, Cardiobacterium, Eikenella, Kingella,* all of which have been associated with human disease and all of which have been isolated from the oral cavity of humans (Fine *et al.*, 2005; van Winkelhoff and Slots, 1999). While these associations suggested that *A.a.* should receive pathogen status, this inference was given more credence when it was discovered that *A.a.* possessed genes that had pathogenic consequences (Fives-Taylor *et al.*, 1999).

Proofs that implicate *A.a.* as a pathogen

A.a. virulence traits from a pathogenesis point of view

From a pathogenesis point of view, the discovery of the fact that *A.a.* possessed a genuine toxin (leukotoxin), coupled with the significant destructive cellular effects of that toxin, provided a biological plausibility to the likelihood that *A.a.* was related to disease (Haraszthy *et al.*, 2000). Baehni and colleagues (1979) first elucidated the role of *A.a.* leukotoxin as a virulence factor that destroyed human PMNs (Baehni *et al.*, 1979). The toxin was first purified in the supernatant of a highly toxic strain of *A.a.* (Tsai *et al.*, 1984). Leukotoxin is classified as a member of the RTX (repeat in toxin) family of toxins found in a number of pathogenic organisms, as exemplified by: the alpha hemolysin of *Escherichia coli* (diarrhoea, meningitis), the adenylate cyclase of *Bordetella petussis* (whopping cough), and the leukotoxin of *Mannheimia haemolytica* (Lally *et al.*, 1999). The family shares the common consensus sequence GGXGXDX(L/I/V/W/Y/F)X, which is repeated in multiples within the structure of these toxins and appears to be critical for its cytotoxic effects. Members of the family form a diversified group of Gram-negative bacterial pathogens, producing cytolytic toxins, metalloproteinases and lipases (Kachlany *et al.*, 2000a). The primary effect of leukotoxin (Ltx) rests in its ability to interfere with the capacity of polymorphonuclear leukocytes (PMNs) to kill bacteria. PMNs provide one of the host's most formidable initial defensive strategies and so interference with PMN activity confers a significant advantage in a microbe's ability to survive. Of interest is the fact that *A.a.* Ltx, unlike all other RTX toxins, was initially thought to be cell-associated (Kachlany *et al.*, 2002). As such, it was proposed that *A.a.* Ltx could only operate if there was close cell contact between *A.a.* and its target cell, the PMNs. This point has recently been clarified and it is now accepted that *A.a.* Ltx, like other RTX toxins, appears in both the cell-associated and secreted forms (Kachlany *et al.*, 2002). Moreover, in addition to its ability to interfere with PMN function, secretion of Ltx will kill PMNs causing them to release a number of cell- and tissue-destructive enzymes that can further dampen host resistance.

The prominence of this toxin can be illustrated by reviewing the earliest temporal events in inflammatory periodontal disease. As in other oral mucosal infections, the first step requires attachment of the offending oral microorganism, in this case *A.a.*, to the tooth-tissue interface. Following this initial event, the provoking bacteria can migrate along the tooth surface to gain access to the space below the gum line, known as the sulcular space (Fig. 6.1). The sulcular space is a 'U'- or 'V'-shaped space that is bordered on one side by the tooth, on the other side by the sulcular epithelium and at its base by the junctional epithelium. Bacteria such as *A.a.* can then migrate in a apical or downward and lateral direction, eventually coming in contact with the junctional and/or sulcular epithelium.

This bacterial/host cell surface contact results in 'cross-talk' between *A.a.* and the epithelium, stimulating the host cells to produce defensins, chemokines and so on. Bacteria, in response to host cells or as part of their physiological survival mechanism, produce a number of toxic products that can injure the epithelium (Feng and Weinberg, 2006). In addition, the collective presence of these bacterial substances along the basement membrane forms an unfavourable osmotic gradient, encouraging the emigration of host fluids or products from the underlying connective tissue outward into the sulcus. Among these substance, chemokines, tissue destructive enzymes and complement eventually can be found to leak into the sulcus or the sulcular space (Feng and Weinberg, 2006). Further interaction with the epithelium can evoke additional epithelial cross-talk, possible epithelial damage, and penetration of organisms, such as *A.a.*, through the weakened epithelial basement membrane (Fives-Taylor *et al.*, 1999). Once bacterial cells or their products cross the basement membrane barrier and pass into the connective tissue they are then in a position to come in contact with the endothelial vessels directly below the basement membrane. The basement membrane of these vessels then sends off signals that cause leukocytes to collect on the inner side of its vessel wall, a process called margination. Further signalling by substances on the inner vessel wall results in 'diapedesis', or migration of PMNs through these vessel walls in an outward direction into the connective tissue surrounding the vasculature (Tonetti *et al.*, 1998). PMNs follow the signalling gradient, created by bacterial products, that can be traced back to the tooth surface. Alternatively, the tissue reaction to these products can cause the PMNs to reposition themselves by ameboid movement, or chemotaxis, to the area of danger (Madrianos *et al.*, 1997). This directed migration causes the PMNs to follow the gradient toward the connective tissue side of the junctional epithelial basement membrane that borders the tooth. Over time, the PMNs then transmigrate the epithelium following the 'scent' of chemotaxin signals, to emerge from the underlying connective tissue into the sulcular space where they form a band of protective PMNs (Schroeder and Attstrom, 1980) (Fig. 6.2). This wall of PMNs attempts to divert the bacteria from further penetrating the epithelial basement membrane whose function is to prevent any foreign substances from gaining access to the underlying connective tissue space. The principal role of the wall of PMNs, called 'professional phagocytes', is to engulf, degrade, and destroy the foreign bacteria. However, if the PMNs fail to accomplish their goal and the bacteria succeed in killing PMNs, not only is their protective strategy thwarted, but PMN death results in release of numerous

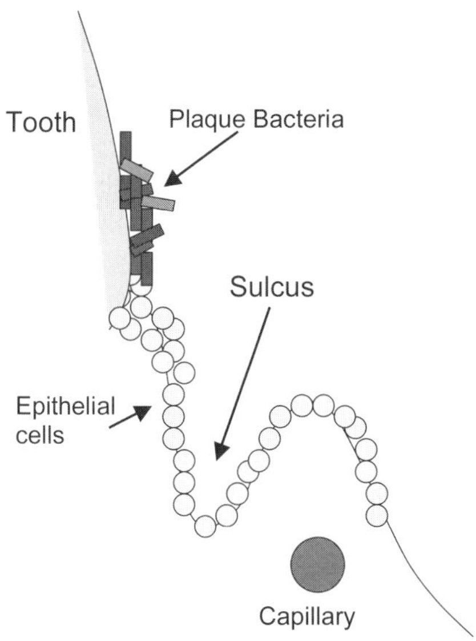

Figure 6.1 Illustration of the gingival sulcus. The junctional epithelium (labelled 'epithelial cells') extends from the tooth surface and forms a 'U'- or 'V'-shaped *cul de sac*. The junctional epithelium is seen on the tooth side and the sulcular epithelium on the opposing side. Note that the vessels (labelled 'capillary') lie directly below the basement membrane of the protective epithelium. Plaque collects on the tooth surface prior to extension and entry into sulcular space.

tissue-damaging enzymes such as, collagenase and hyaluronidase, which will further weaken the epithelial basement membrane barrier (Hooper and Gordon, 2001). Moreover, if the PMNs fail in their mission to destroy the potential invaders, the bacteria continue in their pathogenic journey toward the underlying connective tissue, stimulating lymphatic hyper-reactivity that can lead to damage of the supporting bone (Schroeder and Attstrom, 1980).

Proofs that implicate *A.a.* as a pathogen: virulence traits from a molecular point of view: genes that have been studied extensively

Leukotoxin

From a molecular point of view, identification of the Ltx gene by Kolodrubetz *et al* (1989) and Lally *et al* (1989) brought *A.a.* into the molecular era and provided added credibility to the pathogenic potential of this reputed pathogen.

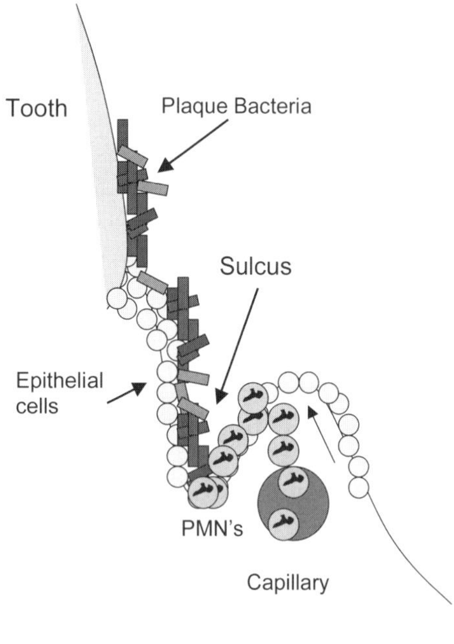

Figure 6.2 Illustrates how plaque on the tooth surface migrates into the sulcus and evokes a response from PMNs that migrate through the connective tissue from the vasculature directly below the sulcus. PMNs form a wall of defensive cells that attempt to protect the underlying connective tissue from the bacterial onslaught.

It is proposed that Ltx interacts with cell surface integrin, CD11a-CD18 (LFA-1) in two distinctly different ways. At low Ltx concentrations, its interaction with LFA-1 causes cells to exhibit changes associated with cell apoptosis. Evidence supporting this pathway demonstrate mitochondrial perturbations occurring in proportion to the concentration of toxin added (Lally *et al.*, 1999). On the other hand, at high concentrations of Ltx, cell necrosis, is associated with the formation of fused transmembrane pores leading to rapid destruction of the PMN cell membrane, is seen (Lally *et al.*, 1999). This form of cell destruction appears to be independent of mitochondrial alteration. As indicated above, apoptosis seems to be the method of cell death; it results in the severe reduction or elimination of host-defence cells, allowing bacterial colonization of the subgingival domain.

Ltx A is a 113 000 mw active protein with glycine repeats at its C-terminus (Fig. 6.3) which appears to be responsible for calcium-binding at this, the toxic portion of the molecule (Haraszthy *et al.*, 2000). However, it has been shown that Ltx increases under anaerobic conditions of growth while downregulation occurs in the presence of an excess of fructose, suggesting that Ltx expression can be regulated by environmental conditions (Inoue *et al.*, 2001; Spitznagel *et al.*, 1995). However, it has also been shown that Ltx is constitutively regulated. In this case, two Ltx promoters, JP2 and 652, have been proposed. The JP2 promoter contains a 530 base-pair deletion in the 3′ end of the promoter sequence. It is thought that this portion of the promoter binds to a repressor protein and thus its absence removes the effect of the repressor, causing increased levels of Ltx secretion (Brogan *et al.*, 1994). This is a reasonable explanation for the fact that those cells with the JP2 promoter appear to secrete more Ltx compared to those that have the promoter region intact, as in the case of the 652 promoter.

As already mentioned, Ltx occurs in two forms, a secreted and cell-bound form. Cell-associated Ltx may be more important as *A.a.* travels through the blood and lymph where it can affect 'pavementing' PMNs and lymphocytes or red blood cells. (Ltx has also been shown to have haemolytic capability). Moreover, the cell-

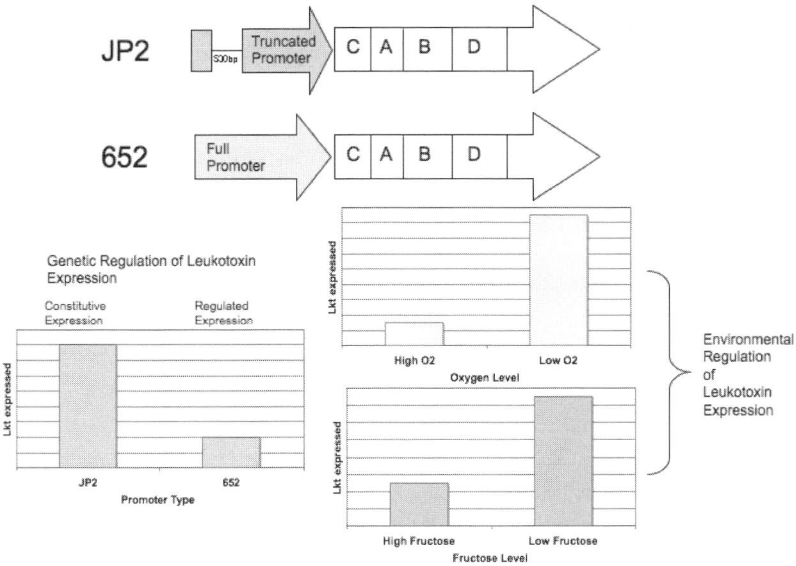

Figure 6.3 Diagram of Leukotoxin (Ltx), its promoter region and the purported regulation of its expression. The Ltx operon consists of four genes (ltx, C, A, B and D, in that order). The first gene, ltxC, encodes an acyltransferase, responsible for modification of the toxin. LtxA is responsible for the toxin structure, while LtxB and LtxD encode proteins required for secretion of the toxin across the bacterial membrane. These genes require post-translational modification to become biologically active. The upper section of the figure shows the gene with its two promoter regions and the bar graphs below illustrate ltx regulation. The bar graph on the lower left panel shows reputed constitutive levels of ltx expression by JP2 and 652 promoter type. Two bar graphs on the right hand side show reputed environmental influences on the expression of ltx. It is proposed that high levels of oxygen and/or fructose signal cells to produce lower levels of ltx expression. (Figure derived from Spitznagel *et al.*, 1995, and Inoue *et al.*, 2001.)

associated form may be more active when *A.a.* is internalized by phagocytes. Secreted Ltx may be more important in connective tissue and when the protein confronts PMNs and macrophages in the sulcus, where Ltx would have a major impact on the first line of defence against professional phagocytes.

Cytolethal distending toxin

To add to the perception that *A.a.* is a periodontal pathogen, a second toxin, the cytolethal distending toxin (Cdt) was subsequently discovered, further implicating *A.a.* as a disease-provoking bacterium. Initially it was hypothesized the Cdt toxin could be responsible for a number of host-cell anomalies including cell distension, growth arrest, nuclear swelling and chromatin fragmentation (Kamen, 1983; Shenker *et al.*, 1982a). While the genes responsible for this pathology were identified in the late 90s (Sugai *et al.*, 1998), the effects of *A.a.* on disturbance of the host cell cycle and growth was discovered in the early 1980s by Shenker *et al.* (1982b). Once again, only when the genes responsible for this effect were identified was this virulence trait considered as a genuine piece of evidence implicating *A.a.* in disease.

Cdt was first discovered in *E. coli* and *Camplylobacter*, in conjunction with diarrhoeal disease (Pickett and Whitehouse, 1999). Cdt toxins are also attributed to a number of bacterial pathogens such as *E. coli*, *C. jejuni*, *Shigella* spp. and *Haemophilus ducreyi* (Pickett and Whitehouse, 1999).

As a tripartite unit complex, CdtA and CdtC form a heterodimeric umbrella or shell that covers or protects the B unit and delivers it to the cell cytoplasm as it enters the cell from the exterior (Fig. 6.4). Once inside the cell, the B unit migrates from the cytoplasm to the nucleus, where it attacks chromatin bodies and acts as a nuclease and/or phosphatase, intoxicating the nucleus and causing cell arrest at the G2/M transition phase. It is postulated that the B-subunit acts to induce breaks in the double strands of DNA, followed by histone phosphorylation (Young *et al.*, 2000). As such, Cdt induces apoptosis for T-cell and

B-cells but not for fibroblasts and is thought to 'hijack' the controls of these cells, leading to the classification of Cdt as 'cytostatins' or 'cyclomodulins' (Ohara *et al.*, 2004). Recently, it has been shown that a mutation in a single amino acid in CdtB significantly affects the activity of *A.a.* Cdt, however this remains controversial. Concerns exist as to the cell types used, bacterial strains used and conditions chosen for analysis (Fine *et al.*, 1999b). These concerns are not unique to the study of Cdt and are true for any and all investigations of bacterial-host interactions because *in vivo* conditions are difficult to replicate *in vitro*.

It has been proposed that the immunopathogenesis of periodontal disease is mediated by a predominance of either Th1 or Th2 cells or a combination thereof. Supporters of Th1 theory of dominance claim that Th1 cells specific for *A.a.* outer membrane protein (Omp 29) induce bone loss (Komatsuzawa *et al.*, 2002). Alternatively, those who support the Th2 theory of dominance claim that *A.a.*-induced Th2 cells secrete cytokines that act on B-cells resulting in non-productive antibody that cause continuous connective tissue and bone destruction (Gemmel and Seymour, 2004). Moreover, and more in keeping with the effect of Cdt, it has been proposed that T-cell receptors, particularly the gamma, delta, T cell receptors populate certain epithelial tissues which constitute one of the earliest lines of defence in the gingiva. In this scenario Cdt can affect these T-cells which play a role in preventing pathogen-induced cytotoxicity of infected and stressed epithelial cells (Lundqvist and Hammarstrom, 1993).

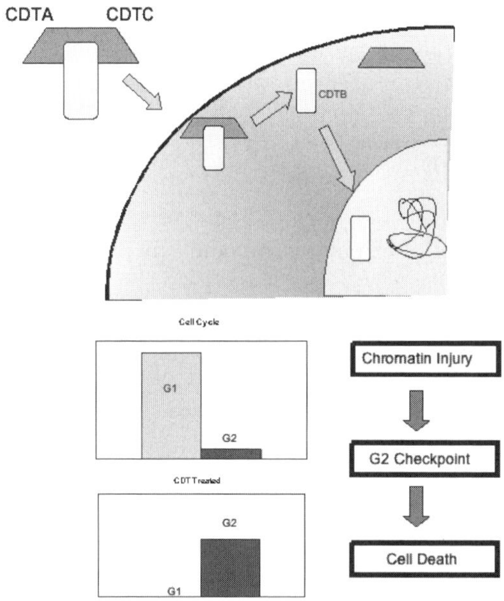

Figure 6.4 Diagram of the effects of Cytolethal distending toxin (Cdt). The *A.a.* Cdt toxin consists of a three-gene operon, as follows: cdtABC. While it is proposed that cdt A and C encode proteins that interact with the host cell surface, cdtB (retangular insert between cdtA and cdtC) encodes the actual toxin. Thus, it is proposed that the actions of A and C facilitate the entry of B into the cell, as seen in the upper left hand corner. As the active B subunit is released in the cytoplasm of the cell, it enters the nucleus where it exhibits DNase activity, resulting in DNA damage leading to G2 arrest and cell distention. (Redrawn from Ohara *et al.*, 2004.)

In summary, the combination of *A.a.* Ltx and Cdt overwhelm the early efforts by the host to incapacitate *A.a.* and thus allow this organism to survive and flourish in the connective tissue environment. It is proposed that *A.a.* can effect both arms of the immune surveillance system. Thus, it can affect the innate arm by means of the interaction of Ltx with LFA-1 receptors carried on cells, such as PMNs and macrophages, that are part of the innate system (Lally *et al.*, 1999). In addition, by interfering with cell-cycle events, *A.a.* Cdt affects the ability of lymphocytes to undergo clonal expansion. By reducing clonal expansion of T cells, Cdt severely compromises the ability of an authentic immune response to be mounted against *A.a.* and thus enables the organism's survival, for extended periods of time, in susceptible host tissue (Shenker *et al.*, 2004). Taken together, these virulence traits, Ltx and Cdt, supply additional ammunition for those who believe that *A.a.* acts as a true pathogen.

Of interest and potential relevance with respect to the relationship of *A.a.* to periodontal disease, it has been proposed that the organism participates in disease in two ways: (1) as an opportunistic pathogen it appears as a member of the resident flora that overgrows when the opportunity arises and then provokes disease in a slowly progressive manner; or (2) as a frankly exogenous pathogen that, when present, can cause rapid and aggressive bone loss. This second scenario is thought to be due to the highly toxic JP2 clonal type, seen in children of African descent; whereas the disease seen in Caucasians is thought to be opportunistic and related to *A.a.* of any clonal type other than the JP2 clone (Haraszthy *et al.*, 2000; Haubek *et al.*, 2001). While stated as a possible explanation for *A.a.* pathogenesis, so far all organisms, other than *A.a.*, that contain either Ltx and/or Cdt toxins have been associated with either human or animal infections (Fabris *et al.*, 2002). The biological nature of these toxins and their close association with other infectious diseases thus provides convincing molecular evidence that *A.a.* is a candidate-pathogen associated with a specific form of periodontal disease, LAP (Haraszthy *et al.*, 2000).

A.a. biofilm and attachment genes

WCI: flp-rcp-tad complex

In addition to the two toxins, Ltx and Cdt, a great deal of interest has been focused on the ability of *A.a.* to attach to and invade tissues. Perhaps the first evidence showing that *A.a.* possesses a characteristic attachment phenotype was pointed out by Scannapieco *et al.*, (1987). Confusion about this characteristic was prompted by the fact that, on subculture in the laboratory, most *A.a.* strains could rapidly lose this phenotype and become smooth and non-adherent. These complications, in addition to the fact that the rough, adherent phenotype was difficult to study in the laboratory, coupled with the fact that there were no genetic tools available to study this phenomenon, put this area of study on hold for several years. The first evidence implicating a gene responsible for the adherence phenotype was identified in 1999 when Inoue and coworkers published the nucleotide sequence of the *flp-1*gene which they derived from the N-terminal protein sequence of isolated and purified fibrillar preparations. Then, Haase and co-workers (1999) by comparing the profiles of the rough and smooth colonies of *A.a.*, isolated a band that was unique to rough colonies; and then, using N-terminal sequencing, identified the rough colony proteins A and B and the *rcpAB* genes associated with these proteins. Almost at the same time, the *A.a. tad* (tight adherence) genes were identified using a random mutagenesis strategy that uncovered a gene locus responsible for non-specific adherence and biofilm formation (Kachlany *et al.*, 2000b). This locus was ultimately recognized as part of the *flp-rcp-tad* genes, organized into what was termed the 'widespread colonization island' (WCI); this was subsequently found in many bacteria and all the *archaea* sequenced to date (Planet *et al.*, 2003). This 14-gene locus, consisting of *flp-1, flp-2, tadV, rcpCAB-tadZABCDEFG*, was shown to be responsible for the adherence, auto-aggregation and biofilm formation of *A.a.* (Planet *et al.*, 2003).

Flp pili are visualized as bundles of long thin fibrils of 50 -70^0 A that extend from the lateral surface of the cell at its polar ends. The *flp-1* gene is responsible for fibril structure while *flp2* appears to have no effect on structure or function (Kachlany *et al.*, 2001). Flp-1 prepilin protein is believed to be cleaved by a prepilin peptidase produced by TadV. Tad G appears to be essential for Flp pilus assembly since mutations in *tad*G fail to produce assembled pilus protein (Planet *et al.*, 2003). TadG may also serve as a membrane anchor for Flp. Tad E and F are processed by Tad V prepilin peptidase and they probably function as transmembrane domains. Tad B and C appear to localize to the inner membrane while rcp A and rcp B are outer membrane components (Fine *et al.*, 2005).

The evolutionary importance of this operon should be considered in light of the fact that adherence to abiotic and biological surfaces is often critical for the survival of microbes in the harsh and turbulent environment in which they live. Extensive phylogenetic analyses have provided significant evidence demonstrating that the earliest bacteria, the *Archaea*, possessed genes in this operon. Moreover, the GC content of the tad loci in many of the 'modern' bacteria we study have significantly different GC content as compared to their own DNA. Therefore, it is highly likely that the WCI, an integrated operon, has been passed from *Archaea* to other bacteria by horizontal gene transfer (Planet *et al.*, 2003). We suggest that this newly acquired mobile genomic island has played a vital role in survival of its host bacteria. The hypothesis has been presented (Planet *et al.*, 2003) that *A.a.* and other members of the family of *Pasteurellaceae,* as well as the genus *Yersinia*, have acquired the WCI from ancestors of *Rhizobiaceae*, most likely by horizontal transfer, a phenomenon dealt with in Chapter 3 of this book.

A.a. has been instrumental in unravelling the secrets of this so called 'pathogenicity island'. It appears that of all the organisms possessing the WCI, *A.a.* has the most complete set of genes in this operon (Planet *et al.*, 2003). With respect to function, it appears that the flp genes, and *flp-1* in particular, are responsible for the structure of the bundled pili seen at the polar ends of *A.a.* (Planet *et al.*, 2003). Thus, when the flp-1 gene is insertionally inactivated, fibril formation ceases, as does attachment to abiotic surfaces. In addition, the Tad A protein is thought to provide the energy, through ATPase activity, required for secretion and assembly of the fibrillar structure (Kachlany *et al.*, 2001).

The strength of attachment of *A.a.* to abiotic and biological surfaces can be used to illustrate its prominent role in colonization. For example, when a fresh isolate of *A.a.* is grown in a tissue culture vessel, set on its side in the laboratory, the entire inoculum attaches and forms a glazed sandy covering that obscures the translucency of the vessel wall (Fine *et al.*, 1999a). When the vessel is stood upright and a sample of broth withdrawn from the centre of the vessel, inoculation of that aliquot onto an enriched agar plate yields no colonies after incubation. This indicates that the broth was sterile, and that all the growth of *A.a.* occurred on the walls of the vessel (Fine *et al.*, 1999b). This remarkable feature of attachment avidity was tested using a shaking water bath assay in which *A.a.* was bound to hydroxyapatite beads and its removal compared to that of *Streptococcus gordonii*, considered to be among the prominent pioneer plaque colonizers (Fine *et al.*, 1999a). In this assay, beads containing *A.a.* and *S. gordonii* were then placed into a larger vessel that was subjected to constant shaking at 200 rev/min in a shaking water bath. One hundred μl aliquots were withdrawn from the vessel supernatant once an hour for 6 hours to determine the number of organisms per microlitre that appeared in the supernatant and were thus detached from the beads as a result of shaking. Over the observation period, *S. gordonii* reached a plateau after 3 hours, while none of the *A.a.* were seen in the supernatant indicating that the shaking did not disturb the latter's binding (Fine *et al.*, 1999a).

Many different methods were used in effort to remove *A.a.* from the beads and neither, high salt, lithium chloride, acid, protease nor heat was successful (Fine *et al.*, 1999a). In contrast, periodate was very effective in detaching *A.a.* from the SHA surface, suggesting that attachment was carbohydrate-dependent and highlighting the biofilm nature of the binding of *A.a.* to abiotic surfaces (Fine *et al.*, 1999a). Difficulty in removing *A.a.* from tooth surfaces has been noted and recently the importance in this process of PGA, a carbohydrate with an N-acetyl galatosamine backbone, has been reported (Kaplan *et al.*, 2001). It is curious that, in spite of the avidity of *A.a.* binding in a relatively non-selective manner, evidence is mounting that initial binding may be quite specific, and dependent on adhesins outside of the WCI, enabling the organism to bind in both a species- and tissue-specific manner (Fine *et al.*, 2005).

The autotransporters

The mouth is a hazardous environment and organisms that survive in this habitat are faced with many extremes of nature that include wide physico-chemical variations. These include temperature (freezing to over 50°C); pH (2.3 to 9.0 depending on the food and drink ingested); currents (depending on flows from saliva, drinking etc.); and physical assaults (due to mastication and tooth-cleaning). Clearly, only the hardiest, most adaptable, and strongest attachers will survive. When the sequence of the *A.a.* genome became available, it was possible to use *in silico* methods to interrogate a region found in *H. influenzae* which was responsible for its binding to epithelium and which was homologous with a gene sequence in *A.a.* Interestingly, the binding of *H. influenzae* to epithelium was tested by examining its binding to buccal epithelial cells, a method developed for studying the binding of oral bacteria to oral mucosa. The binding function of the *hap* adhesin in *H. influenzae* was tested against human buccal epithelial cells. This, coupled with the fact that *Haemophilus spp and Actinobacillus spp* shared many genetic similarities, suggested that the genes for the *hap* adhesins may be present in *A.a.* In the event, the *hap* adhesin region did have homology to a gene sequence in *A.a.* transposon mutagenesis was used to inactivate the autotransporter gene in *A.a.* that shared homology with the *hap* autotransporter gene in *H. influenzae* and thus identified the *aae* gene in *A.a.* (Rose *et al.*, 2003).

Autotransporters consist of three components: (a) a signal or leader sequence, (b) a passenger domain, and (c) a β domain (Henderson *et al.*, 1998). Initially, a pre-adhesin polypeptide exists, in its full tripartite form, in the cytoplasm. This pre-adhesin is directed, by its leader or signal sequence, to contact the inner leaflet of the inner cell membrane of Gram-negative species which contains a sec apparatus (Henderson *et al.*, 2001). A signal peptidase is then directed to cleave the polypeptide which now locates itself at the outer leaflet of the inner membrane. The modified peptide is now released into the periplasm as an intermediate protein consisting of a two-component protein; a β domain and a passenger domain (Henderson *et al.*, 2001). The β domain orients itself in the inner leaflet of the outer membrane forming a β barrel structure which directs a pore opening in the membrane. This barrel structure enables the passenger domain to extrude itself through the pore and exit at the bacterial surface for interaction with host cells (Henderson *et al.*, 2001). The *H. influenzae hap* adhesin, an autotransporter similar to Aae, is formed as a type V secretion and translocation of the hap passenger domain creates an uncleaved adhesin extending from the outer membrane into the external environment for interaction with epithelial cells (St Geme *et al.*, 1994).

Currently, there are five major secretion pathways for Gram-negative bacteria (Henderson *et al.*, 1998). Type I is composed of three proteins that span the inner and outer membrane. Type II is also a multimeric protein but, in this case, requires sec-dependent machinery to move the protein across the inner membrane. Type III is composed of a complex of proteins that also spans both inner and outer membranes and, once again, relies on the sec apparatus. However, in this case, some of the effector molecules cross the inner membrane in a sec-independent manner. Type IV secretion involves 9 proteins interacting with the inner and outer membrane and localized within the periplasm and cytoplasm. Type V, the autotransporter system, is a highly simplified secretion system thought to be the most primitive of the other four secretion systems. All proteins that are secreted in the type V system possess the standard three-component unified structure as described above (Henderson *et al.*, 1998).

At present, *A.a.* is known to have 3 autotransporter proteins; Aae, EmaA, and ApiA (previously known as Omp 100) (Li *et al.*, 2004; Mintz, 2004). Aae differs from the others in that it forms a complete β barrel. In contrast, EmaA and ApiA trimerize four β strands of 70 amino acids each to complete the β barrel unit structure. Translocation of the passenger domain occurs through the β barrel pore. Conventional autotransporters are cleaved by proteases on passage through the outer membrane. This is not true for EmaA and ApiA, since the trimerized translocator domain is covalently anchored in the membrane and so the passenger domain remains uncleaved. While ApiA and Aae appear to mediate attachment of *A.a.* to epithelial cells, EmaA appears to form strong bonds with collagen, which may help to protect *A.a.* from engulfment by macrophages and Langerhans cells in the connective tissue region (Fine *et al.*, 2005).

Expression of the *aae* gene in *E. coli* confirms the functional specificity of this gene as seen in wild type *A.a.* Using this strategy, it could be concluded that this gene, *aae*, allows *A.a.* to bind in a species-specific manner to BECs derived from humans and Old-World monkeys. Moreover, this approach provided conclusive evidence that *A.a.* demonstrated tissue tropism and thus bound to BECs, as opposed to epithelial surfaces obtained from skin, alveolar, nasal tongue or vaginal epithelium. In addition, a similar transposon mutagenesis protocol was used to demonstrate that ApiA, another *A.a.* surface adhesin, had the same species-specificity and that a double knock-out of *aae* and *apiA* completely obliterated the ability of *A.a.* to bind to BEC (Fine *et al.*, 2005).

From an evolutionary point of view, it is of interest that *A.a.* possesses three proteins, Ltx, Aae and ApiA showing a high level of specificity for receptors in humans and Old-World monkeys. Also, *A.a.* is found in humans and Old-World monkeys but not in other mammalian species tested, including New-World monkeys.

Expression of these proteins appears to be related to a high degree of species-specificity and tissue tropism (Fine *et al.*, 2005). These findings suggest, but do not prove, that *A.a.* could have evolved 30 000 000 years ago, at a time when it is proposed that New- and Old-World monkeys diverged. However, this proposed evolutionary timeline pales in comparison to the discovery that the WCI dates back to the most primitive bacteria, the *archaea*. Therefore, the flp-rcp-tad complex appears to have been present when bacteria had to adhere to surfaces or be lost in the primeval ooze. In either scenario, a case can be made that *A.a.* is not a newcomer and has been around for quite a while.

Pil A: Type IV pili

Pil A genetics has provided an improved understanding of the natural competence of *A.a.* and, as such, has afforded many new pieces of information about *A.a.* genetics. With regard to competence it has been shown that only certain strains of *A.a.* are capable of natural transformation (Wang *et al.*, 2002). Serotypes a, d and e are competent while, b, c and f have not been shown to take up DNA (Wang *et al.*, 2002). Mutations in the *pilABCD* genes obliterated the ability of competent cells to take up DNA but had no effect on their colonial morphology or adhesive fibrils. These genes then are responsible for the Type IV secretion system in *A.a.* and, as such, PilA is perceived to be a prepilin protein, Pil B and C have been shown to be transporter proteins whereas Pil D has a prepilin peptidase activity. *A.a.* biofilm is sensitive to DNase which suggests that PilA (Fig. 6.5) may provide the surface exposed binding domain for extracellular DNA (Wang *et al.*, 2003).

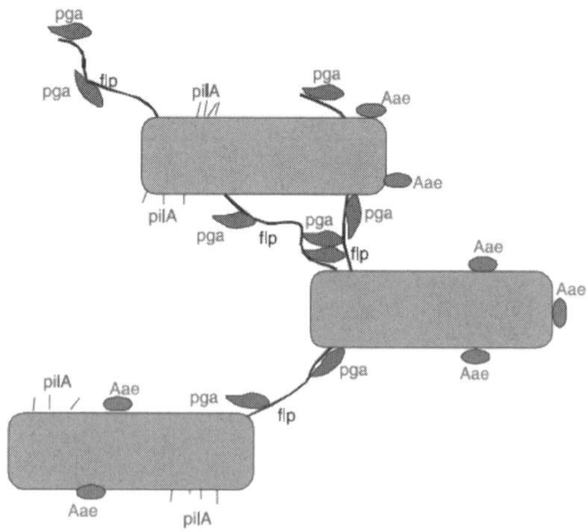

Figure 6.5 Complex interactions related to *A.a.* attachment. Flp designates the Flp fibrils that extend from the *A.a.* surface and appear to modulate binding of *A.a.* to abiotic surfaces in a non-specific manner. Aae represent the adhesins identified as modulating binding to buccal epithelial cells in a species-specific manner. Thus, *A.a.* appears to bind to human and Old World monkey BECs but not to BECs from other mammalian species. ApiA appears to be a second adhesin that is surface-related and is responsible for binding to BECs obtained from human and Old World monkeys. PGA is a β, 1–6 linked hexosamine that appears to interact with Flp, causing cross-linking of organisms through auto-aggregative associations. PilA appears to bind DNA and also appears to modulate auto-aggregation. The pil genes are also involved in the ability of *A.a.* to undergo natural transformation which may also reflect their ability to bind DNA. (Reprinted from Fine *et al.*, 2005.)

PGA: The glue that holds cells together

In addition to fibrils, adhesins, and pili, it appears that extracellular carbohydrates and DNA are also involved in the complex interactions that form the basis for the adherence and biofilm formation of *A.a.* PGA is a β-1–6-linked N-acetyl-D-glucosamine that forms a sticky polymer, the production of which is mediated by a four-gene operon, *pgaABCD* (Kaplan *et al.*, 2003; Wang *et al.*, 2004). *Pga* C and D are subunits of a glycosyltransferase that functions by catalysing PGA synthesis. *Pga* C appears to act as a deacetylase and mediates cell to cell aggregation but appears not to be involved with cell surface attachment. It also appears that extracellular DNA, in association with PGA, is involved with biofilm development. In this scenario, cell lysis results in the release of DNA fragments which then associate with the cell surface of *A.a.* to enable increased cell to cell associations. The exact mechanism of this interaction is still unclear but the action of DNase causes significant reduction in biofilm attachment to surfaces, strongly suggesting the involvement of DNA, probably modulated by

PilA, in association with PGA (Kaplan *et al.*, 2004).

The real world of attachment: complex interactions of surface molecules

Our current knowledge about *A.a.* binding can be summarized as follows: The *flp-rcp-tad* locus appears to be responsible for binding to abiotic and biological surfaces, such as the tooth surface (Planet *et al.*, 2003; Wang and Chen, 2005). Adhesins, such as Aae and ApiA, appear to provide *A.a.* with highly selective adhesins that are specific for tissues and for BECs of Old-World monkeys and humans (Fine *et al.*, 2005). PGA is a polysaccharide that interacts with *flp* and *tad* and appears to act as the cement that helps in auto-aggregation of *A.a.* PilA, in association with extracellular DNA, augment the cell to cell binding (Wang *et al.*, 2003). We also know that *A.a.* can bind to BECs derived from rats and cows but via adhesins that differ from Aae and ApiA (Fine *et al.*, 2005). Overall, it is thought that *A.a.* adherence is dependent on a group of interacting genes and gene products that allow *A.a.* to form tight attaching colonies on a variety of oral tissues.

In summary, and examined from another perspective, *A.a.* may not require the typical non-shedding surfaces, such as the tooth, to colonize. Indeed, it can be found in the mouths of predentate children and is seen in and on the buccal mucosal tissue (Rudney *et al.*, 2005). These findings lead us to believe that buccal mucosal surfaces may possibly be the initial site of *A.a.* colonization. Further since *A.a.* appears to be found in the internal cytoplasm of BECs it has been suggested that BECs provide a protective domain that shields the organism from the hazards of the oral environment (Rudney *et al.*, 2005). While interesting, this data is far from conclusive but does suggest that *A.a.* binding initially relies on a high level of specificity. This point is still controversial because of the avid non-specific binding of *A.a.* to abiotic and biological surfaces, mediated by the *tad* genes in conjunction with PGA, would appear to overwhelm the specific adhesins. However, at present it appears that initial binding of *A.a.* requires host and tissue tropism, mediated by autotransporters Aae and ApiA, but that permanency and survival on the tooth surface is more than likely dependent on the *flp-rcp-tad*-PGA complex that is critical for biofilm formation and survival (Fine *et al.*, 2005).

Genetic methods used to explore *A.a.* genes of importance

Background

The elegance of molecular biology relates to the fact that it provides scientists with the opportunity to isolate a particular gene, study it and then to confirm its function by deletion techniques. The wild type organism can be examined, compared to the strain containing the mutated gene, so that the trait of interest can be assessed. This strategy provides the investigator with an almost ideal experimental protocol design; that is, identify a gene responsible for a virulence trait, knock the gene out and test to see the effect of the knock-out in a testable model system.

The era of molecular biology has provided many previously unexplored possibilities for understanding mechanisms of disease pathogenesis. Ever since it became clear that microorganisms were implicated in human diseases, researchers have made efforts to understand host/bacterial interactions. The person credited with formalizing guidelines for providing science with a strategy that examined the cause and effect relationship between microorganisms and infectious diseases was Robert Koch, the German microbiologist who established Koch's postulates which he used to study *Mycobacterium tuberculosis* and *Bacillus anthracis* (Falkow, 1988).

While the shortcomings of Koch's postulates have been reviewed and revised to better fit our understanding of polymicrobial infections, they nevertheless still prove to be useful (Falkow, 1988). Falkow saw the value of adapting these postulates for studies of 'virulence genes' in suspected pathogens. Thus, as stated by Falkow: (1) the phenotype of the organism under investigation should be associated with virulence in the pathogenic species or strains, as compared to the phenotype seen in non-pathogenic strains or species; (2) Inactivation of the gene responsible for the phenotypic trait under observation should lead to a measurable decrease in the virulence or pathogenecity of the organism; (3) Restoration of the phenotype, by replacement of the mutated gene, should lead to an increase in the virulence of the organism. Falkow's adaptation of Koch's postulates has been of tremendous value in identifying genes of interest and, if used in the appropriate bacteria/host environment, can provide important insights into disease. However, a major shortcoming of this approach is that knowledge of the gene of interest is a pre-requisite in experimental designs. In the following section, methods of gene mobilization used to identify genes of prominence in *A.a.*, are described.

To detect and examine genes of pathogenic importance a molecular toolbox is required to enable deletion of genes and their subsequent restoration to study the trait of interest. However, if the gene is unknown, random mutagensis can be used by mobilizing a transposon into the organism of interest so that it can randomly 'jump' around in its new host. The expectation is that the transposon will ultimately insert itself into a potential virulence gene and inactivate its function. The gene can then be explored in the context of the pathogenic properties of the organism. Over the years we have learned that the identified trait is more than likely, only one of many virulene traits required for a successful pathogen to survive. Indeed, it is clear that pathogenic species require multiple, well-orchestrated, strategies to survive and thrive in the host environment.

Constructs of *A.a.* mutants

There are several systems for constructing mutants in *A.a.* (Table 6.1). DNA has been introduced into the organism via conjugation, chemical transformation, and electroporation (Wang *et al.*, 2003). Conjugation is more efficient than either transformation or electroporation, but requires a donor cell (*E. coli*) and the use of a mobilizable plasmid. Transformation and electroporation are especially inefficient in *A.a.* and efficiency rates vary widely among the many strains that are used (Fives-Taylor *et al.*, 1999). In addition, recombination has not been well studied in *A.a.*, but low rates of recombination may further decrease the efficiency of mutant generation. However, to overcome some of these problems, a highly competent strain, D7S, has been identified and a system for making most strains of *A.a.* competent for DNA uptake has recently been developed (Wang *et al.*, 2003). As already noted, if the gene in question is known, targeted mutagenesis can be carried out. Two methods for interrupting or deleting a gene have been used. Using standard cloning methods, restriction sites within the gene can be used to insert or replace the gene with a resistance gene marker. In addition, a highly efficient method that does not require cloning procedures utilizes an *in vitro* transposition system (Crosby and Kachlany, 2007; Fine *et al.*, 2005; Kaplan *et al.*, 2004; Rhodes *et al.*, 2005). In this method, a plasmid containing the gene of interest is mixed with the transposon and transposase and incubated for several hours. The DNA mixture is then transformed into *E. coli* and cells are selected based on the antibiotic to which the transposon confers resistance. Insertion of the transposon into the gene of interest is easily carried out using PCR. The advantage of the *in vitro* transposon system over the direct cloning method is that the insertion of the transposon occurs randomly; and so it is possible to obtain mutations at different locations within a gene without having to devise many different cloning strategies.

When a genetic screen or selection can be used to identify potentially unknown genes, random mutagenesis is a very powerful tool. Random mutagenesis is useful if the gene to be studied is not known *a priori*. Random mutagenesis is carried out using a transposon system where the transposase can be induced. In *A.a.*, pVJT128, whose transposon confers kanamycin resistance, and pLOF/Sp, whose transposon confers spectinomycin resistance, have been used successfully (Mintz, 2004). When a mutant exhibiting a desired phenotype (defect in biofilm formation, for example) is isolated, the interrupted gene can be identified using inverse-PCR and further characterized.

Isolation of spontaneous mutants has been carried out by selecting for *A.a.* cells that are resistant to manganese (Haraszthy *et al.*, 2006). While there is really no extra step to generate mutants (since it relies on the natural error rate of DNA replication), a strong genetic selection

Table 6.1 Systems for constructing mutants in *A. actinomycetemcomitans*

Mutagenesis strategy	Transfer mechanism	Antibiotic resistance cassette	DNA used	Gene	Function	Reference
Targeted						
	Conjugation					
		Spectinomycin				
			pGP704	*imp*	Inner membrane protein	(Mintz and Fives-Taylor, 2000)
			pVT1460	*orf859*	*In vivo* induced antigen	(Cao *et al*., 2004)
			pVT1460	*orf1402*	*In vivo* induced antigen	(Cao *et al*., 2004)
			pVT1460	*vapA*	*In vivo* induced antigen	(Cao *et al*., 2004)
			pVT1461	*recA*	DNA recombination	(Mintz, 2004)
			pBluescript II KS	*pilABCD*	Natural transformation	(Wang *et al*., 2003)
			pVT1460	*msrA*	Peptide methionine sulfoxide reductase	(Mintz *et al*., 2002)
		Kanamycin				
			pVT1461	*dam*	DNA adenine methylation	(Wu *et al*., 2006)
			pVT1563	*aae*	Binding to KB cells	(Rose *et al*., 2003)
	Transformation					
		Spectinomycin				
			DNA fragment	*cdtABC*	Immune evasion	(Nalbant *et al*., 2003)
		Kanamycin				
			pMB78	*tdeA*	Toxin and drug export	(Crosby and Kachlany, 2007)
			pMB78	*macB*	Antimicrobial resistance	(Crosby and Kachlany, 2007)
			pMB78	*ltxA*	Immune evasion	(Crosby and Kachlany, 2007)
			pMB78	*flp-2*	Adherence	(Perez *et al*., 2006)
			pMB78	*tadV*	Adherence	(Perez *et al*., 2006)

Table 6.1 Continued

Mutagenesis strategy	Transfer mechanism	Antibiotic resistance cassette	DNA used	Gene	Function	Reference
			pMB78	*rcpB*	Adherence	(Perez *et al.*, 2006)
			pMB78	*pgaC*	PGA synthesis	(Kaplan *et al.*, 2004)
			pMB78	*afeABCD*	Iron acquisition	(Rhodes *et al.*, 2005)
			pMB78	*aae*	Binding to human BECs	(Fine *et al.*, 2005)
			DNA fragment	*aae*	Binding to human BECs	(Rose *et al.*, 2003)
		Tetracycline				
			DNA fragment	*ltxA*	Immune evasion	(Nalbant *et al.*, 2003)
			DNA fragment	*flp-tad*	Adherence	(Wang and Chen, 2005)
	Electroporation					
		Spectinomycin				
			pVT736–1	*orf2*	Plasmid stability	(Galli and LeBlanc, 1997)
			pUS19	*ltxBD*	Export of LtxA	(Guthmiller *et al.*, 1995)
			DNA fragment	*apiA* (*omp100*)	Binding and invasion of KB cells, fibronectin binding, serum resistance	(Asakawa *et al.*, 2003)
			pCR-Script	*apiA* (*omp100*)	Binding and invasion of KB cells, fibronectin binding, serum resistance	(Li *et al.*, 2004)
			pUS19	*ltxA*	Immune evasion	(Kolodrubetz *et al.*, 1995)
			pUS19	*cagE*	Apoptosis inducer	(Teng and Hu, 2003)
		Kanamycin				
			pGEM-T	*luxS*	Quorum sensing	(Fong *et al.*, 2001)
			pGEM-T	*rbsB*	Ribosome binding protein	(James *et al.*, 2006)
			pGEM-T	*arcB*	Sensor kinase	(Fong *et al.*, 2003)

		Ampicillin				
			pGEM	*htpG*	Heat shock protein	(Winston *et al.*, 1996)
Random						
	Conjugation					
		Kanamycin				
			pVJT128	*ltxA*	Immune evasion	(Balashova *et al.*, 2006)
			pVJT128	*flp*	Adherence	(Kachlany *et al.*, 2001)
			pVJT128	*tad*	Adherence	(Kachlany *et al.*, 2000b)
			pVJT128	*wzt/dspA*	OPS (serotype f)	(Kaplan *et al.*, 2001)
			pVJT128	*orf10*	OPS (serotype a)	(Suzuki *et al.*, 2000)
			pVJT128	*dspB*	Biofilm dispersal	(Kaplan *et al.*, 2003)
			pVJT128	*katR*	Catalase	(Thomson *et al.*, 1999)
		Spectinomycin				
			pLOF/Sp	*emaA*	Binding to collagen	(Mintz, 2004)

method is required and considerable effort is involved to identify the mutated genes, if they are not already known. In addition, it is nearly impossible to know if more than one mutation occurs within the same cell.

Heterologous expression of *A.a.* genes in *E. coli*

Broad-host-range plasmids that replicate in *A.a.* have proven to be useful tools for studying *A.a.* gene function and regulation (Thomson *et al.*, 1999). Typically, *A.a.* genomic DNA fragments are ligated into a plasmid vector and the plasmid is then introduced into *A.a.* cells by conjugation or electroporation. These plasmid-harbouring cells can then be used to map genes and promoters, to study gene expression and regulation, and to perform genetic complementation tests (Thomson *et al.*, 1999). The major limitation to these studies is the fact that plasmids exist in multiple copies within each cell (usually 10 or more copies), as opposed to the one or two copies of the chromosome that are present in the cell at any given time. The presence of multiple copies of a gene may result in increased gene expression. In addition, the presence of multiple promoter sequences within the cell can titrate out a small number of transcription factors, thereby altering gene expression. These factors must be considered when using any plasmid-based expression system (Kachlany *et al.*, 2001; Kaplan *et al.*, 2004; Thomson *et al.*, 1999).

In some cases – for example, when large amounts of pure protein are needed for structural or functional studies – it may be necessary or desirable to express *A.a.* genes in a heterologous host, usually *E. coli*. This organism is a good choice for this purpose because its physiology and genetics are well characterized, and numerous expression vectors and protein purification methods have been developed (Fine *et al.*, 2005; Kaplan *et al.*, 2004). In addition, *E. coli* grows rapidly (generation time ~20 min) and can be cultured in defined growth medium using large-scale fermenters. One of the problems associated with the expression of foreign genes in *E. coli* is that foreign promoters do not always function in that organism. This is less of a problem when expressing *A.a.* genes in *E. coli* because *A.a.* and *E. coli* are closely related, phylogenetically, both being members of the class *Gammaproteobacteria* (Planet *et al.*, 2003). Nevertheless, studies in which *A.a.* genes have been expressed at high levels in *E. coli* usually utilize high copy number plasmids that contain strong, inducible *E. coli* promoters. *A.a.* proteins that have been successfully purified from over-expressing strains of *E. coli* are listed in Table 6.2.

Another reason for expressing *A.a.* genes in *E. coli* is to simplify the study of gene structure and function. For example, Fine *et al.* (2005) expressed *A.a. aae*, which encodes a surface-exposed autotransporter protein, in *E. coli* (Fine *et al.*, 2005). They showed that the Aae protein is properly expressed on the surface of *E. coli* cells, and that the Aae protein confers upon *E. coli* the ability to bind to buccal epithelial cells of humans and Old-World primates. By performing these studies in *E. coli*, the authors avoided the problems associated with the strong *A.a.* autoaggregation phenotype, which can confound the results of epithelial cell binding assays. Other researchers have used *E. coli*-based systems to confirm the function of cloned *A.a.* genes. For example, Rhodes *et al.* (2005) showed that expression of the *A.a. afeABCD* operon in *E. coli* confers upon *E. coli* cells the ability to grow under iron-chelated conditions; this confirmed that the *A.a. afeABCD* operon encodes an iron acquisition system (Rhodes *et al.*, 2005). Similarly, Fong *et al.* (2001) showed that cell-free conditioned medium from an *E. coli* strain expressing the *A.a. luxS* gene induced luminescence in *Vibrio harveyi*; this confirmed that *A.a. luxS* encodes the production of the quorum -sensing signal autoinducer II (AI-2) (Fong *et al.*, 2001; Fong *et al.*, 2003). Cao *et al.* (2006) took advantage of the ease with which plasmids can be manipulated in *E. coli* to construct and analyse a series of 16 point mutations in the *A.a.* cytolethal distending toxin genes *cdtA* and *cdtC* (Cao *et al.*, 2006). These studies enabled the authors to identify amino acid residues critical for disulfide bond formation and toxin stability. The *A.a.* genes that have been functionally expressed in *E. coli* are listed in Table 6.3.

Heterologous expression of genes in *E. coli* can also be a useful method for screening

Table 6.2 *A. actinomycetemcomitans* proteins purified from over-expressing strains of *E. coli*

Protein	Function	Reference
Dispersin B (poly-N-acetylglucosamine hydrolase)	Biofilm dispersal	(Kaplan *et al.*, 2003)
Cytoletal distending toxin	Cell-cycle modulating toxin	(Akifusa *et al.*, 2001)
dTDP-6-deoxy-L-lyxo-4-hexulose reductase	O polysaccharide synthesis	(Nakano *et al.*, 2000)
GDP-α-d-mannose-4,6-dehydratase	O polysaccharide synthesis	(Suzuki *et al.*, 2000)
GDP-4-keto-6-deoxy-mannise reductase	O polysaccharide synthesis	(Suzuki and Nakano, 2003)
GDP-6-deoxy-D-talose synthetase	O polysaccharide synthesis	(Maki *et al.*, 2003)

A.a. plasmid libraries for genes that encode a desired function. For example, Li *et al.* (2004) transformed *E. coli* with a plasmid library containing *A.a.* genomic DNA fragments and used an immunoscreening procedure to isolate *E. coli* clones that expressed *A.a.* surface antigens (Li *et al.*, 2004). This procedure enabled the authors to associate several *A.a.* genes involved in epithelial cell invasion. *A.a.* invasion of epithelial cells was first discovered by Meyer *et al.*, 1996; however the specific gene or genes responsible to invasion have yet to be disclosed (Meyer *et al.*, 1996). Similarly, Song *et al.* (2002) screened *E. coli* transformants using the *in vivo*-induced antigen technology (IVIAT) procedure. This facilitated the identification of previously unknown *in vivo*-induced genes that are likely to be involved in virulence (Song *et al.*, 2002). Such genes may be useful targets for diagnostic and therapeutic strategies.

Recent genetic approaches that may pave the way for new discoveries

Microbial genetics has provided a wealth of information regarding disease pathogenesis. Many of the findings relate to the earliest stages in the disease process; namely, attachment and the evasion of host defences in the initial stages of the infection. This focus is primarily due to the fact that many of these early stages can be studied *in vitro* using individual cell models. These have become available since cells can be isolated, maintained and used to determine how the bacteria or bacterial trait of interest can affect the host cell under investigation. Pure cultures of bacteria can now be tested in the presence of a specific host cell to determine the effect of the bacteria on many of the host cell's attributes. In this manner it is possible to assess the way in which bacteria can undermine cell integrity. While this approach has provided many pieces of useful information, the effect that these microbes display in the complicated environment of the host still remains a mystery. To address these more complex issues several novel methods have evolved to enable analysis of *in vivo* -induced genes for the study of bacterial virulence. Three prominent techniques that have permitted exploration of the influence of the host on microbial gene expression are described below in the order of their discovery: (1) *in vivo* expression technology (IVET), (2) signature-tagged mutagenesis (STM), and (3) *in vivo*-induced antigen technology (IVIAT). Microarray technology and *in silico* identification of genes of importance will also be discussed.

IVET

In the IVET strategy, genes turned on in the host environment, but not *in vitro*, are identified by comparing gene expression in the two different environments (Mahan *et al.*, 1993). IVET uses a promoter-trap strategy, and homologous recombination to produce the construct. In the IVET strategy a promoter is chosen on the grounds that it will only induce gene expression under conditions indicating that bacterial cells expressing particular genes are doing so because they favour the host environment. However, several issues need to be taken into account in interpreting the

Table 6.3 *A. actinomycetemcomitans* genes that have been functionally expressed in *E. coli*

Gene	Function	Reference
aae	Buccal epithelial cell adhesion	(Fine *et al.*, 2005)
afeABCD	Iron transport system	(Rhodes *et al.*, 2005)
cdtABC	Cytoletal distending toxin	(Cao *et al.*, 2006; Mayer *et al.*, 1999; Saiki *et al.*, 2004; Shenker *et al.*, 2004; Sugai *et al.*, 1998; Ueno *et al.*, 2006)
fur	Iron-regulated transcription factor	(Haraszthy *et al.*, 2006; Haraszthy *et al.*, 2002)
ltxCABD	Leukotoxin production	(Ebersole *et al.*, 1990; Kolodrubetz *et al.*, 1989; Lally *et al.*, 1991; Lally *et al.*, 1989)
luxS	Synthesis of quorum-sensing signal AI-1	(Fong *et al.*, 2001)
omp100	Outer membrane protein	(Asakawa *et al.*, 2003)
O-PS	Serotype-specific O-polysaccharide	(Nakano *et al.*, 1998; Yoshida *et al.*, 1999; Yoshida *et al.*, 1998)
pgaC	Poly-N-acetylglucosamine synthetase	(Kaplan *et al.*, 2004)

results from these experiments. For example, the animal model used and the environment within that animal can, and usually does, differ from the human host-initiated infection. Also, since the comparison is related to the *in vitro* growth of the microbe, the laboratory media used for that growth can also influence the gene expression. Nevertheless, new genes have been discovered using this technology and recently the technique of Differential Fluorescence Induction (DFI) has been developed to address some of these problems. In DFI, a green fluorescent protein (gfp), combined with Fluorescence Activated Cell Sorting (FACS), is used to identify bacterial proteins that 'light up' *in vivo*. The method does not require *in vitro* growth since microbes with green fluorescence highlight genes expressed *in vivo* (Valdivia and Falkow, 1997).

STM

In contrast to IVET, Signature tagged mutagenesis (STM) uses comparative hybridization methods to isolate mutants unable to be expressed in the host environment (Hensel *et al.*, 1995). The end-result is that the microbe containing the tagged gene does not survive in the host's environment. A series of transposon tags that do not cross-hybridize are generated; pools of separate tags are developed and then injected into an animal. Thus, of the pooled and injected tags the ones of interest are those that do not survive in the host, indicating that the gene deletion is important in bacterial survival *in vivo*. In the case of STM, differential hybridization of host-expressed genes are analysed and those essential for survival in the host are highlighted in that they are discovered only when the tagged bacteria do not survive. For example, using STM, intraperitoneal challenge with *Salmonella typhimurium* resulted in identification of a 32-gene pathogenicity island, SP12, that encodes a type III secretion system required for replication of the organism in macrophages (Hensel *et al.*, 1995). While the random transposon mutagenesis protocol is standard and straightforward, the negative selection of the significant silenced *in vivo* gene is laborious and can be misleading.

IVIAT

In efforts to apply these methods to the study of *A.a.*, a unique and novel method for *in vivo*-induced genes was developed and named In Vivo Antigen Induced Technology (IVIAT) (Handfield *et al.*, 2000). In this technique, sera from patients with Localized Aggressive Periodontitis (LAP) is obtained and used to react with *in vitro*-expressed antigens from a virulent strain of *A.a.* The antibody absorbs out any antigens that are reactive *in vitro* so that the antibody remaining in the patient sera after absorption *in vitro* should react only with undetected *A.a.* antigens. Genomic DNA is prepared from the same strain of *A.a.* to develop an expression library for propagation in an *E. coli* host which can be probed with the absorbed antibody using a colony-blot methodology. In this manner, protein antigens that are theoretically produced in the human host during an *A.a.* infection can be identified and differentiated from *in vitro* expressed antigens. The *A.a.* DNA responsible for antigen expression is extracted and purified from the subcultured *E. coli* clones and the cloned DNA is sequenced to identify the *A.a.* genes of interest. This approach has several shortcomings. For example, not all genes are immunogenic; *E. coli* may not process genes equally as well as the native *A.a.* host; and conditions of gene expression may vary and not be optimal. Nevertheless, at least 10 new ORFs have been identified using this technology; five of them homologous to proteins from *H. influenzae* and three homologous to *Pasteurella multocida*. While the relationship of these genes to infection still needs to be proven, this system comes closest to identifying gene expression in the host.

One other relatively recent system used for gene identification relies on methods requiring *in silico* gene searches. Since the genomes of many human and animal pathogens are available, it is possible to perform database searches to identify homologous regions in a particular microbe and compare its genome to other available genomes. In this manner, gene sequences can be created by PCR simulation but it is also possible to create interrupted genes and to use random or directed mutagenesis protocols as mentioned previously to determine gene function in that particular organism.

Microarray technology

As pointed out by Asikainen and Karched in Chapter 1 of this book, DNA microarray can be used in the detection and identification of bacteria in mixed infections, although it was originally used as a tool in gene expression profiling. Indeed, it is most useful for investigating the gene expression profile of a particular microbe when it undergoes distinct environmental challenges. Microarray technologies use mRNA from the microbe to hybridize with the immobilized DNA from which it arose. Typically, when used to evaluate host responses, the target mammalian cDNA is printed or spotted onto a matrix and RNA isolated from challenged vs. unchallenged tissue is used for hybridization. This method permits the analysis of gene expression during different stages in the infectious cycle and to evaluate microbial as well as host responses to stress. For example, one would expect *A.a.* gene expression during attachment to tissue to differ from its expression during invasion of tissue, or on contact with host PMNs. To explore these differences, a target cDNA library is generated from the fully sequenced *A.a.* genome and printed onto a glass slide. Sample RNA is isolated from *A.a.* living in either a control environment (healthy tissue surface) or in an experimental environment (environment with a high level of inflammation). RNA is used as a template to transcribe cDNA which is labelled with fluorescein in the case of the control, or rhodamine in the case of the experimental microbe. The two labelled samples are mixed and then incubated with the printed substrate for hybridization. Samples in which mRNA is elevated appear red (if active), green (if inactive), or yellow (if partially active). Studies utilizing this technology can be invaluable in establishing host influences on bacterial expression or *vice versa* (Feng and Weinberg, 2006). These methods can be enhanced by examining a collection of all genes transcribed (termed the microbial 'transcriptome') during the infectious cycle. Efforts to assess the *A.a.* transcriptome are still in their infancy but should provide insight into many new bacterial and host co-existence strategies (Handfield *et al.*, 2005). With all of the methods described above, a great deal of oversight and methodological control is required to avoid misinterpretation. However, these methods offer great promise because of the broad range of insights that they can provide.

Conclusions

It is clear that bacteria do not act in a vacuum and that most infections occur in a polymicrobial matrix where bacteria compete with each other for nutrients and for the host's attention in a time-dependent manner. *A.a.* is no exception and so the best way to study these interactions would be to place *A.a. in vivo* in its natural habitat, allowing it to compete with its community partners at different points in the infectious cycle. Measurement of bacterial survival under these conditions of stress should be made at different intervals during the infectious cycle. It is clear that certain bacterial genes will be turned on at different times so that the organism can adapt to the ebbs and flows of the host and microbial environment in an energy-efficient manner. Currently, we know that *A.a.* can attach, invade, avoid defence cells and cause tissue destruction, which strongly suggests that it has attained pathogen status (Fine *et al.*, 2005). Our understanding of the power of *A.a.* to perform these various functions has been reinforced by molecular methods that have clearly demonstrated *A.a.* genes of interest. Moreover, deleting these genes have confirmed that these functions are implicated in virulence. In some

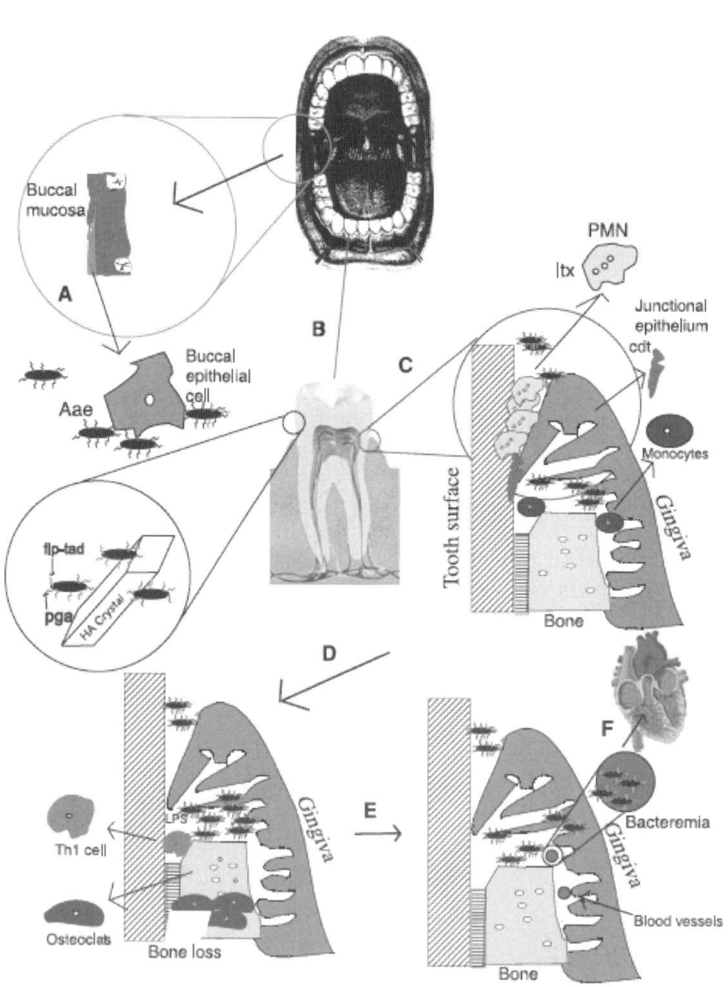

Figure 6.6 Illustration of the relationship of *A.a.* to oral and systemic disease. Panel A shows a proposed way in which *A.a.* first colonizes the mouth via the BEC. Panel B shows how Flp-Tad and PGA all provide *A.a.* with its ability to move from the BEC to the tooth where it encounters a non-shedding surface for a more permanent colonization in the oral cavity. Panel C shows how *A.a.* migrates into the sulcular space and is confronted by PMNs. It is proposed that *A.a.* Ltx can neutralize the PMNs' ability to destroy this foreign invader. This panel also illustrates how Cdt (red arrow-like symbol) is in a position to interact with monocytes in the sub-epithelial space. Then, in conjunction with *A.a.* LPS, is in a position to affect lymphocytes which can then stimulate (as seen in panel D) affected lymphocytes to stimulate precursor cells, initiating bone loss and impairing bone repair. In panel E, it is proposed that *A.a.* can gain access to the bloodstream, collect on heart valves and contribute to an increased risk for heart disease in susceptible subjects. (Reprinted from Fine *et al.*, 2005.) A colour version of this figure is located in the plate section at the back of the book.

cases this has led to discovery of the structure of the protein responsible for this action. In other situations, regulation and modification of gene expression has been explored and defined. This path of discovery, starting at a definition of the responsible gene, its regulation, protein structure and associated function, has provided a wealth of information that would otherwise have been inaccessible, had it not been for the 'molecular microscope' provided by genetic manipulations.

While significant discoveries have been made in the last two decades, it is certain that there is a great deal that still needs to be learned. In the continuing efforts to explore host/parasite interactions and their role in infectious diseases, the next breakthrough will perhaps occur with the development of systems and models allowing scientists to probe genes of interest in environments that are more representative of the interactions seen in the real world. It has become patently clear that microorganisms with pathogenic potential can adapt to a variety of habitats. In this context, *A.a.* has been isolated from many different sites in the body (Fig. 6.6; van Winkelhoff and Slots, 1999). Once a better understanding of normal physiological events is achieved, computer science may be in a position to help in the generation of models that will permit extrapolation from simple to more complex systems. Alternatively, exaggerated and staggered model systems can be developed, allowing the study of disease in its early, middle and late phases.

Since the era of Koch, Pasteur and Ehrlich we have searched for the 'magic bullet' that could eradicate 'undesirable' microbes. In efforts to achieve these ambitions we have often resorted to a 'scorched earth' policy to eliminate 'friendly' bacteria, as well as perceived pathogens. In spite of our successes, we are still working on ways to discover how the latter resist our efforts to eradicate them. At a time when it is manifestly evident that ecological principals are critical to the survival of the planet, maybe it is time to understand that these bacteria are probably part of our own intimate ecosystem and have co-evolved with us. Perhaps it may be time for us to gain a better understanding of the interaction of these 'foreign' cells with our own cells and tissues; and be asking why these targeted bacteria have adapted so well with so little interference with in our own basic ecology. The environmental disturbances caused by these microbes pale in comparison to the environmental disruptions resulting from our own sophisticated energy-dependent societies. Overall, the lessons learned by treating these microscopic creatures as harmful enemies may have limited our ability to learn about our own adaptation, development and survival. Instead of defining ourselves as 'Microbe Hunters' we may want to redirect our 'molecular microscope' through a window that fosters our exploration of microbial survival, in the context of human survival. In this manner we may put ourselves in a better position to understand our past development with an eye toward our future survival.

References

Akifusa, S., Poole, S., Lewthwaite, J., Henderson, B., and Nair, S.P. (2001) Recombinant *Actinobacillus actinomycetemcomitans* cytolethal distending toxin proteis are required to interact to inhibit human cell cycle progression and to stimulate human leukocyte cytokine synthesis. Infect. Immun. *69*, 5925–5930.

Asakawa, R., Komatsuzawa, H., Kawai, T., Yamada, S., Goncalves, R.B., Izumi, S., Fujiwara, T., Nakano, Y., Suzuki, N., Uchida, Y., *et al.* (2003) Outer membrane protein 100, a versatile virulence factor of *Actinobacillus actinomycetemcomitans*. Mol. Microbiol. *50*, 1125–1139.

Baehni, P., Tsai, C.C., McArthur, W.P., Hammond, B.F., and Taichman, N.S. (1979) Interaction of inflammatory cells and oral microorganisms. VIII. Detection of leukotoxic activity of a plaque-derived gram-negative microorganism. Infect. Immun. *24*, 233–243.

Balashova, N.V., Crosby, J.A., Al Ghofaily, L., and Kachlany, S.C. (2006) Leukotoxin confers beta-hemolytic activity to *Actinobacillus actinomycetemcomitans*. Infect. Immun. *74*, 2015–2021.

Brogan, J.M., Lally, E.T., Poulsen, K., Kilian, M., and Demuth, D.R. (1994) Regulation of *Actinobacillus actinomycetemcomitans* leukotoxin expression: analysis of the promoter regions of leukotoxic and minimally leukotoxic strains. Infect. Immun. *62*, 501–508.

Cao, L., Volgina, A., Korostoff, J., and DiRienzo, J.M. (2006) Role of intrachain disulfides in the activities of the CdtA and CdtC subunits of the cytolethal distending toxin of *Actinobacillus actinomycetemcomitans*. Infect. Immun. *74*, 4990–5002.

Cao, S.L., Progulske-Fox, A., Hillman, J.D., and Handfield, M. (2004) *In vivo* induced antigenic determinants of *Actinobacillus actinomycetemcomitans*. FEMS Microbiol. Lett. *237*, 97–103.

Crosby, J.A., and Kachlany, S.C. (2007) TdeA, a TolC-like protein required for toxin and drug export in *Aggregatibacter (actinobacillus) actinomycetemcomitans*. Gene *388*, 83–92.

Ebersole, J., Kraig, E., Bauman, G., Spitznagel, J.K., and Kolodrubetz, D. (1990) Molecular approaches to leucotoxin as a virulence component in *Actinobacillus actinomycetemcomitans*. Arch. Oral Biol. 35, 69S-78S.

Fabris, A.S., DiRienzo, J.M., Wikstrom, M., and Mayer, M.P. (2002) Detection of cytolethal distending toxin activity and cdt genes in *Actinobacillus actinomycetemcomitans* isolates from geographically diverse populations. Oral Microbiol. Immunol. *17*, 231–238.

Falkow, S. (1988) Molecular Koch's postulates applied to microbial pathogenicity. Rev. Infect. Dis. *10*, S274-S276.

Feng, Z., and Weinberg, A. (2006) Role of bacteria in health and disease of periodontal tissues. Periodontol. 2000 *40*, 50–76.

Fine, D.H., Furgang, D., Kaplan, J., Charlesworth, J., and Figurski, D.H. (1999a). Tenacious adhesion of *Actinobacillus actinomycetemcomitans* strain CU1000 to salivary-coated hydroxyapatite. Arch. Oral Biol. *44*, 1063–1076.

Fine, D.H., Furgang, D., Schreiner, H.C., Goncharoff, P., Charlesworth, J., Ghazwan, G., Fitzgerald-Bocarsly, P., and Figurski, D.H. (1999b). Phenotypic variation in *Actinobacillus actinomycetemcomitans* during laboratory growth: implications for virulence. Microbiol. *145 (Pt 6)*, 1335–1347.

Fine, D.H., Velliyagounder, K., Furgang, D., and Kaplan, J.B. (2005) The *Actinobacillus actinomycetemcomitans* autotransporter adhesin *Aae* exhibits specificity for buccal epithelial cells from humans and old world primates. Infect. Immun. *73*, 1947–1953.

Fives-Taylor, P.M., Meyer, D.H., Mintz, K.P., and Brissette, C. (1999) Virulence factors of *Actinobacillus actinomycetemcomitans*. Periodontol. 2000 *20*, 136–167.

Fong, K.P., Chung, W.O., Lamont, R.J., and Demuth, D.R. (2001) Intra- and interspecies regulation of gene expression by *Actinobacillus actinomycetemcomitans* LuxS. Infect. Immun. *69*, 7625–7634.

Fong, K.P., Gao, L., and Demuth, D.R. (2003) luxS and arcB control aerobic growth of *Actinobacillus actinomycetemcomitans* under iron limitation. Infect. Immun. *71*, 298–308.

Galli, D.M., and LeBlanc, D.J. (1997) Identification of a maintenance system on rolling circle replicating plasmid pVT736-1. Mol. Microbiol. *25*, 649–659.

Gemmel, E., and Seymour, G. (2004) Immunoregulatory control of Th1/Th2 cytikine profiles in periodontal disease. Periodontol. 2000 35, 21–41.

Guthmiller, J.M., Kolodrubetz, D., and Kraig, E. (1995) Mutational analysis of the putative leukotoxin transport genes in *Actinobacillus actinomycetemcomitans*. Microb. Pathog. *18*, 307–321.

Haase, E.M., Zmuda, J.L., and Scannapieco, F.A. (1999) Identification and molecular analysis of rough-colony-specific outer membrane proteins of *Actinobacillus actinomycetemcomitans*. Infect. Immun. *67*, 2901–2908.

Handfield, M., Brady, L., Progulske-Fox, A., and Hillman, J.D. (2000) IVIAT: a novel method to select for bacterial genes induced specifically in human infections. Trends Microbiol. *8*, 336–339.

Handfield, M., Progulske-Fox, A., and Hillman, J.D. (2005) *In vivo* induced genes in human diseases. Periodontol. 2000 38, 123–134.

Haraszthy, V.I., Hariharan, G., Tinoco, E.M., Cortelli, J.R., Lally, E.T., Davis, E., and Zambon, J.J. (2000) Evidence for the role of highly leukotoxic *Actinobacillus actinomycetemcomitans* in the pathogenesis of localized juvenile and other forms of early-onset periodontitis. J. Periodontol. *71*, 912–922.

Haraszthy, V.I., Jordan, S.F., and Zambon, J.J. (2006) Identification of Fur-regulated genes in *Actinobacillus actinomycetemcomitans*. Microbiol. *152*, 787–796.

Haraszthy, V.I., Lally, E.T., Haraszthy, G.G., and Zambon, J.J. (2002) Molecular cloning of the fur gene from *Actinobacillus actinomycetemcomitans*. Infect. Immun. *70*, 3170–3179.

Haubek, D., Ennibi, O.K., Poulsen, K., Poulsen, S., Benzarti, N., and Kilian, M. (2001) Early-onset periodontitis in Morocco is associated with the highly leukotoxic clone of *Actinobacillus actinomycetemcomitans*. J. Dent. Res. *80*, 1580–1583.

Henderson, I.R., and Nataro, J.P. (2001) Virulence functions of autotransporter proteins. Infect. Immun. *69*, 1231–1243.

Henderson, I.R., Novarro-Garcia, F., and Nataro, J. (1998) The great escape: structure and function of the autotransporter proteins. Trends Microbiol. *6*, 370–378.

Hensel, M., Shae, J., Gleason, C., Jones, M., Dalton, E., and Holden, D. (1995) Simultaneous identification of bacterial virulence genes by negative selection. Science *269*, 400–403.

Hooper, L., and Gordon, J. (2001) Commensal host-bacterial relationships in the gut. Science *292*, 1115–1118.

Inoue, T., Tanimoto, I., Tada, T., Ohashi, T., Fukui, K., and Ohta, H. (2001) Fermentable-sugar-level-dependent regulation of leukotoxin synthesis in a variably toxic strain of *Actinobacillus actinomycetemcomitans*. Microbiolology *147*, 2749–2756.

James, D., Shao, H., Lamont, R.J., and Demuth, D.R. (2006) The *Actinobacillus actinomycetemcomitans* ribose binding protein RbsB interacts with cognate and heterologous autoinducer 2 signals. Infect. Immun. *74*, 4021–4029.

Kachlany, S.C., Fine, D.H., and Figurski, D.H. (2000a). Secretion of RTX leukotoxin by *Actinobacillus actinomycetemcomitans*. Infect. Immun. *68*, 6094–6100.

Kachlany, S.C., Fine, D.H., and Figurski, D.H. (2002) Purification of secreted leukotoxin (LtxA) from *Actinobacillus actinomycetemcomitans*. Protein Expr. Purif. *25*, 465–471.

Kachlany, S.C., Planet, P.J., Bhattacharjee, M.K., Kollia, E., DeSalle, R., Fine, D.H., and Figurski, D.H. (2000b). Nonspecific adherence by *Actinobacillus actinomycetemcomitans* requires genes widespread in bacteria and archaea. J. Bacteriol. *182*, 6169–6176.

Kachlany, S.C., Planet, P.J., Desalle, R., Fine, D.H., Figurski, D.H., and Kaplan, J.B. (2001) flp-1, the first representative of a new pilin gene subfamily, is required for non-specific adherence of *Actinobacillus actinomycetemcomitans*. Mol. Microbiol. *40*, 542–554.

Kamen, P.R. (1983) Inhibition of keratinocyte proliferation by extracts of Actinobacillus actinomycetemcomitans. Infect. Immun. *42*, 1191–1194.

Kaplan, J.B., Perry, M.B., MacLean, L.L., Furgang, D., Wilson, M.E., and Fine, D.H. (2001) Structural and genetic analyses of O polysaccharide from *Actinobacillus actinomycetemcomitans* serotype f. Infect. Immun. *69*, 5375–5384.

Kaplan, J.B., Ragunath, C., Ramasubbu, N., and Fine, D.H. (2003) Detachment of *Actinobacillus actinomycetemcomitans* biofilm cells by an endogenous beta-hexosaminidase activity. J. Bacteriol. *185*, 4693–4698.

Kaplan, J.B., Velliyagounder, K., Ragunath, C., Rohde, H., Mack, D., Knobloch, J.K., and Ramasubbu, N. (2004) Genes involved in the synthesis and degradation of matrix polysaccharide in *Actinobacillus actinomycetemcomitans* and *Actinobacillus pleuropneumoniae* biofilms. J. Bacteriol. *186*, 8213–8220.

Kolodrubetz, D., Dailey, T., Ebersole, J., and Kraig, E. (1989) Cloning and expression of the leukotoxin gene from *Actinobacillus actinomycetemcomitans*. Infect. Immun. *57*, 1465–1469.

Kolodrubetz, D., Phillips, L.H., Ezzo, P.J., and Kraig, E. (1995) Directed genomic integration in *Actinobacillus actinomycetemcomitans*: generation of defined leukotoxin-negative mutants. Infect. Immun. *63*, 2780–2784.

Komatsuzawa, H., Asakawa, R., Kawai, T., Ochiai, K., Fujiwara, T., Taubman, M.A., Ohara, M., Kurihara, H., and Sugai, M. (2002) Identification of six major outer membrane proteins from *Actinobacillus actinomycetemcomitans*. Gene *288*, 195–201.

Lally, E.T., Golub, E.E., Kieba, I.R., Taichman, N.S., Decker, S., Berthold, P., Gibson, C.W., Demuth, D.R., and Rosenbloom, J. (1991) Structure and function of the B and D genes of the *Actinobacillus actinomycetemcomitans* leukotoxin complex. Microb. Pathogen. *11*, 111–121.

Lally, E.T., Hill, R.B., Kieba, I.R., and Korostoff, J. (1999) The interaction between RTX toxins and target cells. Trends Microbiol. *7*, 356–361.

Lally, E.T., Kieba, I.R., Demuth, D.R., Rosenbloom, J., Golub, E.E., Taichman, N.S., and Gibson, C.W. (1989) Identification and expression of the *Actinobacillus actinomycetemcomitans* leukotoxin gene. Biochem. Biophys. Res. Comm. *159*, 256–262.

Li, L., Matevski, D., Aspiras, M., Ellen, R.P., and Lepine, G. (2004) Two epithelial cell invasion-related loci of the oral pathogen *Actinobacillus actinomycetemcomitans*. Oral Microbiol. Immunol. *19*, 16–25.

Lundqvist, C., and Hammarstrom, M.-L. (1993) T-cell receptor gamma-delta intraepithelial lymphocytes are present in normal and chronically inflamed human gingiva. Immunology *79*, 38–45.

Madrianos, P., Papapanou, P., and Sandros, J. (1997) *Porphyromonas gingivalis* infection of oral epithelium inhibits neutrophil transepithelial migration. Infect. Immun. 65, 3983–3990.

Mahan, M., Slauch, J., and Mekalanos, J. (1993) Selection of bacterial virulence genes that are specifically induced in host tissues. Science *259*, 686–688.

Maki, M., Jarvinen, N., Rabina, J., Maaheimo, H., Mattila, P., and Renkonen, R. (2003) Cloning and functional expression of a novel GDP-6-deoxy-D-talose synthetase from *Actinobacillus actinomycetemcomitans*. Glycobiol. *13*, 295–303.

Mandell, R.L. (1984) A longitudinal microbiological investigation of *Actinobacillus actinomycetemcomitans* and *Eikenella corrodens* in juvenile periodontitis. Infect. Immun. *45*, 778–780.

Mayer, M.P., Bueno, L.C., Hansen, E.J., and DiRienzo, J.M. (1999) Identification of a cytolethal distending toxin locus and features of a virulence-associated region in *Actinobacillus actinomycetemcomitans*. Infect. Immun. *67*, 1227–1237.

Meyer, D., Lippmann, J.E., and Fives-Taylor, P. (1996) Invasion of epithelial cells by *Actinobacillus actinomycetemcomitans*: a dynamic, multistep process. Infect. Immun. *64*, 2988–2997.

Mintz, K.P. (2004) Identification of an extracellular matrix protein adhesin, EmaA, which mediates the adhesion of *Actinobacillus actinomycetemcomitans* to collagen. Microbiol. *150*, 2677–2688.

Mintz, K.P., and Fives-Taylor, P.M. (2000) impA, a gene coding for an inner membrane protein, influences colonial morphology of *Actinobacillus actinomycetemcomitans*. Infect. Immun. *68*, 6580–6586.

Mintz, K.P., Moskovitz, J., Wu, H., and Fives-Taylor, P.M. (2002) Peptide methionine sulfoxide reductase (MsrA) is not a major virulence determinant for the oral pathogen *Actinobacillus actinomycetemcomitans*. Microbiology *148*, 3695–3703.

Nakano, Y., Yoshida, Y., Suzuki, N., Yamashita, Y., and Koga, T. (2000) A gene cluster for the synthesis of serotype d-specific polysaccharide antigen in *Actinobacillus actinomycetemcomitans*. Biochim. Biophys. Acta *1493*, 259–263.

Nakano, Y., Yoshida, Y., Yamashita, Y., and Koga, T. (1998) A gene cluster for 6-deoxy-L-talan synthesis in *Actinobacillus actinomycetemcomitans*. Biochim. Biophys. Acta *1442*, 409–414.

Nalbant, A., Chen, C., Wang, Y., and Zadeh, H.H. (2003) Induction of T-cell apoptosis by *Actinobacillus actinomycetemcomitans* mutants with deletion of ltxA and cdtABC genes: possible activity of GroEL-like molecule. Oral Microbiol. Immunol. *18*, 339–349.

Newman, M.G., Socransky, S.S., Savitt, E.D., Propas, D.A., and Crawford, A. (1976) Studies of the microbiology of periodontosis. J. Periodontol. *47*, 373–379.

Ohara, M., Oswald, E., and Sugai, M. (2004) Cytolethal distending toxin: a bacterial bullet targeted to nucleus. J. Biochem. (Tokyo) *136*, 409–413.

Perez, B.A., Planet, P.J., Kachlany, S.C., Tomich, M., Fine, D.H., and Figurski, D.H. (2006) Genetic analysis of the requirement for flp-2, tadV, and rcpB in *Actinobacillus actinomycetemcomitans* biofilm formation. J. Bacteriol. *188*, 6361–6375.

Pickett, C.L., and Whitehouse, C.A. (1999) The cytolethal distending toxin family. Trends Microbiol. *7*, 292–297.

Planet, P.J., Kachlany, S.C., Fine, D.H., DeSalle, R., and Figurski, D.H. (2003) The Widespread Colonization Island of *Actinobacillus actinomycetemcomitans*. Nat. Genet. *34*, 193–198.

Rhodes, E.R., Tomaras, A.P., McGillivary, G., Connerly, P.L., and Actis, L.A. (2005) Genetic and functional analyses of the *Actinobacillus actinomycetemcomitans* AfeABCD siderophore-independent iron acquisition system. Infect. Immun. *73*, 3758–3763.

Rose, J.E., Meyer, D.H., and Fives-Taylor, P.M. (2003) *Aae*, an autotransporter involved in adhesion of *Actinobacillus actinomycetemcomitans* to epithelial cells. Infect. Immun. *71*, 2384–2393.

Rudney, J.D., Chen, R., and Sedgewick, G.J. (2005) *Actinobacillus actinomycetemcomitans*, *Porphyromonas gingivalis*, and *Tannerella forsythensis* are components of a polymicrobial intracellular flora within human buccal cells. J. Dent. Res. *84*, 59–63.

Saiki, K., Gomi, T., and Konishi, K. (2004) Deletion and purification studies to elucidate the structure of the *Actinobacillus actinomycetemcomitans* cytolethal distending toxin. J. Biochem. *136*, 335–342.

Scannapieco, F.A., Millar, S.J., Reynolds, H.S., Zambon, J.J., and Levine, M.J. (1987) Effect of anaerobiosis on the surface ultrastructure and surface proteins of *Actinobacillus actinomycetemcomitans (Haemophilus actinomycetemcomitans*). Infect. Immun. 55, 2320–2323.

Schroeder, H., and Attstrom, R. (1980) Pocket formation: an hypothesis., Vol II (London,: Academic Press).

Shenker, B.J., Besack, D., McKay, T., Pankoski, L., Zekavat, A., and Demuth, D.R. (2004) *Actinobacillus actinomycetemcomitans* cytolethal distending toxin (Cdt): evidence that the holotoxin is composed of three subunits: CdtA, CdtB, and CdtC. J. Immunol. *172*, 410–417.

Shenker, B.J., Kushner, M.E., and Tsai, C.C. (1982a). Inhibition of fibroblast proliferation by *Actinobacillus actinomycetemcomitans*. Infect. Immun. *38*, 986–992.

Shenker, B.J., McArthur, W.P., and Tsai, C.C. (1982b). Immune suppression induced by *Actinobacillus actinomycetemcomitans*. I. Effects on human peripheral blood lymphocyte responses to mitogens and antigens. J. Immunol. *128*, 148–154.

Slots, J. (1976) The predominant cultivable organisms in juvenile periodontitis. Scand. J. Dent. Res. *84*, 1–10.

Song, Y., Kozarov, E., Walters, S.L., Cao, L., Handfield, M., Hillman, J.D., and Progulske-Fox, A. (2002) Genes if periodontopathogens expressed during disease. Annals Periodontol. *7*, 38–42.

Spitznagel, J., Jr., Kraig, E., and Kolodrubetz, D. (1995) The regulation of leukotoxin production in *Actinobacillus actinomycetemcomitans* strain JP2. Adv. Dent. Res. *9*, 48–54.

St Geme, J.W., III, de la Morena, M.L., and Falkow, S. (1994) A Haemophilus influenzae IgA protease-like protein promotes intimate interaction with human epithelial cells. Mol. Microbiol. *14*, 217–233.

Sugai, M., Kawamoto, T., Peres, S.Y., Ueno, Y., Komatsuzawa, H., Fujiwara, T., Kurihara, H., Suginaka, H., and Oswald, E. (1998) The cell cycle-specific growth-inhibitory factor produced by *Actinobacillus actinomycetemcomitans* is a cytolethal distending toxin. Infect. Immun. 66, 5008–5019.

Suzuki, N., and Nakano, Y. (2003) The production of 6-deoxy-L-talan from *Actinobacillus actinomycetemcomitans* via bacterial coupling *in vitro*. Oral Microbiol. Immunol. *18*, 401–404.

Suzuki, N., Nakano, Y., Yoshida, Y., Nakao, H., Yamashita, Y., and Koga, T. (2000) Genetic analysis of the gene cluster for the synthesis of serotype a-specific polysaccharide antigen in *Actinobacillus actinomycetemcomitans*. Biochim. Biophys. Acta *1517*, 135–138.

Teng, Y.T., and Hu, W. (2003) Expression cloning of a periodontitis-associated apoptotic effector, cagE homologue, in *Actinobacillus actinomycetemcomitans*. Biochem. Biophys. Res. Commun. *303*, 1086–1094.

Thomson, V.J., Bhattacharjee, M.K., Fine, D.H., Derbyshire, K.M., and Figurski, D.H. (1999) Direct selection of IS903 transposon insertions by use of a broad-host-range vector: isolation of catalase-deficient mutants of *Actinobacillus actinomycetemcomitans*. J. Bacteriol. *181*, 7298–7307.

Tonetti, M., Imboden, M., and Lang, N. (1998) Neutrophil migration into the gingival sulcus is associated with transepithelial gradients of interleukin-8 and ICAM -1. J. Periodont. Res. *29*, 386–392.

Tsai, C.C., Shenker, B.J., DiRienzo, J.M., Malamud, D., and Taichman, N.S. (1984) Extraction and isolation of a leukotoxin from *Actinobacillus actinomycetemcomitans* with polymyxin B. Infect. Immun. *43*, 700–705.

Ueno, Y., Ohara, M., Kawamoto, T., Fujiwara, T., Komatsuzawa, H., Oswald, E., and Sugai, M. (2006) Biogenesis of the *Actinobacillus actinomycetemcomitans* cytolethal distending toxin holotoxin. Infect. Immun. *74*, 3480–3487

Valdivia, R., and Falkow, S. (1997) Fluorescence-based isolation of bacterial genes expressed within host cells. Science *277*, 2007–2011.

van Winkelhoff, A.J., and Slots, J. (1999) *Actinobacillus actinomycetemcomitans* and *Porphyromonas gingivalis* in nonoral infections. Periodontology 2000 *20*, 122–135.

Wang, X., Preston, J.F., 3rd, and Romeo, T. (2004) The pgaABCD locus of *Escherichia coli* promotes the synthesis of a polysaccharide adhesin required for biofilm formation. J. Bacteriol. *186*, 2724–2734.

Wang, Y., and Chen, C. (2005) Mutation analysis of the flp operon in *Actinobacillus actinomycetemcomitans*. Gene 3*51*, 61–71.

Wang, Y., Goodman, S.D., Redfield, R.J., and Chen, C. (2002) Natural transformation and DNA uptake signal sequences in *Actinobacillus actinomycetemcomitans*. J. Bacteriol. *184*, 3442–3449.

Wang, Y., Shi, W., Chen, W., and Chen, C. (2003) Type IV pilus gene homologs pilABCD are required for natural transformation in *Actinobacillus actinomycetemcomitans*. Gene *312*, 249–255.

Winston, J.L., Toth, S.I., Roe, B.A., and Dyer, D.W. (1996) Cloning and characterization of the *Actinobacillus actinomycetemcomitans* gene encoding a heat-shock protein 90 homologue. Gene *179*, 199–204.

Wu, H., Lippmann, J.E., Oza, J.P., Zeng, M., Fives-Taylor, P., and Reich, N.O. (2006) Inactivation of DNA adenine methyltransferase alters virulence factors in *Actinobacillus actinomycetemcomitans*. Oral Microbiol. Immunol. *21*, 238–244.

Yoshida, Y., Nakano, Y., Suzuki, N., Nakao, H., Yamashita, Y., and Koga, T. (1999) Genetic analysis of the gene cluster responsible for synthesis of serotype e-specific polysaccharide antigen in *Actinobacillus actinomycetemcomitans*. Biochim. Biophys. Acta *1489*, 457–461.

Yoshida, Y., Nakano, Y., Yamashita, Y., and Koga, T. (1998) Identification of a genetic locus essential for serotype b-specific antigen synthesis in *Actinobacillus actinomycetemcomitans*. Infect. Immun. 66, 107–114.

Young, V., Knox, K., and Schauer, D. (2000) Cytolethal distending toxin sequence and activity in enterhepatic pathogen *Helicobacter hepaticus*. Infect Immun 68, 184–191.

Zambon, J.J. (1985) *Actinobacillus actinomycetemcomitans* in human periodontal disease. J. Clin. Periodontol. *12*, 1–20.

The Molecular Biology of *Porphyromonas gingivalis* 7

Margaret J. Duncan

Abstract

Porphyromonas gingivalis is a Gram-negative oral anaerobe strongly associated with chronic adult periodontitis. The bacterium is relatively easy to grow, produces a number of well-characterized virulence factors, and can be manipulated genetically. The availability of the genome sequence will enhance our understanding of the biology of the *P. gingivalis*, and how it interacts with the environment, other bacteria, and the human host. Beginning with a brief description of the genome, this chapter proceeds with a discussion on strain variability, and the relationship of *P. gingivalis* to other oral and enteric *Bacteroidetes*. The genetic methods we use with *P. gingivalis* are based on those for the enteric *Bacteroides*, and the development and use of these systems is discussed. Among the most studied *P. gingivalis* genes are those that code for functions that either interact directly with the human host cells or protect the bacterium from host activities. Several recent studies are presented to illustrate the technologies and the scope of this research. Finally, new studies to determine the molecular mechanisms of gene regulation are described.

Introduction

Porphyromonas gingivalis is the best-studied Gram-negative oral anaerobe for several reasons. Chief among these is that it has only moderately fastidious growth requirements, and can be easily cultured in the laboratory on blood plates or similarly rich media supplemented with haem and vitamin K (Gibbons and McDonald, 1960). In addition, the organism produces several proteolytic enzymes with activities that are relatively simple to assay, and the initial biochemical characterization of the organism advanced rapidly. Nevertheless, for several years the number of these activities appeared to be limitless and interrelated, however, with the development of genetic systems for the organism this complexity was resolved using reverse genetics and DNA sequencing. Similarly confusing were the multiple names attributed to the bacterium, and the early history of the taxonomy and nomenclature has been recounted by Mayrand and Holt (1988). The organism that became known as *Bacteroides gingivalis* underwent a further name change to *Porphyromonas gingivalis* after physiological and chemical evidence clearly indicated that the organism was sufficiently different from other *Bacteroides* species to warrant reclassification as a new genus (Shah and Collins, 1988). The designation 'Porphyromonas', from the Greek porfura (purple), alludes to the black/purple appearance of colonies growing on blood plates due to the accumulation of haem on the surface of cells. A valuable and comprehensive set of reviews of earlier work with *P. gingivalis* can be found in 'Biology of the Species *Porphyromonas gingivalis*' edited by Haroun Shah, Denis Mayrand, and Robert Genco (1993). Most recently the availability of the complete genome sequence for at least one strain of *P. gingivalis*, has shed further light on pathogenic mechanisms, strain variations in genetic complement, and gene function, and has moved the organism into the realm of modern molecular biology.

The genome

Microbial genomes are now sequenced exclusively by the 'bottom-up, random shot-gun' strategy first used to derive the genomic sequence *H. influenzae* (Fleischmann *et al.*, 1995). Typically, a genome project consists of four phases as shown in Fig. 7.1: first, the preparation of a genomic library of small fragments (ca. 2-kb) obtained by mechanically shearing chromosomal DNA isolated from the organism of interest. Mechanical fragmentation minimizes the biases often observed in libraries made from restriction enzyme-digested DNA due to some enzyme sites being more susceptible to digestion than others. The small DNA fragments are ligated to a high copy number bacterial plasmid, and transformed into *Escherichia coli*. Concurrently, larger insert libraries may be prepared in bacteriophage lambda or plasmid vectors to be used later in the genome assembly phase as backup DNA templates to close sequence and physical gaps. In the second phase, the ends of all the small fragments are sequenced using vector-specific sequencing primers. The goal is to achieve approximately 8-fold coverage of the genome with these small random sequences and for example, for the *P. gingivalis* genome of 2.4 Mb there were approximately 35 000 sequencing runs with an average of 500 readable bases per run. In the third phase, the large number of small random sequences is assembled, usually beginning with 30-bp overlaps, until increasingly larger fragments are built that ultimately can be joined to yield one contiguous sequence. During the final phase, open reading frames of at least 100 amino acids are identified, first by automated annotation based on the BLASTP algorithm (Altschul *et al.*, 1990), and finally by visual inspection.

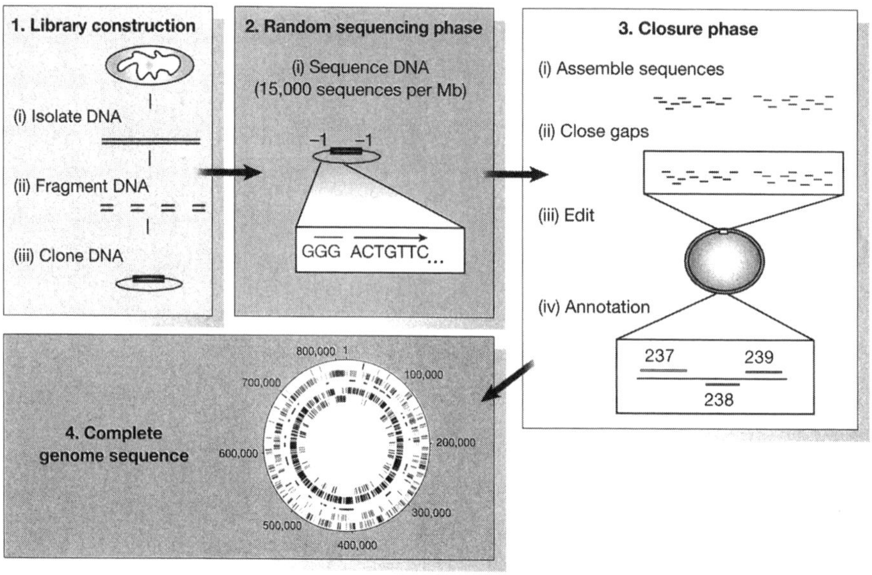

Figure 7.1 The four phases of a whole-genome sequencing project. From Fraser *et al.*, (2000); reprinted with permission.

The genome of the sequenced strain (W83) is approximately 2.34 Mbp and contains a normal complement of rRNA operons (four), and fifty-three tRNAs (Nelson, *et al.*, 2003) and approximately 85% of the genome is coding sequence (information from the Comprehensive Microbial Resource at http://cmr.tigr.org/tigr-scripts/CMR). Because of homologies to open reading frames (ORF) in other organisms, the genome of *P. gingivalis* is relatively well annotated, and as of November 2006, only 201 genes (10% of the genome) were classified as having unknown function. The genome of strain W83 was sequenced at The Institute for Genomic Research (TIGR) and two independent annotations were carried out by TIGR and by the Database Team at the Biosciences Division of the Los Alamos National Laboratory, and are available at: http://cmr.tigr.org/tigr-scripts/CMR and http://www.oralgen.lanl.gov, respectively. The two annotations can be viewed simultaneously at the Bioinformatics Resource for Oral Pathogens (BROP) at http://www.brop.org/ that was developed by Tsute Chen at The Forsyth Institute (Chen *et al.*, 2005).

A surprising feature was the significant amount of the genome (approximately 7%) that contained repeat sequences. Of these, there are 117 copies of insertion sequences (IS) existing as either intact transposases or fragments thereof. Two regions of the genome contain open reading frames that are related to conjugative transposons previously identified in other *Bacteroides* species (Nikolich *et al.*, 1994; Bacic *et al*, 2005) Tabulated information on the conjugative transposons in *P. gingivalis* and the other oral *Bacteroidetes* is available at http://www.oralgen.lanl.gov. The presence of both types of transposon in *P. gingivalis* has numerous implications including their potential to orchestrate the transfer of genetic material between strains, although, as yet, there is no experimental evidence for this. There is evidence, however, for the role of at least one IS in naturally occurring genetic rearrangements in *P. gingivalis* that result in changes in expression of the lys-gingipain cysteine proteinase, a known virulence factor (Lewis and Macrina, 1998).

The DNA homologies provided by repeat sequences make them substrates for intrachromosomal recombination, and there is evidence of homologous recombination between several copies of IS*Pg1* in the genome of W83 (Dong *et al.*, unpublished data). A comparison of IS*Pg1* loci in strains W83 and ATCC 33277 suggested that synteny within the core genome is very similar, but the distribution of ISs is different (Dong *et al.*, 1999). A comparison of genome sequences from more *P. gingivalis* strains will yield greater insights into the extent of genomic rearrangements.

Strain variations

With the goal of correlating characteristics of specific strains to virulence, at least twenty-five phenotypic and genotypic traits of *P. gingivalis* have been examined at the molecular level (Kilian *et al.*, 2006; Koehler *et al.*, 2003). Studies based on multilocus DNA sequencing suggest that the species is panmictic and no one clonal type is associated with disease. Sequencing data indicated hotspots for nucleotide polymorphisms in most of the genes tested (ten) in approximately sixty strains, although the majority of changes were neutral at the protein level (Koehler *et al.*, 2003; Enersen *et al.*, 2006).

A study that compared *P. gingivalis* strains isolated from one hundred and thirty periodontitis patients with those from one hundred and eighty-one healthy subjects used a DNA heteroduplex-based method to detect polymorphisms within the ribosomal internal spacer region (Griffen *et al.*, 1999). There were significant associations between disease and specific heteroduplex types, and that associated with strain W83 was the most dominant. These findings correlated well with animal studies that showed W83 as highly virulent (Genco *et al.*, 1991), and leads to the question of which characteristics of the strain are responsible for virulence. One contender is the polysaccharide capsule that surrounds cells of strains such as W83 that cause a spreading infection in mouse models, compared to the localized abscesses induced by non-capsulated strains (Laine and van Winkelhoff, 1998). A recent comparison of the capsule polysaccharide locus of strain W50, a close relative of strain W83, and non-capsulated strain 381 revealed multiple differences that could potentially lead to loss of capsule production (Aduse-Opoku *et al.*, 2006). Interestingly, strain W83 makes very poor biofilms *in vitro*, yet a mutation in the first gene in this capsule region simultaneously caused a loss of the ability to produce capsule and increased biofilm formation (Davey and Duncan, 2006). A speculation from these findings is that capsule production and poor biofilm growth may be related to the disseminating infection observed with encapsulated strains.

Additional functions potentially associated with the virulence of strains W83 and W50 are encoded in receptor antigen A and B (*ragAB*) genes (Frandsen *et al.*, 2001). Antibodies to these outer membrane proteins were present in the sera of periodontitis patients, while sera from healthy control subjects showed only weak activity (Curtis *et al.*, 1999). Furthermore, *ragAB* genes were detected only in strains that produced spreading infection, e.g. W50, and not in strains 381 and ATCC 33277 (Hanley *et al.*, 1999; Frandsen *et al.*, 2001). In a subsequent study, variant homologues of *ragAB* were detected in virulent and also avirulent strains (Hall *et al.*, 2005), however, while the original alleles (*rag-1*) generally sorted with virulence in animal studies, capsule and fimbrial genotypes also appeared to be factors.

A considerable body of evidence indicates that the genotype of the *fimA* gene, encoding fimbrilin, the subunit protein of *P. gingivalis* major fimbriae, is an important indicator of virulence, and the correlation of type II fimbriae with human infection is highly significant (reviewed by Amano *et al.*, 2004). In a recent elegant genetic study, the type II *fimA* allele was exchanged for the type I allele of the 'avirulent' type strain ATCC 33277, and in the reciprocal exchange the type I allele was incorporated into the virulent strain, hence the effects of *fimA* genotype could be assessed independently of bacterial background (Kato *et al.*, 2006). A number of differences observed in the exchange mutants correlated with their new *fimA* gene, and those associated with the type II gene included a short and peritrichous fimbrial phenotype; increased adherence and invasion of epithelial cells; inhibition of epithelial cell migration and proliferation (potentially related to increased degradation of paxillin); and increased avidity for $\alpha_5\beta_1$ integrin.

Microarrays based on the ORFS of strain W83 are provided by TIGR through the CMR. A microarray-based comparative analysis was used to assess the genetic divergence between strain W83 and the type strain ATCC 33277 at the whole genome level (Chen *et al.*, 2004). The early microarrays used in this study comprised PCR amplicons of median size 461 bp, and were better suited for this type of analysis than the more recent arrays that contain 70 bp amplicons. The protocol is depicted in Fig. 7.2, and briefly, genomic DNA was isolated from each strain (designated A and B), digested with a restriction enzyme, and PCR amplified using random primers. Each sample was labelled with a specific Cy-dye and then mixed for competitive hybridization to the microarrays. After scanning the arrays, normalized signal ratios for each gene were evaluated and ranked with respect to sequence divergence in the two strains. The data indicated that 93% of the genes in both strains showed no significant divergence, and among the remaining 7% (encoding 154 open reading frames) were genes that were known to be different, e.g. *ragB*, IS*Pg4* and IS*Pg5*. Fig. 7.3A presents the distribution of ORFs across the genome of strain W83, and Fig.

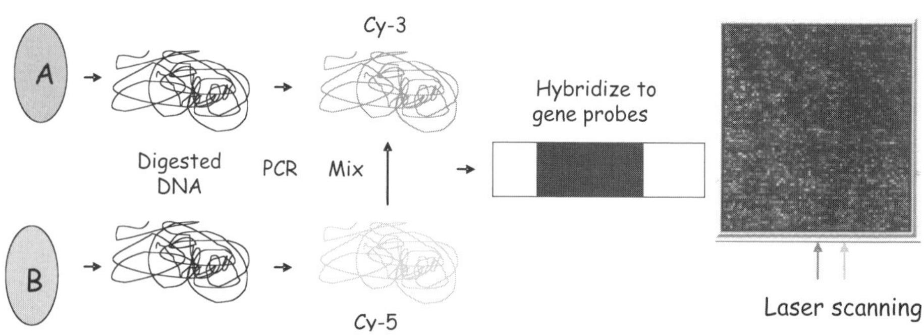

Figure 7.2 Comparison of strains by 'genomotyping'. To compare the gene content and diversity between strains A and B, fragments of their genomic DNA are differentially labelled with Cy3 andCy5 dyes, mixed and hybridized to microarrays. Fluorescent signals are scanned, normalized and scored for divergence. A colour version of this figure is located in the plate section at the back of the book.

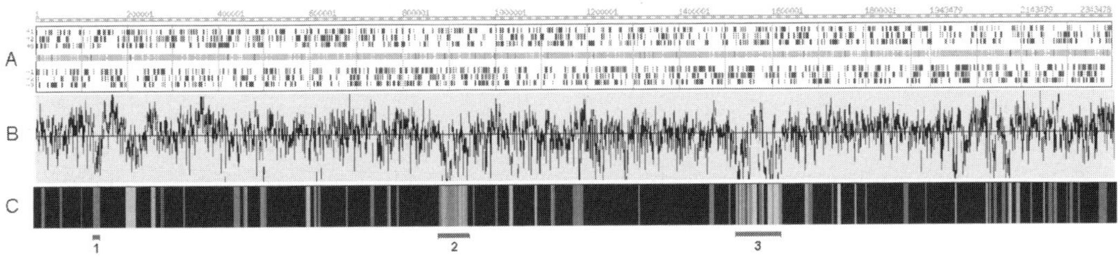

Figure 7.3 Distribution of divergent genes in the genome of strain W83. A. W83 genome and six-frame distribution of predicted genes. B. GC distribution based on G+C content of 500-bp windows across the genome. C. Simulated head map of the distribution of genes classified as slightly (yellow) to highly (red) divergent between strains W83 and ATCC 33277. The black background indicates areas where genes are present in both genomes. Printed with permission (Chen *et al.*, 2004). A colour version of this figure is located in the plate section at the back of the book.

7.3B shows the GC distribution in which several low G+C regions were observed. According to the microarray data, highly divergent regions 1, 2, and 3 in Fig. 7.3C coincided with those of low G+C. Region 1 contained the cluster of genes that encode the putative capsule polysaccharide locus described earlier. Regions 2 and 3 in Fig. 7.3C are approximately 28 and 18 kb, respectively, are paralogues in W83 and are highly divergent in ATCC 33277. Besides having a low G+C content, they are bounded on one side by *Bacteroides* transposon Tn*5520* and on the other by tRNAs, favoured sites for integration of foreign DNA. The regions contain homologues of genes found in *Bacteroides* conjugative transposons and genes encoding conserved hypothetical and proteins of unknown function.

Relationship of *P. gingivalis* to other *Bacteroidetes*

The 16S rRNA-based phylogenetic tree in Fig. 7.4 shows the relationship of *P. gingivalis* to the small number of other *Bacteroidetes* species that have been sequenced. According to the tree, *P. gingivalis* is more closely related to *Tannerella forsythia* than to *Prevotella intermedia* and the enteric *Bacteroides*, and certain biochemical properties substantiate this relatedness, as discussed in more detail by Tanner and Izard (2006). Presented in Table 7.1 is a comparison of the genome content of *P. gingivalis* and the other species, based on the number of genes in each functional category, and the percentage of the genome occupied by each class that potentially reflects the importance of their functions in the biology of the organisms. As noted above, one notable functional category for *P. gingivalis* is that of mobile and extrachromosomal functions which comprises 146 genes occupying 7.34% of genomic space, while in the other species these functions occupy an average 3.85% of the genome. Related to this, *P. gingivalis* is the only organism containing disrupted reading frames caused by the transposition of ISs into structural genes. Interestingly, a single sequence related to IS*Pg4* was found in the *P. intermedia* genome, while relatives of IS*Pg1*, IS*Pg2*, IS*Pg3*, and IS*Pg4* were found in *T. forsythia*. One can only speculate on the origin of these elements within co-inhabitants of the subgingival plaque biofilm, but their increased number and diversity in *T. forsythia* substantiates its closer relationship to *P. gingivalis*. Incidentally,

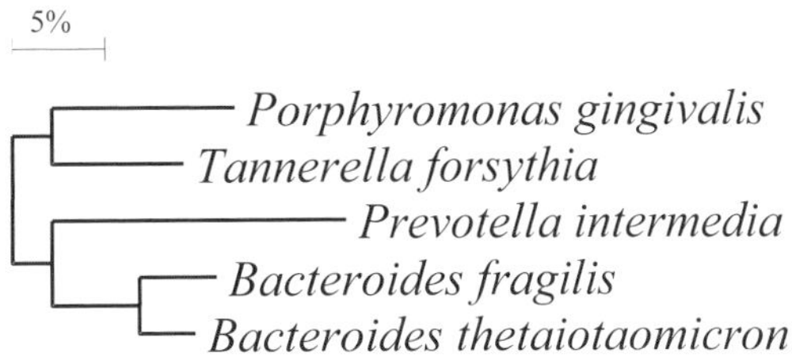

Figure 7.4 Phylogenetic tree of selected *Bacteroidetes* species. Comparative sequence analysis were performed with 16S rRNA sequences. The similarity matrices were corrected for multiple base changes at single positions by the method of Jukes and Cantor (Jukes and Cantor, 1969) and the phylogenetic tree was constructed by the neighbour-joining method of Saitou and Nei (1987). TREECON, a software package for the Microsoft Windows environment, was used for the construction and drawing of evolutionary trees (Van de Peer and De Wachter, 1994). Marker bar represents a 5% difference in nucleotide sequences.

Table 7.1 Comparison of the genome content of oral and enteric *Bacteroidetes* species

	Pg[a] ATCC 3277		*Pi*[b] 17		*Tf*[c] ATCC 43037		*Bf*[d]YCH46		*Bf* NCTC9343		*Bt*[e] VPI-5482	
Genome Size (no. of chromosomes)	2 343 476 bp (1)		2 699 437 bp (2)		3 405 543 bp (1)		5 277 274 bp (1)		5 241 700 bp (2)		6260 361 bp (1)	
Gene Role Category	No. of genes	% of Genes	No. of Genes	% of Genes	No. of Genes	% of Genes	No. of Genes	% of Genes	No. of Genes	% of Genes	No. of Genes	% of Genes
Amino acid biosynthesis	19	0.95	23	0.81	58	1.79	92	2.04	94	2.18	98	2.05
Biosynthesis of cofactors, prosthetic groups, and carriers	85	4.27	63	2.23	92	2.84	104	2.3	108	2.5	89	1.86
Cell envelope	140	7.04	230	8.16	171	5.28	599	13.2	621	14.4	750	15.6
Cellular processes	46	2.31	61	2.16	121	3.73	118	2.61	135	3.13	140	2.93
Central intermediary metabolism	23	1.15	36	1.27	41	1.27	97	2.15	127	2.94	115	2.4
Disrupted reading frame	44	2.21	0	0	0	0	0	0	0	0	0	0
DNA metabolism	81	4.07	91	3.23	91	2.8	153	3.39	169	3.92	154	3.22
Energy metabolism	129	6.48	151	5.36	204	6.3	349	7.74	378	8.77	468	9.79
Fatty acid and phospholipid metabolism	16	0.8	25	0.88	44	1.36	53	1.17	49	1.13	55	1.15
Hypothetical proteins	515	25.9	1089	38.6			324	7.18	710	16.4	214	4.47
Hypothetical proteins-conserved	177	8.9	302	10.7			1144	25.3	565	13.1	1114	23.3

Mobile and extrachromosomal element functions	146	7.34	114	4.04	189	5.83	114	2.52	104	2.41	191	3.99
Protein fate	77	3.87	95	3.37	103	3.18	157	3.48	149	3.45	143	2.99
Protein synthesis	118	5.93	129	4.58	127	3.92	147	3.26	132	3.06	129	2.69
Purines, pyrimidines, nucleosides, and nucleotides	44	2.21	52	1.84	59	1.82	71	1.57	71	1.64	76	1.59
Regulatory functions	43	2.16	66	2.34	47	1.45	165	3.65	223	5.17	238	4.98
Signal transduction	12	0.6	12	0.42	30	0.93	23	0.51	27	0.62	39	0.81
Transcription	30	1.5	36	1.27	78	2.4	89	1.97	106	2.46	100	2.09
Transport and binding functions	118	5.93	109	3.87	239	7.38	300	6.65	348	8.07	333	6.96
Unclassified	0	0	63	2.23	154	4.73	395	8.76	120	2.78	239	5
Unknown function	201	10.1	183	6.49	804	24.81	331	7.34	357	8.28	374	7.82
Viral functions	0	0	0	0	0	0	2	0.04	0	0	0	0

[a]*Pg: P. gingivalis*

[b]*Pi: P. intermedia*

[c]*Tf: T. forsythia*. Data obtained from http:www.oralgen.lanl.gov

[d]*Bf: B. fragilis*

[e]*Bt: B. thetaiotaomicron*

all the oral species contain either complete or fragments of *Bacteroidetes* conjugative transposons.

Another interesting category in the comparison is the class of regulatory functions that deals with signal transduction, i.e. two-component (signal transduction) systems that regulate the expression of other genes in response to environmental changes. Both *P. gingivalis* and *P. intermedia* contain a relatively modest number of such genes, approximately twelve each. In contrast, *T. forsythia* contains thirty genes, on a par with the enteric *Bacteroides* which contain up to thirty-nine; this difference among the oral species raises interesting questions. It has been argued that enteric *Bacteroides*, like *B. thetaiotaomicron*, need a large number of sensing two-component systems, as well as other regulators, to regulate the extensive repertoire of genes involved in utilizing a wide variety of dietary carbohydrates, i.e. the 'glycobiome' (Xu *et al.*, 2004). However, the three oral species occupy the same ecological niche, and one could reason that only a limited number of two-component systems are necessary to contend with this somewhat restricted environment. Therefore, why does *T. forsythia* have more genes? Are they functional in the niche? Do they give the organism a nutritional advantage? On a lighter note, it has been suggested that the number of signal transduction proteins in an organism is its 'bacterial IQ', reflecting metabolic versatility in dealing with complex environments (Galperin, 2005), and also determined whether the organism was an introvert (like *P. gingivalis?*) that was more involved in maintenance of its intracellular environment compared to an extrovert (*T. forsythia?*) that is more astute in sensing and responding to the extracellular milieu.

Tools for molecular genetics

The development of genetic systems for *B. fragilis* and *B. thetaiotaomicron*, together with generosity of the Abigail Salyers laboratory (University of Illinois) in sharing strains and plasmids, were crucial factors that contributed to the development of molecular biology and genetics with *P. gingivalis*. The first *E. coli-Bacteroides* shuttle plasmid was based on the broad host range plasmid RK2 that could be transferred by mating from *E. coli* to *B. fragilis*, followed by selection for antibiotic resistant *Bacteroides* transconjugants (Guiney *et al.*, 1984). In the meantime, the transfer of antibiotic resistance genes between *Bacteroides* species was documented by several groups (Welch *et al.*, 1979; Malamy and Talley, 1981; Smith *et al.*, 1982; Mays *et al.*, 1982) and subsequently indigenous plasmids from the species were developed as shuttle vectors for cloning and targeted mutatgenesis (Guthrie and Salyers, 1986). Next in development, were methods for random mutagenesis based on *Bacteroides* transposons (Shoemaker *et al.*, 1985; Salyers *et al.*, 1987). The translation of these tools and methods to *P. gingivalis* began with the demonstration that *Bacteroides* plasmids could be transferred to and replicate in the organism (Dyer *et al.*, 1992). Subsequently, there followed a flurry of studies to disrupt genes for the already recognized virulence factors, fimbriae (fimbrilin) and gingipain/haemagglutinins (Hamada *et al.*, 1994; Fletcher *et al.*, 1995; Progulske-Fox *et al.*, 1995; Lepine *et al.*, 1996). As in mutagenesis protocols for *Bacteroides*, *P. gingivalis* target genes were cloned into a *Bacteroides* shuttle vector for amplification in *E. coli*. The cloned genes were disrupted with a *Bacteroides* erythromycin resistance cassette, then transferred from *E. coli* to *P. gingivalis* by mating.

Recognizing the utility of random mutagenesis in identifying new potential virulence factors, systems based on the *Bacteroides* transposons Tn*4351* and Tn*4400* were used to generate mutant libraries. Screens for mutants defective in formation of fimbriae (Hoover *et al.*, 1992; Watanabe-Kato *et al.*, 1998; Hayashi *et al.*, 2000), production of surface polysaccharide (Genco *et al.*, 1995; Davey and Duncan, 2006), and pigment production (Chen *et al.*, 2000a; Chen *et al.*, 2000b) have all contributed to the recognition and understanding of new functions that play a role in virulence, gene regulation, and biofilm formation.

While the publication of the genome sequence for *P. gingivalis* strain W83 has made the cloning of genes by PCR amplification efficient and routine, it has also underscored strain differences at the DNA sequence level that range from regions of diversity that contain different genes to single nucleotide polymorphisms (SNPS). The sequencing of at least one more strain, and access to the sequence, will be a valuable and practical resource to the research community.

Most-studied genes

It goes without saying that among the most studied genes are those that encode activities consistently linked to interactions with host cells and pathogenicity. One of the initial steps in *P. gingivalis* infection is colonization of and growth in the gingival sulcus. Within the last decade numerous studies have implicated fimbriae and gingipain-related adhesins in interactions with gingival junctional epithelium.

Studies with either native fimbriae or recombinant fimbrilin (FimA) have focused on its pathogenic effects, its host cell receptors, and ensuing host signalling. Bone resorption in mouse embryo calvariae was induced by native fimbriae and inhibited by fibronectin, suggesting that fimbriae used β_1 integrins as host cell receptors (Kawata *et al.*, 1997). Further supporting this suggestion, the adherence of recombinant type II fimbrilin-coated microspheres to HEp-2 cells was inhibited by anti-$\alpha_5\beta_1$ antibodies in a dose-dependent manner (Nakagawa *et al.*, 2002). The association between recombinant fimbrilin and β_1 integrin from primary gingival epithelial cells was demonstrated by direct binding assays (Yilmaz *et al.*, 2002) and subsequent fimbriae–integrin signalling was documented (Yilmaz *et al.*, 2002; Yilmaz *et al.*, 2003). These data were strengthened with the demonstration that *P. gingivalis* cells and fimbriae competed with extracellular matrix protein ligands for integrin receptors (Nakagawa *et al.*, 2005). Pathways independent of fimbriae-integrin binding were also postulated since *fimA* mutants still showed (reduced) invasion of gingival epithelial cells (Yilmaz *et al.*, 2002), and alternate receptors for fimbriae have been detected (Weinberg *et al.*, 1997).

Gingipains, a family of cysteine proteinases produced by *P. gingivalis*, are present on the cell surface, in extracellular vesicles, and in culture supernatants. The gingipain family comprises four proteins, three of which have catalytic activity. Arg-gingipains A and B (RgpA and RgpB) cleave proteins after arginine residues, while lys-gingipain (Kgp) cleaves after lysine. In addition to their catalytic domains, RgpA and Kgp also possess adhesin domains, while the fourth related protein, haemagglutinin A (HagA), contains only repeats of the adhesin domain. Gingipains also interact with host epithelial cells and several reports have characterized the damaging effects of their proteolytic activities on the integrity of epithelial cell monolayers (Katz *et al.*, 2000; Chen *et al.*, 2001b; Katz *et al.*, 2002). A key study suggested that gingipain adhesin domains may be involved in targeting their associated proteolytic activities to cell surface receptors, and colocalization of the arg-gingipain (RgpA) adhesin domain to fibronectin and its $\alpha_5\beta_1$ integrin receptor was demonstrated (Scragg *et al.*, 1999). The demonstration that this adhesin domain, both native and recombinant, was internalized by oral epithelial cells indicated that the molecular modules required for uptake reside within this sequence (Scragg *et al.*, 2002). Genetic and biochemical approaches have shown that adhesin domains of RgpA and Kgp, and possibly the related domains of HagA, bind to epithelial cells (Chen *et al.*, 2001a). Furthermore, binding of both gingipain adhesins and *P. gingivalis* cells to HEp-2 monolayers was blocked by anti-adhesin domain antibody (Chen and Duncan, 2004).

How an anaerobe protects itself on exposure to air is one of the most intriguing problems in microbiology. As an anaerobe, *P. gingivalis* is somewhat aerotolerant as evidenced by its survival in air in experiments that measure adherence and invasion of epithelial cells *in vitro*. The genes and their products that protect the bacterium in oxidative stress conditions have been studied by genetic methods, and more recently by transcription profiling. Through the activity of superoxide dismutase (Sod) the organism is able to convert reactive oxygen species to hydrogen peroxide (Amano *et al.*, 1990; Nakayama, 1990), and *sod* mutants showed increased sensitivity to air and rapidly lost viability compared to the wild type strain (Nakayama, 1994). However, the organism does not possess catalase so another activity must be responsible for the detoxification of hydrogen peroxide. While NADH oxidase was one candidate for this function, in strain W50 the purified activity was associated with 4-hydroxybutyryl-CoA dehydratase (AbfD), an enzyme involved in glutamate metabolism, and utilization of NADH by the purified enzyme actually generated hydrogen peroxide (Diaz *et al.*, 2004). A database search identified another activity, alkyl hydroperoxide reductase, encoded by the *ahpCF* genes, that reduces alkyl hydroperoxides to non-toxic alcohols using either NADH or NADPH as reducing agent, and Northern analyses suggested that these genes were expressed in strain W50 grown under oxidative conditions (Diaz *et al.*, 2004).

Most recently, mutant analyses have been complemented with transcription profiling to verify activities induced by oxidative stress and identify new stress-responsive activities and their regulators. The basic method for transcriptional profiling is as depicted in Fig. 7.2, except that instead of DNA targets, competitive hybridizations to the microarray slides are carried out with cDNA samples. A recent study investigated *P. gingivalis* genes that were induced during coculture with HEp-2 epithelial cells in air (experimental culture) compared to control bacteria that were incubated alone under the same conditions (Hosogi and Duncan, 2005). In these experiments total RNA was isolated from *P. gingivalis* that adhered to epithelial cells (experimental), and treated to remove contaminating eukaryotic 18 and 23S ribosomal and polyadenylated mRNA species, and enrich for prokaryotic RNA. Experimental and control RNA samples were reverse transcribed to cDNA, and after purification coupled with Cy3 and Cy5 dyes for hybridization to the microarrays. Alkyl hydroperoxide reductase was among the genes induced when *P. gingivalis* was cocultured with epithelial cells under aerobic conditions, supporting a role in removal of peroxide during the oxidative response (Hosogi and Duncan, 2005). In *Bacteroides fragilis* there was also a considerable increase in expression of *ahpCF* on exposure to peroxide or oxygen, and *ahpCF* mutants were more sensitive to peroxide (Rocha and Smith, 1999). The relatedness of *P. gingivalis* and *B. fragilis*, together with the significant amino acid homology between their AhpCF enzymes, suggests not only that they have analogous functions but also that, as in *B. fragils*, the *P. gingivalis* genes may be positively regulated by OxyR (Rocha *et al.*, 2000). This suggestion is supported by several lines of evidence: first, in an *oxyR* mutant there was no detectable expression of *ahpCF* transcripts under anaerobic or aerobic growth conditions, indicating that OxyR was a positive regulator of these genes (Diaz *et al.*, 2006). The data also implied that OxyR expression was constitutive and necessary for anaerobic growth. This work was complemented by a comparative transcriptional profile analysis of anaerobically grown wild type and *oxyR* mutant strains that showed that several genes including *ahpCF*, *sod*, and those encoding oxidative-stress responsive proteins ferritin and thioredoxin, were either directly or indirectly regulated by OxyR (Diaz *et al.*, 2006). Direct binding of recombinant OxyR to the promoter regions of *ahpC* and *sod* was demonstrated by electromobility shift assays, confirming the direct and positive role of the regulator in the expression of these genes (Ohara *et al.*, 2006). In the same study two apparent transcription starts were identified in the *sod* promoter by primer extension; both transcripts were present in the wild type strain, however, one persisted in the *oxyR* mutant suggesting alternate control mechanisms that also coordinate the response to oxidative stress.

Because maintaining viability in the presence of reactive oxygen species is so important to *P. gingivalis*, back-up mechanisms must be available to sense and destroy toxic products, and, not surprisingly, there are many candidates for this role. Increased expression of rubrerythrin (Rbr) in *P. gingivalis* cultures after exposure to peroxide or oxygen suggested a direct or indirect role in survival through its ability to convert hydrogen peroxide to water (Sztukowska *et al.*, 2002). A subsequent study showed that Rbr was important for the growth of *P. gingivalis* in animal models, and afforded protection against reactive nitrogen species that could be generated by monocytes and macrophages (Mydel *et al.*, 2006). Several other proteins that were shown to be protective against hydrogen peroxide included the DNA-binding protein, Dps (Ueshima *et al.*, 2003); a universal stress protein (Chen *et al.*, 2006); and potential regulators included a novel protein, UstA, that was upregulated in stationary phase and may repress the expression of superoxide dismutase and other oxidative stress-responsive activities (Kikuchi *et al.*, 2005); and most recently heat shock protein HtrA was implicated in tolerance to both heat and oxidative stress (Roy *et al.*, 2006).

Gene regulation

Little is known at the molecular level about how *P. gingivalis* senses environmental changes, and how such information is transmitted to the cell. Early experiments showed that while *P. gingivalis* could grow normally at 39°C, the expression of *fimA* and the production of fimbriae decreased (Amano *et al.*, 1994), and a follow up study noted that expression of *fimA* at the mRNA and protein levels was repressed up to 54% at the higher temperature that maybe found in inflamed periodontal pockets (Amano *et al.*, 2001).

Bacteria use two-component signal transduction systems to sense and respond to environmental changes (reviewed by Stock *et al.*, 2000). The first component, a histidine kinase (HK), responds to a specific stimulus by autophosphorylating a designated histidine residue in its sequence. Information is relayed by transfer of activated phosphate from the HK histidine to a designated aspartate residue in the cognate response regulator (RR). It has been established that two-component systems (TCS) play a role in the regulation of virulence in both Gram-negative and -positive bacteria, e.g. the ToxR regulon of *Vibrio cholerae* (Skorupski and Taylor, 1997); the PhoP/PhoQ regulon in *Salmonella* (Garcia *et al.*, 2004); the CovRS system in Group A Streptococcus (Graham *et al.*, 2002); and the PhoP-PhoS system in *Enterococcus faecalis* (Teng *et al.*, 2002). Thus, in relation to disease control by therapeutic intervention, it has been argued that the conserved domains within HKs and RRs represent ideal targets for inhibition of virulence functions (Stephenson and Hoch *et al.*, 2002). Because of these conserved regions, homology searches with the *P. gingivalis* genome sequence revealed a total of four HK-RR pairs, one orphan HK, two orphan RRs, and one HK-RR chimera.

The first TCS in *P. gingivalis* was discovered recently. In a screen for fimbriae-deficient mutants in a random transposon mutant library, eight independent insertions mapped outside the structural gene for fimbrilin to two genes, *fimS* and *fimR*, that encoded sensor histidine kinase and response regulator homologues (Hayashi *et al.*, 2000). The role of FimR as a positive regulator of *fimA* expression was confirmed by *in trans* complementation of the *fimR* mutant with the wild type gene (Nishikawa and Yoshimura, 2001). Microarray-based comparative transcription profiling of parent and *fimR* mutant strains showed that a limited number of genes, clustered around *fimA*, were downregulated in the mutant (Nishikawa *et al.*, 2004). Chromatin immune precipitation (ChIP) with anti-FimR antibody of cross-linked DNA-protein complexes that were formed *in vivo* revealed that FimR could bind to the promoter of *fimX*, the first gene in the *fimA* cluster, and this binding was confirmed in gel shift assays. Thus far, this work shows that the production of fimbriae, which at first glance appeared straightforward, is controlled by a regulation cascade of multiple steps triggered by the activation of FimS-FimR.

Because *P. gingivalis* is a component of the subgingival biofilm it is subjected to changes in nutrient availability and the challenges of living with other organisms in that niche. Thus it was logically anticipated that many functions of *P. gingivalis* would be regulated by quorum sensing mechanisms. The production of extracellular autoinducer (AI-2) by *P. gingivalis* was established using the *Vibrio harveyi* luciferase reporter system (Frias *et al.*, 2001; Chung *et al.*, 2001; Surette and Bassler, 1998). AI-2 production was dependent on the presence of the *luxS* gene, and *luxS* mutants had a pleiotropic phenotype with altered expression of proteins involved in haemin acquisition, gingipain production, and stress responses (Chung *et al.*, 2001; Burgess *et al.*, 2002). Comparative transcriptional profiling of wild type and *luxS* mutant strains indicated that stress-responsive genes were upregulated in the mutants, and possibly responsible for their increased survival, compared to the parent strain, to high pH and oxidative stress conditions (Yuan *et al.*, 2005). Production of A1-2 appears to be growth phase-dependent, with maximal expression occurring at either early or mid-exponential phase (Burgess *et al.*, 2002; James *et al.*, 2006) possibly related to whether the cultures were grown in complex or defined medium. In the latter case, A1-2 appeared to be present and stable in growth medium well into stationary phase, after *luxS* expression had decreased (Burgess *et al.*, 2002). A comparison of wild type and *luxS* mutant strains by differential display PCR suggested that LuxS signalling may modulate the expression of proteins involved in acquisition of haemin (Chung *et al.*, 2001). A more recent report indicates that LuxS-mediated regulation is complex with both induction and repression of proteins involved in iron transport and sequestration, and haemin uptake (James *et al.*, 2006). The topic of cell signalling in oral bacteria is discussed in detail in Chapter 4 of this book.

Conclusions

This chapter focused on the molecular biology of *P. gingivalis*, and clearly we have progressed to the point where the organism can now be manipulated and studied almost as well as *E. coli*. Inevitably, the more we learn about the bacterium the more we are humbled by the realization of how little we understand about the biology of *P. gingivalis* from physiology to gene regulation.

Acknowledgements.

Thanks are due to Dr Bruce Paster for the *Bacteroidetes* phylogenetic tree. Work in the Duncan laboratory was supported by NIH grants DE 10510 and DE 015931.

References

Aduse-Opoku, J., Slaney, J.M., Hashim, A., Gallagher, A., Gallagher, R.P., Rangarajan, M., Boutaga, K., Laine, M. L, Van Winkelhoff, A.J., and Curtis, M.A. (2006) Identification and characterization of the capsular polysaccharide (K-antigen) locus of *Porphyromonas gingivalis*. Infect. Immun. *74*, 449–460.

Altschul, S.F., Gish, W., Miller, W., Myers, E.W., and Lipman, D.J. (1990) Basic local alignment search tool. J. Mol. Biol. *215*, 403–410.

Amano, A., Nakagawa, I., Okahashi, N., and Hamada, N. (2004) Variations of *Porphyromonas gingivalis* fimbriae in relation to microbial pathogenesis. J. Periodont. Res. *39*, 136–142.

Amano, A., Premaraj, T., Kuboniwa, M., Nakagawa, I., Shizukuishi, S., Morisaki, I., and Hamada, S. (2001) Altered antigenicity in periodontitis patients and decreased adhesion of *Porphyromonas gingivalis* by environmental temperature stress. Oral Microbiol. Immunol. *16*, 124–128.

Amano, A., Sharma, A., Sojar, H.T., Kuramitsu, H.K., and Genco, R.J. (1994) Effects of temperature stress on expression of fimbriae and superoxide dismutase by *Porphyromonas gingivalis*. Infect. Immun. *62*, 4682–4685.

Amano, A., Shizukuishi, S., Tamagawa, H., Iwakura, K., Tsunasawa, S., and Tsunemitsu, A. (1990) Characterization of superoxide dismutases purified from either anaerobically maintained or aerated *Bacteroides gingivalis*. J. Bacteriol. *172*, 1457–1463.

Bacic, M., Parker, A.C., Stagg, J., Whitley, H.P., Wells, W.G., Jacob, L.A., and Smith, C.J. (2005) Genetic and structural analysis of the *Bacteroides* conjugative transposon *CTn341*. J Bacteriol. *187*, 2858–2869.

Burgess, N.A., Kirke, D.F., Williams, P., Winzer, K., Hardie, K.R., Meyers, N.L., Aduse-Opoku, J., Curtis, M.A., and Camara, M. (2002) LuxS-dependent quorum sensing in *Porphyromonas gingivalis* modulates protease and haemagglutinin activities but is not essential for virulence. Microbiology. *148*, 763–772.

Chen, T., Abbey, K., Deng, W.J., and Cheng, M.C. (2005) The bioinformatics resource for oral pathogens. Nucleic Acids Res. 33 (Web server issue), W734-W740.

Chen, T., and Duncan, M.J. (2004) Gingipain adhesin domains mediate *Porphyromonas gingivalis* adherence to epithelial cells. Microb. Pathog. *36*, 205–209.

Chen, T., Dong, H., Yong, R, and Duncan, M.J. (2000a). Pleiotropic pigmentation mutants of *Porphyromonas gingivalis*. Microb. Pathog. *28*, 235–247.

Chen, T., Dong, H., Tang, Y.P., Dallas, M.M., Malamy, M.H., and Duncan, M.J. (2000b). Identification and cloning of genes from *Porphyromonas gingivalis* after mutagenesis with a modified Tn*4400* transposon from *Bacteroides fragilis*. Infect. Immun. *68*, 420–423.

Chen, T., Hosogi, Y., Nishikawa, K., Abbey, K., Fleischmann, R.D., Walling, J., and Duncan, M.J. (2004) Comparative whole-genome analysis of virulent and avirulent strains of *Porphyromonas gingivalis*. J. Bacteriol. *186*, 5473–5479.

Chen, T., Nakayama, K., Belliveau, L., and Duncan, M.J. (2001a) *Porphyromonas gingivalis* gingipains and adhesion to epithelial cells. Infect. Immun. *69*, 3048–3056.

Chen, Z., Casiano, C.A., and Fletcher, H.M. (2001b) Protease-active extracellular protein preparations from *Porphyromonas gingivalis* W83 induce N-cadherin proteolysis, loss of cell adhesion, and apoptosis in human epithelial cells. J. Periodontol. *72*, 641–650.

Chen, W., Honma, K., Sharma, A., and Kuramitsu, H.K. (2006) A universal stress protein of *Porphyromonas gingivalis* is involved in stress responses and biofilm formation. FEMS Microbiol. Lett. *264*, 15–21.

Chung, W.O., Park, Y., Lamont, R.J., McNab, R., Barbieri, B., and Demuth, D.R. (2001) Signaling system in *Porphyromonas gingivalis* based on a LuxS protein. J. Bacteriol. *183*, 3903–3909.

Curtis, M.A., Hanley, S.A., and Aduse-Opoku, J. (1999) The *rag* locus of *Porphyromonas gingivalis*: a novel pathogenicity island. J. Periodont. Res. *34*, 400–405.

Davey, M.E., and Duncan. M.J. (2006) Enhanced biofilm formation and loss of capsule synthesis: deletion of a putative glycosyl-transferase in *Porphyromonas gingivalis*. J. Bacteriol. *188*, 5510–5523.

Diaz, P.I., Slakeski, N., Reynolds, E.C., Morona, R., Rogers, A.H., and Kolenbrander, P.E. (2006) Role of *oxyR* in the oral anaerobe *Porphyromonas gingivalis*. J. Bacteriol. *188*, 2454–2462.

Diaz, P.I., Zilm, P.S., Wasinger, V., Corthals, G.L., and Rogers, A.H. (2004) Studies on NADH oxidase and alkyl hydroperoxide reductase produced by

Porphyromonas gingivalis. Oral Microbiol. Immunol. *19*, 137–143.

Dong, H., Chen, T., Dewhirst, F.E., Fleischmann, R.D., Fraser, C.M., and Duncan, M.J. (1999) Genomic loci of the *Porphyromonas gingivalis* insertion element IS*1126*. Infect. Immun. *67*, 3416–3423.

Dyer, D.W., Bilalis, G., Michel, J.H., and Malek, R. (1992) Conjugal transfer of plasmid and transposon DNA from *Escherichia coli* into *Porphyromonas gingivalis*. Biochem. Biophys. Res. Commun. *186*, 1012–1019.

Enersen, M., Olsen, I., van Winkelhoff, A.J., and Caugant, D.A. (2006) Multilocus sequence typing of *Porphyromonas gingivalis* strains from different geographic origins. J. Clin. Microbiol. *44*, 35–41.

Fleischmann, R.D., Adams, M.D., White, O., Clayton, R.A., Kirkness, E.F., Kerlavage, A.R., Bult, C.J., Tomb, J.F., Dougherty, B.A., Merrick, J. M, *et al.* (1995) Whole-genome random sequencing and assembly of *Haemophilus influenzae* Rd. Science. *269*, 496–512.

Fletcher, H.M., Schenkein, H.A., Morgan, R.M., Bailey, K.A., Berry, C.R., and Macrina, F.L. (1995) Virulence of a *Porphyromonas gingivalis* W83 mutant defective in the *prtH* gene. Infect. Immun. *63*, 1521–1528.

Frandsen, E.V., Poulsen, K., Curtis, M.A., and Kilian, M. (2001) Evidence of recombination in *Porphyromonas gingivalis* and random distribution of putative virulence markers. Infect. Immun. *69*, 4479–4485.

Fraser, C.M., Eisen, J.A., and Salzberg, S.L. (2000) Microbial genome sequencing. Nature. *406*, 799–803.

Frias, J., Olle, E., and Alsina, M. (2001) Periodontal pathogens produce quorum sensing signal molecules. Infect. Immun. *69*, 3431–3434.

Galperin, M.Y. (2005) A census of membrane-bound and intracellular signal transduction proteins in bacteria: bacterial IQ, extroverts and introverts. BMC Microbiol. 5, 35.

Garcia Vescovi, E., Soncini, F.C., and Groisman, E.A. (1994) The role of the PhoP/PhoQ regulon in *Salmonella virulence*. Res. Microbiol. *145*, 473–480.

Genco, C.A., Cutler, C.W., Kapczynski, D., Maloney, K., and Arnold, R.R. (1991) A novel mouse model to study the virulence of and host response to *Porphyromonas (Bacteroides) gingivalis*. Infect. Immun. *59*, 1255–1263.

Genco, C.A., Schifferle, R.E., Njoroge, T., Forng, R.Y., and Cutler, C.W. (1995) Resistance of a Tn*4351*-generated polysaccharide mutant of *Porphyromonas gingivalis* to polymorphonuclear leukocyte killing. Infect. Immun. *63*, 393–401.

Gibbons, R.J., and MacDonald, J.B. (1960) Hemin and vitamin K compounds as required factors for the cultivation of certain strains of *Bacteroides melaninogenicus*. J. Bacteriol. *80*, 164–170.

Graham M.R., Smoot, L.M., Migliaccio, C.A., Virtaneva, K., Sturdevant, D.E., Porcella, S.F., Federle, M.J., Adams, G.J., Scott, J.R., and Musser, J.M. (2002) Virulence control in group A Streptococcus by a two-component gene regulatory system: global expression profiling and *in vivo* infection modeling. Proc. Natl. Acad. Sci. USA. *99*, 13855–13860.

Griffen, A.L., Lyons, S.R., Becker, M.R., Moeschberger, M.L., and Leys, E.J. (1999) *Porphyromonas gingivalis* strain variability and periodontitis. J. Clin. Microbiol. *37*, 4028–4033.

Guiney, D.G., Hasegawa, P., and Davis, C.E. (1984) Plasmid transfer *from Escherichia coli* to *Bacteroides fragilis*: differential expression of antibiotic resistance phenotypes. Proc. Natl. Acad. Sci. USA. *81*, 7203–7206.

Guthrie, E.P., and Salyers, A.A. (1986) Use of targeted insertional mutagenesis to determine whether chondroitin lyase II is essential for chondroitin sulfate utilization by *Bacteroides thetaiotaomicron*. J. Bacteriol. *166*, 966–971.

Hall, L.M., Fawell, S.C., Shi, X., Faray-Kele, M.C., Aduse-Opoku, J., Whiley, R.A., and Curtis, M.A. (2005) Sequence diversity and antigenic variation at the *rag* locus of *Porphyromonas gingivalis*. Infect. Immun. *73*, 4253–4262.

Hamada, N., Watanabe, K., Sasakawa, C., Yoshikawa, M., Yoshimura, F., and Umemoto, T. (1994) Construction and characterization of a *fimA* mutant of *Porphyromonas gingivalis*. Infect. Immun. *62*, 1696–1704.

Hanley, S.A., Aduse-Opoku, J., and Curtis, M.A. (1999) A 55-kilodalton immunodominant antigen of *Porphyromonas gingivalis* W50 has arisen via horizontal gene transfer. Infect. Immun. *67*, 1157–1171.

Hayashi, J., Nishikawa, K., Hirano, R., Noguchi, T., and Yoshimura, F. (2000) Identification of a two-component signal transduction system involved in fimbriation of *Porphyromonas gingivalis*. Microbiol. Immunol. *44*, 279–282.

Hoover, C.I., Abarbarchuk, E., Ng, C.Y., and Felton, J.R. (1992) Transposition of Tn*4351* in *Porphyromonas gingivalis*. Plasmid. *27*, 246–250.

Hosogi, Y., and Duncan, M.J. (2005) Gene expression in *Porphyromonas gingivalis* after contact with human epithelial cells. Infect. Immun. *73*, 2327–2335.

James, C.E., Hasegawa, Y., Park, Y., Yeung V, Tribble, G.D., Kuboniwa, M., Demuth, D.R., and Lamont, R.J. (2006) LuxS involvement in the regulation of genes coding for hemin and iron acquisition systems in *Porphyromonas gingivalis*. Infect. Immun. *74*, 3834–3844.

Jukes, T.H., and Cantor, C.R. (1969) Evolution of protein molecules. In Mammalian protein metabolism, H. N. Munro, ed. (Academic Press, Inc., New York, NY) 3, 21–132.

Kato, T., Kawai, S., Nakano, K., Inaba, H., Kuboniwa, M, Nakagawa, I., Tsuda, K., Omori, H., Ooshima, T., Yoshimori, T., and Amano, A. (2007) Virulence of *Porphyromonas gingivalis* is altered by substitution of fimbria gene with different genotype. Cell. Microbiol. 9753–765.

Katz, J., Sambandam, V., Wu, J.H., Michalek, S.M., and Balkovetz, D.F. (2000) Characterization of *Porphyromonas gingivalis*-induced degradation of epithelial cell junctional complexes. Infect. Immun. *68*, 1441–1449.

Katz, J., Yang, Q.B., Zhang, P., Potempa, J., Travis, J., Michalek, S.M., and Balkovetz, D.F. (2002) Hydrolysis of epithelial junctional proteins by *Porphyromonas gingivalis* gingipains. Infect. Immun. *70*, 2512–2418.

Kawata, Y., Iwasaka, H., Kitano, S., and Hanazawa, S. (1997) *Porphyromonas gingivalis* fimbria-stimulated bone resorption is inhibited through binding of the fimbriae to fibronectin. Infect. Immun. *65*, 815–817.

Kikuchi, Y., Ohara, N., Sato, K., Yoshimura, M., Yukitake, H., Sakai, E., Shoji, M., Naito, M., and Nakayama, K. (2005) Novel stationary-phase-upregulated protein of *Porphyromonas gingivalis* influences production of superoxide dismutase, thiol peroxidase and thioredoxin. Microbiology *151*, 841–853.

Kilian, M., Frandsen, E.V., Haubek, D., and Poulsen, K. (2006) The etiology of periodontal disease revisited by population genetic analysis. Periodontol. 2000 *42*, 158–179.

Koehler, A., Karch, H., Beikler, T., Flemmig, T.F., Suerbaum, S., and Schmidt, H. (2003) Multilocus sequence analysis of *Porphyromonas gingivalis* indicates frequent recombination. Microbiology. *149*, 2407–2415.

Laine, M.L., and van Winkelhoff, A.J. (1998) Virulence of six capsular serotypes of Porphyromonas gingivalis in a mouse model. Oral Microbiol. Immunol. *13*, 322–325.

Lepine, G., Ellen, R.P., and Progulske-Fox, A. (1996) Construction and preliminary characterization of three hemagglutinin mutants of *Porphyromonas gingivalis*. Infect. Immun. *64*, 1467–1472.

Lewis, J.P., and Macrina, F.L. (1998) IS*195*, an insertion sequence-like element associated with protease genes in *Porphyromonas gingivalis*. Infect. Immun. *66*, 3035–3042.

Malamy, M.H., and Tally, F.P. (1981) Mechanisms of drug-resistance transfer in *Bacteroides fragilis*. J. Antimicrob. Chemother. *8*, (Suppl D):59–75.

Mayrand, D, and Holt, S.C. (1988) Biology of asaccharolytic black-pigmented *Bacteroides* species. Microbiol. Rev. 52, 134–152.

Mays, T.D., Smith, C.J., Welch, R.A., Delfini, C., and Macrina, F.L. (1982) Novel antibiotic resistance transfer in *Bacteroides*. Antimicrob. Agents Chemother. *21*, 110–118.

Mydel, P., Takahashi, Y., Yumoto, H., Sztukowska, M., Kubica, M., Gibson, F. C. 3rd, Kurtz, D. M. Jr, Travis, J, Collins, L.V., Nguyen, K.A., Genco, C.A., and Potempa, J. (2006) Roles of the host oxidative immune response and bacterial antioxidant rubrerythrin during *Porphyromonas gingivalis* infection. PLoS Pathog. *2*, e76. 712–723.

Nakagawa, I., Amano, A., Inaba, H., Kawai, S., and Hamada, S. (2005) Inhibitory effects of *Porphyromonas gingivalis* fimbriae on interactions between extracellular matrix proteins and cellular integrins. Microbes Infect. *7*, 157–163.

Nakagawa, I., Amano, A., Kuboniwa, M., Nakamura, T., Kawabata, S., and Hamada S. (2002) Functional differences among FimA variants of *Porphyromonas gingivalis* and their effects on adhesion to and invasion of human epithelial cells. Infect. Immun. *70*, 277–285.

Nakayama, K. (1990) The superoxide dismutase-encoding gene of the obligately anaerobic bacterium *Bacteroides gingivalis*. Gene. *96*, 149–150.

Nakayama, K. (1994) Rapid viability loss on exposure to air in a superoxide dismutase-deficient mutant of *Porphyromonas gingivalis*. J. Bacteriol. *176*, 1939–1943.

Nelson, K.E., Fleischmann, R.D., DeBoy, R.T., Paulsen, I.T., Fouts, D.E., Eisen, J.A., Daugherty, S.C., Dodson, R.J., Durkin, A.S., Gwinn, M., Haft, D.H., Kolonay, J.F., Nelson, W.C., Mason, T., Tallon, L., Gray, J., Granger, D, Tettelin, H., Dong, H., Galvin, J.L., Duncan, M.J., Dewhirst, F.E., and Fraser, C.M. (2003) Complete genome sequence of the oral pathogenic bacterium *Porphyromonas gingivalis* strain W83. J. Bacteriol. *185*, 5591–5601.

Nikolich, M.P., Shoemaker, N.B., Wang, G.R., and Salyers, A.A. (1994) Characterization of a new type of *Bacteroides* conjugative transposon, Tcr Emr 7853. J. Bacteriol. *176*, 6606–6612.

Nishikawa, K., and Yoshimura, F. (2001) The response regulator FimR is essential for fimbrial production of the oral anaerobe *Porphyromonas gingivalis*. Anaerobe *7*, 255–262.

Nishikawa, K., Yoshimura, F., and Duncan, M.J. (2004) A regulation cascade controls expression of *Porphyromonas gingivalis* fimbriae via the FimR response regulator. Mol. Microbiol. 54, 546–560.

Ohara, N., Kikuchi, Y., Shoji, M., Naito, M., and Nakayama, K. (2006) Superoxide dismutase-encoding gene of the obligate anaerobe *Porphyromonas gingivalis* is regulated by the redox-sensing transcription activator OxyR. Microbiology. *152*, 955–966.

Progulske-Fox, A., Tumwasorn, S., Lepine, G., Whitlock, J., Savett, D., Ferretti, J.J., and Banas, J.A. (1995) Cloning, expression and sequence analysis of a second *Porphyromonas gingivalis* gene that codes for a protein involved in hemagglutination. Oral Microbiol. Immunol. *10*, 311–318.

Rocha, E.R., and Smith C.J. (1999) Role of the alkyl hydroperoxide reductase (*ahpCF*) gene in oxidative stress defense of the obligate anaerobe *Bacteroides fragilis*. J Bacteriol. *181*, 5701–5710.

Rocha, E.R., Owens, G. Jr, and Smith C.J. (2000) The redox-sensitive transcriptional activator OxyR regulates the peroxide response regulon in the obligate anaerobe *Bacteroides fragilis*. J Bacteriol. *182*, 5059–5069.

Roy, F., Vanterpool, E., and Fletcher, H.M. (2006) HtrA in *Porphyromonas gingivalis* can regulate growth and gingipain activity under stressful environmental conditions. Microbiology. *152*, 3391–3398.

Saitou, N., and Nei, M. (1987) The neighbor-joining method: a new method for reconstructing phylogenetic trees. Mol. Biol. Evol. *4*, 406–425.

Salyers, A.A., Shoemaker, N.B., and Guthrie, E.P. (1987) Recent advances in *Bacteroides* genetics. Crit. Rev. Microbiol. *14*, 49–71.

Scragg, M.A., Alsam, A., Rangarajan, M., Slaney, J.M., Shepherd, P., Williams, D.M., and Curtis, M.A. (2002) Nuclear targeting of *Porphyromonas gingivalis* W50 protease in epithelial cells. Infect. Immun. *70*, 5740–5750.

Scragg, M.A., Cannon, S.J., Rangarajan, M., Williams, D.M., and Curtis, M.A. (1999) Targeted disruption of fibronectin-integrin interactions in human gingival fibroblasts by the RI protease of *Porphyromonas gingivalis* W50. Infect. Immun. *67*, 1837–1843.

Shah, H.N., and Collins, M.D. (1988) Proposal for reclassification of *Bacteroides asaccharolyticus*, *Bacteroides gingivalis*, and *Bacteroides endodontalis* in a new genus *Porphyromonas*. Int. J. Syst. Bacteriol. *38*, 128–131.

Shah, H.N., Mayrand, D, and Genco, R.J. (1993) The biology of the species *Porphyromonas gingvalis*. (CRC Press Inc.).

Shoemaker, N.B., Guthrie, E.P., Salyers, A.A., and Gardner, J.F. (1985) Evidence that the clindamycin-erythromycin resistance gene of *Bacteroides* plasmid pBF4 is on a transposable element. J. Bacteriol. *162*, 626–632.

Skorupski, K., and Taylor, R.K. (1997) Control of the ToxR virulence regulon in *Vibrio cholerae* by environmental stimuli. Mol. Microbiol. *25*, 1003–1009.

Smith, C.J., Welch, R.A., and Macrina, F.L. (1982) Two independent conjugal transfer systems operating in *Bacteroides fragilis* V479–1. J. Bacteriol. *151*, 281–287.

Stephenson, K., and Hoch, J.A. (2002) Two-component and phosphorelay signal-transduction systems as therapeutic targets. Curr. Opin. Pharmacol. *2*, 507–512.

Stock, A.M., Robinson, V.L., and Goudreau, P.N. (2000) Two-component signal transduction. Ann. Rev. Biochem. *69*, 183–215.

Surette, M.G., and Bassler, B.L. (1998) Quorum sensing in *Escherichia coli* and *Salmonella typhimurium*. Proc. Natl. Acad. Sci. USA. *95*, 7046–7050.

Sztukowska, M., Bugno, M., Potempa, J., Travis, J., and Kurtz, D. M. Jr. (2002) Role of rubrerythrin in the oxidative stress response of *Porphyromonas gingivalis*. Mol. Microbiol. *44*, 479–488.

Tanner, A.C., Izard, J. (2006) *Tannerella forsythia*, a periodontal pathogen entering the genomic era. Periodontol. 2000. *42*, 88–113.

Teng, F., Wang, L., Singh, K.V., Murray, B.E., and Weinstock, G.M. (2002) Involvement of PhoP-PhoS homologs in *Enterococcus faecalis* virulence. Infect. Immun. *70*, 1991–1996.

Ueshima, J., Shoji, M., Ratnayake, D.B., Abe, K., Yoshida, S., Yamamoto, K., and Nakayama, K. (2003) Purification, gene cloning, gene expression, and mutants of Dps from the obligate anaerobe *Porphyromonas gingivalis*. Infect. Immun. *71*, 1170–1178.

Van de Peer, Y., and De Wachter, R. (1994) TREECON for Windows: a software package for the construction and drawing of evolutionary trees for the Microsoft Windows environment. Comput. Appl. Biosci. *10*, 569–570.

Watanabe-Kato, T., Hayashi, J.I., Terazawa, Y., Hoover, C.I., Nakayama, K., Hibi, E., Kawakami, N., Ikeda, T., Nakamura, H., Noguchi, T., and Yoshimura, F. (1998) Isolation and characterization of transposon-induced mutants of *Porphyromonas gingivalis* deficient in fimbriation. Microb. Pathog. *24*, 25–35.

Weinberg, A., Belton, C., Park, M.Y., and Lamont, R.J. (1997) Role of fimbriae in *Porphyromonas gingivalis* invasion of gingival epithelial cells. Infect. Immun. *65*, 313–316.

Welch, R.A., Jones, K.R., and Macrina, F.L. (1979) Transferable lincosamide-macrolide resistance in *Bacteroides*. Plasmid *2*, 261–268.

Xu, J., Chiang, H.C., Bjursell, M.K., and Gordon, J.I. (2004) Message from a human gut symbiont: sensitivity is a prerequisite for sharing. Trends Microbiol. *12*, 21–28.

Yilmaz, O., Watanabe, K., and Lamont, R.J. (2002) Involvement of integrins in fimbriae-mediated binding and invasion by *Porphyromonas gingivalis*. Cell. Microbiol. *4*, 305–314.

Yilmaz, O., Young, P.A., Lamont, R.J., and Kenny, G.E. (2003) Gingival epithelial cell signaling and cytoskeletal responses to *Porphyromonas gingivalis* invasion. Microbiology. *149*, 2417–2426.

Yuan, L., Hillman, J.D., and Progulske-Fox, A. (2005) Microarray analysis of quorum-sensing-regulated genes in *Porphyromonas gingivalis*. Infect. Immun. *73*, 4146–4154.

The Molecular Biology of the Survival and Virulence of *Treponema denticola* 8

Marie-Claude Jobin, Mohsen Amin and Richard P. Ellen

Abstract

The levels of oral spirochetes are elevated in patients with periodontal diseases. Among this group, *Treponema denticola* is the most studied and is considered as one of the main etiological bacteria of periodontitis. The environment in which this bacterium thrives has shaped it into the motile and highly proteolytic bacterium that we know today. Application of new molecular biology tools and genomic analysis has contributed greatly to the understanding of this fastidious bacterium. Mutagenesis of specific genes and the complete genome sequence of the Type strain, ATCC 35405, have given new insights into its intrinsic survival mechanisms. This chapter presents a review of the latest reports on *T. denticola* genes and proteins implicated in motility, growth and virulence.

Introduction

Treponema is one of the five genera of the spirochete phylum which includes *Borrelia*, *Brachyspira*, *Leptospira*, and *Spirochaeta*. It has been found in a variety of environments such as the human mouth, hooves and digestive tract of cattle, and termite guts (Paster and Canale-Parola, 1985; Edwards *et al.*, 2003; Trott *et al.*, 2003; Graber *et al.*, 2004; Wyss *et al.*, 2004). Treponemes usually have a spiral/helical shape and demonstrate a great diversity in length and diameter. Another characteristic of treponemes is the possession of cytoplasmic filaments and periplasmic flagella which originate from each end of the bacterium, varying in number from 2 to 8 (Masuda and Kawata, 1989). These flagella are known to contribute to cell shape integrity and motility. Aside from their shape, the surface structure and physiological traits of treponemes are similar to that of Gram-negative bacteria. They possess an outer membrane (outer sheath) containing glycolipids and proteins, a periplasmic space, peptidoglycan, and a cytoplasmic membrane that envelops the cytoplasmic cylinder.

There are at least 57 phylotypes of oral treponemes (Dewhirst *et al.*, 2000), ~80% of which have not yet been cultivated. Treponemes constitute between 20 to 50% of the bacterial flora of periodontally diseased patients as observed by microscopy of periodontal pockets (Dewhirst *et al.*, 2000). They are considered to be opportunistic pathogens since they are almost beyond detection in healthy patients (Riviere and DeRouen, 1998). So far, less than a dozen species have been successfully cultivated and maintained in the laboratory: *Treponema denticola*, *T. pectinovorum*, *T. socranskii*, *T. maltophilum*, *T. putidum*, *T. amylovorum*, *T. lecithinolyticum*, *T. medium*, *T. parvum* and *T. vincentii* (Wyss, 1992; Wyss *et al.*, 1996; Umemoto *et al.*, 1997b; Wyss *et al.*, 1997, 1999, 2001). *T. denticola* is considered to be a member of the proteolytic (so-called «red») complex, together with *Porphyromonas gingivalis* and *Tannerella forsythia*, as defined and updated by Socransky and coworkers (2005). The exact relation between these species is still unclear but studies have demonstrated a strong relationship between *T. denticola* and *P. gingivalis* for growth, biofilm formation and co-localization in subgingival plaque samples (Nilius *et al.*, 1993; Kigure *et al.*, 1995; Hashimoto *et al.*, 2003; Yamada *et al.*, 2005). The concentration of these three bacterial species in the periodontal pocket varies directly with the severity of the disease and they are of very low population density in healthy subjects. *T. denticola*'s link to periodontal diseases has been demonstrated in a number of studies using microscopy (Riviere *et al.*, 1997) and molecular biology techniques (Dewhirst *et al.*, 2000; Paster *et al.*, 2001; Gomes *et al.*, 2006; Haffajee *et al.*, 2006).

The importance of this periodontal pathogen in disease has led to a number of studies to determine the mechanisms and requirements for its establishment in the human mouth and its role in the pathogenesis of periodontal diseases. Though many such mechanisms have been reported, the specific contribution of each to *T. denticola*'s pathogenicity is still unknown. In recent years, research has intensified with improvements in cultivation techniques and the creation of novel molecular biology tools for detection and gene function analysis. Also, the completion of the genomic sequence of *T. denticola* ATCC 35405 has produced a huge amount of data and the possibilities of providing a better understanding of the physiology of this pathogen. This chapter focuses on the characterization of *T. denticola* and the use of molecular tools to determine the implication of newly reported virulence factors and genes in *T. denticola*'s growth, motility and virulence.

Treponema denticola

T. denticola was recognized only recently as a valid and distinct species, and both reference (ATCC 35404 and 33520) and type (ATCC 35405) strains were designated by Chan and co-workers (1993). It is a spiral shaped bacterium (Fig. 8.1), like the other members of the spirochete phylum, and measures between 6 to 16 μm in length and 0.15 to 0.2 μm in width and is considered to be a small spirochete (Chan and McLaughlin, 2000). The outer sheath of *T. denticola* is composed of proteins, glycoproteins, lipids and lipoproteins (Yotis *et al.*, 1991; Yotis *et al.*, 1993; Ishihara and Okuda, 1999). The main lipid of the outer sheath is analogous to lipopolysaccharides (LPS) from Gram-negative bacteria, because of its similar proportions of various fatty acids and its ability to form micelles when extracted. It also bears similarities to lipoteichoic acid (LTA)

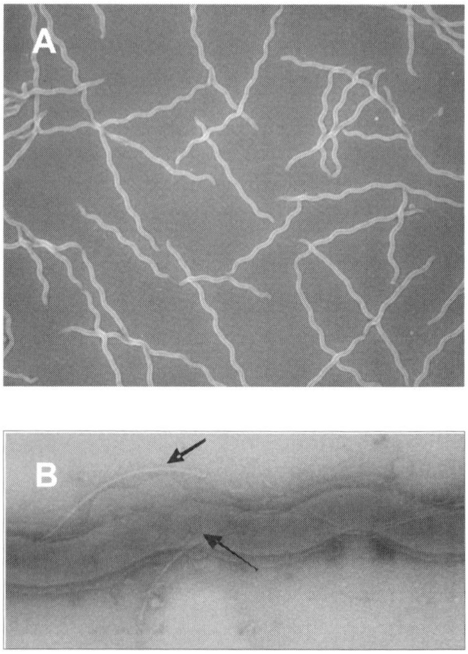

Figure 8.1 Electron photomicrographs of *T. denticola*. (A), Scanning electron photomicrograph taken after adhesion of *T. denticola* cells to a fibronectin-coated surface; (B), Transmission electron micrograph of *T. denticola*; arrows indicate periplasmic flagella. (Reprinted from Ellen *et al.*, 1998. Copyright 1998, with permission from Elsevier.)

produced by Gram-positive bacteria because of the outer membrane anchor which makes this lipid much more mobile than 'normal' LPS (Schultz *et al.*, 1998). However, *T. denticola* does not synthesize a true LPS among its outer sheath glycolipids. This asaccharolytic bacterium expresses strong and varied enzymatic activities that probably function in the acquisition of the required nutrients for growth (Uitto *et al.*, 1988; Chu and Holt, 1994) and self-protection from the proteinaceous molecules of the host defence system (Grenier, 1996; Miyamoto *et al.*, 2006; Yamazaki *et al.*, 2006). Such enzymes are found on the surface of the bacterium and the extracellular vesicles, and some are secreted into the growth medium (Grenier *et al.*, 1990; Rosen *et al.*, 1995; Gazi *et al.*, 1997). Apart from being highly proteolytic, *T. denticola* binds to a great variety of substrates (Fenno *et al.*, 1996; Xu *et al.*, 2001), cells (Baehni *et al.*, 1992; Haapasalo *et al.*, 1996; Ikegami *et al.*, 2004), and other bacterial species (Yao *et al.*, 1996; Hashimoto *et al.*, 2003), which probably helps foster its survival and proliferation. *T. denticola*'s phenotype was most likely shaped by the environment in which it has established its ecological niche, and it has evolved to produce highly specific strategies to survive in the harsh habitat that is the periodontal pocket.

Environment

The periodontal pocket, in which *T. denticola* is found, is a retentive gap located between the tooth and the gingiva, immediately below the gingival margin (Paster *et al.*, 2001). This environment provides bacteria with unique conditions to which they have to adapt or be eliminated. The crevicular fluid, which permeates the periodontal pocket, contains a wide array of proteins, serum, little sugar and almost no oxygen, which are not the optimum conditions in which most bacteria would thrive. Because of the restrictions dictated by this hostile environment, a laboratory medium developed to sustain the growth of such a bacterium must be highly specific, which demands the use of some atypical nutrients. A classic broth for the cultivation of *T. denticola* would include peptides, volatile fatty acids, serum and vitamins (Van Horn and Smibert, 1983), as shown in Table 8.1. Also, growth and multiplication require an anaerobic environment, although *T. denticola* has been shown to grow in the presence of a small amount of oxygen (Syed *et al.*, 1993; Caldwell and Marquis, 1999). A more defined medium was developed by Wyss (1992) and was shown to support the growth of *T. denticola* and other oral treponemes previously thought to be uncultivable. However, it is still a challenge, even for experienced laboratories, to cultivate *T. denticola* on an ongoing basis.

The periodontal pocket and gingival tissue also provide *T. denticola* exposure to several human cell types, such as epithelial cells, which normally form the protective barrier; fibroblasts, responsible for maintaining the extracellular matrix integrity; and cells of innate immunity (neutrophils, dendritic cells and monocytes/macrophages). The presence of these cells can be considered another source of nutrients for *T. denticola* and/or a threat, considering their immunoprotective functions when activated. These interactions have probably shaped the manner in which *T. denticola* senses and reacts to its environment, colonizes the host, and proliferates. Studies in our laboratory and others have demonstrated the ability of *T. denticola* to interact with host cells and to interfere with their functions. For example, reports on neutrophils (Amin *et al.*, 2004; Puthengady Thomas *et al.*, 2006) and epithelial cells (De Filippo *et al.*, 1995; Ellen *et al.*, 2000) have demonstrated dysregulation of the actin network when these mammalian cells are exposed to *T. denticola* and its outer sheath components. In order to function adequately, fibroblasts need an intact cytoskeleton and the capacity to phagocytose collagen, which depends on physiologic properties of their actin filament network. It has been demonstrated that *T. denticola*, and its outer sheath components, perturb these functions. Actin filament assembly and the binding to extracellular matrix via integrins in fibroblasts is perturbed when fibroblasts are challenged with *T. denticola* (Baehni *et al.*, 1992; Yang *et al.*, 1998; Amin *et al.*, 2004); this can have a major impact on actin-dependent cellular functions. These are only a few examples of how *T. denticola* interacts with eukaryotic cells (Ellen, 2002). Interactions of *T. denticola* and its environment (cells, bacteria, immune system) are most likely mediated by a variety of molecules accessible on the bacterial

Table 8.1 Chemical compounds found in *T. denticola* growth media (Van Horn and Snibert, 1983).

Category	Chemical compound
Peptides (or amino acids)	
	Brain heart infusion
	Trypticase/peptone
	Yeast extract
	Cysteine
	Asparagine
Fatty acids	
	Isobutyric acid
	Metholybutyric acid
	Isovaleric acid
	Valeric acid
Vitamins and growth factors	
	Thiamine (Vitamin B1)
	Vitamin K1
	Rabbit serum
	Haemin
Other	
	Sodium thioglycollate
	Resazurin

surface and/or those shed into the extracellular environment.

Major outer sheath protein

The major outer sheath protein, Msp, is found in the outer membrane of *T. denticola*. In its native state, this protein is an oligomer, composed of ~53 kDa polypeptides, with a hexagonal shaped distribution, as shown by electron microscopy (Cockayne *et al.*, 1989; Umemoto *et al.*, 1989; Haapasalo *et al.*, 1992; Egli *et al.*, 1993). The protein complex is thought to form a porin in the bacterial outer membrane and has been shown to form permeable ion channels in *in vitro* models of black lipid bilayers (Egli *et al.*, 1993), and *in vivo* with epithelial cells (Mathers *et al.*, 1996). Msp complex is usually detected in the outer membrane (Weinberg and Holt, 1991) but its principal location was questioned in a report by Caimaino and coworkers (1999), who concluded that Msp is mostly periplasmic. Experiments using a similar protocol were conducted in our laboratory, leading to a different conclusion. We used a procedure to maintain the outer membrane of spirochetes intact throughout the steps of immunofluorescence developed by Cox and coworkers (1995) to study the surface of *Treponema pallidum*. Briefly, *T. denticola* cells were embedded in liquid low-melting-temperature agarose to envelop them in microdroplets. The microdroplets were submitted to an indirect immunofluorescence (IF) protocol using high titre polyclonal rabbit anti-native Msp complex antiserum as the primary antibody. The microdroplets were treated with Triton X-100 to disrupt the outer membrane, exposing antigens in the periplasm. A comparison of the IF reactions provided evidence that Msp is actually an integral membrane protein in the outer membrane of the bacterium with some epitopes exposed to the surface. Labelling was clearly greater on the intact (Fig. 8.2A) than Triton X-100-treated cells (Fig. 8.2B). The major difference between this procedure and the one that was described by Caimaino and coworkers (1999) is that the primary antibody used by our laboratory was raised against native Msp rather than recombinant Msp (rMsp) (Fenno *et al.*, 1996). Although few reports have confirmed the true function of Msp, it is generally accepted that its large pore is sufficient to be a channel for nutrients and ions (Egli *et al.*, 1993; http://www.tcdb.org/) connecting the extracellular environment to the periplasmic space. However, knockout mutations of Msp do not seem to hinder the growth of *T. denticola* (Fenno *et al.*, 1998b); so, alternative ways to transport compounds vital for growth must be present.

Pore-forming capability may well be related to the cytoxicity of Msp for a number of cell types (Fenno *et al.*, 1998a). Not only is Msp potentially toxic to host cells but it also causes major fluctuations in intracellular calcium, which is an important mediator of cell homeostasis and response to stress. Msp induces acute Ca^{2+} transients in fibroblasts, within seconds after exposure, by promoting the entry of extracellular calcium into the cytosol. The cytosolic Ca^{2+} concentration often fluctuates above the baseline levels for the following 30 min (Wang *et al.*, 2001). In addition to these calcium fluxes, or perhaps due to them, Msp induces the disassembly of ventral actin filaments. Coincidently, it promotes the assembly of subcortical actin filaments, which causes the cell to round up (Amin *et al.*, 2004) and uncouples store-operated Ca^{2+} channels from the endoplasmic reticulum (Wang *et al.*, 2001). In fibroblasts, subcortical actin assembly, induced by Msp, is also closely related to the inhibition of collagen binding by plasma membrane β_1 integrins (Batista da Silva *et al.*, 2004). Similar events are also observed with neutrophils, where Msp diminishes chemotaxic and phagocytic abilities through the modification of the cytoskeleton (Puthengady Thomas *et al.*, 2006). A recent study in murine neutrophils has shown that Msp inhibits fMLP-induced activation of the small GTPase Rac1 and its downstream regulation of cell polarity (Magalhães *et al.*, 2007). Thus, Msp has the ability to induce severe changes in cell homeostasis. Notably, both fibroblasts and neutrophils that undergo extensive cytoskeletal reorganization in response to Msp remain viable.

Another example is the addition of Msp to epithelial cells, which induces the depolarization of the cell membrane, by forming ion channels and causing cell death (Mathers *et al.*, 1996). The conduct of these studies requires the use of Msp obtained by extracting a highly enriched native Msp complex from *T. denticola* (Egli *et al.*, 1993; Wang *et al.*, 2001). There is a risk of contamination with minor outer membrane components during the enrichment procedure and this is why sensitivity to absorption with anti-Msp antibodies and to heating at 60°C are done to confirm that contaminants have no influence on the observations made. Even with these tests, the development of novel methods for purification and molecular cloning of Msp,

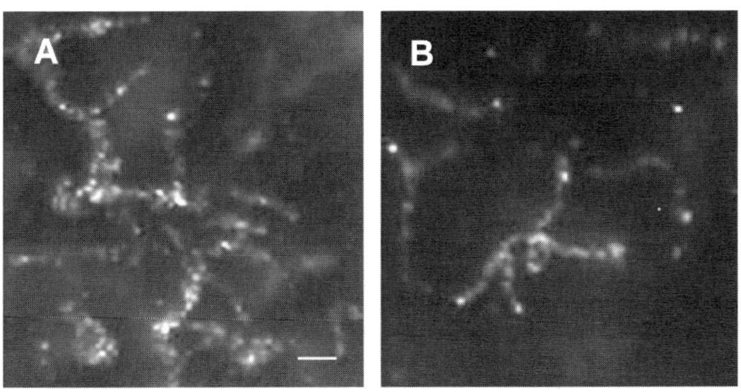

Figure 8.2 Encapsulation of *Treponema denticola* in gel microdroplets followed by indirect immunofluorescence staining of major outer sheath protein (Msp). Encapsulated cells were stained in a two-step process in the absence (A) and presence (B) of Triton X-100 by the method of Cox and coworkers (1995). (Bar, 10 μm).

without losing the functional characteristics of Msp, are of utmost importance. rMsp was successfully expressed in *Escherichia coli* (Haapasalo *et al.*, 1992; Fenno *et al.*, 1996) and was shown to possess the same binding properties, membrane depolarizing properties, and cytoxicity as the native Msp complex enriched from *T. denticola* (Haapasalo *et al.*, 1992; Fenno *et al.*, 1998a). The lower yield of recombinant Msp complex has been an obstacle to detailed study, but Edwards and coworkers (2005) found that cloning the Msp gene with a short N-terminal deletion yielded sufficient amounts of functional rMsp for their experiments. Further studies to fully characterize rMsp are needed to confirm that using rMsp is equivalent to using native Msp.

Msp is an important protein for *T. denticola*, for its porin function and for its interactions with host cells and the extracellular environment. Msp-like proteins have also been detected in other oral treponemes (Fenno *et al.*, 1997) such as *T. maltophilum* (Heuner *et al.*, 1999) and *T. lecithinolyticum* (Park *et al.*, 2002). Apart from its function as a porin, Msp is considered to be an important adhesin for *T. denticola* (Haapasalo *et al.*, 1992; Edwards *et al.*, 2005), suggesting putative roles in *T. denticola* establishment and survival.

Adhesins

Adhesion to host cells and/or proteins and other matrix components is an important function for bacterial colonization and acquisition of nutrients. Attachment to a surface helps establish a bacterium in its optimum ecological location (Gibbons, 1989). Adhesion increases the efficiency of the enzymes produced by the bacterium by bringing them closer to their substrates. Moreover, binding probably has a protective effect for the pathogen by camouflaging itself with host proteins to evade the defence system (Wilson *et al.*, 2002). Proteins found in the plasma and those of the extracellular matrix have different binding properties and biological activities; both are targets for bacterial adhesins (Virkola *et al.*, 1996; Mosesson *et al.*, 2001). Retention of potent proteolytic activities by the bacterial surface may also increase the proteolytic and invasion potential of a pathogen (Hu *et al.*, 1995; Darenfed *et al.*, 1999).

T. denticola probably produces a number of adhesins, but Msp is evidently among the main adhesins for the recognition of components of the extracellular matrix, such as heparin sulphate, laminin, collagen type I and IV, fibronectin, and fibrinogen (Haapasalo *et al.*, 1991, 1992; Umemoto *et al.*, 1993; Umemoto and Namikawa, 1994; Fenno *et al.*, 1996; Edwards *et al.*, 2005); and epithelial cells and fibroblasts (Weinberg and Holt, 1991; Keulers *et al.*, 1993; Fenno *et al.*, 1998a). The full extent of *T. denticola*'s binding capabilities has probably not yet been fully determined.

Fibronectin was the target ligand in several studies because of the numerous proteins from *T. denticola* that bind to this protein (Haapasalo *et al.*, 1992; Umemoto *et al.*, 1993). This was confirmed with inhibition studies using fibronectin. Even if Msp was blocked by using specific antibodies, *T. denticola* could still bind fibronectin, suggesting that other adhesins with an affinity for fibronectin are exposed on the surface (Umemoto *et al.*, 1993; Edwards *et al.*, 2005). Edwards and coworkers (2005) have determined the region of the Msp molecule responsible for fibronectin binding by producing truncated recombinant Msp fragments (N-terminal fragment, central region, and C-terminal fragment). The N-terminal region was found to function in fibronectin binding and it also functioned in the binding of keratin, laminin, collagen type I, fibrinogen, hyalyronic acid and heparin. Another fibronectin binding protein, OppA, was identified; it is accessible at the surface and it also binds plasminogen (Fenno *et al.*, 2000). OppA is involved in peptide uptake in several bacteria and the genes encoding for OppB, -C, -D and -F were identified downstream of *oppA* in *T. denticola* (Fenno *et al.*, 2000). Mutants lacking OppA did not exhibit significant phenotypic differences (growth, cytotoxicity, Msp assembly, chymotrypsin-like protease activity, fibronectin binding) from the wild type strain, which may be explained by the presence of several *opp* genes (Fenno *et al.*, 2000). Localization of *T. denticola* adhesins for fibronectin, by ligand-gold electron microscopy, revealed adhesins distributed over the surface of the bacterium. Rearrangement of these adhesins toward the tip of *T. denticola* was stimulated by binding to the immobilized ligand on an inert surface (Dawson and Ellen, 1990; Dawson and Ellen, 1994). This was a clear demonstration of mobility of an ECM-binding adhesin in *T. denticola*'s fluid outer sheath.

Adhesins for proteins other than fibronectin are present on the *T. denticola* surface and have different properties and roles in bacterial adhesion. Binding to epithelial cells is mediated by a leucin rich repeat (LrrA) protein which is necessary for tissue penetration by *T. denticola* (Ikegami *et al.*, 2004). For epithelial cells and other cells dependent on anchorage for proliferation, the protein FHL-1 is present on their surface and also found in the extracellular matrix. FHL-1 is recognized by H-like protein B (FhbB) of *T. denticola* (McDowell *et al.*, 2005). This protein was cloned, and it might be important for *T. denticola* since it is constitutively produced by several strains of *T. denticola* grown under different conditions (McDowell *et al.*, 2007). Both LrrA and FhbB may be important for binding to the bacterial cell surface and invasion of the tissues. Types I, IV and V collagen are recognized by a collagen-binding peptide (CBP) that is located in the outer sheath (Umemoto *et al.*, 1997a). Similar CBPs were detected in other oral spirochetes such as *T. socranskii* subsp. *buccale* and *T. medium* (Umemoto *et al.*, 1997a). Binding of matrix proteins probably fosters the retention and tissue penetration of *T. denticola* and may, therefore, be important in pathogenicity. Additional surface proteins are crucial for scavenging and uptake of other required nutrients, such as iron.

Uptake of iron in *T. denticola* and a number of other periodontal pathogens is essential for growth. *T. denticola* binds lactoferrin (Staggs *et al.*, 1994) and haemin (Scott *et al.*, 1993; Chu *et al.*, 1994b); the haemin-binding proteins (HbpA and HbpB) are found in the outer sheath, and have similar amino acid sequences (Xu *et al.*, 2001). In complete medium, mutants of Hbp proteins (A and/or B) of *T. denticola* have a growth rate similar to that of the wild type but they grow more slowly under iron-restricted conditions (Xu and Kolodrubetz, 2002). These mutants demonstrate the necessity for the presence of receptors responsible for iron uptake when the bacterium does not have access to free-iron as in the periodontal pocket, where iron is sequestered by transferrin and lactoferrin (Hentze *et al.*, 2004). Acquisition of iron is also achieved through the outer membrane sheath (OMS) receptor which has a haemin-binding capacity (Scott *et al.*, 1996). These proteins, Hbps and OMS, have probably evolved to counter the effect of iron limitation, and they may be important for *T. denticola* virulence (Chu *et al.*, 1994b). Binding of ECM and iron-sequestrating proteins increases the number of interactions between *T. denticola* and the host. By diminishing the distance between the pathogen and the cell (or the source of nutrients), the bacterium will be in a better position to establish and to modulate the host response. Substrate recognition domains of surface enzymes probably play a dual role in adhesion and localization of enzymatic activity.

Enzymatic activities

Proteases and peptidases

Proteases are considered major virulence factors in the pathogenesis of infections because of their wide range of action and their implication in bacterial growth, colonization, degradation of the host tissues, and perturbation of the host defences (Grenier and Mayrand, 1993). Proteases can be further characterized by the substrates that they cleave or by the peptide sequence that they recognize. For *T. denticola*, proteases are crucial for growth, since this asaccharolytic bacterium relies mostly on peptides to obtain the necessary nutrients and energy.

Dentilisin

Dentilisin (PrtP; 72 kDa) is the major protease produced by *T. denticola*. It is bound in the outer sheath, and it cleaves proteins at a phenylalanine residue, which classifies it as a chymotrypin-like protease (Uitto *et al.*, 1988; Rosen *et al.*, 1999). Dentilisin hydrolyses a wide range of native proteins including transferrin, fibrinogen, immunoglobulins and synthetic peptides containing phenylalanine. Its activity is optimal in reduced environments, similar to that of the periodontal pocket (Uitto *et al.*, 1988; Deng *et al.*, 2001; Miyamoto *et al.*, 2006; Yamazaki *et al.*, 2006). It is produced in an inactive form and is most likely activated as it integrates in the outer membrane (Fenno *et al.*, 1998b; Bian *et al.*, 2005). This protease is known to be part of a complex, named 'the chymotrypsin-like protease complex' (CTLP) which includes the accessory proteins PrcA1 and PrcA2 (43 and 38 kDa, respectively). These two proteins are required for PrtP activity (Que and Kuramitsu, 1990; Arakawa and Kuramitsu, 1994; Ishihara *et al.*, 2004). The accessory proteins are cleavage products of the polypeptide translated from the *prcA* transcript. An insertional mutation in *prcA*, or in *prtP*, will create a mutant defective in functional CTLP activity (Ishihara *et al.*, 1998; Lee *et al.*, 2002). Distribution of the *prtP-prcA* locus was characterized in a number of oral *Treponema* species and found to be widely distributed, even in strains that did not exhibit a chymotrypsin-like activity (Heuner *et al.*, 2001; Correia *et al.*, 2003; Bian *et al.*, 2005). Bian and coworkers (2005) characterized a gene, immediately upstream of *prcA*, that is required for dentilisin expression.

A dentilisin-deficient mutant (Ishihara *et al.*, 1998) was found to grow in very large aggregates compared with the wild type (Ellen *et al.*, 2000), suggesting that dentilisin is important in determining the expression or integration of other proteins in the outer sheath and the growth patterns of *T. denticola*. For example, interactions between Msp and CTLP are probably important in determining the structure of the outer sheath since mutation of dentilisin by allelic replacement results in the absence of hexagonal arrays of Msp or reduced expression of Msp. Similarly, Msp-deficient mutants have defects in CTLP expression (Fenno *et al.*, 1998b; Ishihara and Okuda, 1999; Bian *et al.*, 2005). The role of dentilisin extends beyond its proteolytic activity (modification of Msp expression, degradation of proteins/peptides for nutrition); it is also considered an adhesin. The mechanisms involved in biofilm formation and the role of each bacterial component is still unknown, but a mutant defective in dentilisin activity has shown an increased ability to form biofilms (Vesey and Kuramitsu, 2004). *T. denticola* co-localizes with *P. gingivalis* in biofilms *in vivo* (Kigure *et al.*, 1995) and forms biofilms with *P. gingivalis in vitro*. This function is probably highly specific since it was not observed with two other Gram-negative periodontal pathogens, *T. forsythia* and *Fusobacterium nucleatum* (Yamada *et al.*, 2005). One set of coadhesion molecules that bind *T. denticola* and *P. gingivalis* has been identified as the CTLP binding to the *P. gingivalis* fimbriae (Hashimoto *et al.*, 2003). Biofilm formation may present an opportunity for the bacterial species to derive peptides from their combined proteolytic activities, to protect each other from host defences or to exchange genetic material (Wang *et al.*, 2002). *T. denticola* binding properties were also observed with epithelial cells through the attachment of hyaluronan to dentilisin (Haapasalo *et al.*, 1996). Adhesion and potent proteolytic activity, such as mediated by dentilisin, probably fosters disruption of epithelial barriers and invasion of the epithelium by *T. denticola* (Grenier *et al.*, 1990; Haapasalo *et al.*, 1996; Ellen *et al.*, 2000; Chi *et al.*, 2003). Dentilisin has also been proven to be of great importance in the degradation of complement factor 3 (C3) protein (Yamazaki *et al.*, 2006), degradation of the inflammatory cytokines (Deng *et al.*, 2001; Miyamoto *et al.*, 2006) and it is highly cytotoxic for several cell lines (Uitto *et al.*, 1995; Fenno *et al.*, 1998a). The diverse functions that can be traced to dentilisin may provide *T. denticola* with the necessary nutrients and protection to grow and survive in the periodontal environment.

Trypsin-like activity

Several peptidases have been described but only a few have been studied in detail (Mikx *et al.*, 1992). Nevertheless, several studies have been conducted to determine the proteolytic profile of *T. denticola*. In the early 1980s, detection of peptidolytic activity was an important tool for bacterial characterization, and a trypsin-like activity produced by *T. denticola* and other periodontal pathogens was, indeed, found (Laughon *et al.*, 1982). This activity was later used as a prognostic aid to detect the presence of periodontal pathogens and to relate periodontal sites that yielded positive tests to the severity of periodontal lesions (Schmidt *et al.*, 1988; Loesche *et al.*, 1990). This trypsin-like activity was characterized for *T. denticola* (Mäkinen *et al.*, 1995; Gazi *et al.*, 1997) and later found to be mediated by a protein encoded by the ObdB gene (Fenno *et al.*, 2001). It is only recently that *opdB* was analysed (Fenno *et al.*, 2001) and was expressed in *E. coli* (Lee and Fenno, 2004). This activity is required for growth, as demonstrated with an Opd-deficient mutant (Fenno *et al.*, 2001). The expression of OpdB by *E. coli* has demonstrated the same substrate specificity as the native OpdB of *T. denticola*, and this clone will undoubtedly facilitate further characterization of the enzyme (Lee and Fenno, 2004).

Non-proteolytic enzymes

Cystalysin

Cystalysin was first characterized as a haemolytic protein capable of haemoxidizing haemoglobin and causing lysis of human erythrocytes (Chu *et al.*, 1994a). It was first purified in the mid 90s by Chu and coworkers (1994) and then subject to several structural and mechanistic studies (Krupka *et al.*, 2000; Bertoldi *et al.*, 2002; Cellini *et al.*, 2006). It is found in the cytoplasm of *T. denticola*, and it has been characterized as a cystein desulphydrase, responsible for the removal of the sulphydryl and amino acid groups from sulphur compounds to produce pyruvate (an energy source), ammonia (NH_3) and hydrogen sulphide (H_2S), which is highly toxic to eukaryotic cells (Chu *et al.*, 1997). This activity may be implicated in the metabolism of glutathione, which is available in periodontal pockets and may be important for bacterial growth (Chu *et al.*, 2002) because H_2S is produced later in the metabolism of glutathione. The production of H_2S is one of the mechanisms by which cystalysin induces lysis of erythrocytes (Kurzban *et al.*, 1999). Another possibility is that the enzyme binds to the erythrocyte membrane, alters the cell membrane by degrading spectrin (20–30% of the protein of the red blood cell membrane), and chemically oxidizes haemoglobin, resulting in the lysis of the cell (Chu *et al.*, 1999). These events, simultaneously active, probably contribute to the acquisition of iron by *T. denticola*, promoting host cell damage and, thus, providing additional nutrients.

Other enzymes

Other enzymatic activities have been detected in *T. denticola*: a dipeptidyl peptidase IV activity (Gazi *et al.*, 1997), an endo-acting proline-specific oligopeptidase (Mäkinen *et al.*, 1994), a proline iminopeptidase (Mäkinen *et al.*, 1996), a γ-glytlutamyl transferase (Mäkinen and Mäkinen, 1997), an acid phosphatase (Norton Hughes and Yotis, 1990), a glycine reductase (Rother *et al.*, 2001) and an endopeptidase active on bradykinin (Mäkinen *et al.*, 1992). All of these activities may promote bacterial growth, the disruption of the host immune mechanisms or shield *T. denticola* against elimination. *T. denticola* also produces a NADH oxidase which is known to function in oxygen metabolism and may be implicated in resistance to oxygen toxicity (Syed *et al.*, 1993; Caldwell and Marquis, 1999).

It was recently reported that *T. denticola* metabolizes steroids and other cholesterol-containing hormones (Clark and Soory, 2006). These lipids are of nutritional value for *T. denticola* but they may also participate in the dysregulation of the immune and inflammatory response (Clark and Soory, 2006). Cholesterol is not the only lipid important for *T. denticola*. A pathway leading to the production of phosphatidylcholine, the major phospholipid of *T. denticola* membranes, has been detected and characterized (Kent *et al.*, 2004). The gene *licCA*, coding for choline kinase and CDP:phosphocholine cytidylyltransferase, was identified in the *T. denticola* genome, based on the similarity of the predicted *licCA* of *T. pallidum*. The gene was successfully cloned and the recombinant enzyme was active. In addition, a viable mutant was created by inserting an erythromycin cassette in *licCA*. The gene disruption completely blocked the incorporation of choline or phosphatidylcholine by *T. denticola* (Kent *et al.*, 2004). Though the *licCA* was blocked, *T. denticola* compensated for this loss by incorporating phosphatidylethanolamine, another phospholipid, which might demonstrate the ability of this bacterium to modulate its phospholipid expression in response to environmental conditions (Kent *et al.*, 2004). Environmental control of the production of cellular lipids in *T. denticola* was further demonstrated by the use of media lacking specific fatty acids (Wyss, 2007). The comparison of the fatty acid profiles of several strains of *T. denticola* grown in two different media clearly showed their adaptation and their capacity to synthesize long chain fatty acids (Wyss, 2007). Even though *T. denticola* produces a great number of enzymes that contribute to meeting its metabolic needs, other oral bacteria that co-localize in the same environment as *T. denticola* may contribute to its growth and survival.

P. gingivalis, which colocalizes with *T. denticola* (Kigure *et al.*, 1995), has been recognized to promote *T. denticola* growth by providing it with some growth factors (Nilius *et al.*, 1993). Also, co-culture of *T. denticola* with other periodontal pathogens, or the addition of spent media from the growth of these bacteria, increases the yield of cells obtained in *T. denticola* cultures (ter Steeg and van der Hoeven, 1990). These bacteria (*Prevotela intermedia, Eubacterium nodatum, Veillonella parvula and F. nucleatum*) can provide *T. denticola* with growth factors or smaller peptides from cleaved proteins that are not usually used by *T. denticola* (ter Steeg and van der Hoeven, 1990). The possibility that *T. denticola* utilizes the cleavage and waste products of other oral bacteria, or secreted proteins, would be a major advantage in the organism's establishment. Together with these enzymatic activities and binding properties, the motility of *T. denticola* could be of great benefit in scavenging for nutrient sources or for evading toxic compounds.

Motility and chemotaxis

T. denticola typically produces four flagella [two originating at each end] that are intertwined around the cytoplasmic cylinder (Fig. 8.1B). Some regions are exposed and others are buried in the periplasm and outer sheath, and they most often overlap in the middle section of the bacterium. Flagella have proven to be of utmost importance for *T. denticola* structural integrity (Ruby *et al.*, 1997) and, of course, motility (Li *et al.*, 1996). Cytoplasmic filaments contribute to the cell's structural integrity by creating a ribbon-like structure around the protoplasmic cylinder (Izard *et al.*, 2004). These cytoplasmic filaments are found in several treponemes (Masuda and Kawata, 1989) and display a high percentage of conserved amino acid identity (Izard *et al.*, 1999). They are located close to the cytoplasmic membrane of the treponemes and they sprout at each end of the bacterium at a place different to that of the periplasmic flagella (Izard *et al.*, 2004). They are the product of the gene *cfpA*, which was first characterized in *T. pallidum* (You *et al.*, 1996). CfpA is implicated in cell division, since the mutation of the encoding gene leads to the observation of long chains of undivided cells (Izard *et al.*, 2001). These filaments do not contribute to movement, as their mutation has no effect on the expression of periplasmic flagella (Izard *et al.*, 2001). The motility of treponemes is similar to that of a screw into wood, and it is the flagella which provide the necessary traction for the movement to occur (Charon and Goldstein, 2002). The speed of the bacterium depends on the viscosity of the medium; the more viscous the environment, the more efficient the movement is because of the increased traction (Pietrantonio *et al.*, 1988; Klitorinos *et al.*, 1993). Motility is directed by a chemotaxis system similar to that expressed by other motile bacteria. These components enable the bacterium to seek and move toward an optimum environment.

Because of the importance of motility and chemotaxis in bacterial invasion, several studies have been done with *T. denticola* to determine the genes implicated in these functions. Basal bodies, with associated flagellar motors, are present in *T. denticola* at each end of the bacterium and their coordination is required for efficient movement toward or away from a chemical gradient (Charon and Goldstein, 2002). It has been determined that if one motor turns clockwise and the other counterclockwise, linear movement will be observed. If both motors turn in the same direction, the cell will tumble. For a spirochete, this means that the cell will be uncoordinated and flexion will be observed. Several genes implicated in the assembly of the flagella have been characterized for *T. denticola*. One of the first genes to be studied was *flgE*, coding for a flagellar hook protein. The phenotype of *flgE* mutants includes the loss of bacterial motility and lack of intact flagella (Li *et al.*, 1996). Using electroporation, Li and coworkers (1996) constructed a mutation in *flgE* by the insertion of an erythromycin cassette. The mutant lost its motility and showed a perfect helical morphology which contrasted with the irregular helical morphology of the wild type strain (Li *et al.*, 1996). This phenomenon was also observed in a spontaneous mutant lacking motility (Ruby *et al.*, 1997). The system used by Li and coworkers (1996) to create this *flgE* mutant was the first successful mutagenesis methodology with *T. denticola*.

FliG, a major protein implicated in flagellum assembly, rotation and chemotaxis in enteric bacteria, was detected in a *T. denticola* DNA bacteriophage library following its screening with sera from patients suffering from periodontal diseases (Heinzerling *et al.*, 1995). It was later found to be a flagellar switch component (Heinzerling *et al.*, 1997). The DNA sequences of the region flanking *fliG* were analysed for the presence of other genes related to flagellar structure. A gene homologous to *fliF*, which has been implicated in flagellum assembly in other bacteria, was detected by Heinzerling *et al.* (1997). The characterisation of these genes led the discovery of a total of seven genes related to filament proteins, including *flgB*, *flgC*, *fliE*, *fliF*, which encode the basal body constituents of the flagella; and *fliH* and *fliI*, which encode flagellar export proteins (Heinzerling *et al.*, 1997). The flagellar operon of *T. denticola*, *fla*, was identified and 11 genes were analysed. They included *tap1*, *flgD*, *flgE*, *orf4*, *motB*, *fliM*, *fliY*, *orf10* and *fliP*, which are all believed to be related to the motility and environment-sensing functions of *T. denticola* (Limberger *et al.*, 1999; Stamm and Bergen, 1999). The *fla* operons of *T. denticola* and *T. pallidum* are highly conserved and demonstrate the phylogenetic relatedness of these two spirochetes (Stamm and Bergen, 1999). Two other genes, *mglA* and *mglB*, probably part of an ABC transporter system, were later sequenced in *T. denticola* and shown to have a high degree of homology with the flagellar export apparatus of *Bacillus subtilis* (Lépine and Ellen, 2000).

Though motility is a crucial function of spirochetes, it is chemotactic sensing that directs the bacterium to move directionally toward positive stimuli and away from noxious stimuli. The similarity of the loci responsible for the expression of methyl-accepting chemotaxis proteins (MCPs) in other organisms, such as *T. pallidum*, *Leptospira illini* and *Treponema phagedenis* was sought and a probe made from a consensus region was used to screen a *T. denticola* DNA library (Greene and Stamm, 1997). The *T. denticola mcpA* was characterized by Greene and Stamm (1997). At the same time, the gene *dmcA* was discovered when the sequence upstream of a gene responsible for the expression of a chymotrypsin-like protease, *prtB* (Arakawa and Kuramitsu, 1994), was analysed (Kataoka *et al.*, 1997). Insertional mutants of DmcA (Kataoka *et al.*, 1997) or DmcB (Li *et al.*, 1999) were produced and found to be defective in chemotaxis for some nutrients. In addition to the surface receptor for chemotaxis, several proteins are needed to relay the message from receptor to the flagellar motors to give direction and energize movement. The gene cluster responsible for the chemotaxis of *T. denticola* contains *cheA*, *-W*, *-Y*, and *-X* (Greene and Stamm, 1999). One of the key proteins is CheA, which is part of a putative 2-component phosphorelay system implicated in the orientation of the cells (Lux *et al.*, 2002). A mutant of *cheA* was achieved by insertional mutagenesis with an ErmF cassette (Lux *et al.*, 2002). The mutant had impaired responses to chemotactic stimuli while retaining coordinated cell movements which resulted in non-directional motility. Protein-protein interaction assays were done with CheA and CheX, another member of the sophisticated chemotaxis signal transduction system, to characterize the mechanism and interactions between the implicated proteins. They were found to interact with one another (Sim *et al.*, 2005). CheX was further characterized by testing its interactions with receptor proteins, other Che proteins and motor proteins. With these new data, Sim and colleagues (2005) were able to determine a more detailed mechanism of *T. denticola* motility/chemotaxis protein interaction. In their model, CheX relays the message from CheA, which is associated with the MCP complex at the cell surface, to the Fli proteins, considered to be the motor proteins of the flagella. Other proteins, like CheY, could have similar functions to CheX (Sim *et al.*, 2005). Further studies are required to determine the exact mechanisms responsible for the control of motility/chemotaxis and the analysis of the proteins implicated in the process.

Chemotaxis and motility are important for *T. denticola* in terms of tissue invasion and mixed biofilm formation with *P. gingivalis*. A CheA-deficient mutant did not penetrate oral epithelial cell layers, unlike the wild type strain, which invaded readily (Lux *et al.*, 2001). Other mutants defective in DmcA and DmcB were also tested and exhibited a similar behaviour to the CheA mutant, suggesting that motility alone is not sufficient for invasion by *T denticola*. Indeed, both motility and chemotaxis need to be fully operational for effective invasion (Lux *et al.*, 2001). The genetic analysis of biofilm formation has been studied with mutants defective for *cfpA*, *flgE*, and *dmcA* (Vesey and Kuramitsu, 2004). DmcA did not appear to play a role in biofilm formation but little or no biofilm was observed with mutants targeting the flagellum genes, *cfpA* and *flgE* (Vesey and Kuramitsu, 2004). Motility and chemotaxis are highly related to one another, and both are important for *T. denticola*'s response to a variety of stimuli. Although they influence one another, their functions are highly specific.

Conclusion

As one of the members of the proteolytic (red) complex, so designated by Socransky and co-workers (2005), *T. denticola* evidently plays a major role in periodontal diseases. However, its pathogenic mechanisms are still unclear because there is no reliable animal model system of experimental chronic periodontitis in which *T. denticola* has been established and studied. Its interactions with the host and other oral bacteria are extensive because of the diverse environment of the periodontal pocket. *T. denticola* obtains nutrients from the degradation of host proteins into peptides or it imports the cleavage products or growth factors produced by neighbouring bacteria. Its interactions with *P. gingivalis* are those most studied because of their close association *in vivo*. *T. denticola* produces a number of adhesins to bind substrates and to attach itself to the host, which may help its invasion and protection against the host immune system. Also, because of its motility, *T. denticola* can invade tissues to gain access to fresh sources of nutrients or to evade host defences. Years after its initial description, the scientific community is finally gaining a better understanding of the molecular basis of this fastidious bacterium's physiology and virulence. Our ability to transform and mutate genes in *T. denticola*, accomplished 10 years ago in Kuramitsu's laboratory, and the sequencing of the *T. denticola* ATCC 35405 genome, have greatly advanced our knowledge. Nevertheless, we are a long way from fully understanding the mechanisms implicated in the complex interactions between *T. denticola* and its environment.

Apart from furthering our understanding of the biology of an important periodontal pathogen, studies with *T. denticola* have assisted the study of *T. pallidum*, which causes syphilis and which still cannot be continuously cultivated *in vitro*. These two pathogens share a number of proteins that may contribute to virulence; indeed, the comparison of their genomes has shown that around 200 genes share homology. More than 800 genes are also similar among *T. pallidum*, *Borrelia burgdorferi* and/or *Leptospira interrogans* – other spirochetes whose genomes have been sequenced (Seshadri *et al.*, 2004). The findings from studies of *T. denticola* have also had some beneficial impact on the study of other treponemes, such as those implicated in bovine digital dermatitis; these include *T. medium*, *T. vincentii*, *T. phagedenis*, *T. brennaborense*, and a *T. denticola*-like spirochete identified in papillomatous digital dermatitis. Even with recently intensified research on *T. denticola*, details of its pathogenic mechanisms and its role in periodontitis are still obscure.

References

Amin, M., Ho, A.C., Lin, J.Y., Batista da Silva, A.P., Glogauer, M., and Ellen, R.P. (2004) Induction of de novo subcortical actin filament assembly by *Treponema denticola* major outer sheath protein. Infect. Immun. *72*, 3650–3654.

Arakawa, S., and Kuramitsu, H.K. (1994) Cloning and sequence analysis of a chymotrypsinlike protease from *Treponema denticola*. Infect. Immun. *62*, 3424–3433.

Baehni, P.C., Song, M., McCulloch, C.A., and Ellen, R.P. (1992) *Treponema denticola* induces actin rearrangement and detachment of human gingival fibroblasts. Infect. Immun. *60*, 3360–3368.

Batista da Silva, A.P., Lee, W., Bajenova, E., McCulloch, C.A., and Ellen, R.P. (2004) The major outer sheath protein of *Treponema denticola* inhibits the binding step of collagen phagocytosis in fibroblasts. Cell. Microbiol. *6*, 485–498.

Bertoldi, M., Cellini, B., Clausen, T., and Voltattorni, C.B. (2002) Spectroscopic and kinetic analyses reveal the pyridoxal 5′-phosphate binding mode and the catalytic features of *Treponema denticola* cystalysin. Biochemistry *41*, 9153–9164.

Bian, X.L., Wang, H.T., Ning, Y., Lee, S.Y., and Fenno, J.C. (2005) Mutagenesis of a novel gene in the *prcA-prtP* protease locus affects expression of *Treponema denticola* membrane complexes. Infect. Immun. *73*, 1252–1255.

Caimano, M.J., Bourell, K.W., Bannister, T.D., Cox, D.L., and Radolf, J.D. (1999) The *Treponema denticola* major sheath protein is predominantly periplasmic and has only limited surface exposure. Infect. Immun. *67*, 4072–4083.

Caldwell, C.E., and Marquis, R.E. (1999) Oxygen metabolism by *Treponema denticola*. Oral Microbiol. Immunol. *14*, 66–72.

Cellini, B., Bertoldi, M., Montioli, R., Laurents, D.V., Paiardini, A., and Voltattorni, C.B. (2006) Dimerization and folding processes of *Treponema denticola* cystalysin: the role of pyridoxal 5′-phosphate. Biochemistry *45*, 14140–14154.

Chan, E.C., and McLaughlin, R. (2000) Taxonomy and virulence of oral spirochetes. Oral Microbiol. Immunol. *15*, 1–9.

Chan, E.C., Siboo, R., Keng, T., Psarra, N., Hurley, R., Cheng, S.L., and Iugovaz, I. (1993) *Treponema denticola* (ex Brumpt 1925) sp. nov., nom. rev., and identification of new spirochete isolates from periodontal pockets. Int. J. Syst. Bacteriol. *43*, 196–203.

Charon, N.W., and Goldstein, S.F. (2002) Genetics of motility and chemotaxis of a fascinating group of bacteria: the spirochetes. Annu. Rev. Genet. *36*, 47–73.

Chi, B., Qi, M., and Kuramitsu, H.K. (2003) Role of dentilisin in *Treponema denticola* epithelial cell layer penetration. Res. Microbiol. *154*, 637–643.

Chu, L., Dong, Z., Xu, X., Cochran, D.L., and Ebersole, J.L. (2002) Role of glutathione metabolism of *Treponema denticola* in bacterial growth and virulence expression. Infect. Immun. *70*, 1113–1120.

Chu, L., Ebersole, J.L., and Holt, S.C. (1999) Hemoxidation and binding of the 46-kDa cystalysin of *Treponema denticola* leads to a cysteine-dependent hemolysis of human erythrocytes. Oral Microbiol. Immunol. *14*, 293–303.

Chu, L., Ebersole, J.L., Kurzban, G.P., and Holt, S.C. (1997) Cystalysin, a 46-kilodalton cysteine desulfhydrase from *Treponema denticola*, with hemolytic and hemoxidative activities. Infect. Immun. *65*, 3231–3238.

Chu, L., and Holt, S.C. (1994) Purification and characterization of a 45 kDa hemolysin from *Treponema denticola* ATCC 35404. Microb. Pathog. *16*, 197–212.

Chu, L., Kennell, W., and Holt, S.C. (1994a). Characterization of hemolysis and hemoxidation activities by *Treponema denticola*. Microb. Pathog. *16*, 183–195.

Chu, L., Song, M., and Holt, S.C. (1994b). Effect of iron regulation on expression and binding-binding function of outer-sheath proteins from *Treponema denticola*. Microb. Pathog. *16*, 321–335.

Clark, D.T., and Soory, M. (2006) The metabolism of cholesterol and certain hormonal steroids by *Treponema denticola*. Steroids *71*, 352–363.

Cockayne, A., Sanger, R., Ivic, A., Strugnell, R.A., MacDougall, J.H., Russell, R.R., and Penn, C.W. (1989) Antigenic and structural analysis of *Treponema denticola*. J. Gen. Microbiol. *135*, 3209–3218.

Correia, F.F., Plummer, A.R., Ellen, R.P., Wyss, C., Boches, S.K., Galvin, J.L., Paster, B.J., and Dewhirst, F.E. (2003) Two paralogous families of a two-gene subtilisin operon are widely distributed in oral treponemes. J. Bacteriol. *185*, 6860–6869.

Cox, D.L., Akins, D.R., Porcella, S.F., Norgard, M.V., and Radolf, J.D. (1995) *Treponema pallidum* in gel microdroplets: a novel strategy for investigation of treponemal molecular architecture. Mol. Microbiol. *15*, 1151–1164.

Darenfed, H., Grenier, D., and Mayrand, D. (1999) Acquisition of plasmin activity by *Fusobacterium nucleatum* subsp. *nucleatum* and potential contribution to tissue destruction during periodontitis. Infect. Immun. *67*, 6439–6444.

Dawson, J.R., and Ellen, R.P. (1990) Tip-oriented adherence of *Treponema denticola* to fibronectin. Infect. Immun. *58*, 3924–3928.

Dawson, J.R., and Ellen, R.P. (1994) Clustering of fibronectin adhesins toward *Treponema denticola* tips upon contact with immobilized fibronectin. Infect. Immun. *62*, 2214–2221.

De Filippo, A.B., Ellen, R.P., and McCulloch, C.A. (1995) Induction of cytoskeletal rearrangements and loss of volume regulation in epithelial cells by *Treponema denticola*. Arch. Oral Biol. *40*, 199–207.

Deng, Q.D., Han, Y., Xia, X., and Kuramitsu, H.K. (2001) Effects of the oral spirochete *Treponema denticola* on interleukin-8 expression from epithelial cells. Oral Microbiol. Immunol. *16*, 185–187.

Dewhirst, F.E., Tamer, M.A., Ericson, R.E., Lau, C.N., Levanos, V.A., Boches, S.K., Galvin, J.L., and Paster, B.J. (2000) The diversity of periodontal spirochetes by 16S rRNA analysis. Oral Microbiol. Immunol. *15*, 196–202.

Edwards, A.M., Dymock, D., and Jenkinson, H.F. (2003) From tooth to hoof: treponemes in tissue-destructive diseases. J. Appl. Microbiol. *94*, 767–780.

Edwards, A.M., Jenkinson, H.F., Woodward, M.J., and Dymock, D. (2005) Binding properties and adhesion-mediating regions of the major sheath protein of *Treponema denticola* ATCC 35405. Infect. Immun. *73*, 2891–2898.

Egli, C., Leung, W.K., Müller, K.H., Hancock, R.E., and McBride, B.C. (1993) Pore-forming properties of the major 53-kilodalton surface antigen from the outer sheath of *Treponema denticola*. Infect. Immun. *61*, 1694–1699.

Ellen, R.P., Wikström, M., Grove, D.A., Song, M., and Elwing, H. (1998) Polar adhesion of *Treponema denticola* on wettability gradient surfaces. Colloids Surf. B Biointerfaces *11*, 177–186.

Ellen, R.P. (2002) Adhesion of oral spirochaetes to host cells and its cytopathogenic consequences. In Bacterial adhesion to host tissues: mechanisms and consequences, M. Wilson, ed. (Cambridge, United Kingdom, Cambridge University Press), pp. 247–276.

Ellen, R.P., Ko, K.S., Lo, C.M., Grove, D.A., and Ishihara, K. (2000) Insertional inactivation of the *prtP* gene of *Treponema denticola* confirms dentilisin's disruption of epithelial junctions. J. Mol. Microbiol. Biotechnol. *2*, 581–586.

Fenno, J.C., Hannam, P.M., Leung, W.K., Tamura, M., Uitto, V.J., and McBride, B.C. (1998a). Cytopathic effects of the major surface protein and the chymotrypsinlike protease of *Treponema denticola*. Infect. Immun. *66*, 1869–1877.

Fenno, J.C., Lee, S.Y., Bayer, C.H., and Ning, Y. (2001) The *opdB* locus encodes the trypsin-like peptidase activity of *Treponema denticola*. Infect. Immun. *69*, 6193–6200.

Fenno, J.C., Müller, K.H., and McBride, B.C. (1996) Sequence analysis, expression, and binding activity of recombinant major outer sheath protein (Msp) of *Treponema denticola*. J. Bacteriol. *178*, 2489–2497.

Fenno, J.C., Tamura, M., Hannam, P.M., Wong, G.W., Chan, R.A., and McBride, B.C. (2000) Identification of a *Treponema denticola* OppA homologue that binds host proteins present in the subgingival environment. Infect. Immun. *68*, 1884–1892.

Fenno, J.C., Wong, G.W., Hannam, P.M., and McBride, B.C. (1998b). Mutagenesis of outer membrane virulence determinants of the oral spirochete *Treponema denticola*. FEMS Microbiol. Lett. *163*, 209–215.

Fenno, J.C., Wong, G.W., Hannam, P.M., Müller, K.H., Leung, W.K., and McBride, B.C. (1997) Conservation of *msp*, the gene encoding the major outer membrane protein of oral *Treponema* spp. J. Bacteriol. *179*, 1082–1089.

Gazi, M.I., Cox, S.W., Clark, D.T., and Eley, B.M. (1997) Characterization of protease activities in *Capnocytophaga* spp., *Porphyromonas gingivalis*, *Prevotella* spp., *Treponema denticola* and *Actinobacillus actinomycetemcomitans*. Oral Microbiol. Immunol. *12*, 240–248.

Gibbons, R.J. (1989) Bacterial adhesion to oral tissues: a model for infectious diseases. J. Dent. Res. *68*, 750–760.

Gomes, B.P., Jacinto, R.C., Pinheiro, E.T., Sousa, E.L., Zaia, A.A., Ferraz, C.C., and Souza-Filho, F.J. (2006) Molecular analysis of *Filifactor alocis*, *Tannerella forsythia*, and *Treponema denticola* associated with primary endodontic infections and failed endodontic treatment. J. Endodont. 32, 937–940.

Graber, J.R., Leadbetter, J.R., and Breznak, J.A. (2004) Description of *Treponema azotonutricium* sp. nov. and *Treponema primitia* sp. nov., the first spirochetes isolated from termite guts. Appl. Environ. Microbiol. *70*, 1315–1320.

Greene, S.R., and Stamm, L.V. (1997) Identification, sequence, and expression of *Treponema denticola mcpA*, a putative chemoreceptor gene. FEMS Microbiol. Lett. *157*, 245–249.

Greene, S.R., and Stamm, L.V. (1999) Molecular characterization of a chemotaxis operon in the oral spirochete, *Treponema denticola*. Gene 232, 59–68.

Grenier, D. (1996) Degradation of host protease inhibitors and activation of plasminogen by proteolytic enzymes from *Porphyromonas gingivalis* and *Treponema denticola*. Microbiology *142*, 955–961.

Grenier, D., and Mayrand, D. (1993) Potential roles of proteinases. In Biology of the species *Porphyromonas gingivalis*, H. N. Shah, D. Mayrand, and R. J. Genco, eds. (Boca Raton, CRC Press), pp. 234–238.

Grenier, D., Uitto, V.J., and McBride, B.C. (1990) Cellular location of a *Treponema denticola* chymotrypsinlike protease and importance of the protease in migration through the basement membrane. Infect. Immun. 58, 347–351.

Haapasalo, M., Hannam, P., McBride, B.C., and Uitto, V.J. (1996) Hyaluronan, a possible ligand mediating *Treponema denticola* binding to periodontal tissue. Oral Microbiol. Immunol. *11*, 156–160.

Haapasalo, M., Müller, K.H., Uitto, V.J., Leung, W.K., and McBride, B.C. (1992) Characterization, cloning, and binding properties of the major 53-kilodalton *Treponema denticola* surface antigen. Infect. Immun. *60*, 2058–2065.

Haapasalo, M., Singh, U., McBride, B.C., and Uitto, V.J. (1991) Sulfhydryl-dependent attachment of *Treponema denticola* to laminin and other proteins. Infect. Immun. 59, 4230–4237.

Haffajee, A.D., Teles, R.P., and Socransky, S.S. (2006) Association of *Eubacterium nodatum* and *Treponema denticola* with human periodontitis lesions. Oral Microbiol. Immunol. *21*, 269–282.

Hashimoto, M., Ogawa, S., Asai, Y., Takai, Y., and Ogawa, T. (2003) Binding of *Porphyromonas gingivalis* fimbriae to *Treponema denticola* dentilisin. FEMS Microbiol. Lett. *226*, 267–271.

Heinzerling, H.F., Olivares, M., and Burne, R.A. (1997) Genetic and transcriptional analysis of *flgB* flagellar operon constituents in the oral spirochete *Treponema denticola* and their heterologous expression in enteric bacteria. Infect. Immun. *65*, 2041–2051.

Heinzerling, H.F., Penders, J.E., and Burne, R.A. (1995) Identification of a *fliG* homologue in *Treponema denticola*. Gene *161*, 69–73.

Hentze, M.W., Muckenthaler, M.U., and Andrews, N.C. (2004) Balancing acts: molecular control of mammalian iron metabolism. Cell *117*, 285–297.

Heuner, K., Bergmann, I., Heckenbach, K., and Göbel, U.B. (2001) Proteolytic activity among various oral *Treponema* species and cloning of a *prtP*-like gene of *Treponema socranskii* subsp. *socranskii*. FEMS Microbiol. Lett. *201*, 169–176.

Heuner, K., Choi, B.K., Schade, R., Moter, A., Otto, A., and Göbel, U.B. (1999) Cloning and characterization of a gene (*mspA*) encoding the major sheath protein of *Treponema maltophilum* ATCC 51939^{T}. J. Bacteriol. *181*, 1025–1029.

Hu, L.T., Perides, G., Noring, R., and Klempner, M.S. (1995) Binding of human plasminogen to *Borrelia burgdorferi*. Infect. Immun. *63*, 3491–3496.

Ikegami, A., Honma, K., Sharma, A., and Kuramitsu, H.K. (2004) Multiple functions of the leucine-rich repeat protein LrrA of *Treponema denticola*. Infect. Immun. *72*, 4619–4627.

Ishihara, K., Kuramitsu, H.K., Miura, T., and Okuda, K. (1998) Dentilisin activity affects the organization of the outer sheath of *Treponema denticola*. J. Bacteriol. *180*, 3837–3844.

Ishihara, K., Kuramitsu, H.K., and Okuda, K. (2004) A 43-kDa protein of *Treponema denticola* is essential for dentilisin activity. FEMS Microbiol. Lett. *232*, 181–188.

Ishihara, K., and Okuda, K. (1999) Molecular pathogenesis of the cell surface proteins and lipids from *Treponema denticola*. FEMS Microbiol. Lett. *181*, 199–204.

Izard, J., McEwen, B.F., Barnard, R.M., Portuese, T., Samsonoff, W.A., and Limberger, R.J. (2004) Tomographic reconstruction of treponemal cytoplasmic filaments reveals novel bridging and anchoring components. Mol. Microbiol. *51*, 609–618.

Izard, J., Samsonoff, W.A., Kinoshita, M.B., and Limberger, R.J. (1999) Genetic and structural analyses of cytoplasmic filaments of wild-type *Treponema phagedenis* and a flagellar filament-deficient mutant. J. Bacteriol. *181*, 6739–6746.

Izard, J., Samsonoff, W.A., and Limberger, R.J. (2001) Cytoplasmic filament-deficient mutant of *Treponema denticola* has pleiotropic defects. J. Bacteriol. *183*, 1078–1084.

Kataoka, M., Li, H., Arakawa, S., and Kuramitsu, H. (1997) Characterization of a methyl-accepting chemotaxis protein gene, *dmcA*, from the oral spirochete *Treponema denticola*. Infect. Immun. *65*, 4011–4016.

Kent, C., Gee, P., Lee, S.Y., Bian, X., and Fenno, J.C. (2004) A CDP-choline pathway for phosphatidylcholine biosynthesis in *Treponema denticola*. Mol. Microbiol. *51*, 471–481.

Keulers, R.A., Maltha, J.C., Mikx, F.H., and Wolters-Lutgerhorst, J.M. (1993) Attachment of *Treponema denticola* strains to monolayers of epithelial cells of different origin. Oral Microbiol. Immunol. *8*, 84–88.

Kigure, T., Saito, A., Seida, K., Yamada, S., Ishihara, K., and Okuda, K. (1995) Distribution of *Porphyromonas gingivalis* and *Treponema denticola* in human subgingival plaque at different periodontal pocket depths examined by immunohistochemical methods. J. Periodont. Res. *30*, 332–341.

Klitorinos, A., Noble, P., Siboo, R., and Chan, E.C. (1993) Viscosity-dependent locomotion of oral spirochetes. Oral Microbiol. Immunol. *8*, 242–244.

Krupka, H.I., Huber, R., Holt, S.C., and Clausen, T. (2000) Crystal structure of cystalysin from *Treponema denticola*: a pyridoxal 5′-phosphate-dependent protein acting as a haemolytic enzyme. EMBO J. *19*, 3168–3178.

Kurzban, G.P., Chu, L., Ebersole, J.L., and Holt, S.C. (1999) Sulfhemoglobin formation in human erythrocytes by cystalysin, an L-cysteine desulfhydrase from *Treponema denticola*. Oral Microbiol. Immunol. *14*, 153–164.

Laughon, B.E., Syed, S.A., and Loesche, W.J. (1982) API ZYM system for identification of *Bacteroides* spp., *Capnocytophaga* spp., and spirochetes of oral origin. J. Clin. Microbiol. *15*, 97–102.

Lee, S.Y., Bian, X.L., Wong, G.W., Hannam, P.M., McBride, B.C., and Fenno, J.C. (2002) Cleavage of *Treponema denticola* PrcA polypeptide to yield protease complex-associated proteins Prca1 and Prca2 is dependent on PrtP. J. Bacteriol. *184*, 3864–3870.

Lee, S.Y., and Fenno, J.C. (2004) Expression of *Treponema denticola* oligopeptidase B in *Escherichia coli*. Curr. Microbiol. *48*, 379–382.

Lépine, G., and Ellen, R.P. (2000) *MglA* and *mglB* of *Treponema denticola*; similarity to ABC transport and *spa* genes. DNA Seq. *11*, 419–431.

Li, H., Arakawa, S., Deng, Q.D., and Kuramitsu, H. (1999) Characterization of a novel methyl-accepting chemotaxis gene, *dmcB*, from the oral spirochete *Treponema denticola*. Infect. Immun. *67*, 694–699.

Li, H., Ruby, J., Charon, N., and Kuramitsu, H. (1996) Gene inactivation in the oral spirochete *Treponema denticola*: construction of an *flgE* mutant. J. Bacteriol. *178*, 3664–3667.

Limberger, R.J., Slivienski, L.L., Izard, J., and Samsonoff, W.A. (1999) Insertional inactivation of *Treponema denticola tap1* results in a nonmotile mutant with elongated flagellar hooks. J. Bacteriol. *181*, 3743–3750.

Loesche, W.J., Giordano, J., and Hujoel, P.P. (1990) The utility of the BANA test for monitoring anaerobic infections due to spirochetes (*Treponema denticola*) in periodontal disease. J. Dent. Res. 69, 1696–1702.

Lux, R., Miller, J.N., Park, N.H., and Shi, W. (2001) Motility and chemotaxis in tissue penetration of oral epithelial cell layers by *Treponema denticola*. Infect. Immun. 69, 6276–6283.

Lux, R., Sim, J.H., Tsai, J.P., and Shi, W. (2002) Construction and characterization of a *cheA* mutant of *Treponema denticola*. J. Bacteriol. *184*, 3130–3134.

Magalhães, M. A.O., Sun, C.X., Glogauer, M., and Ellen, R.P. (2007) The major outer sheath protein of *Treponema denticola* selectively inhibits Rac1 activation in murine neutrophils. Cell. Microbiol., *in press*.

Mäkinen, K.K., Chen, C.Y., and Mäkinen, P.L. (1996) Proline iminopeptidase from the outer cell envelope of the human oral spirochete *Treponema denticola* ATCC 35405. Infect. Immun. *64*, 702–708.

Mäkinen, K.K., Mäkinen, P.L., Loesche, W.J., and Syed, S.A. (1995) Purification and general properties of an oligopeptidase from *Treponema denticola* ATCC 35405-a human oral spirochete. Arch. Biochem. Biophys. *316*, 689–698.

Mäkinen, K.K., Mäkinen, P.L., and Syed, S.A. (1992) Purification and substrate specificity of an endopeptidase from the human oral spirochete *Treponema denticola* ATCC 35405, active on furylacryloyl-Leu-Gly-Pro-Ala and bradykinin. J. Biol. Chem. *267*, 14285–14293.

Mäkinen, P.L., and Mäkinen, K.K. (1997) γ-Glutamyltransferase from the outer cell envelope of *Treponema denticola* ATCC 35405. Infect. Immun. 65, 685–691.

Mäkinen, P.L., Mäkinen, K.K., and Syed, S.A. (1994) An endo-acting proline-specific oligopeptidase from *Treponema denticola* ATCC 35405: evidence of hydrolysis of human bioactive peptides. Infect. Immun 62, 4938–4947.

Masuda, K., and Kawata, T. (1989) Isolation and characterization of cytoplasmic fibrils from treponemes. Microbiol. Immunol. 33, 619–630.

Mathers, D.A., Leung, W.K., Fenno, J.C., Hong, Y., and McBride, B.C. (1996) The major surface protein complex of *Treponema denticola* depolarizes and induces ion channels in HeLa cell membranes. Infect. Immun. 64, 2904–2910.

McDowell, J.V., Frederick, J., Stamm, L., and Marconi, R.T. (2007) Identification of the gene encoding the FhbB protein of *Treponema denticola*, a highly unique FHL-1 binding protein. Infect. Immun. *75*, 1050–1054.

McDowell, J.V., Lankford, J., Stamm, L., Sadlon, T., Gordon, D.L., and Marconi, R.T. (2005) Demonstration of factor H-like protein 1 binding to *Treponema denticola*, a pathogen associated with periodontal disease in humans. Infect. Immun. *73*, 7126–7132.

Mikx, F.H., Jacobs, F., and Satumalay, C. (1992) Cell-bound peptidase activities of *Treponema denticola* ATCC 33520 in continuous culture. J. Gen. Microbiol. *138*, 1837–1842.

Miyamoto, M., Ishihara, K., and Okuda, K. (2006) The *Treponema denticola* surface protease dentilisin degrades interleukin-1β (IL-1β), IL-6, and tumor necrosis factor alpha. Infect. Immun. *74*, 2462–2467.

Mosesson, M.W., Siebenlist, K.R., and Meh, D.A. (2001) The structure and biological features of fibrinogen and fibrin. Ann. N. Y. Acad. Sci. *936*, 11–30.

Nilius, A.M., Spencer, S.C., and Simonson, L.G. (1993) Stimulation of *in vitro* growth of *Treponema denticola* by extracellular growth factors produced by *Porphyromonas gingivalis*. J. Dent. Res. *72*, 1027–1031.

Norton Hughes, C.A., and Yotis, W.W. (1990) A study of the acid phosphatase of *Treponema denticola*. Zentralbl. Bakteriol. *274*, 195–202.

Park, K.K., Heuner, K., Göbel, U.B., Yoo, Y.J., Kim, C.K., and Choi, B.K. (2002) Cloning and characterization of a major surface protein (MspTL) of *Treponema lecithinolyticum* associated with rapidly progressive periodontitis. FEMS Microbiol. Lett. *207*, 185–192.

Paster, B.J., Boches, S.K., Galvin, J.L., Ericson, R.E., Lau, C.N., Levanos, V.A., Sahasrabudhe, A., and Dewhirst, F.E. (2001) Bacterial diversity in human subgingival plaque. J. Bacteriol. *183*, 3770–3783.

Paster, B.J., and Canale-Parola, E. (1985) *Treponema saccharophilum* sp. nov., a large pectinolytic spirochete from the bovine rumen. Appl. Environ. Microbiol. *50*, 212–219.

Pietrantonio, F., Noble, P.B., Amsel, R., and Chan, E.C. (1988) Locomotory characteristics of *Treponema denticola*. Can. J. Microbiol. *34*, 748–752.

Puthengady Thomas, B., Sun, C.X., Bajenova, E., Ellen, R.P., and Glogauer, M. (2006) Modulation of human neutrophil functions *in vitro* by *Treponema denticola* major outer sheath protein. Infect. Immun. *74*, 1954–1957.

Que, X.C., and Kuramitsu, H.K. (1990) Isolation and characterization of the *Treponema denticola prtA* gene coding for chymotrypsinlike protease activity and detection of a closely linked gene encoding PZ-PLGPA-hydrolyzing activity. Infect. Immun. 58, 4099–4105.

Riviere, G.R., and DeRouen, T.A. (1998) Association of oral spirochetes from periodontally healthy sites with development of gingivitis. J. Periodontol. *69*, 496–501.

Riviere, G.R., DeRouen, T.A., Kay, S.L., Avera, S.P., Stouffer, V.K., and Hawkins, N.R. (1997) Association of oral spirochetes from sites of periodontal health with development of periodontitis. J. Periodontol. *68*, 1210–1214.

Rosen, G., Naor, R., Rahamim, E., Yishai, R., and Sela, M.N. (1995) Proteases of *Treponema denticola* outer sheath and extracellular vesicles. Infect. Immun. *63*, 3973–3979.

Rosen, G., Naor, R., and Sela, M.N. (1999) Multiple forms of the major phenylalanine specific protease in *Treponema denticola*. J. Periodont. Res. *34*, 269–276.

Rother, M., Böck, A., and Wyss, C. (2001) Selenium-dependent growth of *Treponema denticola*: evidence for a clostridial-type glycine reductase. Arch. Microbiol. *177*, 113–116.

Ruby, J.D., Li, H., Kuramitsu, H., Norris, S.J., Goldstein, S.F., Buttle, K.F., and Charon, N.W. (1997) Relationship of *Treponema denticola* periplasmic flagella to irregular cell morphology. J. Bacteriol. *179*, 1628–1635.

Schmidt, E.F., Bretz, W.A., Hutchinson, R.A., and Loesche, W.J. (1988) Correlation of the hydrolysis of benzoyl-arginine naphthylamide (BANA) by plaque with clinical parameters and subgingival levels of spirochetes in periodontal patients. J. Dent. Res. *67*, 1505–1509.

Schultz, C.P., Wolf, V., Lange, R., Mertens, E., Wecke, J., Naumann, D., and Zähringer, U. (1998) Evidence for a new type of outer membrane lipid in oral spirochete *Treponema denticola*. Functioning permeation barrier without lipopolysaccharides. J. Biol. Chem. *273*, 15661–15666.

Scott, D., Chan, E.C., and Siboo, R. (1996) Iron acquisition by oral hemolytic spirochetes: isolation of a hemin-binding protein and identification of iron reductase activity. Can. J. Microbiol. *42*, 1072–1079.

Scott, D., Siboo, I.R., Chan, E.C., Klitorinos, A., and Siboo, R. (1993) Binding of hemin and congo red by oral hemolytic spirochetes. Oral Microbiol. Immunol. *8*, 245–250.

Seshadri, R., Myers, G.S., Tettelin, H., Eisen, J.A., Heidelberg, J.F., Dodson, R.J., Davidsen, T.M., DeBoy, R.T., Fouts, D.E., Haft, D.H., *et al.* (2004) Comparison of the genome of the oral pathogen *Treponema denticola* with other spirochete genomes. Proc. Natl. Acad. Sci. USA *101*, 5646–5651.

Sim, J.H., Shi, W., and Lux, R. (2005) Protein-protein interactions in the chemotaxis signalling pathway of *Treponema denticola*. Microbiology *151*, 1801–1807.

Socransky, S.S., and Haffajee, A.D. (2005) Periodontal microbial ecology. Periodontol. 2000 *38*, 135–187.

Staggs, T.M., Greer, M.K., Baseman, J.B., Holt, S.C., and Tryon, V.V. (1994) Identification of lactoferrin-binding proteins from *Treponema pallidum* subspecies *pallidum* and *Treponema denticola*. Mol. Microbiol. *12*, 613–619.

Stamm, L.V., and Bergen, H.L. (1999) Molecular characterization of a flagellar (*fla*) operon in the oral spirochete *Treponema denticola* ATCC 35405. FEMS Microbiol. Lett. *179*, 31–36.

Syed, S.A., Mäkinen, K.K., Mäkinen, P.L., Chen, C.Y., and Muhammad, Z. (1993) Proteolytic and oxidoreductase activity of *Treponema denticola* ATCC 35405 grown in an aerobic and anaerobic gaseous environment. Res. Microbiol. *144*, 317–326.

ter Steeg, P.F., and van der Hoeven, J.S. (1990) Growth stimulation of *Treponema denticola* by periodontal microorganisms. Anton. van Leeuwen. *57*, 63–70.

Trott, D.J., Moeller, M.R., Zuerner, R.L., Goff, J.P., Waters, W.R., Alt, D.P., Walker, R.L., and Wannemuehler, M.J. (2003) Characterization of *Treponema phagedenis*-like spirochetes isolated from papillomatous digital dermatitis lesions in dairy cattle. J. Clin. Microbiol. *41*, 2522–2529.

Uitto, V.J., Grenier, D., Chan, E.C., and McBride, B.C. (1988) Isolation of a chymotrypsinlike enzyme from *Treponema denticola*. Infect. Immun. 56, 2717–2722.

Uitto, V.J., Pan, Y.M., Leung, W.K., Larjava, H., Ellen, R.P., Finlay, B.B., and McBride, B.C. (1995) Cytopathic effects of *Treponema denticola* chymotrypsin-like proteinase on migrating and stratified epithelial cells. Infect. Immun. *63*, 3401–3410.

Umemoto, T., Li, M., and Namikawa, I. (1997a) Adherence of human oral spirochetes by collagen-binding proteins. Microbiol. Immunol. *41*, 917–923.

Umemoto, T., Nakatani, Y., Nakamura, Y., and Namikawa, I. (1993) Fibronectin-binding proteins of a human oral spirochete *Treponema denticola*. Microbiol. Immunol. *37*, 75–78.

Umemoto, T., Nakazawa, F., Hoshino, E., Okada, K., Fukunaga, M., and Namikawa, I. (1997b) *Treponema medium* sp. nov., isolated from human subgingival dental plaque. Int. J. Syst. Bacteriol. *47*, 67–72.

Umemoto, T., and Namikawa, I. (1994) Binding of host-associated treponeme proteins to collagens and laminin: a possible mechanism of spirochetal adherence to host tissues. Microbiol. Immunol. *38*, 655–663.

Umemoto, T., Namikawa, I., Suido, H., and Asai, S. (1989) A major antigen on the outer envelope of a human oral spirochete, *Treponema denticola*. Infect. Immun. *57*, 2470–2474.

Van Horn, K.G., and Smibert, R.M. (1983) Albumin requirement of *Treponema denticola* and *Treponema vincentii*. Can. J. Microbiol. *29*, 1141–1148.

Vesey, P.M., and Kuramitsu, H.K. (2004) Genetic analysis of *Treponema denticola* ATCC 35405 biofilm formation. Microbiology *150*, 2401–2407.

Virkola, R., Lähteenmäki, K., Eberhard, T., Kuusela, P., van Alphen, L., Ullberg, M., and Korhonen, T.K. (1996) Interaction of *Haemophilus influenzae* with the mammalian extracellular matrix. J. Infect. Dis. *173*, 1137–1147.

Wang, B.Y., Chi, B., and Kuramitsu, H.K. (2002) Genetic exchange between *Treponema denticola* and *Streptococcus gordonii* in biofilms. Oral Microbiol. Immunol. *17*, 108–112.

Wang, Q., Ko, K.S., Kapus, A., McCulloch, C.A., and Ellen, R.P. (2001) A spirochete surface protein uncouples store-operated calcium channels in fibroblasts: a novel cytotoxic mechanism. J. Biol. Chem. *276*, 23056–23064.

Weinberg, A., and Holt, S.C. (1991) Chemical and biological activities of a 64-kilodalton outer sheath protein from *Treponema denticola* strains. J. Bacteriol. *173*, 6935–6947.

Wilson, J.W., Schurr, M.J., LeBlanc, C.L., Ramamurthy, R., Buchanan, K.L., and Nickerson, C.A. (2002) Mechanisms of bacterial pathogenicity. Postgrad. Med. J. *78*, 216–224.

Wyss, C. (1992) Growth of *Porphyromonas gingivalis*, *Treponema denticola*, *T. pectinovorum*, *T. socranskii*, and *T. vincentii* in a chemically defined medium. J. Clin. Microbiol. *30*, 2225–2229.

Wyss, C. (2007) Fatty acids synthesized by oral treponemes in chemically defined media. FEMS Microbiol. Lett. *in press*.

Wyss, C., Choi, B.K., Schüpbach, P., Guggenheim, B., and Göbel, U.B. (1996) *Treponema maltophilum* sp. nov., a small oral spirochete isolated from human periodontal lesions. Int. J. Syst. Bacteriol. *46*, 745–752.

Wyss, C., Choi, B.K., Schüpbach, P., Guggenheim, B., and Göbel, U.B. (1997) *Treponema amylovorum* sp. nov., a saccharolytic spirochete of medium size isolated from an advanced human periodontal lesion. Int. J. Syst. Bacteriol. *47*, 842–845.

Wyss, C., Choi, B.K., Schüpbach, P., Moter, A., Guggenheim, B., and Göbel, U.B. (1999) *Treponema lecithinolyticum* sp. nov., a small saccharolytic spirochaete with phospholipase A and C activities associated with periodontal diseases. Int. J. Syst. Bacteriol. *49 Pt 4*, 1329–1339.

Wyss, C., Dewhirst, F.E., Gmür, R., Thurnheer, T., Xue, Y., Schüpbach, P., Guggenheim, B., and Paster, B.J. (2001) *Treponema parvum* sp. nov., a small, glucoronic or galacturonic acid-dependent oral spirochaete from lesions of human periodontitis and acute necrotizing ulcerative gingivitis. Int. J. Syst. Evol. Microbiol. *51*, 955–962.

Wyss, C., Moter, A., Choi, B.K., Dewhirst, F.E., Xue, Y., Schüpbach, P., Göbel, U.B., Paster, B.J., and Guggenheim, B. (2004) *Treponema putidum* sp. nov., a medium-sized proteolytic spirochaete isolated from lesions of human periodontitis and acute necrotizing ulcerative gingivitis. Int. J. Syst. Evol. Microbiol. *54*, 1117–1122.

Xu, X., Holt, S.C., and Kolodrubetz, D. (2001) Cloning and expression of two novel hemin binding protein genes from *Treponema denticola*. Infect. Immun. *69*, 4465–4472.

Xu, X., and Kolodrubetz, D. (2002) Construction and analysis of hemin binding protein mutants in the oral pathogen *Treponema denticola*. Res. Microbiol. *153*, 569–577.

Yamada, M., Ikegami, A., and Kuramitsu, H.K. (2005) Synergistic biofilm formation by *Treponema denticola* and *Porphyromonas gingivalis*. FEMS Microbiol. Lett. *250*, 271–277.

Yamazaki, T., Miyamoto, M., Yamada, S., Okuda, K., and Ishihara, K. (2006) Surface protease of *Treponema denticola* hydrolyzes C3 and influences function of polymorphonuclear leukocytes. Microbes Infect. *8*, 1758–1763.

Yang, P.F., Song, M., Grove, D.A., and Ellen, R.P. (1998) Filamentous actin disruption and diminished inositol phosphate response in gingival fibroblasts caused by *Treponema denticola*. Infect. Immun. 66, 696–702.

Yao, E.S., Lamont, R.J., Leu, S.P., and Weinberg, A. (1996) Interbacterial binding among strains of pathogenic and commensal oral bacterial species. Oral Microbiol. Immunol. *11*, 35–41.

Yotis, W., Keene, J., Hoerman, K., and Simonson, L.G. (1993) Fatty acid profiles of the outer membrane of ATCC strains 35405, 35404 and 33521 of *Treponema denticola*. J. Periodont. Res. *28*, 387–395.

Yotis, W.W., Sharma, V.K., Gopalsami, C., Chegini, S., McNulty, J., Hoerman, K., Keene, J., Jr., and Simonson, L.G. (1991) Biochemical properties of the outer membrane of *Treponema denticola*. J. Clin. Microbiol. *29*, 1397–1406.

You, Y., Elmore, S., Colton, L.L., Mackenzie, C., Stoops, J.K., Weinstock, G.M., and Norris, S.J. (1996) Characterization of the cytoplasmic filament protein gene (*cfpA*) of *Treponema pallidum* subsp. *pallidum*. J. Bacteriol. *178*, 3177–3187.

The Molecular Basis of Host–Pathogen Interaction in the Oral Cavity

9

Janina P. Lewis

Abstract

The interaction between oral microflora and eukaryotic cells is highly complex and involves active processes allowing both types of partners to co-exist. The microbial encounter with the host is initiated by attachment of the organism to the host; and bacterial surface proteins, termed 'adhesins', mediate this step. Bacteria also secrete mediators that bind or invade the host. The microbial attachment, as well as the encounter with the secreted proteins, serve as signals that are deciphered through multimediator cascades ultimately affecting gene expression in the host. Many receptors recognizing the pathogens and mediating the host response, as well as the variety of microbial molecules triggering the host response, have been demonstrated. Also, the application of high-throughput microarray analyses have revealed that many more players remain to be investigated, as well as providing insight into the complex nature of the host-microbe interaction. The recent advances regarding both the host and microbial players involved in the host-pathogen dialogue, as well as possible mechanisms utilized by pathogenic bacteria to evade the host protective recognition mechanisms, will be discussed. Finally, new directions, with regard to the investigation of the molecular mechanisms of the host-pathogen interaction, are discussed. This includes genomic approaches to investigating host-pathogen interactions on a large scale.

Introduction

Host cells respond to various signals and alter their protein expression accordingly. One of these signals is the encountering of microorganisms by host cells. The mutual co-existence of microorganisms with their hosts is mediated via a molecular dialogue between the host and microbe, resulting in an alteration of protein expression on both sides, ultimately facilitating their mutual adaptation. Although significant evidence that the pathogenic bacteria trigger the host response has accumulated over the years, recent studies have shown that commensal organisms also stimulate the host response.

The oral cavity is inhabited by at least 700 microorganisms, including both pathogenic and commensal species. Oral microorganisms have access to multiple surfaces and cell types in the oral cavity. Firstly, the hard, non-shedding surfaces of the tooth are excellent for plaque deposition. However, host cells, such as the epithelial cells lining the oral cavity, are also colonized by oral microflora. Furthermore, upon injury to the gingival epithelium, the oral microflora gain access to deeper tissues, such as connective tissue, which is composed mainly of fibroblasts. Finally, immune cells are present within the gingival tissues and may also be exposed to bacteria. Microbial cell surface proteins, termed 'adhesins', play a major role in mediating the attachment of the organisms to the host tissue. Recent studies have provided evidence that the microbial contact with host cells is not merely a passive event, but involves a series of interactive responses, involving both the host and the microbe, leading to changes in both partners and ultimately allowing the co-existence of the microbes with the host. The microbial adhesins serve not only to mediate the adherence process, but also have a role as signalling molecules that affect the expression profiles of the host cells. Also, the host is able to respond appropriately, not only to direct contact with the microbe, but also to secreted microbial components. The host response is initiated by the interaction of host cell receptors with the microbial components. Once activated, these receptors transfer signals via multicomponent signalling cascades involving G-proteins, kinases, and ultimately transcription factors, to the nucleus. Significant progress has been made in the identification and characterization of the various host receptors and their expression in gingival tissues. Also, insights into the role of these receptors in response to oral microorganisms have been gained from recent studies. Hundreds of gene products are involved in deciphering the microbial message and mounting an appropriate response. The genomic revolution has allowed for a more comprehensive understanding of the host response to both pathogenic and commensal bacteria. These data, along with the results obtained from the investigation of non-oral organisms, have provided some insight into the possible mechanisms employed by the host to distinguish between commensal and pathogenic bacteria. Importantly, such data are expected to contribute to our understanding of the roles of the commensal and pathogenic microorganisms, in both the health or disease states, and may ultimately serve to identify targets for the design of novel antimicrobial agents.

Microbial complexity of the oral cavity

The human oral cavity is the gateway to our bodies. Approximately two thirds of all microorganisms gain entry through the mouth, and some of these organisms establish themselves in the oral cavity. Approximately 10^{10} bacteria, belonging to 700 bacterial species, inhabit the teeth, gingival crevices, buccal mucosa, and tongue. It is estimated that only 60% of these are cultivable (Paster *et al.*, 2001). Common bacteria found in the oral cavity include (Paster *et al.*, 2001):

Spirochaetes: Treponema medium, T. denticola, T. maltophilum
Fusobacteria: Fusobacterium naviforme, F. animalis, F. nucleatum, Leptotrichia buccalis
Actinobacteria: Actinomyces naeslundii, A. israelii, A. odontolyticus, Rothia dentocriosa, Atopobium, Bifidobacterium dentium, Corynebacterium matruchotii, Propionibacterium propionicus
Firmicutes:
Class 'Bacilli': *Streptococcus oralis, S. mitis, S. gordonii, S. mutans, S. sobrinus, S. sanguis, S. parasanguis, S. salivarius, S. intermedius, S. constellatus, S. anginosus; Abiotrophia adiacens, A. defective, Gemella haemolysans*
Class 'Mollicutes': *Mycoplasma; Solobacterium moorei*
Class 'Clostridia': *Catonella morbid, Dialister, Eubacterium brachy, E. saburreum, Megasphaera, Peptostreptococcus anaerobius, P. micros, Selenomonas, Veillonella dispar, V. parvula, Eubacterium saphenum, Filifactor alocis, Catonella morbid, Megasphaera elsdenii, Dialister pneumocintes, Selenomonas sputigena*
Proteobacteria: Haemophilus parainfluenzae, Campylobacter rectus, C. gracilis, C. concisus, Neisseria mucosa, Desulfobulbus, Aggregatibacter (Actinobacillus) actinomycetemcomitans, Eikenella corrodens, Escherichia coli, Pseudomonas aeruginosa
Bacteroidetes: Porphyromonas gingivalis, P. endodontalis, Tannerella forsythia (previously known as *Bacteroides forsythus*), *Prevotella denticola, P. intermedia, P. oris, P. tannerae, Capnocytophaga ochracea, C. gingivalis*

Besides bacteria, other microorganisms, such as viruses, fungi, and protozoa, are present. *Candida albicans* is one such example. This is a fungus found in 60 percent of the total population and is the causative agent of oral candidiasis (also known as 'oral thrush'), which affects a significant number of young, elderly, and immunocompromised individuals (Cannon and Chaffin, 2001).

The oral cavity is a unique habitat. Both hard, non-shedding surfaces (teeth) and soft tissues are present. These surfaces provide attachment sites for various microorganisms. Also, microorganisms display tissue tropism, and one niche is preferentially occupied by a certain bacterial type that is capable of adhesion and growth at that site. This tropism is mainly dictated by the availability of appropriate receptors for bacterial adhesion, as well as nutrient availability.

Tooth microbiota

The surfaces of the teeth accumulate complex populations of bacteria that exist as biofilms (Kolenbrander, 2000). The main microorganisms involved are streptococci and *Actinomyces* spp. The colonization of freshly cleaned tooth surface is accompanied by the deposition of the saliva-derived pellicle, which provides attachment sites for the microorganisms.

Mucosal microbiota

Gingival, buccal, and palatal tissues are covered by epithelium. One of the characteristics of the epithelium is that desquamation of the surface cells of the epithelium occurs, resulting in the shedding of the microorganisms. Due to the shedding property of the epithelium, the microbial community forming at the mucosal surfaces is not as voluminous as that observed at the tooth surface. Thus, in order to persist at the mucosal surfaces, the microbes must continuously grow and re-attach themselves to the host tissues. Gingival crevices have low oxygen content, and thus are inhabited by large numbers of Gram-negative anaerobes. Among them, are bacteria considered to be involved in the development of periodontal disease; these include *Porphyromonas gingivalis*, *Tannerella forsythia*, and *Prevotella intermedia*. These mucosa-microbe interactions are extremely dynamic and will be the main focus of this chapter.

Tongue microbiota

The tongue is a unique site because it is covered with papillae, affording numerous protected sites that provide sheltered habitats for bacterial colonization. Indeed, with the exception of the teeth and gingival sulcus, the dorsum of the tongue harbours a greater microbial biomass than any other site in the mouth. Relatively few studies dealing with the microbiology of the tongue have been undertaken; however, it appears that the tongue microbiota include bacterial species that also colonize the teeth, as well as those that are selective for the tongue, e.g. *Streptococcus mitis* and *salivarius*.

The oral microbial flora is highly complex, with both pathogenic and commensal species being present in the oral cavity (Ezzo and Cutler, 2003). The major pathogens that have been the focus of most investigations are those underlined in the above microbial list. It is now well established that some oral diseases, such as periodontal diseases, result from the host response to certain bacterial pathogens. How the host distinguishes between pathogenic and non-pathogenic bacteria, and what the host responses are to the pathogenic and non-pathogenic bacteria, are not very clear. In this chapter, an attempt will be made to show what is known regarding the host response to the oral pathogens. The main focus will be on pathogens, as most studies regarding oral bacteria have involved organisms that may play a role in triggering oral diseases. Among those organisms are: *Streptococcus mutans* (initiation and progression of dental caries), *Aggregatibacter* (*Actinobacillus*) *actinomycetemcomitans* (aggressive localized periodontitis), *Porphyromonas gingivalis* (chronic periodontitis), and *Tannerella forsythia* (chronic periodontitis). However, examples indicating that commensals are not merely passive bystanders, but are also responsible for triggering a host response, will also be provided.

Colonization of the oral cavity

The critical first step of the successful establishment of bacteria in the oral cavity is the efficient adherence of the bacteria to the host components or other microorganisms present in the oral cavity. The colonization of the oral cavity is something of a challenge due to the mechanical forces of tongue movement, the flow of saliva and gingival crevicular fluid, the desquamation of cells, and the host immune response. This is further compounded by the ability of salivary molecules to cause the microorganisms to aggregate and subsequently clear them during expectoration or swallowing (Theilade, 1990). Therefore, successful oral colonizers must possess efficient attachment mechanisms. There are a variety of surfaces available in the oral cavity to which bacteria can attach. For example, *P. gingivalis* has been shown to attach to saliva and crevicular fluid components (proline rich proteins and glycoproteins, statherin, histatins, lysozyme and lactoferrin), host cells and matrix components (epithelial cells, fibroblasts, erythrocytes, fibrinogen, collagen and laminin), and microbial cells (*Streptococcus* spp., *Actinomyces* spp., *F. nucleatum*, *T. forsythia*, *T. denticola*, *Veillonella* spp. *and Capnocytophaga* spp.).

Microbial establishment is initiated by low-affinity interactions, such as electrostatic and hydrophobic attraction. However, these forces only enable the relatively weak attachment of the microorganisms and so bacteria can be easily removed from the surfaces. For a more permanent attachment, higher affinity adhesion mechanisms are required. These stronger attachment mechanisms involve bacterial cell surface structures, known as adhesins, that specifically bind to host receptors. These interactions are highly specific; for example, specific adhesins bind to specific receptors on the host surfaces. Therefore, the expression of an adhesin with certain specificity will dictate the type of host cell that has a complementary receptor for that adhesin. Thus, specific bacteria can only inhabit certain niches; for instance, *S. sanguis* is found mainly on the tooth surface, while *S. salivarius* is present on the surface of the tongue.

Host surfaces

A professionally cleaned tooth is first coated by the pellicle, which is composed of salivary proteins; and it is to the pellicle, rather than to the surface of the tooth, that bacteria attach themselves. Among the first colonizers are the Gram-positive streptococci and a fungus-like bacterium, *A. naeslundii*. Together with their extracellular products, these organisms provide a niche for the establishment of secondary colonizers. As the tooth surface is a hard, non-shedding surface, these communities can grow and result in the large masses of accumulated microorganisms known as dental plaque.

The adhesion surface of the host epithelial cells is markedly different. These cells are continuously shed, along with any attached bacteria. They do express receptors to which bacteria can bind directly; however, they are also embedded in an extracellular matrix (ECM). The ECM components include collagens, laminins, elastin, proteoglycans, hyauluronan, fibronectin and fibrinogen; and, at this point, a description of these components is relevant:

Collagens are the major structural components of the eukaryotic ECM. At least 25 different collagen gene products have been identified but the most prevalent (>90%) of the human body collagens are the fibril-forming collagens types I, II and III. Collagen type IV forms basement membrane networks. The triple-helical conformation of collagen is achieved through processing and the post-translational modification of collagens. Bacteria can interact with various forms of collagen both through the recognition of the triple-helical and denatured forms of collagen in a conformation-independent manner, triple-helix-dependent manner, and in a manner that requires fibrillar structures. There are multiple sites along the triple helix to which bacteria can attach themselves. As other ECM proteins, such as fibronectin, vitronectin, laminin, and proteoglycans, can tightly associate with collagens to form multimolecular aggregates, the bacterial interaction with the host is not solely dependent on the collagen component.

Laminins are the major constituent of basement membranes; they interact specifically with type IV collagen, proteoglycans, and other ECM components, as well as with various cell types, thus playing a central role in basement membranes. Laminin is a 900-kDa glycoprotein composed of three chains, alpha, beta, and gamma, linked to each other by disulfide bridges. Varying chain compositions allow for the protein to assume up to 10 different isoforms. Laminin is exposed to bacteria following tissue damage or inflammation. Such tissue damage in the periodontal pocket may be the result of the action of the combination of bacterial and host proteases.

Elastin and fibrillins form fibres that are embedded into the ECM and account for the elasticity of blood vessels, in particular. The main component (90%) of fibrils is elastin, a polymer of tropoelastin. Fibrillins form the microfibrils found in the elastic fibres. Bacteria containing elastin-binding proteins can colonize the elastin-containing organs, such as blood vessels, and thus have a role in the systemic complications resulting from oral diseases.

Proteoglycans are the most abundant, heterogeneous, and functionally versatile group of non-fibrillar ECM components and are found both in the interstitial ECM, as well as in basement membranes. They are composed of polypeptides to which glycosaminoglycan side chains are attached and it is these side chains that account for the proteoglycan's high functional versatility.

Hyaluronan is found in the interstitial ECM in most tissues. It is a high molecular weight polysaccharide with a high hydration capacity. It binds to a variety of proteoglycan members, such as the sulphated-proteoglycan, aggrecan, and contributes to the viscoelastic properties of tissues. The hyaluronan covering gingival epithelial cells serves as ligand for the periodontal pathogen *T. denticola*.

Fibronectin is a ubiquitous matrix protein that is indispensable for the adhesion of most cell types. It contains the Arg-Gly-Asp (RGD) motif that is recognized by receptors present on the host cell surface and designated integrins (see below for a description of integrins). Fibronectin plays a major role in the adherence of a variety of pathogens, and specific fibronectin-binding adhesins have been demonstrated in both Gram-positive and Gram-negative bacteria. In addition, several oral pathogens, including *P. gingivalis*, *P. intermedia*, *S. mutans*, and *T. forsythia*, have been shown to bind to fibronectin.

The precursor of fibrinogen is fibrin, a 350-kDa plasma glycoprotein that associates with other adhesion proteins during the formation of a fibrin clot, the ECM network required for wound healing. A variety of bacteria, including oral species, such as *P. intermedia*, *P. gingivalis*, and *T. forsythia*, contain fibrinogen-binding proteins that may mediate their establishment in the host.

As the ECM is indispensable for tissue integrity, the degradation of the ECM is required to allow bacteria to gain access to the host tissues. Certain oral bacteria have developed a means to gain entry into the host by degrading the ECM molecules; e.g. *P. gingivalis* produces potent proteases capable of degrading a variety of proteins, including ECM components.

Bacteria can also establish themselves by binding directly to the receptors present on host cells. Several types of receptors that have evolved to mediate various host processes, such as cellular activation, proliferation, differentiation, and cellular mobility, are present in the host cell and include: integrins, selectins, cadherins, immunoglobulin superfamily cell adhesion molecules (IgCAM), and syndecans. These receptors mediate cell adhesion and engage in signal transduction from the exterior to the interior of the cell. Various signal transduction processes that are involved in the primary cellular processes, and ultimately survival, are also mediated by these receptors. Bacteria do, in fact, directly bind the host receptors as a means of adhering to the host tissues. However, many bacterial pathogens adhere to the host receptors as a means of exploiting the underlying signalling cascade that allows them to gain entry into the cell. In addition, the engagement of the host receptors by bacteria may lead to the loss of important cellular functions. The main host adhesion receptors are described below.

Integrins are family of transmembrane heterodimeric molecules composed of various non-covalently bound alpha- and beta-chains; for example, $\alpha2\beta1$, $\alpha3\beta1$, $\alpha6\beta1$, and $\alpha6\beta4$ are present on gingival epithelial cells. Specific combination between an alpha and beta subunit determines ligand specificity; for instance, $\alpha5\beta1$ is a fibronectin receptor and $\alpha6\beta1$ binds laminin. Since a given integrin heterodimer can bind to several ligands and can show overlapping function with other integrins, such specificity is relative. The integrin family is defined by its beta-chains but both alpha- and beta-chains contribute to the binding of ligands. Integrins mediate cell-ECM, as well as cell–cell, interactions and play a role in various functions such as cell migration, wound healing, and embryogenesis. The $\beta1$-integrins are mainly extracellular matrix receptors, $\beta2$-integrins are present on the surface of leukocytes and mediate cell–cell adhesion processes during inflammation, and $\beta3$-integrins are present on platelets and play a role in blood clotting.

Selectins are a family of lectin receptors and are mainly expressed on the surface of endothelial cells and leukocytes. They play a role in the vascular system, mediating the attachment and the rolling of leukocytes on endothelial tissues during inflammation.

Cadherins are homodimers present at the cell adherent junctions. Three types of cadherins have been identified and classified as P, E, and N. Cadherins mediate cell–cell interactions by binding the cadherins of the adjacent cells. They are also able to interact with the cell cytoskeleton through the cytoplasmic tails of the catenins.

Immunoglobulin Superfamily Cell Adhesion Molecules (IgCAM) mainly mediate cell–cell interactions and consist of receptors with extracellular domains homologous to those of immunoglobulins. They can mediate homotypic (where the interacting molecules present on the two partner cells are identical) and heterotypic interactions (e.g. ICAM1 and vascular cell adhesion molecule 1 [VCAM-1]). They play a role in inflammation as well as regulating the response to growth factors.

Various microbial adhesins use the above adhesion molecules as their cognate receptors. An example is the *P. gingivalis* adhesin, FimA, which interacts with the host alpha5, beta1-integrin receptor.

Microbial adhesins can be classified into several types:

Fimbrial adhesins

For example, fimbrial adhesins are located at the tip of a scaffold-like structure on the bacterial surface (named 'pili' or 'fimbriae') and are present in oral pathogens, such as *S. mutans*, *P. gingivalis*, and *A. actinomycetemcomitans* (*A.a*).

Non-fimbrial adhesins

In contrast, non-fimbrial adhesins are anchored in the bacterial membrane, but have their surfaces exposed. Examples of these are the *P. gingivalis* adhesins, HagA and HagB, as well as the *P. intermedia* adhesin, AdpB.

Examples of known oral microbial adhesins and their targets are listed in Table 9.1.

P. gingivalis *adhesins*

The mechanisms of adhesion in oral Gram-negative, anaerobic bacteria are best exemplified by *P. gingivalis*. The major adherence-mediating determinants in this organism are fimbriae (peritrichous structures present on the cell surface of the bacterium), 0.3 to 3.0 μm long and 5 nm wide. They are composed of proteinaceous subunits, fimbrillin (FimA, 41–49 kDa depending on the strain) and have been shown to mediate the adhesion of *P. ginigvalis* to a variety of substrates present in the oral cavity; for example, to salivary molecules (proline-rich proteins, glycoproteins, statherin), oral epithelial cells, macrophages, fibroblasts, ECM components (fibrinogen and fibronectin), lactoferrin, and other organisms (reviewed in (39)). In addition to the major fimbriae (composed of FimA monomers), *P. gingivalis* contains two other types of minor fimbriae, one composed of a 67 kDa protein and the other of a 72 kDa protein. The precise functions of these fimbriae remain are not known but it has been demonstrated that the short fimbriae (Mfa1 is the structural protein) mediate adhesion of *P. gingivalis* to *Streptococcus gordonii* (Park *et al.*, 2005). Apart from fimbriae, *P. gingivalis* produces multiple haemagglutinins/adhesins, HagA, HagB, and HagC. HagA-like haemagglutinins are also encoded by two protease genes, *rgpA* and *kgp*. These proteins mediate the attachment of the bacteria to red blood cells, which are in turn used as a source of haemin, an indispensable nutrient

Table 9.1 Examples of adhesins and their receptors in oral bacteria

Microorganism	Adhesin	Host adhesion receptor/ECM component
Streptococcus mutans	FimA	Saliva-coated hydroxyapatite fibrin
	AgI/II	Salivary receptors, fibronectin, laminin, collagen type I
Porphyromonas gingivalis	FimA	Salivary proteins, integrin receptors
	Mfa1	*S. gordonii*
	HagA	Host cells, Hb, collagen, fibronectin, fibrinogen
	HagB	To be determined
Treponema denticola	38 kDa (flagellin)	
	53 kDa	
	72 kDa	
	64 kDa	
	Lrr	
Actinobacillus actinomycetemcomitans	EmaA	Collagen
	AaE	
Fusobacterium nucleatum	Fad	Surface proteins of KB cells
Tannerella forsythia	BspA	Fibrinogen, fibronectin
Prevotella intermedia	AdpB	Fibronectin, fibrinogen, laminin
	18 putative adhesins	To be determined

for the growth of the bacteria. In addition to haemagglutinating properties, the HagA sequences have also been shown to participate in binding to ECM components (fibronectin, laminin, fibrinogen) (Pike *et al.*, 1996) and epithelial cells (Chen *et al.*, 2001). As the HagA –like adhesins are associated with protease components capable of degrading a variety of molecules, including ECM proteins such as fibronectin and fibrinogen, the combination of the adherent and protease activities may be a mechanism by which the organism specifically adheres to the host components, and then degrades them to gain access to deeper host tissues. Another cell surface adhesin, HagB, has been shown to play a role in the adherence of *P. gingivalis* to human coronary artery endothelial cells (HCAEC) (Song *et al.*, 2005).

Knowledge of the adhesion process of other Gram-negative anaerobic members of the oral community is currently quite limited.

T. forsythia *adhesins*

T. forsythia adhesins allow the organism to bind to a variety of host components, including oral epithelial cells. Also, haemagglutination activity has been reported in this organism (Lamont and Jenkinson, 1998). BspA, one of the most thoroughly characterized adhesins in this organism, has been demonstrated to bind both fibrinogen and fibronectin (Sharma *et al.*, 1998). Investigation of the genomic sequence of *T. forsythia* identified 6 other genes that encode proteins with similarities to the originally identified BspA protein. The presence of multiple copies of this adhesin may indicate the importance of the protein in the establishment and survival of the bacterium but the exact contribution of those proteins remains to be established. Finally, it is probable that other, yet to be identified, adhesins are present in this bacterium. The genome sequence of *T. forsythia* is available, thus enabling bioinformatic, in addition to functional genomic, approaches to identify other mediators of the *T. forsythia*-host adhesion process.

Prev. intermedia *adhesins*

P. intermedia has been shown to attach to host cells but the host and the microbial components mediating the interaction remain to be clearly defined. *P. intermedia* produces fimbriae of several different types. They are classified depending on their diameter: type C (8-nm-diameter, *P. intermedia* 17), type D (5-nm-diameter, *P. intermedia* 27), and type A (1- to 2-nm-diameter, *P. intermedia* 25611) (Leung *et al.*, 1989). The presence of type C fimbriae in *P. intermedia* 17 may mediate the attachment and invasion of host oral epithelial cells (Dorn *et al.*, 1998) but more studies are required to confirm this possibility. While the fimbriae-associated proteins have been shown to agglutinate red blood cells (Leung *et al.*, 1999), further investigations, involving allelic-exchange mutants, are needed to clearly define the role of these proteins in the haemagglutination process.

P. intermedia has been demonstrated to bind to type I collagen, laminin, fibronectin, and fibrinogen (Grenier, 1996; Alugupalli, 2003; Kalfas *et al.*, 1992). A 62 kDa laminin-binding protein present throughout the whole cell, as well as in the outer membrane extracts, has been reported but the gene and protein sequence remain unknown. A collagen-, fibronectin-, and fibrinogen-binding component, designated AdpB, has recently been identified, and as expected, this protein was also shown to be located at the cell surface (Yu *et al.*, 2006). Of note, the availability of the genomic sequence of *P. intermedia* enabled the bioinformatic investigation of the sequences for similarities with known adhesins from other bacteria. This search identified eighteen ORFs encoding putative adhesins that are present on the genome of *Prev. intermedia* 17. Five genes PI0136, PI0493, PI1564, PI1571, and PI2109, coded for BspA-like adhesin from *T. forsythia*. BspA is a cell surface associated protein that has been shown to bind fibronectin and fibrinogen in a concentration-dependent manner, and also to inhibit the binding of *T. forsythia* to these ECM components (Sharma *et al.*, 1998; Honma *et al.*, 2001). Two genes, PI0137 and pPI0072, encoded homologues of lectin-like adhesin from *Prevotella loescheі* (Manch-Citron *et al.*, 1992). This adhesin, consisting of six subunits of 75 kDa in size, mediates the coaggregation of *P. loeschei* with streptococci and the haemagglutination of neuraminidase-treated erythrocytes (London and Allen, 1990). Another *P. intermedia* haemagglutinin, encoded by pPI0381, has been identified (Okamoto *et al.*, 1999) and this haemagglutinin gene has been demonstrated to be present in

88% of *P. intermedia* isolates. On the genome of *P. intermedia* 17 we also identified an ORF (PI0360) encoding a putative fibrinogen binding protein. This protein is similar to the fibrinogen-binding protein from *Streptococcus equi* (Meehan *et al.*, 1998). Furthermore, two genes, PI0023 and PI1883, coding for internalin-like proteins, are present on the genome of *P. intermedia* 17. Internalins are a family of cell surface proteins containing leucine rich repeats (LRR). While internalin A (InlA) and Internalin B (InlB) have been shown to mediate the entry of *Listeria* into host cells (Ireton and Cossart, 1997), the role of other members of the internalin family remains to be determined (Dramsi *et al.*, 1997). ORF PI1717 encodes AdpB, the second fibronectin-binding adhesin identified in our laboratory using proteomic approaches (see data above). Finally, PI0035 codes for a 160 kDa fibronectin-binding protein (AdpA). Homology searches have shown that the most significant homology (29% identity and 43% similarity over 230 C-terminal amino acids of AdpA) was noted with the haemagglutinin/adhesin protein, HagA, of *P. gingivalis*. Five ORFs, PI0757, PI1073, pPI0626, PI0429, and PI0428, encoding AdpA-like proteins, were identified on the genome of *P. intermedia* 17, indicating that this protein may be indispensable for attachment to and interaction with the host.

T. denticola *adhesins*

T. denticola is a spirochete shown to adhere to various epithelial, fibroblast, and endothelial cells (Sela, 2001). Furthermore, it was shown to adhere directly to host cells, as well as bind various ECM matrix proteins, including laminin and fibronectin. The adherence of the organism is polar and correlates with the polar clustering of adhesins in the outer regions, forming the basis for 'tip adhesion'. 53-kDa and 72-kDa as well as 38-kDa flagellar proteins have been demonstrated to bind human plasma fibronectin (Umemoto *et al.*, 1993). Also, a 64-kDa adhesin has been isolated (Weinberg and Holt, 1991). Finally, a leucine-rich repeat protein, designated LrrA, has also been identified in *T. denticola*, and has been shown to have a role in binding to host cells and to another periodontopathogen, *T. forsythia*. The binding of *T. denticola* to *T. forsythia* is mediated through the specific association of LrrA with BspA (Ikegami *et al.*, 2004). Finally, dentilisin, a chymotrypsin-like protease, has been identified as a *P. gingivalis* fimbria-binding protein, thus serving as one of the attachment mechanisms between oral pathogens (Hashimoto *et al.*, 2003).

A. actinomycetemcomitans (A.a) *adhesins*

A.a. binds and invades a variety of cells. Several adhesins have been identified in this organism, including both fimbrial and afimbrial proteins. The major fimbrial component is a 6.5 kDa peptide, designated Flp (fimbrial low molecular weight protein), and the minor component has a molecular weight of 54 kDa (Inoue *et al.*, 1998). A genomic locus, composed of 14 genes: *flp-1-flp-2-tad*V-*rcp*CAB-*tad*ZABCDEFG, has been identified and is referred to as the *tad* locus. Mutations in this locus have rendered the bacterium unable to induce bone loss, when compared to the parental strain, in a rat model, indicating that this locus is important in the pathogenesis of *A.a.* (Schreiner *et al.*, 2003). Also, an extracellular amorphous material, associated either with the bacterium or released into the milieu, can confer adhesiveness to *A.a.* Furthermore, two genomic loci, *apiA* and *apiBC*, encode proteins binding the host components, and have a role in the invasion of *A.a.* into epithelial cells. ApiA binds human collagen types II, III andV, and fibronectin. ApiBC are involved in binding to collagen type III (Li *et al.*, 2004). In addition, autotransporter adhesin Aae is a surface-exposed outer membrane protein required for binding to buccal epithelial cells (Rose *et al.*, 2003; Fine *et al.*, 2005). Finally, an outer membrane protein, termed 'extracellular matrix protein adhesin', EmaA, was firstly identified as a collagen-binding protein (Mintz, 2004). It shares protein features with another collagen-binding protein, YadA of *Yersinia enterocolitica*. Interestingly, antenna-like structures were demonstrated to form in *A.a.*, depending on the presence of EmaA (Ruiz *et al.*, 2006).

F. nucleatum *adhesins*

Fusobacterium nucleatum is designated as an 'adhesive' organism because it is known to bind a variety of host cells, including epithelial and endothelial cells, polymorphonuclear leukocytes, monocytes, erythrocytes, fibroblasts, and

HeLa cells, as well as salivary macromolecules, extracellular matrix proteins, and human immunoglobulin G (IgG) (Babu *et al.*, 1995; Han *et al.*, 2000; Ozaki *et al.*, 1990; Winkler *et al.*, 1987). It also plays an important role in oral microbial community formation as it co-aggregates with a variety of oral bacteria. As such, the knowledge of the molecular basis of the microorganism's adhesion mechanisms is essential to the understanding of its pathogenesis. Both lectin- and non-lectin adhesins are suggested to be present in this organism. In addition, 40-kDa FomA, 39-kDa and 30-kDa proteins appear to play a role in mediating the interbacterial coaggregation, and a high-molecular-mass component (approximately 300 kDa) has a role in binding to galactose. The role of these proteins in mediating the binding of the bacterium to host cells remains to be determined. A spontaneous lam mutant, lacking the ability to bind erythrocytes and attach to or invade human gingival epithelial cells, has been isolated. Characterization of this mutation, with respect to the affected gene, will aid in understanding the molecular basis of the adhesion process in *F. nucleatum*. Finally, a small 13.6-kDa adhesin, designated FadA (*Fusobacterium* adhesin A), was shown to play a role in the attachment of *F. nucleatum* to oral mucosal cells (Han *et al.*, 2005). Of note, the adhesin is conserved among a variety of closely related *Fusobacterium* species.

S. mutans *adhesins*

Streptococcus mutans can bind salivary proteins, coaggregate with other oral bacteria, and adhere to host cells. A 167-kDa AgI/II adhesin has been demonstrated to bind to salivary proteins, as well as to salivary receptors (Crowley *et al.*, 1993; Hajishengallis *et al.*, 1994), extracellular matrix proteins, such as collagen type I, fibronectin, and laminin. In addition, a 36-kDa lipoprotein, FimA, is associated with *S. mutans* fimbriae. This protein is also found on the surface of other oral streptococci, such as *S. parasanguis*, and other *viridans* streptococci that have been examined to date. A study of FimA has shown that it is responsible for the binding of the bacterium to saliva-coated hydroxyapatite, as well as to fibrin. The latter adhesion property facilitates the adherence of oral streptococci to fibrin, and is believed to be an essential factor in the colonization of heart valves (Kitten *et al.*, 2002). Interestingly, this protein is also implicated in the transport of the essential metal manganese in the streptococcal cell.

Regarding the attachment processes of oral bacteria, much still remains to be clarified but comprehensive approaches are now possible, due to the availability of genomic sequences for a variety of oral organisms. The identification and characterization of receptors mediating the attachment of microorganisms to the host will certainly contribute to a better understanding of the colonization process, the host-microbe interaction and provide new targets for the development of anti-periodontitis strategies.

The adhesion process

The primary colonizers of freshly cleaned teeth are facultative bacteria, mainly streptococci. After the establishment of streptococci, other organisms follow; thus, the tooth is eventually colonized by Gram-negative anaerobes and spirochetes. The tooth surfaces are not the only sites available for colonization by oral bacteria; for example, microorganisms bind to the epithelial cells lining the soft tissue surfaces of the oral cavity. Some oral bacteria are capable of penetrating the host epithelial barrier and transversing deeper into the host gingival tissues, thus reaching the connective tissues which are composed mainly of fibroblasts. Finally, oral microorganisms can gain access to the bloodstream and so interact with endothelial cells or, ultimately, with aortic cells. The hallmark of oral soft tissue infections, such as gingivitis and periodontitis, is the recruitment of immune cells to the site of infection. Oral pathogens thus interact with cells such as macrophages, dendritic cells, and neutrophils. There is, therefore, a variety of host components to which the oral bacteria can attach and gain access (Fig. 9.1A).

Microbial adhesion to the host cells may be direct and may involve host cell-surface receptors (integral host membrane proteins, such as lectins, eukaryotic cell surface proteoglycans, glycolipids, and cell surface adhesion receptors); or, it may be indirect, adhering to extracellular matrix (ECM) components, such as collagen, fibronectin, laminin, hyaluronan (such as is used by *T. denticola* to attach to epithelial cells in periodontal tissue), and glycoproteins (fibronectin, vitronectin and fibrinogen-fibrin) (Fig. 9.1B).

The first cell type encountered during adhesion to the host tissue is the epithelial cell. As the carbohydrates expressed on the apical site of these cells are readily available for bacterial attachment, the association between bacterial sugar-binding adhesins, also known as lectins, is the major adhesion mechanism. As specific host cell receptors, such as cadherins and integrins, are expressed on the basolateral site of the epithelial cells, and are thus not readily available for microbial adhesion, such adhesins must be recruited in order for the organisms to be able to attach. Since factors such as inflammation or mechanical injury can cause receptor redistribution to the apical site, such events are probably required for subsequent microbial colonization. Apart from direct adhesin-host receptor binding interactions, other factors also play a role. The process of attachment in the oral cavity is unique in that oral bacteria live as multispecies communities. These polymicrobial communities establish themselves using various mechanisms, including attachment to each other and metabolic interactions. Bacteria-host and bacteria-bacteria interactions are thus involved in the colonization process in the oral cavity.

Emergent antibiotic resistance, as well as the harmful side reactions due to the cross-reactivities between bacterial surface proteins and host factors, require novel strategies for the prevention and treatment of infectious diseases. Based on the identification of adhesion and invasion mechanisms of pathogens, and the relevant components involved in these processes at different host niches, novel 'anti-adhesion therapies' can be designed. These include various substances, such as soluble bacterial adhesins or host receptor analogues, antibodies against bacterial adhesins, and low-molecular-weight antagonists. Since each microorganism has the ability to use multiple mechanisms of adhesion to initiate infection, effective anti-adhesion drugs should contain cocktails of different inhibitors. These strategies are, however, complicated by the fact that not all bacterial adhesion genes are present in every strain, and that environmental factors can affect the expression of these genes. Thus, knowledge of the strategies used by microorganisms for their tissue and cellular entry may be utilized in the

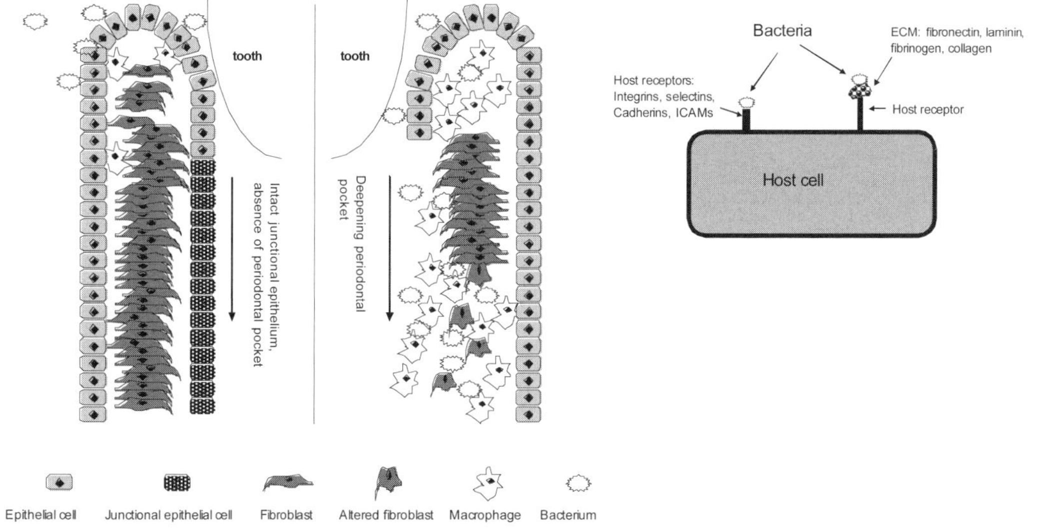

Figure 9.1 Adhesion of microorganisms to host cells. A. Various host cells are available for microbial adhesion in the oral cavity. The host cell types available depend on the health status of the individual. At healthy sites, mainly epithelial cells are available for microbial contact. At diseased sites, epithelial cells and fibroblasts (comprising the bulk of the connective tissue) as well as infiltrating monocytic cells, are available. B. The adherence to host cells can be directly to host receptors and indirectly, mediated through the host extracellular matrix components such as fibronectin, fibrinogen and laminin.

development of beneficial therapeutic regimens. Finally, a further benefit of an understanding of the adhesins may involve the ability of these proteins to elicit a host response, as described below. Thus, intervention with adhesin binding to a host receptor may also lead to reduced host response to the encounter with the microbe.

Invasion of host cells by oral microorganisms

Several oral pathogens are capable not only of attaching to oral cells, but also of invading the cells. This serves as an important virulence mechanism that enables the evasion of host defence responses. Also, such a niche may provide the microorganism with a nutritionally rich environment. Finally, the epithelial cells harbouring the bacteria are able to serve as reservoirs of the pathogens in recurrent infections. The importance of intracellular invasion as a virulence factor for several enteric pathogens, such as *Salmonella* and *Shigella*, is well established, and recent studies show it may also be an important aspect of the pathophysiology of several oral organisms, including *P. gingivalis*, *A. a.*, *T. forsythia*, *P. intermedia*, and *F. nucleatum*. Both bacterial and host components are required for adhesion and invasion. As outlined above, host receptors and ECM components are targets for bacterial binding. However, the bacterial challenge has a more complex set of'consequences' for the host. For example, the invasion of *F. nucleatum* requires the involvement of actins, microtubules, signal transduction, protein synthesis, and energy metabolism of the epithelial cell, as well as protein synthesis by *F. nucleatum*. An indication that specific bacterial components are required for bacterial invasion is the observation that a spontaneous mutant of the bacterium that does not bind red blood cells or host cells is also incapable of invading host cells (Han *et al.*, 2000).

The invasion of phagocytic cells may be passive, involving only the simple phagocytosis of the pathogens. However, the entry of bacteria into non-phagocytic cells is an active process, and involves both the host and bacterial components. Thus, upon binding to the host cells, surface rearrangement of the host cytoskeleton occurs. This is mediated by actin rearrangement, leading to the reorganization of the host cytoskeleton and resulting in membrane ruffling and the subsequent envelopment of bacteria (Fig. 9.2). Studies have demonstrated that both the host and *P. gingivalis* not only remain viable for extended periods of time, but also that *P. gingivalis* multiplies within the host cells. In endothelial cells, *P. gingivalis* is internalized via lipid rafts, and is incorporated into host membrane-bound organelles, known as phagosomes. These phagosomes provide a nutritious niche as the replicating vacuole. The suppression of autophages by 3-methyladenine or wortmannin resulted in the transport of the internalized *P. gingivalis* to the phagolysosome, where the bacterium was degraded and destroyed. Thus, it was concluded that the survival of *P. gingivalis* depends upon the activation of autophagy and the survival of the endothelial host cell (Belanger *et al.*, 2006). However, the mechanism by which *P. gingivalis* initiates autophagy remains unclear. The routes of invasion and intracellular trafficking seem to vary depending on the host cell and so more investigations are needed to determine the fate of *P. gingivalis* in various types of host cells. Studies on the invasion of *P. gingivalis* into epithelial cells have shown that this process is initiated by the adhesion of *P. gingivalis* FimA to beta-1 integrin. This association triggers actin rearrangement, ultimately resulting in membrane ruffling which facilitates bacterial invasion. Of note, *P. gingivalis* induces the formation of actin-mediated membrane protrusions through which the organism migrates to adjacent cells, thus bypassing the need for the release of the bacteria into the extracellular space, and ultimately their exposure to the humoral immune response (Yilmaz *et al.*, 2006). Similarly, the *A.a.* invasion of epithelial cells involves multiple dynamic processes. Bacterial attachment induces the effacement of the microvilli and the bacteria enter through ruffled apertures in the cell membrane vacuole. However, this membrane is soon broken down and the bacteria are released into the cytoplasm where they can then replicate. *A.a.* then spreads into adjacent cells through membrane protrusions, as described for *P. gingivalis*. The cell-to-cell spread results in the microbes gaining access to deeper tissues and thus facilitates the spread of the organisms, as well as possibly allowing the to avoid the immune response. Unlike the 'ruffling' mechanisms employed by *P. gingvalis* and *A. a.*, the invasion of epithelial cells

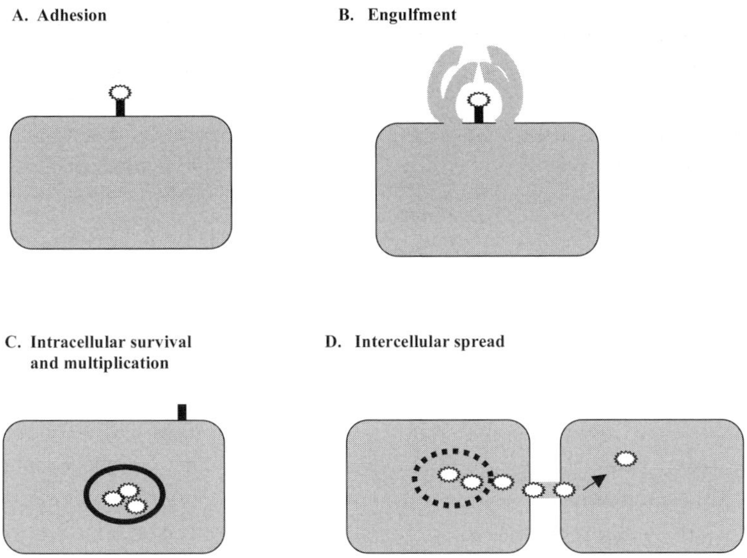

Figure 9.2 Invasion and spread of *P. gingivalis* in epithelial cells. A. *P. gingivalis* binds to an integrin receptor on the host cell surface. B. Bacterial binding triggers a host response leading to microtubule and microfilament rearrangement, ultimately resulting in cytoskeleton alteration which manifests itself in membrane ruffling and engulfment of the pathogen. C. *P. gingivalis* escapes host protective mechanisms by localizing into membrane vacuoles in which it survives and multiplies for prolonged period of time. D. The transfer of the pathogen into neighbouring cells, through protrusions in the host cell is directed by *P. gingivalis* itself. Cell-to-cell spread results in the bacterium gaining access deeper in the tissue and further shields it from the immune response directed to infected cells.

by *F. nucleatum* occurs via a 'zipper' mechanism, which also involves bacteria-induced actin polymerization. Thus, the bacteria-triggered host cell cytoskeleton rearrangement is common for most bacteria that invade the host.

It is of interest that recent findings also show that intracellular oral bacteria are present in human buccal epithelial cells as multispecies communities (Rudney *et al.*, 2005). So far, the polymicrobial communities have been examined for the presence of three oral pathogens *A.a.*, *P. gingivalis*, and *T. forsythia*. Given the polymicrobial nature of the oral communities, it is probable that more bacterial species are present within these communities.

Changes in the host triggered by oral microorganisms

In order to adapt to each other, the microorganisms inhabiting the oral cavity respond to the challenges imposed by the host and the host responds to the presence of the microorganisms. These interactions manifest themselves in changes in gene and protein expression, both in microorganism and host, allowing the co-existence of the two various organisms. The host response to bacterial infection plays a major role in determining whether the bacteria will remain confined or whether they will evade the innate defences of the host, ultimately resulting in tissue destruction.

Firstly, the organisms activate the production of host elements that mediate the adhesion and invasion of the organisms. Then, the host cell is manipulated to allow the survival and spread of the intracellular bacteria.

Furthermore, oral pathogenic bacteria can trigger a host response that leads to host tissues injury. The severe inflammation and tissue destruction seen in periodontal diseases is the result of the host response to periodontal bacteria. In this case, the host either responds disproportionally, or the pathogenic bacteria establish themselves in vast numbers, thus serving as a potent trigger for the host proinflammatory response. As the immune system response plays a major role in the development of oral diseases, such as periodontal diseases, it was the first and most widely investigated response elicited by oral pathogens. Finally, other alterations in the host cell response, including the death response due to either necrosis or apoptosis pathways, are affected by interaction with oral pathogens. Listed below are the major changes observed in host cells upon encountering bacteria.

Cell surface modifications

Upon contact with oral pathogens, host cells undergo morphological changes. The gingival epithelial cells containing internalized *P. gingivalis* round up and detach from the substratum (Belton *et al.*, 1999). Also, fibroblasts, upon contact with *P. gingivalis*, undergo cell rounding and detachment.

The cell rounding and detachment is associated with the degradation of cell adhesion molecules, such as E-cadherin, occludin, beta-1 integrin, ICAM-1, and vascular cell adhesion molecule-1. This degradation is mediated through bacterial proteinases, such as gingipains, produced by *P. gingivalis* (Katz *et al.*, 2000; Tada *et al.*, 2003; Wang *et al.*, 1999).

Apoptosis

Exposure to oral bacteria may also trigger apoptosis in host cells. The oral pathogen *A.a.* triggers cell death through the apoptosis of multiple cell lines (Mans *et al.*, 2006).

In contrast, early infection with *P. gingivalis* induces anti-apoptotic response in infected cells. However, this response changes as the challenge with the bacterium continues and the late-response to *P. gingivalis* involves pro-apoptotic events (Urnowey *et al.*, 2006). *F. nucleatum* has also been shown to stimulate the apoptosis of human peripheral white blood cells (Jewett *et al.*, 2000; Shenker and Datar, 1995). Finally, *T. forsythia* has been shown to induce cytolytic activity in host cells.

Cytokine production

A cytokine response to *P. gingivalis* and *A.a.* was demonstrated in an *in vivo* calvarial mouse model. An elevated expression of mRNA for the proinflammatory cytokines IL1-beta, IL6 and TNF-alpha was demonstrated. Interestingly, the commensal bacterium *S. gordonii* was also able, albeit to a markedly lesser degree than the pathogenic species, to stimulate the proinflammatory cytokine expression in the animal host. These results reveal an active host response to the pathogenic flora, as well as the ability of the host to respond to commensal bacteria (Kesavalu *et al.*, 2002). The cytokine induction was also demonstrated using a variety of *in vitro* models; for example, the ability of *P. gingivalis* to induce the expression of inflammatory cytokines and the activation of lipid inflammatory mediators, has also been demonstrated by a number of investigators (Nair *et al.*, 1983; Sismey-Durrant and Hopps, 1991; Tamura *et al.*, 1992; Wells *et al.*, 1993; Zhou *et al.*, 2005; Seymour and Gemmell, 2001). Also, another bacterium, *F. nucleatum*, stimulates the production of IL-8 from epithelial cells, indicating its ability to induce inflammation (Darveau *et al.*, 1998; Han *et al.*, 2000).

Antimicrobial peptide production

Antimicrobial peptides are small, endogenous, polycationic molecules that play an important role in innate immunity. They exert their antimicrobial activity by interfering with bacterial membranes, ultimately resulting in interference to the bilayer and pore formation (Brogden *et al.*, 2005). Gingival epithelial cells express calprotectin and β-defensins, which have been demonstrated to have a role in protection against *P. gingivalis* infections. Since calprotectin is a cytosolic calcium-binding protein with broad spectrum antimicrobial activity (Steinbakk *et al.*, 1990), it may kill or inhibit the growth of intracellular *P. gingivalis* and other oral bacteria. The elevated expression of calprotectin observed in the gingival crevicular fluid from patients with periodontitis (Nakamura *et al.*, 2000) may hint at its role during infection with periodontopathic bacteria. The peptide human β-defensin (hBD) is mainly found in association with gingival epithelial surfaces and may play a role in restricting intracellular bacterial replication, as well as preventing the physical destruction of the host cell. The production of hBD has been shown to be increased in gingival epithelial cells exposed to *P. gingivalis* and *A.a.* (Feucht *et al.*, 2003; Chung *et al.*, 2004). Interestingly, *F. nucleatum* also induces the production of hBDs in gingival epithelial cells and this induction is likely to suppress the growth of competitive species (Krisanaprakornkit *et al.*,

2000). This is because *F. nucleatum* is resistant to the induced ß-defensins, and other bacteria, such as *P. gingivalis,* are sensitive to and thus will be killed by the elevated levels of the selectins.

Modulation of matrix metalloproteinases (MMP) secretion

Matrix metalloproteinases are zinc-dependent proteinases that play a role in tissue remodelling and cell migration. An abnormal increase in the production of these proteinases may be involved in the pathophysiology of periodontal diseases (Uitto *et al.*, 2003). These enzymes can degrade extracellular matrix proteins, such as collagen types I, III, and IV, fibronectin, elastin, and proteoglycan. Exposure to bacterial components, such as lipopolysaccharide and phospholipase C, can induce MMP secretion in gingival epithelial cells (Birkedal-Hansen, 1993). Also, the *P. gingivalis* cysteine proteases, gingipains, upregulate MMP expression in epithelial cells (DeCarlo *et al.*, 1998; Birkedal-Hansen, 1993).

Other modulations

Several changes in host cells may be induced upon contact with *T. denticola.* This bacterium produces a chymotrypsin-like enzyme that can degrade junctional complexes. This enzyme can also be transported into the cell where it disturbs the actin cytoskeleton. Furthermore, the major surface protein of *T. denticola* inserts itself into the epithelial cell membrane where it forms a conductance ion channel and depolarizes the membrane.

It should be noted that, although pathogenic bacteria were the focus of the majority of the above investigations, commensal bacteria are not merely passive bystanders; indeed, they are capable of triggering host responses, as evidenced by the cytokine stimulation by *S. gordonii.*

Activation of the host response

The host response to the encroaching bacteria is accomplished by a variety of signalling molecules that are secreted or present on the bacterial surface and which bind to receptors expressed by the host cell. The binding of the bacterial signalling molecules (also called ligands) initiates a cascade of intracellular reactions, ultimately affecting the regulation of a variety of aspects of the host cell behaviour, including cytoskeleton rearrangement, movement, metabolism, inflammatory response, differentiation, and survival. Therefore, an understanding of the molecular components of the pathways affected by infection with oral bacteria, and how they are regulated, is of major importance, since the deregulation of the cascades may result in a variety of diseases, such as periodontal diseases, or even more serious complications, such as cancer.

The changes observed in the host upon challenge with bacteria are the result of 'complex' signalling cascades initiated by the binding of a bacterial ligand to a host cell receptor. There are a variety of bacterial ligands, and the major groups are discussed below (see 'bacterial mediators'). The bacterial cell surface ligands bind to receptors expressed on the host cell surface but secreted bacterial ligands may cross the plasma membrane and bind to intracellular receptors. Furthermore, intracellular bacteria may exert their effect by exposing their surface components to the host's intracellular receptors. Host receptors are mainly surface exposed, but intracellular receptors also play a role. Upon binding of the ligand, the receptors transmit signals to intracellular targets, ultimately reaching the nucleus, targeting transcription factors and resulting in changes in gene expression. Multistep and multicomponent signal transduction cascades involving kinases, GTPases, and other proteins are present, and distinct pathways exist for various receptors; however, some components of the pathways are shared. Several host receptors that play a role in the response to oral microbial challenge have been identified, and these, as well as the signalling pathways downstream of these receptors, will be discussed below (Fig. 9.3B).

Host mediators

Signalling through adhesion molecules and lipid rafts

Host adhesion molecules can serve as receptors through which bacteria can signal. Bacterial interaction with adhesion molecules has been shown to play a role in the adhesion and invasion of a variety of pathogens, including *Yersinia spp,* and *Listeria monocytogenes.* These interactions involve the binding of the pathogens to integrins

(*Yersinia*) and E-cadherins (*Listeria*). Of note, similar modes of bacteria-induced phagocytosis are also observed in oral bacteria. For example, *P. gingivalis* FimA adhesin binds the host beta-1 integrin receptor resulting in the activation and rearrangement of the focal protein paxillin and focal adhesion kinase (FAK), thereby facilitating bacterial entry into the cytoplasm (Figs 9.2; 3A,B). The signalling cascade involves tyrosine kinase signalling, the activation of the small GTPase Rac, and actin polymerization. It should be noted that both bacterial and epithelial cells remain viable following the internalization process, indicating that this is a stable relationship.

Once internalized, other intracellular host receptors are likely to be involved, as the down-regulation of neutrophil interleukin IL-8 and the inhibition of apoptosis are the outcomes of the invasion of epithelial cells by *P. gingivalis*. Since IL-8 is required for the recruitment of neutrophils to the site of infection, the observed inhibition probably impairs host detection and the subsequent clearance of the organism. Also, the regulation of matrix metalloproteinase production by gingival epithelial cells is altered following contact with the organism, and this interferes with extracellular matrix repair and reorganization.

A.a. has also been shown to invade oral epithelial cells via binding to integrins. The bacterial integrin-binding protein has been identified, and when this is expressed in a heterologous host, a non-invasive *Escherichia coli* strain, it transforms the non-invasive bacterium into an invasive strain. The invasion of *A.a.* into epithelial cells has also been shown to be mediated via the transferrin receptor. Phosphorylcholine-positive strains of *A.a.* are able to invade endothelial cells, and this invasion is dependent on the platelet-activating factor receptor, thus indicating that various host receptors may play a role in bacterial invasion.

Immune signalling

The first line of the host response to infection is the innate immune response. Microbes are detected by the host pattern-recognition receptors (PRRs) that recognize multiple pathogen-associated molecular patterns (PAMPs), common to a variety of microorganisms but not present on the host tissues. One of these are the Toll-like receptors (TLRs) (Akira and Takeda, 2004) (Fig. 9.3B,C). TLRs are a family of receptors composed of an extracellular region containing leucine-rich repeats, a transmembrane region, and an intracellular Toll/interleukin-1-like domain (TIR). Upon activation, the TIR domain recruits an adaptor molecule, such as the myeloid differentiation factor 88 (MyD88), TIRAP (TIR-domain-containing adaptor protein), TRIF (TIR-domain-containing adaptor protein inducing interferon beta) or TRAM (TRIF-related adaptor molecule), which act downstream of the TLRs. The TLR family includes multiple receptors that recognize specific pathogen molecules, such as lipoproteins and peptidoglycan (TLR1, TLR2, TLR6), double-stranded RNA (TLR3), lipopolysaccharide (TLR4), flagellin (TLR5), and unmethylated CpG DNA (TLR9) (Akira *et al.*, 2006; Imler and Hoffmann, 2003). TLRs are expressed on various cells, such as macrophages, dendritic cells, B cells, fibroblasts, and epithelial cells, and their expression is rapidly modulated in response to bacteria, cytokines, and environmental factors. While some TLRs are expressed on the cell surface (TLRs 1, 2, 4, 5, and 6), others (TLRs 3, 7, 8, and 9) are found almost exclusively in intracellular compartments, such as endosomes. Various microbial PAMPs activate their respective TLRs, which then transmit the signals via a variety of signalling pathways to the nucleus. The expression of various host genes is ultimately affected, producing molecules such as proinflammatory cytokines, co-stimulatory molecules, growth factors, chemokines, and interferon. The different signalling pathways used by TLRs depend on which adaptor molecule is recruited downstream of the TLR. MyD88 can associate with all TLRs, except TLR3, and activates two pathways, NF-kB (nuclear factor kB) and mitogen-activated protein kinase (MAPK). The NF-kB pathways are activated through the recruitment of IRAK family kinase, TRAF6, the IKK complex (inhibitor of kB kinase), ANK (ankyrin repeats domain), RHD (Rel repeats domain), and NF-kB. The MyD88-associated pathway leads to the upregulation of immune response genes, such as proinflammatory cytokines and co-stimulatory molecules. In contrast to MyD88, TRIF binds only a subset of TLRs (TLR3 and 4), and triggers the activation of

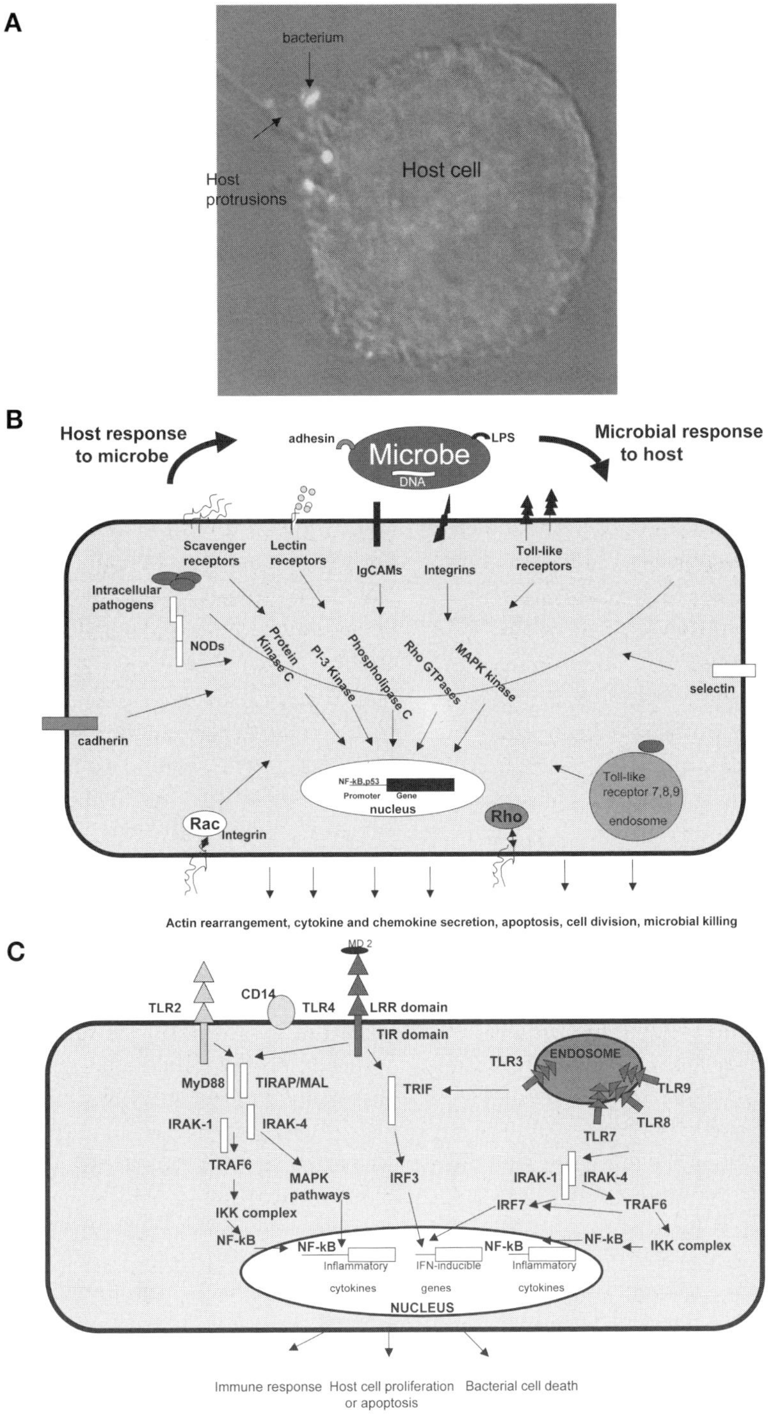

Figure 9.3 Receptor and signalling cascades during host-pathogen interaction. A. Co-culture of *P. gingivalis* with human umbilical vein endothelial cells (HUVECs). Upon exposure to *P. gingivalis* HUVEC forms fingerlike cell protrusions interacting with the microorganism. B. Receptors and signalling cascades are engaged upon microbial contact with the host cell. Multiple receptors can simultaneously recognize microbial components. The binding to receptor triggers intracellular signals that, through various pathways engaging multiple molecules, deliver the message to the nucleus and alter gene transcription in the host cell. This ultimately leads to a host cell response manifesting itself in actin rearrangement, cytokine/chemokine secretion, immune response, apoptosis, cell division, and microbial killing. The host-microbe contact results not only in host response but also in microbial response, thus indicating that this is a highly interactive process. C. TLR pathways. A colour version of this figure is located in the plate section at the back of the book.

interferon regulative factor-3 (IRF3), which in turn induces the transcription of genes coding for interferon α and β. The different pathways utilized by various TLRs explain how different TLR elicitors can trigger different transcription programs in the same immune cell type. The appropriate immune response to the pathogen is the result of the activation of different TLRs in the same cell, resulting in cross talk between intracellular signal transduction pathways (Akira and Takeda, 2004).

Furthermore, since TLRs can detect different pathogen components, the same TLR may be activated by various pathogens, leading to a common response against various microbial species. Also, although different TLRs detect different pathogen components, these TLRs ultimately upregulate many of the same genes, due to activation of shared transcription factors, such as NF-kB. Thus, TLR signalling pathways can converge on a common set of genes and allow the host cell to generate a common response to a variety of pathogens.

NOD receptors

Finally, the nucleotide-binding oligomerization domain (NOD) family of proteins, particularly NOD1 and NOD2, is implicated in recognition of intracellular pathogens (Inohara, 2005; Inohara and Nunez, 2003). NOD1 is a cytoplasmic protein composed of N-terminal caspase activating and recruitment domain (CARD), a central NOD, and a C-terminal leucine-rich repeat (LRR) domain. NOD2 contains two CARDs. NODs interact with CARD domain-containing kinases such as kinase-receptor interacting protein 2 (RIP2) or RICK and activate NF-kB pathway. The LRR domains of NODs recognize various peptidoglycan breakdown products such as muramyl dipeptide MurNAc-L-Ala-D-isoGln (MDP) that is recognized by NOD2 and γ-D-glutamyl-meso-DAP (i.e.-DAP), a dipeptide present in variety of Gram-negative bacteria is detected by NOD1.

Recent findings show that a retinoic-acid-inducible gene-I (RIG-I) may play a role in detection of intracellular dsRNA (mainly virus products) (Yoneyama and Fujita, 2004). This detection mode may play a role in the innate response to viral infection. RIG-I, a cytoplasmic helicase, contains two N-terminal CARDs that transduce signals leading to the activation of IRF3 resulting in type I IFN production and NF-kB activation and ultimately in proinflammatory cytokine secretion.

Co-receptors for TLRs

Recent studies have revealed that the TLR receptors require accessory molecules for their microbial recognition (Miyake, 2006). Lipopolysaccharide (LPS) binding protein (LBP), CD14 and MD-2 were required for LPS recognition. MD-2 directly mediated ligand binding while LBP and CD14 affected the ligand presentation to the receptor complex, TLR/MD-2 ultimately influencing the responses to the microbial ligand. Both TLR2 and TLR4 require similar accessory molecules for recognition. Thus, the innate immune response to microorganisms also depends on the presence of the co-receptors on host cells.

Signalling through lectin receptors

Eukaryotic cells express a variety of surface lectins that mediate detection of both self- and foreign-carbohydrates. These receptors also bind microbial components and, as such, serve in the cooperative detection of microbes. One such lectin is dectin-1, a beta-glucan binding receptor. Another beta-glucan receptor is a high-affinity binding lectin CR3.

Signalling through scavenger receptors

Scavenger receptors were originally defined by their ability to bind and internalize modified lipoproteins. However, these receptors were later demonstrated to also bind various ligands such as polyribonucleotides, lipopolysaccharides, and silica particles (Platt *et al.*, 1999). Two scavenger receptors have been demonstrated to have a role in binding and internalizing microbes (Underhill and Ozinsky, 2002). The scavenger receptor A (SR-A), a transmembrane homo-trimer with an extracellular collagenous triple-helix domain, binds whole bacteria as well as microbial components such as LTA and LPS. MARCO (macrophage receptor with collagenous structure) is the second receptor implicated in phagocytosis. Another receptor, CD91 (also known as alpha-macroglobulin receptor or LRP-1), is a surface molecule involved in endocytosing polyanionic ligands such as low-density lipoproteins (Van Berkel *et al.*, 2005). However, these receptors also have a role in signalling in response to microbial factors.

Cooperation of host receptor signals in host response to microbes

Multiple host receptors play a role in recognition of microbes. It is the combined activation of the various receptors that leads to the either synergistic or antagonistic effects that modulate the ultimate outcome of the host response. Therefore, in order to understand the role of the various host receptors in the response to microbes, a complete understanding of these multiple receptor interactions is required.

Immune signalling receptors on host cells encountered by oral bacteria

Attachment to epithelial cells is the first type of encounter between the oral bacteria and the host. Upon tissue injury, the bacteria can gain access to the underlying connective tissues, which are composed mainly of fibroblasts. As a result of the immune response to oral pathogens, monocytic cells are recruited to the site of infection, where they interact with the bacteria. Furthermore, oral bacteria are able to gain access to the circulatory system, where they can interact with the endothelial cells. During periods of health, a balance is maintained between the host and the oral pathogens, and the main type of host cells engaged in the interaction with the oral bacteria are epithelial cells. However, as the disease status is initiated and progresses, oral bacteria begin to interact with a variety of host cells. Recent studies have demonstrated that gingival epithelial cells and gingival fibroblasts express functional TLRs and NODs. The major role of the epithelial cells lining the gingival crevice was previously believed to be the provision of a mechanical barrier that would physically prevent the bacteria from reaching deeper gingival tissue. However, it has now been established that epithelial cells can also sense and respond to bacteria. This is expected to involve activation through the TLRs or NODs (for an intracellular bacteria), an elaborate communication network that ultimately controls the expression of various host effectors, ultimately conveying information to the bacteria, other epithelial cells, and the underlying immune and inflammatory cells of the mucosa. The outcome of this communication determines the overall health status of the gingival tissues. Interestingly, in contrast to monocytic cells and gingival fibroblasts, epithelial cells do not produce inflammatory cytokines upon stimulation with respective TLR or NOD ligands (Uehara and Takada, 2007; Uehara *et al.*, 2005; Uehara *et al.*, 2001). They do, however, produce antimicrobial factors, such as beta-defensin 2 and peptidoglycan-recognition proteins. Such antimicrobial factor upregulation, without the stimulation of proinflammatory cytokines, is reasonable in view of the fact that the epithelial cells constitutively interact with a variety of microbial species, including commensal species; and inflammatory responses, if elicited by those bacteria, may lead to tissue destruction. However, gingival fibroblasts are isolated from the oral microflora by the epithelial barrier, and only upon tissue injury do microorganisms gain an access to the underlying fibroblasts, resulting in the triggering of an inflammatory response. Such an inflammatory response may explain the periodontal tissue destruction, possibly due to pathogenic bacteria gaining access to deeper host tissues.

As noted above, the cooperation between the various receptors dictates the ultimate response to microbial contact. The various types of cells in the oral cavity may simultaneously express overlapping, but not identical, combinations of TLRs and other PRRs and this may contribute to tissue- or site-specific responses to microbial stimulation. Invading microorganisms are likely to interact with the multiple TLRs and non-TLRs, and so both the magnitude and quality of the subsequent immune response are likely to depend on the distribution of receptors on the cells encountered by the microorganisms and the coordinated sum of the signals induced by these different receptors. The emerging information regarding the interaction of TLRs with each other and with members of the non-TLR PRR families as well as the outcome of such interactions on host resistance to infection is described in recent review (Trinchieri and Sher, 2007). Furthermore, as noted above, the host TLR receptors may require co-receptors for their signalling activity. Since some differences in expression of the co-receptors among various host cells may exist, differences in response to the specific microbial pattern may also appear when examining various host cells. For example, gingival epithelial cells do not express the TLR2 co-receptor CD14 that is required for response to the *P. gingivalis* FimA. This co-receptor is expressed on monocytic cells and so these cells do respond to FimA with elevated cytokine expression (Eskan *et al.*, 2007). These differences may account for the differential response of the various cell lines to *P. gingivalis*; also similar differences in the expression of co-receptors on various cell types can account for differential response to other oral microbes.

Host immune receptors and oral bacteria

Host activation by extracellular bacteria involves the engagement of cell surface receptors. The recognition of various oral microbial patterns has been investigated, and the role of TLR2 in the recognition of FimA adhesin, as well as *P. gingivalis* LPS, has been established. Interestingly, a recent study also demonstrated that the proinflammatory response to *P. gingivalis* is mediated through the TLR2 receptor. These results are in agreement with an *in vivo* mouse infection model demonstrating that inflammation and alveolar bone loss was reduced in animals deficient in the TLR2 receptor.

Intracellular bacteria are also able to manipulate the host signalling cascades, ultimately modulating the host expression for their mutual benefit. The observation of oral bacteria invading host cells is very recent, and little is known regarding either the host response, or the molecular mechanisms mediating these responses. However, as mentioned above, internalized *P. gingivalis* alters the host expression of metalloproteases, IL-8, and apoptosis effectors, thus indicating that the host responds to the internalized bacterium.

Host signalling molecules

Large numbers of signalling molecules are activated during microbial contact with the host and linking the key regulatory molecules will be best suited for computational modelling. However, there are few signalling molecules characteristic of multiple signalling pathways. These include the phosphoinositide 3-kinase (PI 3-kinase), phospholipase C (PLC), Rho GTPase, protein kinase C (PKC), and MAPK (Fig. 9.3B). These molecules play a role in engulfment of microbes as well as in regulation of the inflammatory response to the microbes.

Phosphoinositide 3-kinase

Phosphoinositide 3-kinase (PI 3-kinase or PI3Ks) are a family of enzymes capable of phosphorylating the 3 position hydroxyl group of the inositol ring of phosphatidylinositol (PtdIns) (Vanhaesebroeck *et al.*, 2001). The 3-phosphorylated phosphoinositides then function through a group of signalling proteins containing Phox domains (PX domains), pleckstrin homology domain (PH domains), FYVE domains and other phosphoinositide-binding domains that are recruited to various cellular membranes. One of the PI 3-kinase activated proteins is protein kinase B (PKB also known as AKT) which is recruited to specific membrane regions during microbe contact. Various classes of PI 3-kinases can play a role in modulating the host response during various stages of host-microbe interaction. Thus, membrane extensions are dependent on the presence of class I PI 3-kinase while phagosome maturation requires the presence of class II PI 3-kinase. The PI 3-kinase is also required for inflammatory response and it was shown to be recruited to TLR receptors when host cells were infected with *S. aureus*, ultimately leading to NF-kB-mediated cytokine production (Botelho *et al.*, 2000).

Phospholipase C (PLC)

Phospholipase is an enzyme that cleaves phospholipids into fatty acids and other lipophilic substances. Four classes termed A, B, C, and D are recognized, based on the type of reaction they catalyse. The phospholipase C cleaves before the phosphate and releases diacylglycerol (DAG) and a phosphate-containing group, inositoltriphosphate (IP3). DAG is necessary for further activation of Protein Kinase C while IP3 leads to the release of intracellular calcium. Therefore, phospholipase C (PLC) plays a central role in signal transduction events for many receptors. Its activity is required for microbe-induced pro-inflammatory signalling in macrophages, due to its role in activating PKC (Chen *et al.*, 2000a). Also, inhibition of PLC activity blocks the formation of actin filaments thus preventing microbial internalization (Botelho *et al.*, 2000).

Small GTPases

Small GTPases are 20–25 kDa nucleotide-binding proteins serving as molecular switches for a variety of cellular functions including growth regulation, morphogenesis, cell mobility, axonal guidance, cytokinesis and trafficking. At least 60 different small GTPases have been identified in mammalian cells. The Ras and Rho families are of special interest as they transduce signals from extracellular stimuli to intracellular signal transduction pathways. They cycle between inactive (GDP-bound) and active (GTP-bound) states. Under normal conditions GTP-ases are GDP-bound and are inactive but, when stimulated, they release GDP and bind to GTP. This reaction is mediated by guanine exchange factors (GEFs), while GTP-ase activating proteins (GAPs) inactivate the proteins (Chimini and Chavrier, 2000). Members of the Rho family of GTPases regulate actin cytoskeleton rearrangements during adhesion, membrane ruffling, and stress fibre formation. They also cooperate with the Ras family of GTPases in the regulation of gene expression and proliferation (Bar-Sagi and Hall, 2000). The members of the Rho GTPase family including Cdc42, Rac, and Rho have been shown to play crucial roles in phagocytosis of microbes, as well as other signalling events due to their ability to interact with other signalling pathways. As the Rho GEFs and GAPs are regulated by various signalling pathways including growth factor receptor signalling, adhesion-dependent signalling, and cell cycle progression, many stimuli may affect their function. One of the Rho family GAP is the YopE protein of *Yersinia pseudotuberculosis* that when injected by the pathogen into the cytoplasm of macrophages prevents its phagocytosis (Black and Bliska, 2000). Also, the Rho GTPase-mediated activation of downstream signals may be affected by various coincident signals since, for example, both Rho and Ras can activate MAPK signalling.

Protein kinase C (PKC)

Protein kinase is an enzyme that modifies other proteins by phosphorylating them. This is accompanied by functional change of the target protein, resulting in change of its enzymatic activity, cellular location, or association with other proteins. Kinases are known to regulate variety of proteins and especially those involved in signal transmission within the cell. The human genome contains as many as 500 genes encoding protein kinases. The activity of the kinases involves removal of a phosphate group from ATP and its covalent attachment to one of three amino acids that have a free hydroxyl group; that is, serine, threonine, and tyrosine. Most of the kinases phosphorylate both serine and threonine while others act on tyrosines. Also, kinases with dual specificity are capable of phosphorylating all three amino acids.

The protein kinase C, a family consisting of 12 isozymes, is a calcium-dependent and/or diacylglycerol (DAG), and phospholipid (such as phosphatidylcholine) enzyme. They play a key roles in numerous signalling pathways both to the actin cytoskeleton and to the nucleus, including signals stimulated by hormones, cytokines, and adhesion. Their substrates are MARCKS proteins, MAP kinase, transcription factor inhibitor IκB, Raf kinase, calpain, and the epidermal growth factor receptor.

Mitogen-activated protein (MAP) kinases

Mitogen-activated protein (MAP) kinases are serine/threonine-specific protein kinases that respond to external stimuli and regulate a variety of cellular activities including gene expression, differentiation, and cell survival/apoptosis. The signalling cascade (also known as the 'MAPK cascade') is composed of MAP kinase, MAP kinase kinase (MKK), and MAP kinase kinase kinase (MKKK). MKKK is activated by external stimuli and phosphorylates MKK, which in turn activates MAPK through phosphorylation of its serine and tyrosine residues. To date, four distinct groups of MAPKs have been characterized in mammals: extracellular signal-regulated kinases (ERKs), c-Jun N-terminal kinases (JNKs) (also known as stress-activated protein kinases), p38 isoforms, ERK5 (activated by growth factors and stress stimuli and participates in cell proliferation). Both JNK and p38 are responsive to stress stimuli such as cytokines, heat shock, osmotic shock and mediate cell differentiation and apoptosis.

Bacterial mediators

Bacterial components are the signalling molecules or ligands for host receptors. A variety of bacterial components can serve as signals for the host cell response. Several types of components, as well as the host receptors engaged by the bacterial components, are discussed below.

Cell surface proteins

Bacterial cell surface proteins are the first to come in contact with the host. As such, they may serve as a means of attachment (e.g. adhesins). However, bacterial surface proteins may also trigger the host to respond to the contact with the pathogen. For example, the *P. gingivalis* 41 kDa FimA protein binds to the host cell adhesion receptor integrin and induces a signalling cascade ultimately leading to the uptake of the bacterium. FimA also interacts with other host receptors, such as the TLR receptors, and triggers the release of inflammatory cytokines in a variety of cells. Also, the 67-kDa MfaI protein induces cytokine production in a TLR2-dependent manner (Hiramine *et al.*, 2003). These results demonstrate that bacterial adhesins can trigger signal transduction events in the host cells. The *P. gingivalis* HagA adhesin has also been shown to attach to (Duncan *et al.*, 1996; Chen, 2001) and invade (Scragg, 2002) host epithelial cells. So far, however, the eukaryotic receptor, as well as the signal transduction events mediating the internalization of the HagA molecule, remain to be clarified. The *T. forsythia* BspA protein was described above as an adhesin that promotes the adherence of the organism to ECM components, such as fibronectin and fibrinogen, and to the host cells. Recent studies have demonstrated that it also induces cytokine production from monocytes, and chemokines from osteoclasts (Sharma *et al.*, 2005). This cytokine production was mediated through the interaction of the protein with the host TLR2 receptor and it required the CD14 co-receptor to effect host stimulation (Hajishengallis *et al.*, 2002).

Finally, the *S. mutans* 167-kDa AgI/II adhesin that was shown to bind to salivary proteins, as well as to salivary receptors and ECM components, also stimulates the production of a tumour necrosis factor alpha (TNF-alpha) in monocytic cells, and of IL1-beta in a TLR4-dependent manner (Chatenay-Rivauday *et al.*, 1998; Sciotti *et al.*, 1997; Crowley *et al.*, 1993; Hajishengallis *et al.*, 1994).

Among the bacterial cell surface proteins are those mediating adhesion, as well as others that are not involved in the adherence of the microorganism to the host cells.

There are many examples of oral bacterial surface proteins that stimulate the host to release immune modulators. For instance, the fimbria-associated proteins, such as the 12-kDa lipoprotein associated with *P. gingivalis* fimbriae, induce IL-1 beta release from macrophages in a TLR2-dependent manner. A surface component of *P. intermedia* (a 55 kDa protein) has been shown to induce multiple cytokines (interleukin-1 alpha, -1 beta, -6, –8 and TNF-alpha) in various cell lines, such as peripheral blood mononuclear cells, umbilical vascular endothelial cells, and gingival fibroblast cultures (Matsushita, 1994).

The precise roles of other adhesins and oral bacterial surface proteins in activating signal transduction pathways remain to be determined.

Lipopolysaccharide (LPS)

One of the components of the cell wall of Gram-negative bacteria is lipopolysaccharide (LPS). LPS is a well-established stimulator of an immune response that results in a destructive inflammatory response. It commonly requires TLR4 for its activity. However, the LPS of Gram-negative oral pathogens, such as that of *P. gingivalis*, differ in structure from that of the LPS of *E. coli*. The *P. gingivalis* LPS is heterogeneous. For example, the number of phosphate groups and the number and position of lipid A fatty acids differ, resulting in multiple lipid A structures. Interestingly, the unusual LPS does not elicit a potent inflammatory reactions in animals (Reife *et al.*, 2006), and has the ability to signal through TLR2. Furthermore, the pattern of elicited inflammatory mediators differs from that of *E. coli* LPS. Finally, it has been demonstrated to inhibit *E. coli* LPS signalling through TLR4.

Peptidoglycan

This is a ligand for TLR2; also, it is recognized by the host intracellular receptors NOD1 and 2.

Lipoproteins

Bacterial lipoproteins are commonly recognized by TLR1 and TLR6. However, *P. gingivalis* lipoprotein, which is encoded by an ORF, PG1828, activates the host through the TLR2 receptor. Since it is associated with *P. gingivalis* LPS, it is the principal component that enhances the TLR2-mediated activity of the host.

Capsule

Some oral pathogens, such as *P. gingivalis*, produce capsular polysaccharide. This polysaccharide encases the entire cell surface, shielding the bacterial cells from the host immune response, since it covers some surface components that may serve as signalling molecules. Therefore, the presence of a capsule has mainly been regarded as having a dampening effect on the response of the host to a pathogen. However, recent studies have shown that the host can respond to capsular material from organisms, such as *Bacteroides fragilis* and *Staphylococcus aureus*. Interestingly, capsular polysaccharide serotype K1 from *P. gingivalis* also elicited chemokine production from macrophage cells, as well as facilitating cell migration of polymorphonuclear leukocytes (d'Empaire *et al.*, 2006). So far, however, the composition of the capsule, as well as that of the host receptor for the capsular material, remains to be defined.

Heat shock proteins

GroEL proteins derived from periodontopathic bacteria can stimulate an inflammatory response from host cells. *P. gingivalis* GroEL signals through both TLR2 and TLR4 (Argueta *et al.*, 2006). The homologue of the human heat shock protein-90 (HSP-90), HtpG of *P. gingivalis*, induces the chemokine CXCL8 in human monocytic and endothelial vein cells. HSP-90 is a chaperone altering signal transduction cascades by association with transcription factors, hormone receptors and protein kinases. The receptor for Hsp90 is CD91 (also known as the alpha-macroglobulin receptor or LRP-1). The host receptors for the *P. gingivalis* HtpG protein are both CD91 and TLR4. The induction of the CXCL8 facilitates the chemoattraction of plasma cells and neutrophils and, as such, is an important factor in response to periodontal infection.

DNA

Some bacterial signalling molecules become accessible upon the lysis of the bacteria. Such lysis may occur within the host cells, and ligands for host intracellular receptors, such as DNA, and specifically the CpG motifs in bacterial DNA, are ligands for the TLR9 receptor. Recognition of the CpG motifs by the intracellular host receptor may trigger signalling cascades, ultimately leading to the release of IFN-α.

Host access to ligands

The bacterial signalling molecules may be cell-associated, but cell surface proteins may also be released and act at sites distant to the site of infection. For example, the *P. gingivalis* HagA, as well as the *T. forsythia* BspA, may be secreted. One of the mechanisms of the secretion of effector molecules may be the type III (or IV) secretion system. The effector molecule associated with *A.a.* apoptosis in human gingival epithelial cells seems to be CagE, a component of the type IV secretion system.

Furthermore, oral pathogens can produce outer membrane vesicles that are derived from the outer membrane and have a composition that is very similar to the outer membrane. These vesicles can also spread far from the site of pathogen colonization and activate host signalling cascades.

State of the knowledge regarding ligands of oral microbes

So far, the majority of efforts have centred on the host response to a variety of oral pathogens, such as *P. gingivalis*, and these have mainly been limited to recognition and signalling by *P. gingivalis* lipopolysaccharide (LPS) and fimbriae (Dixon *et al.*, 2004; Bainbridge and Darveau, 2001; Hirschfeld *et al.*, 2001). Several of the virulence factors, including fimbriae, LPS, and cell surface proteins, have been shown to play a role in host activation, leading to the release of cytokines and ultimately tissue destruction (Bainbridge and Darveau, 2001; Kishimoto *et al.*, 2006; Hajishengallis *et al.*, 2004). However, the host usually encounters live microorganisms; thus, in order to understand the host defence mechanisms and the pathogenesis of these bacteria, it is important to examine the host response to live bacterial cells. This statement is further supported by a recent study showing that the host is able to sense live *P. gingivalis* and its components (LPS and FimA protein) differently (Zhou *et al.*, 2005), indicating that the host response to the live organism is the sum of the responses to all inducing structures present in this organism.

Finally, the magnitude of the host response is dependent on the virulence and the number of microbial cells present. Thus, it is the composition and sum total of bacterial components, such as peptidoglycan, teichoic acid, fimbriae, outer membrane proteins, the capsule, and lipopolysaccharide, that can stimulate the development of a host immune reaction capable of causing severe tissue destruction that ultimately determines the outcome of the bacterial challenge on the host.

Host response to oral pathogens and microarray studies

Recent advances in biotechnology have allowed for analyses of host-pathogen interactions on a global level. Using genomic microarrays, it is now possible to define differential gene expression profiles in both the pathogen and the host. For example microarray analysis was used to define alteration of gene expression in fibroblasts infected with *P. gingivalis* at various times post-infection. Upon infection with *P. gingivalis*, fibroblasts undergo morphological changes. Initially, the cells are spindle-shaped but 21h post infection mainly rounded cells are observed (Fig. 9.4A). It is assumed that the drastic morphological changes are accompanied by an alteration in transcriptional profile of the cells. Indeed, using microarrays containing probes for most of the genes expressed by fibroblasts, the changes could be demonstrated (Fig. 9.4B). In agreement with the morphological changes observed in the host cells, some changes in gene expression were observed at 8h; however, those changes were more drastic (as reflected by altered expression patterns in a larger number of genes) at 21h post-infection. Thus, such a simple experimental approach may identify multiple players in the host response to microbes and it is becoming clear that such studies are necessary for the discovery of previously unknown genes in the infection process, which can then be characterized using conventional experimental approaches. In addition, transcriptional analyses allow the identification of specific pathways that are regulated in response to microorganisms or microbial components. When combined with appropriate bioinformatics and conventional approaches, global analyses will provide a more complete understanding of the host-pathogen dialogue.

Global transcriptional analyses have been employed to provide insights into the host response to oral pathogens. For example, the response of epithelial cells to two oral pathogens, *A.a.* and *P. gingivalis*, was investigated (Handfield *et al.*, 2005) (Fig. 9.4C) using microarrays encompassing probes for more than half of the genes present in the human genome. Such a comprehensive study revealed that multiple host pathways were affected by exposure to the bacterial pathogens. Sixteen pathways, which included molecular functions, cellular components, and biological processes, were identified as being affected in response to both pathogens. Interestingly, these studies also revealed that the response to each of the pathogens differed markedly, resulting in 21 pathways specifically affected by exposure to *P. gingivalis*, and 49 pathways associated with *A.a.* infection. One of the affected pathways was the apoptosis pathway. This is in agreement with results of phenotypic analyses, demonstrating the ability of the organisms to affect the apoptosis of host cells (see above). Analysis of the genes that had altered expression provided a more detailed understanding of the mediators of the complex signalling cascades involved in apoptosis manipulation by the oral pathogens. As shown in Fig. 9.4C, most of the affected genes could be linked to the P53 network. This network is involved in cell division and apoptosis, and the P53 protein, known as a tumour suppressor gene, plays a major role in these processes. The major effector molecules, whose expression was induced by infection with *A. a.*, were the pro-apoptotic molecules GADD45A, E2F1 and ATM. Repressed were cMYC and FOX03A, both of which are anti-apoptotic molecules. In contrast, *P. gingivalis* induced the expression of two mitochondrial anti-apoptotic effectors, Bcl-2 and Bfl-1, as well as cMYC and SGK, with the latter playing a role in

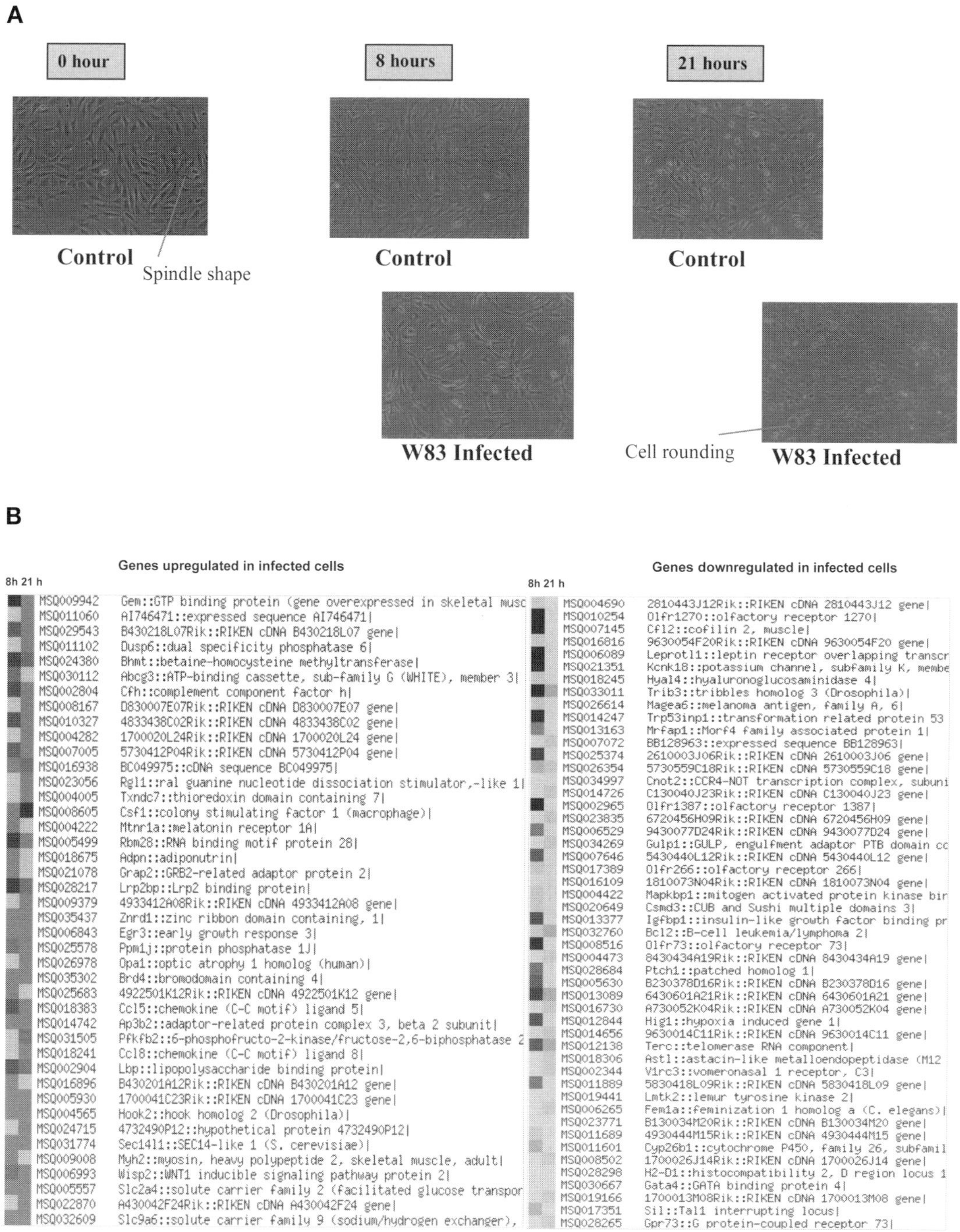

Figure 9.4 Host response to oral bacteria gleaned from genomic approaches. A. Morphological changes occurring in fibroblasts infected with *P. gingivalis* W83 at various times post-infection (8 hours and 21 hours). B. Transcriptional response of the infected fibroblasts at various times post-infection. At 8 hours, some 23 000 genes were detected, of which 182 were regulated. At 21 hours, some 20 000 genes were detected, of which 568 were regulated, at a 1.8-fold change. Among the regulated genes were pro-inflammatory cytokines or cytokine receptors; e.g. Cxcl5, which has recently been shown to play a role in oral carcinoma development. This study sets the stage for determining the *P. gingivalis*-activated transcriptional networks that may play a role not only in the development of periodontal diseases but also in the aetiology of cancer. The red and green colours represent up and down gene regulation, respectively. C (opposite). Apoptotic pathways affected by *A.a.* and *P. gingivalis*. (Continued overleaf.)

C

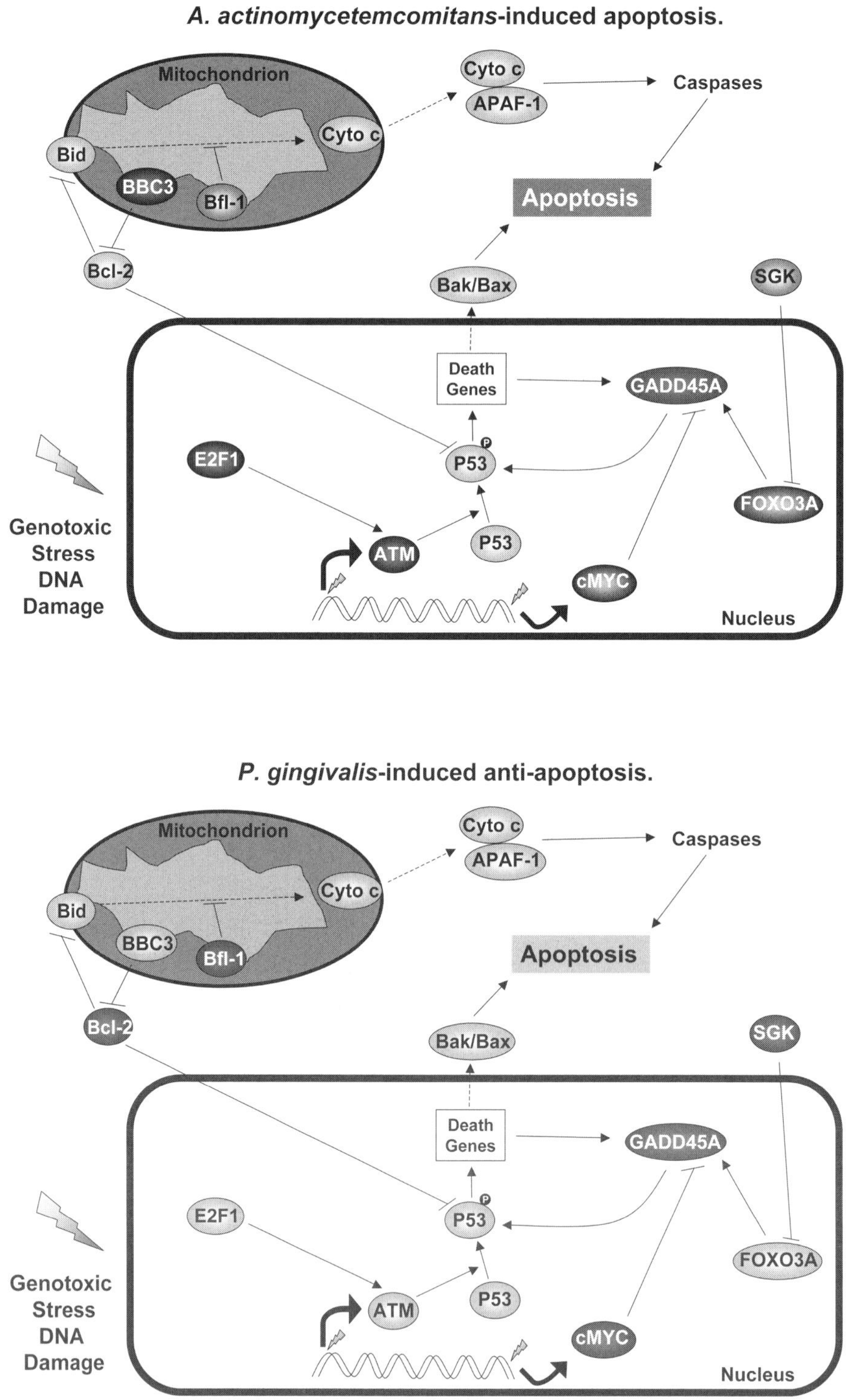
A. actinomycetemcomitans-induced apoptosis.
Mitochondrion
Cyto c
APAF-1
Caspases
Cyto c
Bid
BBC3
Bfl-1
Apoptosis
Bcl-2
Bak/Bax
SGK
Death Genes
GADD45A
E2F1
P53
FOXO3A
Genotoxic Stress DNA Damage
ATM
P53
cMYC
Nucleus
P. gingivalis-induced anti-apoptosis.
Mitochondrion
Cyto c
APAF-1
Caspases
Cyto c
Bid
BBC3
Bfl-1
Apoptosis
Bcl-2
Bak/Bax
SGK
Death Genes
GADD45A
E2F1
P53
FOXO3A
Genotoxic Stress DNA Damage
ATM
P53
cMYC
Nucleus

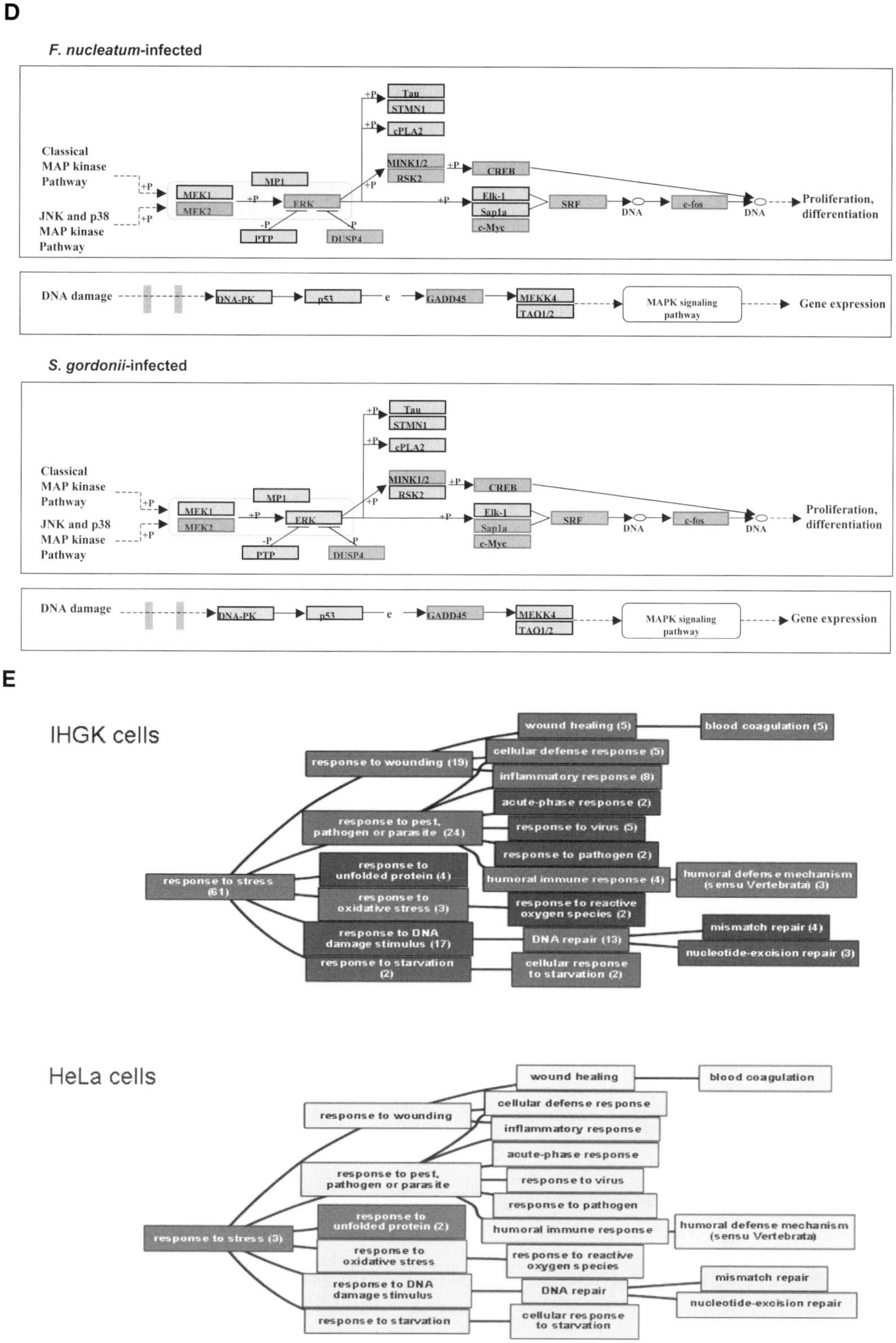

Figure 9.4 Continued. D. Epithelial cell response to commensal (*S. gordonii*) and pathogenic (*P. gingivalis*) microorganisms. E. Cell-line dependent (IHGK and HeLa) host response to *A.a.* Figures adapted from *Cell. Microbiol*. (2005) *7*, 811–823, (C); *Infect Immun*. (2007) *75,* 2540–2547 (D); *Oral Microbiol. Immunol.* (2006), 21261–267 (E) and reprinted by permission. A colour version of this figure is located in the plate section at the back of the book.

cell survival. In addition, to the P53 pathway, other differentially regulated effector molecules playing a role in apoptosis were identified. These data show that multiple apoptotic levels are affected by exposure to the two pathogens. The microarray data are in agreement with phenotypic analyses, which demonstrated that *A.a.* induces apoptosis in HIGK, while infection with *P. gingivalis* had an anti-apoptotic effect. Furthermore, these data are consistent with the different pathophysiologies of the two organisms analysed. While *A.a.* is associated with acute tissue destruction and no inflammation, the disease associated with *P. gingivalis* is chronic and involves inflammatory tissue destruction. Thus, the induction of apoptosis may manifest itself in rapid tissue destruction, while the inhibition of this process may contribute to bacterial persistence in the host, and chronic tissue destruction. In conclusion, these data further confirmed the phenotypic studies and provided invaluable insight into the various players involved in the pathways regulated by infection with oral pathogens.

In another study, the host response to commensal, or 'resident', bacteria was investigated. It is now accepted that these organisms are not simply passive bystanders but engage in an active dialogue with the host. Accordingly, an understanding of the host response to such organisms may provide important clues as to their possible role in maintaining the health status of the host. One of the commensals and early colonizers of the oral cavity is *S. gordonii*. Another commensal also capable of inducing oral pathology is *F. nucleatum*. Microarray analysis was employed to gain insight into the host response to these two bacterial species (Hasegawa *et al.*, 2007). As compared to the host response to oral pathogens, *P. gingivalis* and *A. a.*, the commensal bacteria altered the gingival epithelial cell transcriptome to a lesser degree. Two pathways, the MAP kinase and TLR signalling pathways, were among the most markedly affected. Both of these pathways ultimately effect cytokine expression (Fig. 9.4D). While cytokine IL-6 and IL-8 secretion was upregulated in the epithelial cells stimulated with *F. nucleatum*, downregulation of these proteins was observed in the host cells challenged with *S. gordonii*. The stimulation of the proinflammatory cytokines by the more pathogenic species *F. nucleatum* would be in agreement with its disease-causing potential, while commensal microorganisms, such as *S. gordonii*, may contribute to the maintenance of status of health by inhibiting proinflammatory cytokine production.

As described above, specific bacterial components can elicit specific responses from the host. To gain an insight into the roles of certain bacterial components known to play a role in the pathogenesis of the oral pathogens, the host response to mutants deficient in the components was compared to that of the parental strain. Thus, the role of the gene ORF859 product of *A.a.* was analysed. This gene is induced in humans infected with *A. a.* but its role is not known. Microarray analyses revealed that only a few genes were differentially regulated; notably, the gene encoding IL-27 was upregulated in the parental strain when compared with the mutant deficient in ORF859. Since IL-27 plays a role in the stimulation of host cell-mediated immunity, the ORF859 gene product may contribute to the intracellular adaptation and homeostasis of *A.a.* Such analyses may provide an insight into the role of bacterial products with unknown functions. The second bacterial component analysed was the FimA adhesin of *P. gingivalis*. As described above, this protein plays a role not only in the adhesion and invasion of the bacterium, but also induces a proinflammatory host response. Significant differences in the host response were observed between the parental and the FimA-deficient strains of *P. gingivalis*. As would be expected, based on the role of the protein in the attachment and invasion of *P. ginigvalis*, a large number of genes whose products are involved in cytoskeleton rearrangement and membrane/receptor activity were induced only in the parental strain; this confirmed the role of FimA as a signal in integrin-mediated triggering of the host cytoskeleton rearrangement and the subsequent bacterial invasion. These genes encoded actin binding proteins, LIM protein 1, filamin B, coronin 2A, beta 3, 4 and 6 integrin, alpha V, 3 and 4 integrin, and CD47, (an integrin-associated signal transducer). The apoptosis pathways were not affected by the FimA deficiency, thus indicating this protein plays no role in triggering such pathways. In agreement with the microarray data, the FimA mutant strain had a reduced ability to bind and invade host cells. However, it did have potential, equal to that of the parental strain, in inhibiting the apoptosis of host cells.

In summary, these two examples show that microarray analyses may be used to gain insight into the role of a suspected virulence factor, as well as providing a more detailed understanding of the pathways affected by the microbial components. It should be noted that such data still need confirmation.

The fourth example of the use of a microarray involved the investigation of the role of the difference in the response of various host cell lines to the same pathogen (Mans *et al.*, 2006). Two similar cell types, both of which are epithelial cell lines, HeLa and immortalized human gingival keratinocytes (IHGK), were used for the study. The cells were infected with *A.a.* Comparison of the transcriptional profiles of the infected cells revealed that the expression of most of the genes analysed (80%) remained unaffected, in agreement with the assumption that the processes that are generally important to cell homeostasis would be unaffected by challenge by the bacterium, and would thus be constitutively expressed in all epithelial cell lines. Among the processes unaffected by bacterial infection were RNA synthesis, metabolism, and protein synthesis. However, despite similarities in house-keeping functions and intermediate metabolism, marked differences in the response to infection between the examined cell lines were detected. Only 67 genes were differentially expressed in HeLa cells, while 625 genes were affected by infection with IHGK cells. Furthermore, only 8 genes manifested a common response to the infection (Fig. 9.4E). Among the pathways affected by infection with *A.a.* was the stress response, which included genes involved in response to pathogens, oxidative stress, unfolded proteins, DNA damage, starvation, and wounding. In conclusion, the results of the studies demonstrate that the response to infection of each of the cell lines differed greatly, indicating that the choice of cell lines for the host-pathogen interaction studies should be made very carefully; also that more relevant data can be obtained using host cells that are actually derived from the tissue, as well as a host that provides the naturally occurring niche for the microorganism.

Finally, the fifth microarray-based study was also performed to gain insight into the response of endothelial cells to infection (Chou *et al.*, 2005). In contrast to the previous studies, involving the infection of epithelial cells and analysis of the host response to the total bacterial challenge, this study was undertaken to determine the response of host cells only to intracellular bacteria. Such a study is relevant in the light of the fact that oral bacteria, such as *P. gingivalis*, can invade host cells. Thus, following a 1 hour incubation that allowed for the entry of the bacteria into the host cells, the extracellular bacteria were killed by antibiotic treatment, and the infected host cells were incubated for an additional 5 hours. Sixty-eight genes were found to be regulated in the infected cells when compared to the non-infected cells. The affected pathways included the inflammatory response (cytokine, chemokine, and their receptors, adhesion molecules, and enzymes), angiogenesis, apoptosis (both pro- and anti-apoptotic effectors were differentially regulated), and NF-kB signal transduction. Most of the genes affected were those encoding proteins that mediate the recruitment and trafficking of lymphocytes to sites of vascular inflammation; such as proteins from the chemokine family and adhesion molecules (ICAM and VCAM). These results are in agreement with the observation of the elevated expression of ICAM-1 and VCAM-1 in plaque lesions in aortic tissues derived from animals challenged with parental *P. gingivalis* when compared to tissues derived from uninfected animals, or animals infected with the *P. gingivalis* mutant strain that is unable to invade host cells. Thus, this example of microarray use provides further insight into the complexity of the host response to internalized bacteria, as well as the pathogenic potential of this organism when present within the vascular epithelium.

The major focus of the above studies was the modulation of the innate immune response following challenge with pathogenic bacteria. The studies have confirmed and extended our understanding of the immuno-regulatory and apoptotic pathways. Furthermore, they indicated that many other cellular processes are involved during the host-microbe interactions. These included numerous transcription factors, as well as cell–cell communication factors that were differentially regulated in infected host cells compared to the non-infected controls. As already noted, the recent investigation of the host response to commensal oral bacteria, *F. nucleatum* and *S. gordonii*, revealed that the host responses to the two bacterial species are very similar and differ from the response of pathogen infected cells (Hasegawa *et al.*, 2007). Further investigation of the response may provide insight into the role of the bacteria in promoting homeostasis and its interference with the pathogenic potential of other bacteria. Also, the specific host response to pathogenic bacteria, that may further our understanding of the pathogenesis of the organisms and ultimately the host effectors affected by pathogens, may be used as a target for the development of therapeutic approaches.

Pathogenic and non-pathogenic microbial flora

The innate immune system plays a prominent role as a first line of defence against pathogens. It is based on the recognition of conserved pathogen-associated molecular patterns (PAMPs) that are present on pathogens, but are not found on the host cells. The PAMPs are recognized by pattern recognition receptors (PRRs), such as the TLRs. TLRs sense the presence of microbial infection and are crucial for the initiation of inflammatory and immune defence responses. Of note, the bacterial ligands recognized by TLRs are not unique to pathogens, but rather are shared by entire classes of bacteria, and are also produced, therefore, by commensal bacteria. We do not presently understand whether and how the host distinguishes between pathogenic and commensal bacteria. Moreover, the differences between the two groups of microorganisms are not always clear-cut.

Several mechanisms have been proposed to explain how the host discriminates between pathogens and commensals in order to trigger TLR signalling when appropriate. These include (i) ignorance of commensal PAMPs, (ii) compartmentalized TLR expression, and (iii) the commensal-driven attenuation of proinflammatory signalling (Ramos *et al.*, 2004). Thus, first it has been postulated that both host and commensal bacteria have evolved to co-exist, not activating the acute immune responses through selective interaction with TLRs. One example is the signalling through LPS or lipoteichoic acid (LTA). The LPS-binding protein and the CD14 are accessory proteins that bind LPS and transfer it to TLR receptors thus initiating the signalling cascade ultimately resulting in a host response. However, these proteins can also transfer LPS or LTA to serum proteins that are non-activating and thus remove the bacterial components and avoid activation of the innate immune response (Gutsmann *et al.*, 2001). There is variation in the number of fatty acid side-chains and phosphate groups of LPS isolated from various bacteria. Furthermore, the variation in LPS structure also affects the recognition of the components by LPS-binding proteins and CD14, thus ultimately affecting recognition of these components by the host (Cunningham *et al.*, 1996). Accordingly, one of the mechanisms enabling the host to differentiate between commensals and pathogenic bacteria may be the differential affinity of bacterial components for PRRs. Moreover, TLR may be expressed differentially; for example, the human endothelial cells predominantly express TLR4 and low levels of TLR2. As TLR4 mainly recognizes LPS, while TLR2 responds to lipoproteins, it would be expected that the endothelial cells would predominately respond to LPS. Consistent with this, endothelial cells respond to bacterial LPS but not to the *Mycobacterium tuberculosis* lipoprotein (Faure *et al.*, 2000). Finally, it has been postulated that commensal bacteria down-modulate TLR expression, ultimately preventing their recognition (Melmed *et al.*, 2003). However, such a suggestion contradicts findings that recognition of commensal bacteria by TLR2 and TLR4 is indispensable for induction of cell-protecting factors – such as IL-6, tumour necrosis factor, cytoprotective heat-shock proteins (Hsp) – critical for the maintenance of homeostasis in the gut (Rakoff-Nahoum *et al.*, 2004).

As similar PAMPs are present on both commensals and pathogens, it is probable that the latter differ from the former mainly in terms of the differential expression of cell surface molecules, such as adhesion proteins and other virulence factors that may affect the host (Fischer *et al.*, 2006). A recent, and very attractive, model was proposed by Tournier and Quesnel-Hellmann (2006). This model is a 'pathogen-centric' model of the host recognition of the damage-associated molecular pattern (DAMP) by dendritic cells (DCs) originally proposed by Seong and Matzinger (2004). Tournier and Quesnel-Hellmann further expanded that model by proposing that pathogens encode DAMP-interfering molecules, which they suggested be termed 'negative signals' (NS). These are pathogen components that are not recognized as DAMPs by the host, and that block or modify the intracellular signalling downstream of PRRs, ultimately resulting in tolerance.

Two methods by which virulent bacteria can elicit tolerance to a host's antigen presenting cells (APCs) have been proposed: First, direct interference with the APC pattern recognition-mediated signalling, and second by triggering the apoptosis of cells, which in turn inhibits APCs. TLR-mediated signalling leads to the activation of transcription factors, NF-kB, IFN-regulatory factor (IRF) 3, and IRF7. Examples of direct interference are provided by non-oral organisms. For instance, viruses, such as the vaccinia virus (VV), have several strategies for disrupting TLR-mediated signalling. First, the VV A46R protein is homologous to the Toll-like-interleukin-1 receptor (TIR), and thus targets the host TIR adaptors, MyD88, as well as the MyD88-adaptor-like-TIR-domain-containing adaptor-inducing IFN-beta (TRIF), and TRIF-related adaptor molecule, and thus interferes with the downstream activation of MAPKs, NF-kB, and IRF3 (Stack *et al.*, 2005; Bowie *et al.*, 2000). VV also has a A52R protein that associates with TNF-receptor-associated factor 6 and interleukin-1-associated kinase 2, and blocks the activation of NF-kB (Harte *et al.*, 2003). Finally, VV E3L protein can directly interact with IRF3 and IRF7 and inhibit the phosphorylation of those transcription factors (Smith *et al.*, 2001). Bacteria were shown to act further downstream of the signalling cascade and inhibit MAPK and NF-κB (Finlay and McFadden, 2006). The *Bacillus anthracis* lethal factor (LF), a zinc metalloprotease, can cleave six of the mitogen-activated kinase kinases, and inhibit IRF3 (Dang *et al.*, 2004). The *Yersinia pestis* YopJ and *Y. enterocolitica* YopP proteins inhibit MAPK signalling and NF-κB (Navarro *et al.*, 2005).

Several examples illustrate the death of the infected host cell by pathogen-induced apoptosis. Viruses can extensively manipulate the host machinery, ultimately resulting in apoptosis. The vaccinia virus (VV) can induce the apoptosis of infected macrophages but the molecular mechanism of this process remains unknown. In addition, various bacterial pathogens induce the apoptosis of host cells. For example, the LF protein of *B. anthracis* inhibits p38 MAP kinase, and thus induces apoptosis (Park *et al.*, 2002). Also, the *Yersinia spp* YopP and YopJ proteins inhibit the MAPK and NF-κB pathways, thus disrupting the transcription of anti-apoptotic genes, as well as triggering the apoptotic signalling cascade upstream of tBid (Denecker *et al.*, 2001). The induction of apoptosis has a direct effect as it removes immunostimulatory cells, such as dendritic cells (DCs). The apoptotic cells are rapidly internalized by the phagocytic cells (macrophages and dendritic cells), and this prevents the release, into the environment, of potentially immunogenic components from the dying cells. Also, apoptotic cells release anti-inflammatory and other inhibitory mediators to the neighbouring cells, and these mediators may trigger tolerance under the steady state (Kim *et al.*, 2004; Albert, 2004; Hugues *et al.*, 2002). Collaboratively, pathogens can use apoptosis to attenuate the immune response. The use of apoptosis is a very efficient technique, as apoptotic cells release apoptotic blebs that deliver both negative signals and pathogen antigens, thus inhibiting the recruitment and activation of APCs. Since one apoptotic cell can release a large number of blebs, it can target multiple APCs, and so the effect of immunosuppression can be very severe. An example of such immunosuppressive ability is possessed by the measles virus, which triggers DC apoptosis and causes severe immunosuppression (Hahm *et al.*, 2007).

Oral pathogens can utilize several techniques to induce tolerance. First, the downregulation of TLR expression was observed upon the multiple sequential exposure of the host to bacterial components (e.g. bacterial adhesins, such as the *P. gingivalis* FimA-induced tolerance for the host receptors). Furthermore, one of the oral pathogens' effector molecules that may inhibit the signalling pathways may be the *P. gingivalis* cysteine protease. These proteases have been shown to bind, invade and localize in the perinuclear region of the infected cells (Scragg *et al.*, 2002). The intracellular localization of gingipains may interfere with several signalling pathways, including the focal adhesion molecules FAK/paxillin and MAP kinase, MEK/ERK mediated-regulation. In addition, the *A.a.* leukotoxin or CagA effectors may be internalized and have an effect on the host signalling pathways.

The ability of oral pathogens to induce apoptosis is well-documented for *A. a.*, and is described above. The *P. gingivalis* early response involves an anti-apoptotic response in infected host cells but prolonged infection of host cells with the bacterium (24–48 h) induces a pro-apoptotic effect (Urnowey *et al.*, 2006). Thus, triggering apoptosis by oral pathogens may be an efficient technique of immune signalling, which will then allow for the persistence of the organisms.

Finally, the oral commensals and pathogens induce different types of defensins from host cells. One of the proteins that may have a role in triggering the differential defensin expression is the *P. gingivalis* HagB adhesion molecule (Lu *et al.*, 2006).

While the ability of pathogenic and commensal oral bacteria, such as *P. gingivalis, A. a., F. nucleatum,* and *S. gordonii,* to trigger the host response is documented, a detailed analysis of the host response to oral pathogens and commensals over an extended period of time should provide further insight into the pathogenic strategies used by oral bacteria.

Oral microflora and diseases

Oral microorganisms trigger periodontal diseases; the mechanism involves an exacerbated host response manifesting itself as excessive inflammation, ultimately leading to tissue destruction. The above information, which highlights the ability of oral pathogens to trigger the inflammatory response, may explain why these are the specific species involved in periodontal disease initiation and progression. However, oral organisms may become distributed systemically, leading to more serious complications, such as cardiovascular diseases. This may be explained by the ability of some of the organisms implicated in cardiovascular disease progression, such as *P. gingivalis*, to stimulate the expression of adhesion molecules (ICAM and VCAM). These molecules are required for macrophage recruitment to the site of infection. The macrophages, upon activation by various stimuli, such as oxidized low-density lipoprotein, endothelin, and angiotensin II, kill cells using effector molecules such as Fas-L and nitric oxide. Finally, the potential for bacteria to cause cancer was highlighted by the discovery of *Helicobacter pylori* cancer-associated activity. As oral diseases are also chronic, it is possible that oral microorganisms may also play a role in oral carcinoma development. This is further supported by the earlier described effects of some of the oral microorganisms on the host signalling cascades. For example, *P. gingivalis* can persist intracellularly and suppress apoptosis in infected cells. Also, comprehensive microarray analysis has demonstrated that this organism is a potent chemokine stimulator (Fig. 9.4B). Chemokines, such as chemokine 5 (CCL5), have recently been demonstrated to play a role in oral carcinoma development (Miyazaki *et al.*, 2006).

Microbial response

Bacteria also have the ability to modulate protein expression depending on the environmental stimuli

Typically, genes whose products are required in a given microenvironment are over expressed. Bacteria encounter several challenges when exposed to the host environment. Firstly, they must adhere to the host, and so over expression of the cell surface components required for the establishment of the microorganism in the host would be expected to successfully mediate this step. The multiplicity of adhesins displayed by oral bacteria in turn increases the likelihood and strength of receptor binding, and provides the means by which bacteria can manipulate the host cell-signalling pathways that are stimulated through individual receptor engagements. Secondly, the bacterium must escape the host immune defence mechanisms, including that of ingestion by phagocytic cells, and oxidative and 'nitrosative' stresses that are elevated in host cells upon encountering microorganisms. Next, the bacteria need to proliferate, thus necessitating the acquisition of nutrients for growth. As iron is limited during the host upregulation, iron uptake proteins would be expected to be produced in the bacteria inhabiting the host. However, as the intracellular environment is rich in iron, and since iron is an oxidative stress mediator, the uptake of this nutrient must be limited, and thus the expression of iron uptake mechanisms would be expected to be reduced in organisms that grow intracellularly.

One of the approaches used to identify genes that are required for the adhesion of *P. gingivalis* was the SDS-PAGE analysis of protein profiles derived from the bacteria attached to gingival epithelial cells, and that of free bacteria. Using this method, a 51-kDa protein was detected as being upregulated in epithelial cell-bound bacteria, indicating that this protein plays a role in the attachment of the bacteria to host cells (Agnani *et al.*, 2000). The use of *in vivo* induced antigen technology (IVIAT) identified three *A.a.* genes that were upregulated in cells grown intracellularly in HeLa cells. Mutants deficient in two of the gene products (ORF1402 and ORF859) had significantly reduced abilities to survive in the host cells, thus confirming the predicted role in the intracellular survival of the bacterium (Cao *et al.*, 2004).

The genomic sequences of a variety of oral bacteria are currently known (*P. gingivalis* W83, *A.a.*, *T. denticola*, *P. intermedia* 17, *F. nucleatum*, *and T. forsythia*), allowing for the design of genomic microarrays, which are an excellent resource with which to investigate the global gene expression in these bacteria. Studies involving the effect of host exposure on bacterial transcription were performed in *P. gingivalis*. The first study involved the exposure of *P. gingivalis* to epithelial cells. Several stress related genes, such as *sodB*, *ftn*, *ahpC*, *ahpF*, *trxB*, *groEL*, *dnaK*, and *htrA* were upregulated, in agreement with the stress conditions, such as oxidative and nitrosative stress, imposed by the host on the bacteria (Fig. 9.5). The induction of these pathways indicates that they are indispensable for the survival of the bacterium when it is in contact with epithelial cells. Interestingly, several genes encoding hypothetical proteins, as well as putative regulatory proteins (*tetR* and *rprY*), were upregulated. Finally, capsule synthesis was upregulated upon contact with host cells, indicating the importance of the capsule in host-*P. gingivalis* interactions (Hosogi and Duncan, 2005).

The gene expression in host cell-internalized bacteria was also examined. First, to identify possible *P. gingivalis* factors involved in invasion and survival, genes that are upregulated in internalized *P. gingivalis* were identified by differential display reverse transcription-PCR (DDRT-PCR) (Park *et al.*, 2004). Twenty-five genes upregulated in bacteria present within the GEC were detected. Among them were genes whose product would not normally be suspected of being involved in the *P. gingivalis* invasion process. Also, a functional analysis of three of the genes, encoding endopeptidase PepO, cation-transporting ATPase, and ABC transporter, further confirmed the involvement of these gene products in the invasion process of *P. gingivalis*. Another insight into the gene expression alteration upon invasion of *P. gingivalis* into host cells was gained from studies by Rodrigues and Progulske-Fox (2005). *P. gingivalis* genes differentially expressed during the invasion of human coronary artery endothelial cells included the upregulation of 11 genes and the downregulation of 62 genes.

Among the upregulated were genes whose products may be involved in intracellular trafficking and/or interactions with autophagosomal vesicles, such as PG1682, which encodes a glycosyl transferase, and PG1683, which encodes a homologue of amylases. The glycosyl transferase maybe involved in the attachment of *P. gingivalis* to epithelial cells (glycosyl transferase) (Chen *et al.*, 2000b), while PG1683 may have a role in the co-aggregation of *P. gingivalis* with other oral bacterial species (Ellen *et al.*, 1997; Kamaguch *et al.*, 2001). It is suspected that PG1682 and PG1683 might also play a role in the coaggregation of *P. gingivalis* with cell host intracellular cell membranes (such as the ones encountered in autophagosomes). In addition, a gene encoding putative ABC transporter permease (PG0280) was upregulated, as well as a gene encoding a transporter with unknown function (PG0092). These gene products may play a role in transport in the intracellular bacterium. Two genes, PG1286 and PG1172, encoding ferritin and a putative iron-sulphur cluster binding protein, respectively, are expected to play a role in oxidative stress protection of the intracellular bacterium.

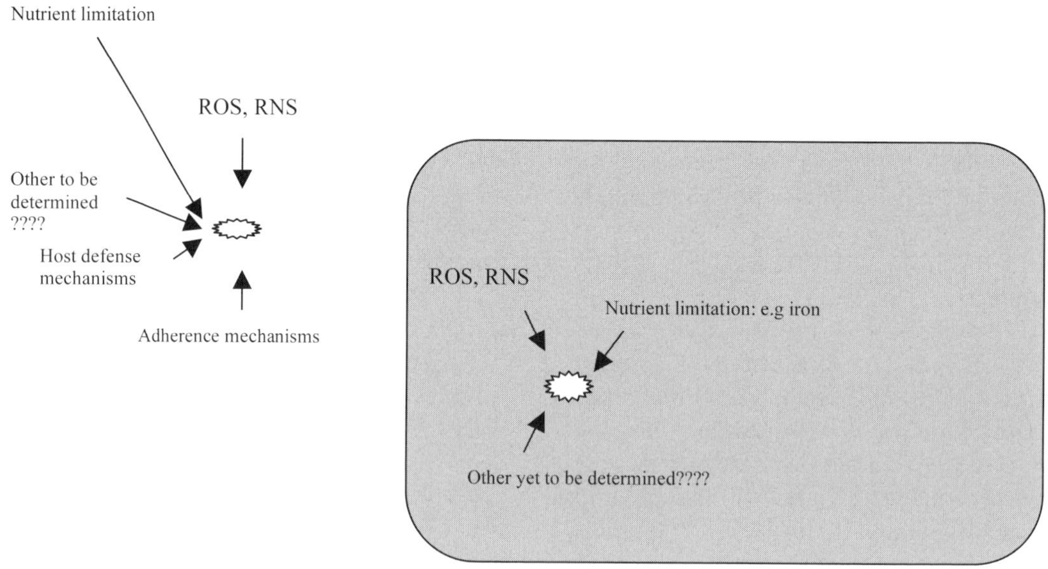

Figure 9.5 Host factors having an effect on microbial gene expression. Microorganisms are sensors of the environment and the microbial transcriptional profile is the read-out of the conditions that they encounter. A variety of host factors can serve to trigger alteration of gene expression in microbial cells upon contact with the host cells. While the majority of the host factors having an effect on microorganisms is universal, some cell-specific responses are also present.

Finally, PG1896 encodes proteins synthesizing *S*-adenosylmethionine (SAM). This is a major methyl donor acting in metabolic processes, and its absence in some bacteria was associated with a defect in cell division (Newman *et al.*, 1998). Since *P. gingivalis* was shown to be able to replicate inside late autophagosomes (Dorn *et al.*, 2001), it is probable that this gene product is involved in intracellular replication or the metabolic processes necessary for the survival of *P. gingivalis* in host cells.

The large group of genes downregulated in the intracellular bacteria included genes that encode hypothetical proteins, and proteins likely to be involved in protein synthesis, transcription, and energy metabolism that may be related to the reduced intracellular bacterial growth rate.

A global approach, using the proteomic comparison of proteins from bacteria grown in media exposed to the host (conditioned media) to bacteria grown in unconditioned media revealed that 479 proteins were differentially expressed upon the exposure of the bacterium to the epithelial cell growth medium (Zhang *et al.*, 2005). As many as 276 of the proteins were hypothetical. Notably upregulated were: a homologue of the *L. monocytogenes* internalin, demonstrated to play a role in the invasion of the bacterium into host cells, and ATP-dependent Clp protease. The role of Clp protease in the invasion of *P. gingivalis* was further confirmed using a protease-deficient mutant that, when compared to the parent strain, had a 50% reduced ability to invade epithelial cells.

The above studies provide insight into the molecular basis of the interaction of the bacteria with the host. Some of the results are expected, based on the host challenges imposed on the bacteria. However, novel genes, whose products were not expected to play a role in the bacterial interaction with the host, were also detected. Future studies, which will be conducted over longer periods of time and using a variety of host cells, are required to gain a more comprehensive understanding of the bacterial factors required for colonization, growth, and survival, once in contact with host cells. Also, although the majority of investigations have centred on the detection and characterizarion of virulence genes, it is possible that the major mechanism differentiating virulent and avirulent microbial strains is alteration in the regulatory systems; such as mutations in major regulatory proteins. Thus, comparison of the gene expression in virulent *versus* avirulent bacterial strains upon exposure to host components is also a promising area for investigating the host effect on microbial response.

Conclusion and future directions

The co-existence of the host and oral microflora is the result of complex host-pathogen interactions resulting in mutual adaptation. The rapid development of molecular biology tools has enabled remarkable progress in the understanding of these relationships. Firstly, multiple microbial cell surface structures mediating the adherence of the organism to host surfaces have been identified. Furthermore, the role of the adhesins in triggering the host response has been demonstrated. In the meantime, new information regarding the host signalling pathways utilized in the response to infection with microorganisms has accumulated. A degree of understanding of the role of the host signalling pathways during infection with oral microflora was possible due to recent investigations involving knockout animals and defined eukaryotic cells expressing receptors of interest. Finally, genome wide approaches involving microarray analysis of the host response to bacterial infection, as well as the bacterial response upon exposure to the host, has afforded some insight into the molecular basis of host-pathogen interactions.

Future efforts to further define the molecular basis of the host-pathogen interactions in the oral cavity will be challenging and will require multidisciplinary approaches, including bioinformatics, as well as transcriptome and proteome analyses conducted simultaneously on the host and bacteria. So far, the majority of the global expression studies have been performed using various cultured cells. Because the oral ecosystem involves dynamic interactions among microorganisms, epithelial cells, and the immune system, an approach that will more accurately elucidate the responses should involve *in vivo* studies using aseptic or gnotobiotic animals. Furthermore, such studies, leading to the identification of bacterial effects on the host should be conducted in defined gnotobiotic animals over a significant time period. Such time-dependent alterations in the host response will provide indispensable information about the dynamic nature of the response to a given pathogen, as well as providing insight into the regulatory processes that underlie it. In addition, approaches to monitor bacterial biochemistry and physiology in the animal models will be required. Examples of host-pathogen interactomes already exist. One such example is the simultaneous investigation of the host and pathogen responses during malaria infection (Lovegrove *et al.*, 2006). This study employed two murine strains of cerebral malaria, susceptible and resistant, and the responses of both the *Plasmodium berghei* ANKA and the mouse were investigated. The results have shown unique strain- and tissue-specific expression profiles for both the parasite and mouse. Such a comprehensive approach, when applied to the investigation of oral host-microbe interaction would be expected to give insight into the gene expression, both microbial and host, during various stages of oral diseases, as well as age.

Since the oral microflora exists as a multispecies community, it is the effect of this community on the host that determines the outcome of the host-microbe interaction. Thus, although it is important to continue investigation of the host response to a single bacterial species, such studies should be undertaken to further our understanding of the host response to the microbial communities in which the microbe of interest resides. Investigation of the host response to polymicrobial communities has yet to be carried out. However, it has been shown that epithelial cells containing internalized *P. gingivalis* do not secrete the neutrophil chemokine IL-8 in response to other oral bacteria, such as *F. nucleatum* (Darveau and Lamont, 1998). Such an inhibition of IL-8 secretion would lead to the inhibition of neutrophil recruitment to the site of infection and, therefore, may have a debilitating effect on the host immune surveillance system, ultimately leading to bacterial overgrowth and disease initiation. This example further reinforces the importance of examining the effect of microbial communities, as opposed to single species, on the host cell response.

Finally, most of the effectors, both eukaryotic and microbial, of the host-microbe 'cross-talk' remain undefined. The complete genome sequence of a variety of oral bacteria offers an unprecedented opportunity for the identification of numerous virulence factors, as well as cell surface-exposed determinants that can be further investigated for their role as signalling effectors. This could be accomplished by examining isogenic mutants, deficient in the effector molecule, for their impact on host response. An alternative approach would be to expose cells to specific microbial proteins in order to identify those generating the host response. The use of mutant hosts, such as cells or animals that are deficient in a specific protein or cellular process could be used to identify the cellular mechanisms that underlie the host response. Another approach would be the usage of activators or specific inhibitors of a specific effector molecule and the subsequent analysis of the host response.

It should be noted that the oral cavity is very accessible to both microbial and host sampling over a long period of time and, as such, represents the ideal system with which to study host-pathogen interactions. Hopefully, data obtained using this system will enrich our understanding, not only of oral diseases but also of other chronic inflammatory diseases.

Summary

Multiple microbial receptors, termed adhesins, mediate the contact with the host target cell.

Many different host receptors recognize microbes and the host response to microbes is mediated simultaneously through multiple receptors

Various microbe-recognition receptors induce different signalling cascades and these signalling cascades may interact cooperatively to mediate the ultimate response.

Pathogenic microorganisms may regulate the signalling cascades to evade destruction by host immune response

References

Agnani, G., Tricot-Doleux, S., Du, L., and Bonnaure-Mallet, M. (2000) Adherence of *Porphyromonas gingivalis* to gingival epithelial cells: modulation of bacterial protein expression. Oral Microbiol. Immunol. *15*, 48–52.

Akira, S. and Takeda, K. (2004) Toll-like receptor signalling. Nat. Rev. Immunol. *4*, 499–511.

Akira, S., Uematsu, S., and Takeuchi, O. (2006) Pathogen recognition and innate immunity. Cell *124*, 783–801.

Albert, M.L. (2004) Death-defying immunity: do apoptotic cells influence antigen processing and presentation? Nat. Rev. Immunol. *4*, 223–231.

Alugupalli KR, Kalfas S Edwardsson S Forsgren A Arnold RR Naidu AS.(1994) Effect of lactoferrin on interaction of *Prevotella intermedia* with plasma and subepithelial matrix proteins. Oral Microbiol. Immunol. 9, 174–179.

Argueta, J.G., Shiota, S., Yamaguchi, N., Masuhiro, Y., and Hanazawa, S. (2006) Induction of *Porphyromonas gingivalis* GroEL signalling via binding to Toll-like receptors 2 and 4. Oral Microbiol. Immunol. *21*, 245–251.

Babu, J.P., Dean, J.W., and Pabst, M.J. (1995) Attachment of *Fusobacterium nucleatum* to fibronectin immobilized on gingival epithelial cells or glass coverslips. J. Periodontol. 66, 285–290.

Bainbridge, B.W. and Darveau, R.P. (2001) *Porphyromonas gingivalis* lipopolysaccharide: an unusual pattern recognition receptor ligand for the innate host defense system. Acta Odontol. Scand. 59, 131–138.

Bar-Sagi, D. and Hall, A. (2000) Ras and Rho GTPases: a family reunion. Cell *103*, 227–238.

Belanger, M., Rodrigues, P.H., Dunn, W.A., Jr., and Progulske-Fox, A. (2006) Autophagy: a highway for *Porphyromonas gingivalis* in endothelial cells. Autophagy. *2*, 165–170.

Belton, C.M., Izutsu, K.T., Goodwin, P.C., Park, Y., and Lamont, R.J. (1999) Fluorescence image analysis of the association between *Porphyromonas gingivalis* and gingival epithelial cells. Cell Microbiol. *1*, 215–223.

Birkedal-Hansen, H. (1993) Role of matrix metalloproteinases in human periodontal diseases. J. Periodontol. 64, 474–484.

Black, D. S. and Bliska, J.B. (2000) The RhoGAP activity of the *Yersinia pseudotuberculosis* cytotoxin YopE is required for antiphagocytic function and virulence. Mol. Microbiol. 37, 515–527.

Botelho, R.J., Teruel, M., Dierckman, R., Anderson, R., Wells, A., York, J.D., Meyer, T., and Grinstein, S. (2000) Localized biphasic changes in phosphatidylinositol-4,5-bisphosphate at sites of phagocytosis. J. Cell Biol. *151*, 1353–1368.

Bowie, A., Kiss-Toth, E., Symons, J.A., Smith, G.L., Dower, S.K., and O'Neill, L.A. (2000) A46R and A52R from vaccinia virus are antagonists of host IL-1 and toll-like receptor signalling. Proc. Natl. Acad. Sci. U. S. A *97*, 10162–10167.

Brogden, K.A., Guthmiller, J.M., and Taylor, C.E. (2005) Human polymicrobial infections. Lancet 365, 253–255.

Cannon, R. D. and Chaffin, W.L. (2001) Colonization is a crucial factor in oral candidiasis. J. Dent. Educ. 65, 785–787.

Cao, S.L., Progulske-Fox, A., Hillman, J.D., and Handfield, M. (2004) In vivo induced antigenic determinants of *Actinobacillus actinomycetemcomitans*. FEMS Microbiol. Lett. *237*, 97–103.

Chatenay-Rivauday, C., Yamodo, I., Sciotti, M.A., Ogier, J.A., and Klein, J.P. (1998) The A and the extended V N-terminal regions of streptococcal protein I/IIf mediate the production of tumour necrosis factor alpha in the monocyte cell line THP-1. Mol. Microbiol. *29*, 39–48.

Chen T, Nakayama K Belliveau L Duncan MJ. *Porphyromonas gingivalis* gingipains and adhesion to epithelial cells. Infect Immun. 69, 3048–3056. 2001.

Chen, C.C., Sun, Y.T., Chen, J.J., and Chiu, K.T. (2000a). TNF-alpha-induced cyclooxygenase-2 expression in human lung epithelial cells: involvement of the phospholipase C-gamma 2, protein kinase C-alpha, tyrosine kinase, NF-kappa B-inducing kinase, and I-kappa B kinase 1/2 pathway. J. Immunol. *165*, 2719–2728.

Chen, T., Dong, H., Yong, R., and Duncan, M.J. (2000b). Pleiotropic pigmentation mutants of *Porphyromonas gingivalis*. Microb. Pathog. *28*, 235–247.

Chimini, G. and Chavrier, P. (2000) Function of Rho family proteins in actin dynamics during phagocytosis and engulfment. Nat. Cell Biol. *2*, E191-E196.

Chou, H.H., Yumoto, H., Davey, M., Takahashi, Y., Miyamoto, T., Gibson, F.C., III, and Genco, C.A. (2005) *Porphyromonas gingivalis* fimbria-dependent activation of inflammatory genes in human aortic endothelial cells. Infect. Immun. *73*, 5367–5378.

Chung, W.O., Hansen, S.R., Rao, D., and Dale, B.A. (2004) Protease-activated receptor signalling increases epithelial antimicrobial peptide expression. J. Immunol. *173*, 5165–5170.

Crowley, P.J., Brady, L.J., Piacentini, D.A., and Bleiweis, A.S. (1993) Identification of a salivary agglutinin-binding domain within cell surface adhesin P1 of Streptococcus mutans. Infect. Immun. *61*, 1547–1552.

Cunningham, M.D., Seachord, C., Ratcliffe, K., Bainbridge, B., Aruffo, A., and Darveau, R.P. (1996) *Helicobacter pylori* and *Porphyromonas gingivalis* lipopolysaccharides are poorly transferred to recombinant soluble CD14. Infect. Immun. *64*, 3601–3608.

d'Empaire, G., Baer, M.T., and Gibson, F.C., III (2006) The K1 serotype capsular polysaccharide of *Porphyromonas gingivalis* elicits chemokine production from murine macrophages that facilitates cell migration. Infect. Immun. *74*, 6236–6243.

Dang, O., Navarro, L., Anderson, K., and David, M. (2004) Cutting edge: anthrax lethal toxin inhibits activation of IFN-regulatory factor 3 by lipopolysaccharide. J. Immunol. *172*, 747–751.

Darveau, R.P., Belton, C.M., Reife, R.A., and Lamont, R.J. (1998) Local chemokine paralysis, a novel pathogenic mechanism for *Porphyromonas gingivalis*. Infect. Immun. *66*, 1660–1665.

DeCarlo, A.A., Grenett, H.E., Harber, G.J., Windsor, L.J., Bodden, M.K., Birkedal-Hansen, B., and Birkedal-Hansen, H. (1998) Induction of matrix metalloproteinases and a collagen-degrading phenotype in fibroblasts and epithelial cells by secretedgingiv *Porphyromonas gingivalis* proteinase. J. Periodont. Res. *33*, 408–420.

Denecker, G., Declercq, W., Geuijen, C.A., Boland, A., Benabdillah, R., van, G.M., Sory, M.P., Vandenabeele, P., and Cornelis, G.R. (2001) *Yersinia enterocolitica* YopP-induced apoptosis of macrophages involves the apoptotic signalling cascade upstream of bid. J. Biol. Chem. *276*, 19706–19714.

Dixon, D.R., Bainbridge, B.W., and Darveau, R.P. (2004) Modulation of the innate immune response within the periodontium. Periodontol. 2000. *35*, 53–74.

Dorn, B.R., Dunn, W.A., Jr., and Progulske-Fox, A. (2001) *Porphyromonas gingivalis* traffics to autophagosomes in human coronary artery endothelial cells. Infect. Immun. *69*, 5698–5708.

Dorn, B.R., Leung, K.L., and Progulske-Fox, A. (1998) Invasion of human oral epithelial cells by Prevotella intermedia. Infect. Immun. *66*, 6054–6057.

Dramsi, S., Dehoux, P., Lebrun, M., Goossens, P.L., and Cossart, P. (1997) Identification of four new members of the internalin multigene family of *Listeria monocytogenes* EGD. Infect. Immun. *65*, 1615–1625.

Duncan, M.J., Emory, S.A., and Almira, E.C. (1996) *Porphyromonas gingivalis* genes isolated by screening for epithelial cell attachment. Infect. Immun. *64*, 3624–3631.

Ellen, R.P., Lepine, G., and Nghiem, P.M. (1997) *In vitro* models that support adhesion specificity in biofilms of oral bacteria. Adv. Dent. Res. *11*, 33–42.

Eskan, M.A., Hajishengallis, G., and Kinane, D.F. (2007) Differential activation of human gingival epithelial cells and monocytes by *Porphyromonas gingivalis* fimbriae. Infect. Immun. *75*, 892–898.

Ezzo, P. J. and Cutler, C.W. (2003) Microorganisms as risk indicators for periodontal disease. Periodontol. 2000. *32*, 24–35.

Faure, E., Equils, O., Sieling, P.A., Thomas, L., Zhang, F.X., Kirschning, C.J., Polentarutti, N., Muzio, M., and Arditi, M. (2000) Bacterial lipopolysaccharide activates NF-kappaB through toll-like receptor 4 (TLR-4) in cultured human dermal endothelial cells. Differential expression of TLR-4 and TLR-2 in endothelial cells. J. Biol. Chem. *275*, 11058–11063.

Feucht, E.C., DeSanti, C.L., and Weinberg, A. (2003) Selective induction of human beta-defensin mRNAs by *Actinobacillus actinomycetemcomitans* in primary and immortalized oral epithelial cells. Oral Microbiol. Immunol. *18*, 359–363.

Fine, D.H., Velliyagounder, K., Furgang, D., and Kaplan, J.B. (2005) The *Actinobacillus actinomycetemcomitans* autotransporter adhesin *Aae* exhibits specificity for buccal epithelial cells from humans and old world primates. Infect. Immun. *73*, 1947–1953.

Finlay, B. B. and McFadden, G. (2006) Anti-immunology: evasion of the host immune system by bacterial and viral pathogens. Cell *124*, 767–782.

Fischer, H., Yamamoto, M., Akira, S., Beutler, B., and Svanborg, C. (2006) Mechanism of pathogen-specific TLR4 activation in the mucosa: fimbriae, recognition receptors and adaptor protein selection. Eur. J. Immunol. *36*, 267–277.

Grenier, D. (1996) Collagen-binding activity of *Prevotella intermedia* measured by a microtitre plate adherence assay. Microbiology *142 (Pt 6)*, 1537–1541.

Gutsmann, T., Muller, M., Carroll, S.F., MacKenzie, R.C., Wiese, A., and Seydel, U. (2001) Dual role of lipopolysaccharide (LPS)-binding protein in neutralization of LPS and enhancement of LPS-induced activation of mononuclear cells. Infect. Immun. *69*, 6942–6950.

Hahm, B., Cho, J.H., and Oldstone, M.B. (2007) Measles virus-dendritic cell interaction via SLAM inhibits innate immunity: Selective signalling through TLR4 but not other TLRs mediates suppression of IL-12 synthesis. Virology 358, 251–257.

Hajishengallis, G., Koga, T., and Russell, M.W. (1994) Affinity and specificity of the interactions between *Streptococcus mutans* antigen I/II and salivary components. J. Dent. Res. 73, 1493–1502.

Hajishengallis, G., Martin, M., Sojar, H.T., Sharma, A., Schifferle, R.E., DeNardin, E., Russell, M.W., and Genco, R.J. (2002) Dependence of bacterial protein adhesins on toll-like receptors for proinflammatory cytokine induction. Clin. Diagn. Lab Immunol. *9*, 403–411.

Hajishengallis, G., Sojar, H., Genco, R.J., and DeNardin, E. (2004) Intracellular signalling and cytokine induction upon interactions of *Porphyromonas gingivalis* fimbriae with pattern-recognition receptors. Immunol. Invest 33, 157–172.

Han, Y.W., Ikegami, A., Rajanna, C., Kawsar, H.I., Zhou, Y., Li, M., Sojar, H.T., Genco, R.J., Kuramitsu, H.K., and Deng, C.X. (2005) Identification and characterization of a novel adhesin unique to oral fusobacteria. J. Bacteriol. *187*, 5330–5340.

Han, Y.W., Shi, W., Huang, G.T., Kinder, H.S., Park, N.H., Kuramitsu, H., and Genco, R.J. (2000) Interactions between periodontal bacteria and human oral epithelial cells: *Fusobacterium nucleatum* adheres to and invades epithelial cells. Infect. Immun. *68*, 3140–3146.

Handfield, M., Mans, J.J., Zheng, G., Lopez, M.C., Mao, S., Progulske-Fox, A., Narasimhan, G., Baker, H.V., and Lamont, R.J. (2005) Distinct transcriptional profiles characterize oral epithelium-microbiota interactions. Cell Microbiol. *7*, 811–823.

Harte, M.T., Haga, I.R., Maloney, G., Gray, P., Reading, P.C., Bartlett, N.W., Smith, G.L., Bowie, A., and O'Neill, L.A. (2003) The poxvirus protein A52R targets Toll-like receptor signalling complexes to suppress host defense. J. Exp. Med. *197*, 343–351.

Hasegawa, Y., Mans, J.J., Mao, S., Lopez, M.C., Baker, H.V., Handfield, M., and Lamont, R.J. (2007) Gingival Epithelial Cell Transcriptional Responses to Commensal and Opportunistic Oral Microbial Species. Infect. Immun. 75, 2540–2547.

Hashimoto, M., Ogawa, S., Asai, Y., Takai, Y., and Ogawa, T. (2003) Binding of *Porphyromonas gingivalis* fimbriae to *Treponema denticola* dentilisin. FEMS Microbiol. Lett. 226, 267–271.

Hiramine, H., Watanabe, K., Hamada, N., and Umemoto, T. (2003) *Porphyromonas gingivalis* 67-kDa fimbriae induced cytokine production and osteoclast differentiation utilizing TLR2. FEMS Microbiol. Lett. *229*, 49–55.

Hirschfeld, M., Weis, J.J., Toshchakov, V., Salkowski, C.A., Cody, M.J., Ward, D.C., Qureshi, N., Michalek, S.M., and Vogel, S.N. (2001) Signalling by toll-like receptor 2 and 4 agonists results in differential gene expression in murine macrophages. Infect. Immun. *69*, 1477–1482.

Honma, K., Kuramitsu, H.K., Genco, R.J., and Sharma, A. (2001) Development of a gene inactivation system for *Bacteroides forsythus*: construction and characterization of a BspA mutant. Infect. Immun. 69, 4686–4690.

Hosogi, Y. and Duncan, M.J. (2005) Gene expression in *Porphyromonas gingivalis* after contact with human epithelial cells. Infect. Immun. *73*, 2327–2335.

Hugues, S., Mougneau, E., Ferlin, W., Jeske, D., Hofman, P., Homann, D., Beaudoin, L., Schrike, C., Von, H.M., Lehuen, A., and Glaichenhaus, N. (2002) Tolerance to islet antigens and prevention from diabetes induced by limited apoptosis of pancreatic beta cells. Immunity. *16*, 169–181.

Ikegami, A., Honma, K., Sharma, A., and Kuramitsu, H.K. (2004) Multiple functions of the leucine-rich repeat protein LrrA of *Treponema denticola*. Infect. Immun. *72*, 4619–4627.

Imler, J.L. and Hoffmann, J.A. (2003) Toll signalling: the TIReless quest for specificity. Nat. Immunol. *4*, 105–106.

Inohara, Chamaillard McDonald C Nunez G.(2005) NOD-LRR proteins: role in host-microbial interactions and inflammatory disease. Annu Rev Biochem. *74*, 355–383.

Inohara, N. and Nunez, G. (2003) NODs: intracellular proteins involved in inflammation and apoptosis. Nat. Rev. Immunol. 3, 371–382.

Inoue, T., Tanimoto, I., Ohta, H., Kato, K., Murayama, Y., and Fukui, K. (1998) Molecular characterization of low-molecular-weight component protein, Flp, in *Actinobacillus actinomycetemcomitans* fimbriae. Microbiol. Immunol. *42*, 253–258.

Ireton, K. and Cossart, P. (1997) Host-pathogen interactions during entry and actin-based movement of *Listeria monocytogenes*. Annu. Rev. Genet. *31*, 113–138.

Jewett, A., Hume, W.R., Le, H., Huynh, T.N., Han, Y.W., Cheng, G., and Shi, W. (2000) Induction of apoptotic cell death in peripheral blood mononuclear and polymorphonuclear cells by an oral bacterium, *Fusobacterium nucleatum*. Infect. Immun. *68*, 1893–1898.

Kalfas, S., Tigyi, Z., Wikstrom, M., and Naidu, A.S. (1992) Laminin binding to *Prevotella intermedia*. Oral Microbiol. Immunol. *7*, 235–239.

Kamaguch, A., Nakayama, K., Ohyama, T., Watanabe, T., Okamoto, M., and Baba, H. (2001) Coaggregation of *Porphyromonas gingivalis* and *Prevotella intermedia*. Microbiol. Immunol. 45, 649–656.

Katz, J., Sambandam, V., Wu, J.H., Michalek, S.M., and Balkovetz, D.F. (2000) Characterization of *Porphyromonas gingivalis*-induced degradation of epithelial cell junctional complexes. Infect. Immun. *68*, 1441–1449.

Kesavalu, L., Chandrasekar, B., and Ebersole, J.L. (2002) In vivo induction of proinflammatory cytokines in mouse tissue by *Porphyromonas gingivalis* and *Actinobacillus actinomycetemcomitans*. Oral Microbiol. Immunol. *17*, 177–180.

Kim, S., Elkon, K.B., and Ma, X. (2004) Transcriptional suppression of interleukin-12 gene expression following phagocytosis of apoptotic cells. Immunity. *21*, 643–653.

Kishimoto, M., Yoshimura, A., Naito, M., Okamoto, K., Yamamoto, K., Golenbock, D.T., Hara, Y., and Nakayama, K. (2006) Gingipains inactivate a cell surface ligand on *Porphyromonas gingivalis* that induces

TLR2- and TLR4-independent signalling. Microbiol. Immunol. *50*, 315–325.

Kitten, T., Munro, C.L., Wang, A., and Macrina, F.L. (2002) Vaccination with FimA from *Streptococcus parasanguis* protects rats from endocarditis caused by other viridans streptococci. Infect. Immun. *70*, 422–425.

Kolenbrander, P.E. (2000) Oral microbial communities: biofilms, interactions, and genetic systems. Ann. Rev. Microbiol. *54*, 413–437.

Krisanaprakornkit, S., Kimball, J.R., Weinberg, A., Darveau, R.P., Bainbridge, B.W., and Dale, B.A. (2000) Inducible expression of human beta-defensin 2 by *Fusobacterium nucleatum* in oral epithelial cells: multiple signalling pathways and role of commensal bacteria in innate immunity and the epithelial barrier. Infect. Immun. *68*, 2907–2915.

Lamont, R. J. and Jenkinson, H.F. (1998) Life below the gum line: pathogenic mechanisms of *Porphyromonas gingivalis*. Microbiol. Mol. Biol. Rev. *62*, 1244–1263.

Leung, K., Nesbitt, W.E., Okamoto, M., and Fukushima, H. (1999) Identification of a fimbriae-associated haemagglutinin from *Prevotella intermedia*. Microb. Pathogen. *26*, 139–148.

Leung, K.P., Fukushima, H., Sagawa, H., Walker, C.B., and Clark, W.B. (1989) Surface appendages, hemagglutination, and adherence to human epithelial cells of *Bacteroides intermedius*. Oral Microbiol. Immunol. *4*, 204–210.

Li, L., Matevski, D., Aspiras, M., Ellen, R.P., and Lepine, G. (2004) Two epithelial cell invasion-related loci of the oral pathogen *Actinobacillus actinomycetemcomitans*. Oral Microbiol. Immunol. *19*, 16–25.

London, J. and Allen, J. (1990) Purification and characterization of a *Bacteroides loeschei* adhesin that interacts with procaryotic and eucaryotic cells. J Bacteriol. *172*, 2527–2534.

Lovegrove, F.E., Pena-Castillo, L., Mohammad, N., Liles, W.C., Hughes, T.R., and Kain, K.C. (2006) Simultaneous host and parasite expression profiling identifies tissue-specific transcriptional programs associated with susceptibility or resistance to experimental cerebral malaria. BMC. Genomics *7*, 295.

Lu, X., Kurago, Z., and Brogden, K.A. (2006) Effects of polymicrobial communities on host immunity and response. FEMS Microbiol. Lett. *265*, 141–150.

Manch-Citron, J.N., Allen, J., Moos, M., Jr., and London, J. (1992) The gene encoding a *Prevotella loescheii* lectin-like adhesin contains an interrupted sequence which causes a frameshift. J Bacteriol. *174*, 7328–7336.

Mans, J.J., Baker, H.V., Oda, D., Lamont, R.J., and Handfield, M. (2006) Distinctive characteristics of transcriptional profiles from two epithelial cell lines upon interaction with *Actinobacillus actinomycetemcomitans*. Oral Microbiol. Immunol. *21*, 261–267.

Matsushita K, Nagaoka S Arakaki R Kawabata Y Iki K Kawagoe M Takada H.(1994) Immunobiological activities of a 55-kilodalton cell surface protein of *Prevotella intermedia* ATCC 25611. Infect Immun. *62*, 2459–2469. 1994.

Meehan, M., Nowlan, P., and Owen, P. (1998) Affinity purification and characterization of a fibrinogen-binding protein complex which protects mice against lethal challenge with *Streptococcus equi* subsp. *equi*. Microbiology *144*, 993–1003.

Melmed, G., Thomas, L.S., Lee, N., Tesfay, S.Y., Lukasek, K., Michelsen, K.S., Zhou, Y., Hu, B., Arditi, M., and Abreu, M.T. (2003) Human intestinal epithelial cells are broadly unresponsive to Toll-like receptor 2-dependent bacterial ligands: implications for host-microbial interactions in the gut. J. Immunol. *170*, 1406–1415.

Mintz, K.P. (2004) Identification of an extracellular matrix protein adhesin, EmaA, which mediates the adhesion of *Actinobacillus actinomycetemcomitans* to collagen. Microbiology *150*, 2677–2688.

Miyake, K. (2006) Roles for accessory molecules in microbial recognition by Toll-like receptors. J. Endotoxin. Res. *12*, 195–204.

Miyazaki, H., Patel, V., Wang, H., Edmunds, R.K., Gutkind, J.S., and Yeudall, W.A. (2006) Down-regulation of CXCL5 inhibits squamous carcinogenesis. Cancer Res. *66*, 4279–4284.

Nair, B.C., Mayberry, W.R., Dziak, R., Chen, P.B., Levine, M.J., and Hausmann, E. (1983) Biological effects of a purified lipopolysaccharide from *Bacteroides gingivalis*. J. Periodont. Res. *18*, 40–49.

Nakamura, T., Kido, J., Kido, R., Ohishi, K., Yamauchi, N., Kataoka, M., and Nagata, T. (2000) The association of calprotectin level in gingival crevicular fluid with gingival index and the activities of collagenase and aspartate aminotransferase in adult periodontitis patients. J. Periodontol. *71*, 361–367.

Navarro, L., Alto, N.M., and Dixon, J.E. (2005) Functions of the *Yersinia* effector proteins in inhibiting host immune responses. Curr. Opin. Microbiol. *8*, 21–27.

Newman, E.B., Budman, L.I., Chan, E.C., Greene, R.C., Lin, R.T., Woldringh, C.L., and D'Ari, R. (1998) Lack of S-adenosylmethionine results in a cell division defect in *Escherichia coli*. J. Bacteriol. *180*, 3614–3619.

Okamoto, M., Maeda, N., Kondo, K., and Leung, K.P. (1999) Hemolytic and hemagglutinating activities of *Prevotella intermedia* and *Prevotella nigrescens*. FEMS Microbiol. Lett. *178*, 299–304.

Ozaki, M., Miyake, Y., Shirakawa, M., Takemoto, T., Okamoto, H., and Suginaka, H. (1990) Binding specificity of *Fusobacterium nucleatum* to human erythrocytes, polymorphonuclear leukocytes, fibroblasts, and HeLa cells. J. Periodontal Res. *25*, 129–134.

Park, J.M., Greten, F.R., Li, Z.W., and Karin, M. (2002) Macrophage apoptosis by anthrax lethal factor through p38 MAP kinase inhibition. Science *297*, 2048–2051.

Park, Y., Simionato, M.R., Sekiya, K., Murakami, Y., James, D., Chen, W., Hackett, M., Yoshimura, F., Demuth, D.R., and Lamont, R.J. (2005) Short fimbriae of *Porphyromonas gingivalis* and their role in coadhesion with *Streptococcus gordonii*. Infect. Immun. *73*, 3983–3989.

Park, Y., Yilmaz, O., Jung, I.Y., and Lamont, R.J. (2004) Identification of *Porphyromonas gingivalis* genes specifically expressed in human gingival epithelial cells by using differential display reverse transcription-PCR. Infect. Immun. 72, 3752–3758.

Paster, B.J., Boches, S.K., Galvin, J.L., Ericson, R.E., Lau, C.N., Levanos, V.A., Sahasrabudhe, A., and Dewhirst,

F.E. (2001) Bacterial diversity in human subgingival plaque. J. Bacteriol. *183*, 3770–3783.

Pike, R.N., Potempa, J., McGraw, W., Coetzer, T.H., and Travis, J. (1996) Characterization of the binding activities of proteinase-adhesin complexes from *Porphyromonas gingivalis*. J. Bacteriol. *178*, 2876–2882.

Platt, N., da Silva, R.P., and Gordon, S. (1999) Class A scavenger receptors and the phagocytosis of apoptotic cells. Immunol. Lett. *65*, 15–19.

Rakoff-Nahoum, S., Paglino, J., Eslami-Varzaneh, F., Edberg, S., and Medzhitov, R. (2004) Recognition of commensal microflora by toll-like receptors is required for intestinal homeostasis. Cell *118*, 229–241.

Ramos, H.C., Rumbo, M., and Sirard, J.C. (2004) Bacterial flagellins: mediators of pathogenicity and host immune responses in mucosa. Trends Microbiol. *12*, 509–517.

Reife, R.A., Coats, S.R., Al-Qutub, M., Dixon, D.M., Braham, P.A., Billharz, R.J., Howald, W.N., and Darveau, R.P. (2006) *Porphyromonas gingivalis* lipopolysaccharide lipid A heterogeneity: differential activities of tetra- and penta-acylated lipid A structures on E-selectin expression and TLR4 recognition. Cell Microbiol. *8*, 857–868.

Rodrigues, P.H. and Progulske-Fox, A. (2005) Gene expression profile analysis of *Porphyromonas gingivalis* during invasion of human coronary artery endothelial cells. Infect. Immun. *73*, 6169–6173.

Rose, J.E., Meyer, D.H., and Fives-Taylor, P.M. (2003) *Aae*, an autotransporter involved in adhesion of *Actinobacillus actinomycetemcomitans* to epithelial cells. Infect. Immun. *71*, 2384–2393.

Rudney, J.D., Chen, R., and Sedgewick, G.J. (2005) *Actinobacillus actinomycetemcomitans*, *Porphyromonas gingivalis*, and *Tannerella forsythensis* are components of a polymicrobial intracellular flora within human buccal cells. J Dent. Res. *84*, 59–63.

Ruiz, T., Lenox, C., Radermacher, M., and Mintz, K.P. (2006) Novel surface structures are associated with the adhesion of *Actinobacillus actinomycetemcomitans* to collagen. Infect. Immun. *74*, 6163–6170.

Schreiner, H.C., Sinatra, K., Kaplan, J.B., Furgang, D., Kachlany, S.C., Planet, P.J., Perez, B.A., Figurski, D.H., and Fine, D.H. (2003) Tight-adherence genes of *Actinobacillus actinomycetemcomitans* are required for virulence in a rat model. Proc. Natl. Acad. Sci. U. S. A *100*, 7295–7300.

Sciotti, M.A., Yamodo, I., Klein, J.P., and Ogier, J.A. (1997) The N-terminal half part of the oral streptococcal antigen I/IIf contains two distinct binding domains. FEMS Microbiol. Lett. *153*, 439–445.

Scragg, M.A., Alsam, A., Rangarajan, M., Slaney, J.M., Shepherd, P., Williams, D.M., and Curtis, M.A. (2002) Nuclear targeting of *Porphyromonas gingivalis* W50 protease in epithelial cells. Infect. Immun. *70*, 5740–5750.

Sela, M.N. (2001) Role of *Treponema denticola* in periodontal diseases. Crit Rev. Oral Biol. Med. *12*, 399–413.

Seong, S.Y. and Matzinger, P. (2004) Hydrophobicity: an ancient damage-associated molecular pattern that initiates innate immune responses. Nat. Rev. Immunol. *4*, 469–478.

Seymour, G.J. and Gemmell, E. (2001) Cytokines in periodontal disease: where to from here? Acta Odontol. Scand. *59*, 167–173.

Sharma, A., Inagaki, S., Honma, K., Sfintescu, C., Baker, P.J., and Evans, R.T. (2005) *Tannerella forsythia*-induced alveolar bone loss in mice involves leucine-rich-repeat BspA protein. J. Dent. Res. *84*, 462–467.

Sharma, A., Sojar, H.T., Glurich, I., Honma, K., Kuramitsu, H.K., and Genco, R.J. (1998) Cloning, expression, and sequencing of a cell surface antigen containing a leucine-rich repeat motif from *Bacteroides forsythus* ATCC 43037. Infect. Immun. *66*, 5703–5710.

Shenker, B.J. and Datar, S. (1995) Fusobacterium nucleatum inhibits human T-cell activation by arresting cells in the mid-G1 phase of the cell cycle. Infect. Immun. *63*, 4830–4836.

Sismey-Durrant, H.J. and Hopps, R.M. (1991) Effect of lipopolysaccharide from *Porphyromonas gingivalis* on prostaglandin E2 and interleukin-1-beta release from rat periosteal and human gingival fibroblasts *in vitro*. Oral Microbiol. Immunol. 6, 378–380.

Smith, E.J., Marie, I., Prakash, A., Garcia-Sastre, A., and Levy, D.E. (2001) IRF3 and IRF7 phosphorylation in virus-infected cells does not require double-stranded RNA-dependent protein kinase R or Ikappa B kinase but is blocked by Vaccinia virus E3L protein. J. Biol. Chem. *276*, 8951–8957.

Song, H., Belanger, M., Whitlock, J., Kozarov, E., and Progulske-Fox, A. (2005) Hemagglutinin B is involved in the adherence of *Porphyromonas gingivalis* to human coronary artery endothelial cells. Infect. Immun. *73*, 7267–7273.

Stack, J., Haga, I.R., Schroder, M., Bartlett, N.W., Maloney, G., Reading, P.C., Fitzgerald, K.A., Smith, G.L., and Bowie, A.G. (2005) Vaccinia virus protein A46R targets multiple Toll-like-interleukin-1 receptor adaptors and contributes to virulence. J. Exp. Med. *201*, 1007–1018.

Steinbakk, M., Naess-Andresen, C.F., Lingaas, E., Dale, I., Brandtzaeg, P., and Fagerhol, M.K. (1990) Antimicrobial actions of calcium binding leucocyte L1 protein, calprotectin. Lancet 336, 763–765.

Tada, H., Sugawara, S., Nemoto, E., Imamura, T., Potempa, J., Travis, J., Shimauchi, H., and Takada, H. (2003) Proteolysis of ICAM-1 on human oral epithelial cells by gingipains. J. Dent. Res. 82, 796–801.

Tamura, M., Tokuda, M., Nagaoka, S., and Takada, H. (1992) Lipopolysaccharides of *Bacteroides intermedius (Prevotella intermedia)* and *Bacteroides (Porphyromonas) gingivalis* induce interleukin-8 gene expression in human gingival fibroblast cultures [published erratum appears in Infect Immun 1993 Jan; 61(1):368]. Infect. Immun. *60*, 4932–4937.

Theilade, E.(1990) Factors controlling the microflora of the healthy mouth. In Human Microbial Ecology, M.J. Hill and P.D. Marsh, eds (Boca Raton, USA: CRC Press, Inc.), pp1–56.

Tournier, J. N. and Quesnel-Hellmann, A. (2006) Host-pathogen interactions: a biological rendez-vous of the infectious nonself and danger models? PLoS. Pathog. *2*, e44.

Trinchieri, G. and Sher, A. (2007) Cooperation of Toll-like receptor signals in innate immune defence. Nat. Rev. Immunol. *7*, 179–190.

Uehara, A., Sugawara, S., Tamai, R., and Takada, H. (2001) Contrasting responses of human gingival and colonic epithelial cells to lipopolysaccharides, lipoteichoic acids and peptidoglycans in the presence of soluble CD14. Med. Microbiol. Immunol. *189*, 185–192.

Uehara, A., Sugawara, Y., Kurata, S., Fujimoto, Y., Fukase, K., Kusumoto, S., Satta, Y., Sasano, T., Sugawara, S., and Takada, H. (2005) Chemically synthesized pathogen-associated molecular patterns increase the expression of peptidoglycan recognition proteins via toll-like receptors, NOD1 and NOD2 in human oral epithelial cells. Cell Microbiol. *7*, 675–686.

Uehara, A. and Takada, H. (2007) Functional TLRs and NODs in Human Gingival Fibroblasts. J. Dent. Res. *86*, 249–254.

Uitto, V.J., Overall, C.M., and McCulloch, C. (2003) Proteolytic host cell enzymes in gingival crevice fluid. Periodontol. 2000. *31*, 77–104.

Umemoto, T., Nakatani, Y., Nakamura, Y., and Namikawa, I. (1993) Fibronectin-binding proteins of a human oral spirochete *Treponema denticola*. Microbiol. Immunol. *37*, 75–78.

Underhill, D.M. and Ozinsky, A. (2002) Phagocytosis of microbes: complexity in action. Annu. Rev. Immunol. *20*, 825–852.

Urnowey, S., Ansai, T., Bitko, V., Nakayama, K., Takehara, T., and Barik, S. (2006) Temporal activation of anti- and pro-apoptotic factors in human gingival fibroblasts infected with the periodontal pathogen, *Porphyromonas gingivalis*: potential role of bacterial proteases in host signalling. BMC. Microbiol. *6*, 26.

Van Berkel, T.J., Out, R., Hoekstra, M., Kuiper, J., Biessen, E., and Van, E.M. (2005) Scavenger receptors: friend or foe in atherosclerosis? Curr. Opin. Lipidol. *16*, 525–535.

Vanhaesebroeck, B., Leevers, S.J., Ahmadi, K., Timms, J., Katso, R., Driscoll, P.C., Woscholski, R., Parker, P.J., and Waterfield, M.D. (2001) Synthesis and function of 3-phosphorylated inositol lipids. Annu. Rev. Biochem. *70*, 535–602.

Wang, P.L., Shinohara, M., Murakawa, N., Endo, M., Sakata, S., Okamura, M., and Ohura, K. (1999) Effect of cysteine protease of *Porphyromonas gingivalis* on adhesion molecules in gingival epithelial cells. Jpn. J. Pharmacol. *80*, 75–79.

Weinberg, A. and Holt, S.C. (1991) Chemical and biological activities of a 64-kilodalton outer sheath protein from *Treponema denticola* strains. J. Bacteriol. *173*, 6935–6947.

Wells, J.M., Wilson, P.W., Norton, P.W., Gasson, M.J., and LePage, R. W.F. (1993) *Lactococcus lactis*: high level expression of tetanus toxin fragment C and protection against lethal challenge. Molec. Microbiol. *8*, 1155–1162.

Winkler, J.R., John, S.R., Kramer, R.H., Hoover, C.I., and Murray, P.A. (1987) Attachment of oral bacteria to a basement-membrane-like matrix and to purified matrix proteins. Infect. Immun. *55*, 2721–2726.

Yilmaz, O., Verbeke, P., Lamont, R.J., and Ojcius, D.M. (2006) Intercellular spreading of *Porphyromonas gingivalis* infection in primary gingival epithelial cells. Infect. Immun. *74*, 703–710.

Yoneyama, M. and Fujita, T. (2004) RIG-I: critical regulator for virus-induced innate immunity. Tanpakushitsu Kakusan Koso *49*, 2571–2578.

Yu, F., Iyer, D., Anaya, C., and Lewis, J.P. (2006) Identification and characterization of a cell surface protein of *Prevotella intermedia* 17 with broad-spectrum binding activity for extracellular matrix proteins. Proteomics. *6*, 6023–6032.

Zhang, Y., Wang, T., Chen, W., Yilmaz, O., Park, Y., Jung, I.Y., Hackett, M., and Lamont, R.J. (2005). Differential protein expression by *Porphyromonas gingivalis* in response to secreted epithelial cell components. Proteomics 5, 198–211.

Zhou, Q., Desta, T., Fenton, M., Graves, D.T., and Amar, S. (2005) Cytokine profiling of macrophages exposed to *Porphyromonas gingivalis*, its lipopolysaccharide, or its FimA protein. Infect. Immun. *73*, 935–943.

The Molecular Basis of Host Defence Mechanisms in Oral Disease 10

Alex B. Berezow, Lijian Jin, and Richard P. Darveau

Abstract

The innate defence status of clinically healthy periodontal tissue is highly orchestrated. For example, histochemical, immunohistochemical, and *in situ* studies examining clinically healthy human periodontal tissue have revealed that neutrophils routinely transit through periodontal tissue from the highly vascularized tissue surrounding the tooth root surface to the gingival crevice located outside of periodontal tissue. Several innate defence response mediators including IL-8, ICAM, and E-selectin, have been shown to be expressed in clinically healthy tissue and have been proposed to facilitate neutrophil trafficking. Both host developmental programs and the host protective response to oral bacterial commensal colonization contribute to the highly organized tissue structure and molecular expression of select innate defence components.

This chapter will provide evidence for the contribution of oral commensal bacteria to the protective state found in clinically healthy tissue. Using bacterial lipopolysaccharide, a major pattern-recognition receptor ligand as an example, we will describe how the innate host response, formerly considered to be non-specific, can display an exquisite specificity. We demonstrate how this specificity is mediated by the binding and transfer characteristics of innate host defence proteins and may facilitate selective responses to oral commensal bacteria. It is then proposed that this specificity-facilitated development of a specific commensal microflora contributes to the protective state found in the host.

The complex oral microbial ecosystem in humans

It has been well established that dental plaque exists as a microbial biofilm characterized by varying areas of high and low bacterial biomass interlaced with aqueous channels (Costerton and Lewandowski, 1997; Costerton *et al.*, 1995; Darveau *et al.*, 1997; Marsh and Bradshaw, 1995). The crucial properties of biofilm bacteria include cell-to-cell communication through quorum sensing or so-called 'cell density-mediated gene expression', gene transfer, regulation of gene expression, and antimicrobial resistance (Tatakis and Kumar, 2005; see also Chapter 4 of this book). It is estimated that at least 10^{14} commensal microbes reside in a form of biofilm on the surfaces of skin, teeth, dentures, dental restorations, prosthetic implants, as well as the mucosal epithelial lining of the oral cavity and the respiratory, gastrointestinal, and urinary tracts (Cohen and Slavkin, 2000). The oral cavity contains over six billion microbes representing 500–700 species (Paster *et al.*, 2001; Socransky and Haffajee, 1994; Wilson *et al.*, 1997). Up to 300 oral bacterial species can be cultured from oral plaque samples. It has been shown that a large number of oral bacterial species are not detectable by using the traditional culture methods. Recent studies, based on acquisition of phylogenetically useful microbial sequences (e.g. 16S rRNA gene), have revealed a great deal of previously unsuspected bacterial and archaeal diversity. In recent years, the human subgingival crevice has been intensely studied as an endogenous microbial niche. Unfortunately, cultivating microorganisms woefully under-represents the true extent of microbial diversity. The best current approach for exploring unidentified species is based on isolating DNA from the target environment, PCR amplifying the rRNA gene, cloning the amplicons into *E. coli*, and sequencing the cloned 16S rRNA gene inserts (Paster *et al.*, 2001). Recently, a number of novel oral phylotypes have been identified from periodontitis samples by molecular phylogenetic methods. These include methanogenic *Archaea*, *Desulfobulbus*, *Deferribacteres*, and *Bacteroides* (Dewhirst *et al.*, 2000; Kroes *et al.*, 1999; Kulik *et al.*, 2001; Kumar *et al.*, 2003; Lepp *et al.*, 2004; Paster *et al.*, 2001).

Innate host defence status in periodontal health

The innate defence status of oral tissues is well organized and regulated as well. It is known that healthy periodontal tissues contain a unique expression of highly selected inflammatory mediators. Early histological studies (Page and Schroeder, 1976) showed that these tissues contained a cellular infiltrate located in juxtaposition to the colonized tooth surface (Kornman *et al.*, 1997). A portion of this cellular infiltrate has been described as forming a 'wall' of neutrophils precisely located between bacteria and junctional epithelium, the epithelial cell surface closest to the dental plaque biofilm (Kornman *et al.*, 1997). The constant migration of neutrophils from the vasculature to the tissue is regulated by different types of adhesion molecules expressed on endothelial cells. These include endothelial cell-adhesion molecule 1 (ECAM-1), vascular cell-adhesion molecule 1 (VCAM-1) and intercellular adhesion molecule 1 (ICAM-1) (Kornman *et al.*, 1997). The migration of neutrophils through the junctional epithelium towards the gingival sulcus is guided by ICAM-1 and IL-8 (a neutrophil chemoattractant) expressed in the junctional epithelium in an apico-coronal direction (Crawford, 1992; Tonetti *et al.*, 1994). IL-8 expression was greatest at the most superficial junctional epithelial cell layers, and the levels of ICAM-1 increased toward areas exposed to bacterial challenges. These data have indicated that neutrophil infiltration and migration into the gingival sulcus are not due to random diffusion; on the contrary, this is a highly organized process that results in the recruitment of neutrophils to the area. Consistent with these observations, it is noteworthy that the connective tissue supporting the junctional epithelium is structurally different from that supporting the oral gingival epithelium, and inflammatory cells, such as polymorphonuclear leukocytes and T-lymphocytes, continually extravasate from the dense capillary and postcapillary venule network and migrate across the junctional epithelium into the gingival sulcus (Nanci and Bosshardt, 2006). Previous studies have supported the notion that innate host defence mediator expression in clinically healthy tissue is vital to the maintenance of periodontal health (Attström and Schroeder, 1979; Carrassi *et al.*, 1989; Hart *et al.*, 1994; MacFarlane *et al.*, 1992; Page *et al.*, 1987; Yoshinari *et al.*, 1994).

More recently, and for the first time, the profiles of common pattern recognition molecules in periodontal tissues have been described (Jin, 2006). Firstly, sCD14 was detected in gingival crevicular fluid (GCF), and its levels were significantly higher than those in serum, indicating that this soluble pattern recognition receptor (PRR) could be produced in periodontal tissues in response to microbial challenge. Furthermore, sCD14 levels in GCF were associated with fewer and shallower periodontal pockets, indicating that sCD14 may play a protective role in local host response to bacterial challenge (Jin and Darveau, 2001). The expression profile of mCD14 in gingival tissues was subsequently investigated in an extended study, which showed that mCD14 mRNA and protein were commonly detected in healthy or diseased periodontal tissues (Jin *et al.*, 2004). Expression of mCD14 was predominantly confined to the gingival epithelium-connective tissue interface. Its expression level in periodontally healthy subjects was significantly higher than in periodontitis patients. Interestingly, within the patients, clinically healthy tissues showed higher levels of mCD14 than periodontal pocket and granulation tissues, suggesting that mCD14 may play a crucial role in mediation of an effective inflammatory response to bacterial challenge. Usually, LPS binding protein (LBP) is known to be an acute-phase reactant, predominantly derived from the hepatocytes, and it may serve to both neutralize LPS and enhance its ability to activate host cells. Recently, it was found that LBP mRNA and protein are expressed in gingival epithelia with a defined expression profile confined to the cytoplasm of granular and keratinized layers of gingival epithelium, spreading from the oral sulcular epithelium to oral epithelium with the expression density decreasing gradually from coronal to apical portion. Furthermore, LBP mRNA and protein were more frequently expressed in healthy tissues than in diseased tissues (Ren *et al.*, 2004). These novel findings imply that local expression of LBP in gingival tissues might contribute to periodontal homeostasis and health through neutralization of LPS. This notion has been supported by our recent *in vitro* study showing that recombinant human LBP (rhLBP) could significantly down-regulate the *E. coli* LPS-induced expression of IL-6 mRNA and protein, as well as suppress the *E. coli* LPS-induced expression of TLR2 and 4 (Ren *et al.*, 2005). It is known that LPS-induced innate host responses are modulated by LBP, CD14, and the TLR superfamily, prior to the subsequent transduction of transmembrane signalling cascades. Our recent study further explored the potential interrelationship of *in vivo* expression of LBP and mCD14 in human gingival tissues as well as the co-expression of TLR2 and 4 in association with periodontal health and disease (Ren *et al.*, 2005). A positive correlation was found in the protein expression levels of two closely related PRRs, LBP and mCD14. The expression profiles of these PRRs seem to differ markedly in periodontally healthy and diseased conditions. In diseased tissues, TLR2 was detected in both periodontal pocket epithelia and macrophage-like cells in connective tissues, and TLR4 was predominantly detected in connective tissues. However, in healthy tissues, only a weak expression of TLR2 was found in gingival epithelia and no obvious TLR4 expression was detected. Furthermore, in diseased tissues, mCD14 was co-detected on CD68-labelled macrophages in the underlying connective tissues of the periodontal pocket epithelium as well as on CD1a-labelled dendritic cells in the interface between the pocket epithelium and connective tissue, whereas a similar expression profile could not be detected in periodontally healthy tissues isolated from periodontitis patients or healthy control subjects. These novel findings, of an altered cellular expression profile of mCD14 and TLR2 and 4 in periodontal pocket tissues, imply that these PRRs may play a crucial role in periodontal homeostasis and pathogenesis.

Expression and regulation of antimicrobial peptides in gingival epithelia

Human gingival epithelium is a stratified squamous epithelium consisting of oral gingival epithelium, oral sulcular epithelium and junctional epithelium. These three categories of epithelia differ in their microanatomy and express distinct cytokeratins and cell surface markers (Crawford, 1992; Mackenzie *et al.*, 1991). Human gingival epithelium, especially the oral sulcular epithelium and junctional epithelium, which are constantly exposed to bacteria and bacterial products, form the first line of host defence in periodontal tissues.

It is believed that gingival epithelium has evolved specific mechanisms to identify and control microbial challenge. Currently, it is well recognized that gingival epithelia serve not merely as physical barriers to microbial challenges but also as active biological barriers by suppressing the propagation of pathogens through production of antimicrobial peptides. These are a diverse group of peptides, which directly kill a broad spectrum of bacteria, fungi, and viruses by acting as natural antibiotics. In addition, they also modify the local inflammatory response and activate host adaptive immunity. Antimicrobial peptides can be divided into several categories on the basis of their structures but most of them maintain certain common structural features, including a cationic charge and the ability to interact with bacterial membranes through hydrophobic amino acids. The antimicrobial mechanisms of these peptides are diverse and they include: fatal depolarization of the normally energized bacterial membrane; creation of physical holes that cause leakage of cellular contents; induction of hydrolases that degrade the cell wall, scrambling the usual distribution of lipids between the leaflets of the bilayer (which results in disturbance of membrane functions); and damaging critical intracellular targets after internalization of the peptides (Zasloff, 2002).

Cathelicidins (e.g. LL-37) and defensins (e.g. human α- and β-defensins) are the major groups of antimicrobial peptides in gingival epithelium; and other peptides in the gingival epithelium include proteinase inhibitors, chemokines, and neutrophil peptides. Constitutive and inducible expression of human cathelicidin, as well as human β-defensin, has been observed in gingival epithelium (Dale et *al.*, 2001). In recent years, human α- and β-defensins have been detected in oral and gingival epithelia, by our group and others, and are considered to be of major importance in innate host defence (Dale, 2002; Lu *et al.*, 2004; Lu *et al.*, 2005). In gingival epithelium, human α-defensins-1, -2, -3 and -4, also termed 'human neutrophil peptides -1, -2, -3 and -4', and LL-37 can be detected in neutrophils that migrate through the junctional epithelium, while human β-defensins (hBDs) are usually detected in keratinocytes of the oral sulcular and gingival epithelia. It is believed that this specific distribution pattern of these antimicrobial proteins in various regions of the gingival epithelium may act to synergize the host defence against microbial challenge (Dale *et al.*, 2001). Recently, we further described the expression patterns of hBDs-1, -2, and -3 in both periodontal health and disease (Lu *et al.*, 2004; Lu *et al.*, 2005). The expression of both hBD-1 and -2 peptides was predominantly confined to the granular and spinous layers of gingival epithelium, while hBD-3 peptides were mainly detected in the basal layer in healthy tissue but their expression did extend from the basal layer to the spinous layers in diseased tissue. These novel findings of hBD-1, -2, and -3 expression patterns in gingival epithelia are significant, indicating the establishment of a functional synergy within the local compartment, with an enhanced innate defence mechanism against microbial challenge. Moreover, hBD-3 peptide is expressed not only in gingival keratinocytes but also in Langerhans' cells and Merkel cells. Additionally, periodontal tissues from healthy controls expressed higher levels of hBD-2 peptides than did clinically healthy tissues from periodontitis patients. Taken together, our findings suggest that appropriate expression of hBDs-1 and -2 may contribute to maintenance of periodontal homeostasis, while hBD-3 peptides contribute to periodontal homeostasis by their antimicrobial activity and promotion of adaptive immune responses (Lu *et al.*, 2004; Lu *et al.*, 2005).

The regulatory mechanism of defensin expression is currently under intensive investigation. Emerging evidence shows that gingival epithelial cells contain bacterial recognition mechanisms that may facilitate discrimination between different bacterial species. It has been shown that commensal bacteria provide host protection by consistently stimulating epithelial surfaces to express antimicrobial peptides at levels that kill opportunistic or pathogenic organisms (Boman, 2000). Interestingly, both a commensal oral bacterium, *Fusobacterium nucleatum*, and a known key periodontopathogen, *Porphyromonas gingivalis*, can elicit β-defensin-2 expression in gingival epithelial cells. However, *F. nucleatum* elicits NF-κB activation, whereas *P. gingivalis* does not (Krisanaprakornkit *et al.*, 2000). Furthermore, another similar study showed that epithelial cells obtained from different body sites utilized two different pathways to facilitate β-defensin-2 secretion, prompting the authors to conclude that 'epithelial cells from different body sites have common signalling mechanisms to distinguish between commensal and pathogenic bacteria' (Chung and Dale, 2004). In agreement with this, IL-8 secretion in response to oral commensal bacteria and *P. gingivalis* is significantly different (Darveau *et al.*, 1998; Huang *et al.*, 2001; Madianos *et al.*, 1997). Although whole commensal bacteria (Huang *et al.*, 2001) or cell wall extracts (Huang *et al.*, 2001; Kornman *et al.*, 1997) elicit secretion of IL-8, *P. gingivalis* does not elicit IL-8 accumulation (Huang *et al.*, 2001; Vankeerberghen *et al.*, 2005). Moreover, it inhibits the IL-8 response to other oral commensal microbes (Darveau *et al.*, 1998; Huang *et al.*, 2001; Madianos *et al.*, 1997),a phenomenon now known as 'local chemokine paralysis' (Darveau *et al.*, 1998). All of this evidence further underscores the ability of epithelial cells to distinguish between commensal and pathogenic bacteria.

The induction of hBD expression in many tissues is mediated by TLRs. Activation of TLR2 by bacterial lipopeptide on tracheobronchial epithelial cells induces hBD-2 mRNA and peptide accumulation (Hertz *et al.*, 2003). Both TLR2 and TLR4 have been linked to inducible hBD-2 expression in intestinal epithelial cells stimulated with peptidoglycan or LPS (Vora *et al.*, 2004) and in keratinocytes stimulated with *Propionibacterium acnes* (Nagy *et al.*, 2005), suggesting that TLRs are involved in monitoring epithelial surfaces to enhance host defence by inducing the production of hBDs. Our recent pilot study showed that *P. gingivalis* LPS could enhance the *in vitro* expression of hBD mRNAs in gingival epithelia, probably through the involvement of TLR2 and 4; however, *in vivo*, *P. gingivalis* might evade innate defence surveillance by interfering with hBD-2 peptide production (Jin *et al.*, 2006). Interestingly, in an extended study on the expression of hBDs-1, -2, and -3 during experimental oral candidiasis, we found that oral epithelia possess mechanisms for sensing the early invasion of *Candida albicans* through recognition of the yeast's hyphae and proteinases. The oral epithelia respond to the insult by producing antimicrobial peptides (Lu *et al.*, 2006). Moreover, protease-activated receptors, which belong to the seven-transmembrane domain G protein-coupled receptor family, are involved in regulation of hBD-2. The induction of hBD-2 by *P. gingivalis* proteases in cultured gingival epithelial cells may be partially mediated by protease-activated receptor-2 and the resultant signalling pathway (Chung *et al.*, 2004).

Furthermore, epithelial cells obtained from sites in contact with commensal bacteria (including gingival epithelial cells) do not secrete inflammatory mediators in response to TLR2 and TLR4 ligands, such as lipotechoic acid (LTA) and LPS (Abreu *et al.*, 2001; Bocker *et al.*, 2003; Hedlund *et al.*, 1999; Melmed *et al.*, 2003; Uehara *et al.*, 2005). While epithelial cells do not produce inflammatory mediators in response to these microbial-associated molecular patterns (MAMPs), gingival epithelial cells utilize TLR2 and TLR4, as well as a newly discovered class of peptidoglycan recognition receptor, to induce β-defensin-2 secretion in response to MAMPs such as LTA and LPS (Sugawara *et al.*, 2006; Uehara *et al.*, 2005). Therefore, the ability of gingival epithelial cells to discriminate between commensal and pathogenic bacteria, combined with their unresponsiveness to MAMPs (with respect to inflammatory mediator secretion), is consistent with a 'screening' function that contributes to the highly orchestrated expression of select innate defence components in clinically healthy periodontal tissue. However, it has been shown that the expression levels of both TLR2 and TLR4 are significantly increased in the gingival epithelium tissue obtained from periodontitis sites (Ren *et al.*, 2005; Sugawara *et al.*, 2006), which may significantly alter gingival epithelial cell responses. For example, intestinal epithelial cells have also been shown to be unresponsive to the major MAMPs (e.g. LTA and LPS) in that these bacterial components do not elicit inflammatory mediator secretion (Abreu *et al.*, 2001; Melmed *et al.*, 2003). However, transient transfection of intestinal epithelial cells with either TLR2 (Melmed *et al.*, 2003) or TLR4 plus MD-2 (Abreu *et al.*, 2001) to increase their TLR expression levels has been shown to significantly alter epithelial cell responses, resulting in NF-κB activation in response to LTA (Melmed *et al.*, 2003) or LPS (Abreu *et al.*, 2001). It is, therefore, possible that the 'screening' characteristic of healthy gingival epithelial cells may be altered when the TLR expression level is increased.

The innate host response to LPS is specific

The innate immune response, previously and perhaps erroneously referred to as 'non-specific immunity', in fact relies upon host recognition of specific, conserved bacterial features. Examples of these conserved bacterial structures, commonly referred to as MAMPs (microbial-associated molecular patterns), include lipopolysaccharide (LPS), lipopeptides, flagellin, and double-stranded RNA (Kawai and Akira, 2005). These conserved structures are common among large groups of bacteria, including both Gram-positive and Gram-negative bacteria, as well as viruses. The reason for this structural conservation is due to the fact that these structures are essential for the microorganism to remain viable. Therefore, attempts to modify them generally result in microbial death or loss of virulence. Because of this, the host is able to recognize many different microbes by using a relatively small number of pattern recognition receptors (PRRs), and this idea forms the basis of the 'pattern recognition hypothesis' (Janeway, 1992). This hypothesis provides the conceptual framework for understanding specificity in the innate host response. Although certain microbial structures are invariant due to their essential nature for viability of the microorganism, bacteria have acquired mechanisms to modify MAMPs structurally, such that they still function for the microorganism, yet elicit significantly different innate host responses. The most studied class of PRRs are the Toll-like receptors (TLRs), which are discussed in more detail below. To demonstrate the specificity of the innate immune response, this section will focus upon the recognition of LPS by TLR4.

Lipopolysaccharide (LPS) structure

Lipopolysaccharide (LPS) is a large, complex molecule located in the outer leaflet of the outer membrane of Gram-negative bacteria. It can be divided into three general subcomponents: (1) O-Antigen; (2) Outer and inner core; and (3) Lipid A. The O-Antigen is a highly variable polysaccharide region that determines the serotype of any particular Gram-negative bacterial species. It consists of between 0 and ~50 repeating subunits, each one consisting of between one and eight glycosyl residues (Erridge *et al.*, 2002). The O-Antigen of some bacteria provides special advantages. For instance, in *Helicobacter pylori*, the etiologic agent of gastric ulcers, a phenomenon known as 'phase variation' allows the bacterium to alter the content of its O-Antigen, presumably as a way of allowing the organism to adapt to the microenvironments encountered in the host (Appelmelk *et al.*, 1998). In other bacteria, such as *Salmonella*, the O-Antigen prevents formation of the complement membrane activation complex (Erridge *et al.*, 2002). Comparatively, the outer core is more variable than the inner core but both core structures are much less variable than the O-Antigen. The outer core is often made up of hexoses, such as glucose and galactose, while the inner core consists of more unusual sugars, such as 3-deoxy-D-manno-octulosonic acid (Kdo) and heptoses (Erridge *et al.*, 2002). In general, the core region (as well as the Lipid A region) is considered essential to bacterial viability. Lipid A, also known as endotoxin, is the biologically active component of LPS, responsible for inducing septic shock. It is also the least variable subcomponent of the LPS molecule. Lipid A is composed of a glucosamine disaccharide backbone, to which fatty acids are attached via amide- and ester-linkages. The 'canonical' lipid A structure, represented by lipid A isolated from *Escherichia coli*, consists of a hexa-acylated, 1,4′-biphosphorylated glucosamine disaccharide (Dixon and Darveau, 2005). Of the six fatty acids, four are directly attached to the disaccharide backbone (primary linkage), while the other two are attached to hydroxyl groups on the primary fatty acids (secondary linkage) (Dixon and Darveau, 2005). However, not all lipid A structures are identical to that isolated from *E. coli*. Lipid A structures show variability in the number, length, and composition of the attached fatty acids, as well as variability in the level of phosphorylation and number and types of substituted groups found attached to the phosphates (Erridge *et al.*, 2002). While these slight alterations may seem trivial, they can have a profound impact on the host's ability to recognize and respond to LPS.

Host recognition of LPS

The ability of the host to properly clear Gram-negative bacterial infections largely depends upon the host's recognition of LPS and production of an inflammatory response. A host lacking the ability to respond to LPS is, therefore, extremely susceptible to infection by certain Gram-negative pathogens. As an example, the LD_{50} for *Salmonella* infection is 1000-fold less for TLR4-deficient mice as compared to wild-type mice (Roy *et al.*, 2006).

The first step in recognition of LPS involves a serum protein called LPS-binding protein (LBP) (Fig. 10.1). This 60-kDa glycoprotein is produced mainly by the liver (Zweigner *et al.*, 2006). The folded structure of LBP is similar to a 'boomerang,' and each side of the 'boomerang' contains a lipid-binding pocket (Beamer *et al.*, 1999). In other words, the N-terminal portion of LBP (one side of the 'boomerang') and the C-terminal portion of LBP (the other side of the 'boomerang') can each bind lipid moieties. Furthermore, the tip of the N-terminal domain of LBP contains a cluster of positively charged amino acids (Zweigner *et al.*, 2006). Evidence such as this has led to the conclusion that the N-terminal domain binds LPS (Beamer *et al.*, 1999). This model is consistent with the structure of lipid A, the subcomponent of LPS that LBP specifically binds. The disaccharide backbone of lipid A is negatively charged due to the presence of phosphate groups, and presumably, the cluster of positively charged amino acids in the N-terminal domain of LBP would therefore bind to it. The non-polar fatty acids covalently linked to the disaccharide backbone would fit into the lipid-binding pockets in the N- and C-terminal domains. The transfer of LPS to the next molecule in the recognition pathway is performed by the C-terminal domain of LBP, since it was discovered that LBP truncated at Ile-197 was able to bind LPS but was unable to transfer

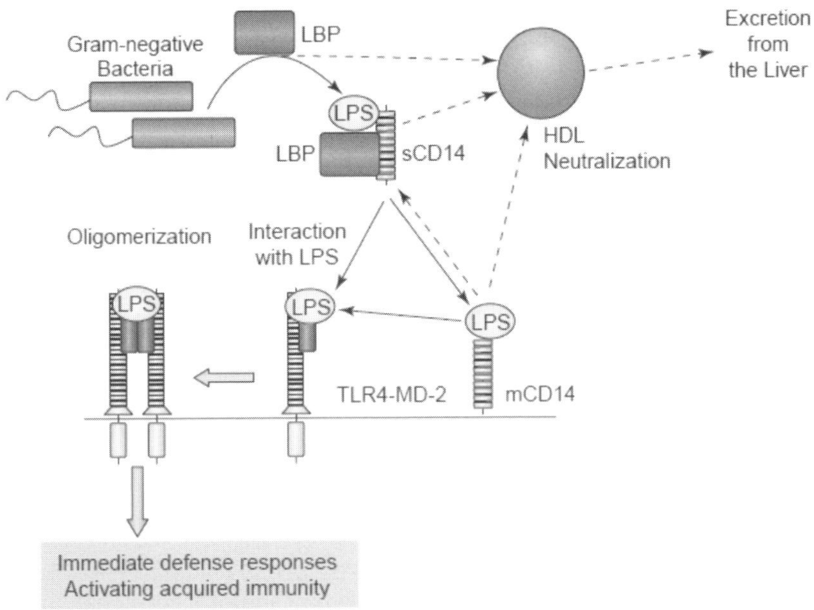

Figure 10.1 Lipopolysaccharide (LPS) processing, signalling and clearance. LPS in the outer membrane of Gram-negative bacteria has to be processed by LPS-binding protein (LBP) and CD14. LBP removes LPS from the outer membrane and forms complexes consisting of LPS, LBP and soluble CD14 (sCD14). These complexes deliver LPS to membrane CD14 for LPS signalling. LPS processing for signalling is depicted by solid lines. LPS is neutralized by its transfer to high-density lipoprotein (HDL) from LBP or CD14. LPS clearance is shown by dotted lines. LPS bound to CD14 is transferred to Toll like receptor (TLR)4-MD-2, which undergoes oligomerization and triggers a signal. The signal via TLR activates immediate and acquired immune responses. (Taken from Miyake, 2004, with permission.) A colour version of this figure is located in the plate section at the back of the book.

it to CD14 (Beamer *et al.*, 1999; Han *et al.*, 1994).

The role of LBP, therefore, is to bind LPS and transfer it to CD14, a protein that can be found either in the serum (soluble CD14, sCD14) or bound to the outer membrane of primarily monocytes and neutrophils via a glycosylphosphatidylinositol anchor (membrane CD14, mCD14). CD14 appears to be extremely specific for smooth LPS (i.e. LPS containing an O-Antigen), as it has been shown that LPS, truncated in its carbohydrate region, was less able to stimulate TNF-α secretion from CD14$^{+/+}$ macrophages, when compared to smooth LPS (Gangloff *et al.*, 2005). Interestingly, CD14$^{-/-}$ macrophages were not only less sensitive to LPS stimulation but also lost their specificity for smooth LPS (Gangloff *et al.*, 2005). It has also been speculated that less-immunostimulatory LPS, such as that isolated from *Porphyromonas gingivalis* and *Helicobacter pylori,* is due to an inability to bind CD14 as efficiently as *E. coli* LPS (Cunningham *et al.*, 2000). Not only do these LPS structures differ in their carbohydrate regions but they also show considerable variation in their lipid A moieties. It is likely that these alterations contribute to the decreased potency of *P. gingivalis* and *H. pylori* LPS (Cunningham *et al.*, 2000).

In order to elucidate the regions of CD14 responsible for binding LPS, single-amino acid replacement mutations were constructed (Cunningham *et al.*, 2000). It was found that replacement of N-terminal hydrophilic residues (in particular, arginine, glutamate, and aspartate) resulted in a dramatic decrease in the ability of CD14 to stimulate an immune response as measured by IL-8 secretion from human umbilical vein endothelial cells (HUVECs) (Cunningham *et al.*, 2000). Additionally, it was also discovered that these CD14 mutant proteins were less able to bind LPS (Cunningham *et al.*, 2000). Recently, CD14 was crystallized and the structural data validated the findings of the mutation experiments. The crystal structure of CD14 showed that a hydrophobic pocket located in the N-terminus of the protein was primarily responsible for binding LPS (Kim *et al.*, 2005). However, surrounding this hydrophobic pocket were several hydrophilic regions, each of which was also important for LPS binding (Cunningham *et al.*, 2000; Kim *et al.*, 2005). Thus, it appears that CD14 is able to 'capture' LPS through both polar and non-polar interactions. Furthermore, mutagenesis experiments defined another region close to this hydrophobic pocket responsible for transferring LPS from CD14 to the TLR4/MD-2 complex (Kim *et al.*, 2005).

Once transferred from CD14, the binding of LPS to the TLR4/MD-2 complex initiates the inflammatory cascade. Evidence points toward MD-2 as being the protein that binds LPS. While many believe that TLR4 is capable of directly interacting with LPS, experimental proof is lacking (Visintin *et al.*, 2006). However, it has been shown that not only does MD-2 bind LPS, it is also able to discriminate between immunostimulatory (agonistic) LPS and non-immunostimulatory (antagonistic) LPS (Muroi and Tanamoto, 2006). This was discovered following the observation of species-specific differences in MD-2. For instance, murine MD-2 can bind and trigger a strong immune response to compound 506 (typical, agonistic lipid A) and also to lipid IVa (tetra-acylated, antagonistic lipid A). However, human MD-2 triggers an immune response to compound 506 but not to Lipid IVa. Further research indicated that the ability of murine MD-2 to initiate an immune response to both lipid A structures was due to three amino acids (Thr-57, Val-61, and Glu-122), and that murine MD-2 lost the ability to stimulate an immune response to lipid IVa if these amino acids were replaced with the corresponding human MD-2 sequence (Muroi and Tanamoto, 2006). Thus, these three amino acids are pivotal in the recognition of a given lipid A structure as agonistic or antagonistic by MD-2 (Muroi and Tanamoto, 2006). In agreement with this, other research has shown that under-acylated LPS, such as that from *P. gingivalis*, is able to antagonize *E. coli* LPS by competing for the same binding site on human MD-2 (Coats *et al.*, 2007).

As stated above, evidence supporting a direct interaction between TLR4 and LPS is scarce. However, it has recently been shown that the hypervariable region of TLR4 may play a role in LPS recognition. By using chimeric TLR4 constructs, made from different regions of human and murine TLR4, the 82-amino acid hypervariable region of human TLR4 appears to be responsible for discriminating between hexa- and penta-acylated *Pseudomonas aeruginosa* lipid A (Hajjar *et al.*, 2002). Whether or not this differential recognition is due to direct binding of TLR4 to LPS or binding of TLR4 to MD-2 is still unclear (Hajjar *et al.*, 2002).

It is widely accepted that TLR4 cannot transmit an LPS signal in the absence of MD-2 (Nishitani *et al.*, 2006). Thus, the ability of TLR4 to bind MD-2 is vital for proper signalling, and it was shown that the region of TLR4 interacting with MD-2 is Glu-24–Lys-47 (Nishitani *et al.*, 2006). It is believed that the binding of LPS to the TLR4/MD-2 complex triggers receptor clustering. In support of this, evidence indicates that while one region of MD-2 is responsible for binding LPS, another is responsible for regulating ligand-induced receptor clustering (Kobayashi *et al.*, 2006). Clustering of the TLR4/MD-2 complexes then initiates the intracellular signal that results in inflammation.

Much research has shown that LPS structures derived from different bacterial species exhibit varying abilities to stimulate an inflammatory response via TLR4/MD-2. For instance, it was shown that *E. coli* LPS was able to stimulate a strong E-selectin response from HUVECs, while *P. gingivalis* LPS was not (Coats *et al.*, 2003). It has been speculated that the reason for this differential response is due to the varying abilities of different lipid A structural forms to induce TLR4/MD-2 receptor clustering (Chen *et al.*, 2007). For example, it is suggested that *E. coli* lipid A is able to strongly induce TLR4/MD-2 dimerization, thus allowing for a strong, frequent signal to induce the inflammatory cascade. However, it is believed that lipid A isolated from *P. gingivalis* only causes weak TLR4/MD-2 dimerization and, as a result, only a weak induction of the inflammatory cascade ensues. In support of this view, it was recently shown using DNA microarray analysis of HUVECs stimulated with various LPS preparations, that under-acylated, less potent LPS species (e.g. *P. gingivalis* LPS) activated only a subset of genes that were activated by the fully acylated, highly potent LPS species (e.g. *E. coli* LPS) (Chen *et al.*, 2007). This observed 'hierarchy' of gene activation lends substantial credibility to the hypothesis that the reason under-acylated LPS species are less potent is due to their inability to stimulate strong TLR4/MD-2 dimerization, and subsequently, a selective inflammatory response representing a subset of the more potent *E. coli* LPS response is found (Chen *et al.*, 2007).

LPS clearance and modulation of the immune response

While inflammation is essential for eradication of an invading pathogen, prolonged inflammation can be detrimental to the health of the host. In particular, the influx of immune cells, such as neutrophils and activated macrophages, which release toxic substances that not only kill foreign invaders but also harm host cells, can significantly damage surrounding tissue. Therefore, it is in the best interests of the host to limit inflammation and attenuate the immune response once the threat has been successfully neutralized. One mechanism used by the host to achieve this is to bind and remove LPS without further stimulating the inflammatory pathway.

High-density lipoprotein (HDL) is a complex, macromolecular assembly of various lipid and protein moieties. Colloquially referred to as 'good cholesterol', HDL not only plays a role in prevention of atherosclerosis, but also has an important function in immune modulation. It appears the primary role of HDL is 'reverse cholesterol transport', in which excess lipids and cholesterol from the diet are bound and taken to the liver for degradation and excretion in bile (Murch *et al.*, 2007). LPS, which contains a lipid A moiety, can also be bound by HDL and excreted from the liver (Miyake, 2004). However, elimination of LPS by this mechanism does not stimulate an innate immune response via TLR4.

The molecular mechanisms underlying LPS clearance are complex and poorly understood. However, certain well-characterized molecules in the LPS recognition pathway, including both LBP and CD14, are known to play a role in LPS clearance as well. The current paradigm is that LBP can directly transfer LPS to HDL or it can transfer LPS indirectly to HDL by first transferring it to sCD14 (Murch *et al.*, 2007). Strangely, it appears that the indirect method of LPS clearance is favoured over the direct one (Wurfel *et al.*, 1995). Once bound by HDL, LPS loses its potent toxicity and, subsequently, the innate immune response begins to diminish (Dixon *et al.*, 2004). As a result, the host is able to strike a fine balance between inflammation to eliminate a microbial threat and prevention of unnecessary tissue damage.

In addition to binding and removing LPS, HDL also appears to have other immune modulating functions. These include attenuating expression of adhesion molecules necessary for neutrophil recruitment, reducing LPS-induced TNF-α expression, and reducing the transmigration of monocytes into tissue (Murch *et al.*, 2007). Again, all of these functions serve to modulate the immune response, thereby attempting to keep host tissue damage to a minimum. Therefore, transfer of LPS to non-inflammatory removal systems or entry of LPS into the inflammatory cascade will depend upon both the structural type of LPS and the relative amounts of host proteins locally available for interaction with the LPS. In this fashion, the innate host response also demonstrates remarkable specificity.

In conclusion, innate host recognition of LPS by three key innate host defence proteins (LBP, CD14, and TLR4) demonstrates a remarkable ability to differentiate different lipid A structures. For example, LBP and CD14 can transfer LPS to serum proteins, resulting in neutralization of the LPS without inducing an inflammatory response, or these molecules may facilitate LPS transfer to the TLR4/MD-2 complex for subsequent cellular activation. Binding affinities to LBP and transfer rates to CD14 are significantly altered by lipid A structure and this, combined with the relative amounts of these proteins and their substrates (serum HDL or cellular TLR4), will determine the extent of LPS neutralization or cellular activation. This represents one mechanism by which the innate host response displays specificity based upon lipid A structure. Exquisite specificity is also observed at the level of TLR4 / MD-2 recognition of LPS. Lipid A structures that are poorly recognized by this complex, for example those containing fewer fatty acids, elicit a limited subset of inflammatory mediators which represents a hierarchical expression pattern for TLR4 activation. These examples of innate host specificity in LPS recognition clearly demonstrate that the 'non-specific' innate host response can elicit highly specific responses.

Bacteria can modify their lipid A structure in response to environmental conditions and therefore modulate innate host responses

Several different examples of how bacteria modify their lipid A structure in response to environmental conditions have appeared in the literature. These examples underscore the ability of bacteria to capitalize on the innate host specificity described in the previous section.

For example, bacteria contain mechanisms to sense local environmental changes (Mekalanos, 1992), such as the PhoP-PhoQ sensor kinase and transcriptional activator system (Miller *et al.*, 1989), which have been shown to be an important component of bacterial pathogenesis. Indeed, in *Salmonella typhimurium* it has been demonstrated that a constitutive *S. typhimurium* mutant designated $PhoP^c$, which partially mimics the intracellular regulation state of the bacteria, contains a modified lipid A species composition (Guo *et al.*, 1997). This renders both whole bacteria and isolated LPS less potent activators of E-selectin and TNF-α when compared to wild type or another $PhoP^-$ mutant (Guo *et al.*, 1997). This was the first description of genetic regulation resulting in lipid A alterations, in turn modifying the innate host response. In addition, although the genetic basis for the modification is not yet completely understood, an environmental modification of the lipid A species composition has been found in *Yesinia pestis*, the causative agent of the bubonic plague (Kawahara *et al.*, 2002; Rebeil *et al.*, 2004). This bacterium, when incubated at 27° C (environmental temperature of its flea vector), displayed lipid A heterogeneity, containing a mix of tri-, tetra-, penta-, and hexa-acylated lipid A species. Incubation at 37° C (mammalian host temperature) significantly altered the lipid A species composition in that the hexa-acylated lipid A species was missing and the penta-acylated species was significantly reduced. These underacylated *Y. pestis* lipid A species resulted in less potent innate defence activation, consistent with the structural studies cited above with the closely related *E. coli* lipid A, and with this bacterium's strategy to evade host defences (Kawahara *et al.*, 2002).

One microenvironmental condition that may alter the interactions of *P. gingivalis* with the innate host defence system is the haemin concentration (Olczak *et al.*, 2005; Smalley *et al.*, 1996). Haemin binds host iron and represents the major iron-acquisition system for *P. gingivalis* (Olczak *et al.*, 2005). Apparently contrasting results, which may depend upon the strains examined, have been obtained (Olczak *et al.*, 2005). It is clear, nevertheless, that the concentration of haemin in the growth medium can regulate the expression of several virulence factors, including gingipains (Olczak *et al.*, 2005) and extracellular vesicle formation (Olczak *et al.*, 2005), as well as increasing the virulence of *P. gingivalis* in a mouse model of infection (Marsh *et al.*, 1994). It has been proposed that *P. gingivalis* obtains iron almost exclusively through haemin uptake *in vivo* (Olczak *et al.*, 2005). Although not experimentally verified, it is believed that the local concentration of haemin (in the form of haemoglobin) can vary considerably, depending on the state of vascular permeability, as measured by bleeding upon probing. We have recently reported that the concentration of haemin in the growth medium can significantly affect the lipid A structural composition of *P. gingivalis* LPS (Al-Qutub *et al.*, 2006). LPS isolated from *P. gingivalis* is highly heterogeneous, containing numerous lipid A types that differ in their content of lipid A fatty acids and phosphate groups (Fig. 10.2). The lipid A content of *P. gingivalis*, after growth in a medium containing 1 μg/ml haemin, consists of a single major monophosphoryl penta-acylated lipid A cluster (centred at *m/z* 1690). In contrast, the lipid A content of bacteria incubated with 10 μg/ml haemin showed a significantly reduced amount of this lipid A structure and a significant increase in both monophosphoryl tetra-acylated lipid A structures (*m/z* 1435 and 1449) and a diphosphoryl penta-acylated lipid A cluster (Fig. 10.3).

Furthermore, it appears that *P. gingivalis* can 'sense' the amount of accumulated haemin and, when a threshold level is reached, alterations in the lipid A structural content occur. Evidence

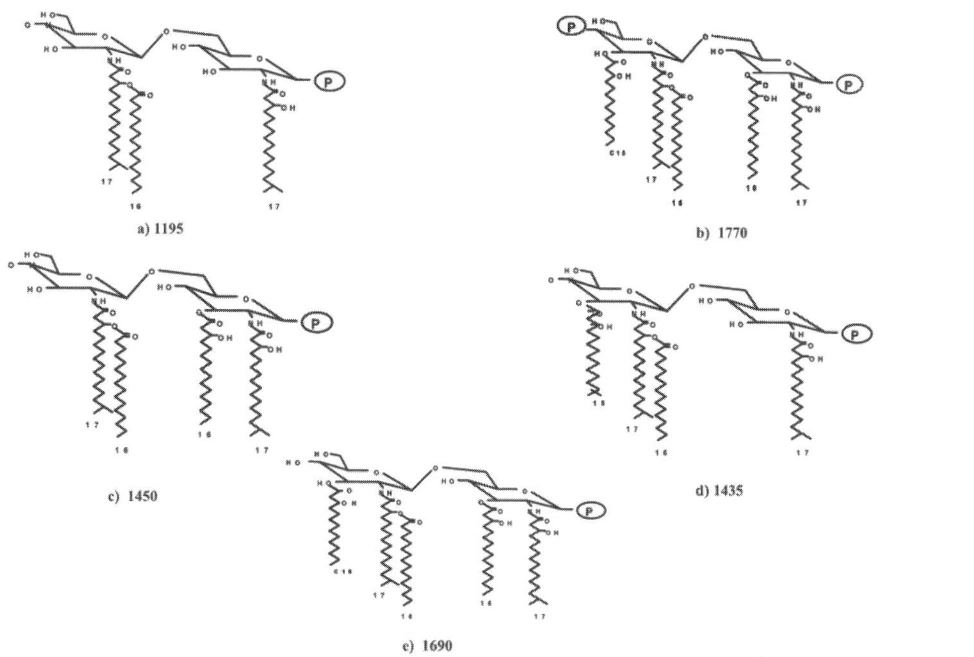

Figure 10.2 Structures of previously characterized *P. gingivalis* lipid A mass ions. *P. gingivalis* LPS contains several different lipid A structures that differ in their number of phosphate groups and fatty acids. In this figure, the MALDI-TOF lipid A peaks that have been characterized with respect to their structure are presented. (A) Structure determined by Ogawa (1993);(B to E) structures elucidated by Kumada *et al.*(1995). (Taken from Al-Qutub *et al.*, 2006, with permission.)

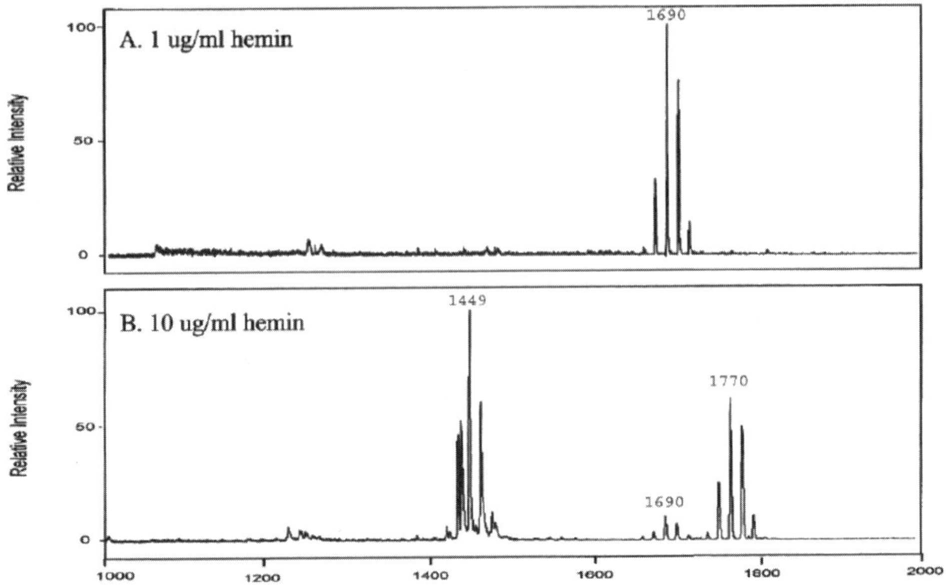

Figure 10.3 Effect of haemin concentration on lipid A content of *P. gingivalis* strain 33277. *P. gingivalis* strain 33277 was incubated in TYHK medium with either 1 (A) or 10 (B) μg/ml haemin. *P. gingivalis* LPS was obtained by the Tri reagent procedure (Yi and Hackett, 2000), and lipid A was cleaved and separated from the LPS as described by Caroff *et al.*(1988). MALDI-TOF analysis was performed as previously described (Guo *et al.*, 1997). All values given are average masses rounded to the nearest whole number for singly charged deprotonated molecules. (Taken from Al-Qutub *et al.*, 2006, with permission.)

for this comes from experiments examining the lipid A structural composition of strain WS15 and its isogenic parent strain A7436 (Simpson *et al.*, 2004). Strain WS15 demonstrates impaired haemin binding due to mutations in two haemin uptake proteins, one of which is a well-characterized protease designated Kgp, and the other, designated HmuR, is involved in haemin transport across the periplasmic space (Simpson *et al.*, 2004). It was found that, in contrast to the parental strain, the mutant strain, WS15, contained a single penta-acylated lipid A cluster centred around *m/z* 1690 when incubated with 1 or 10 μg/ml haemin and tetra-acylated lipid A structures when incubated with 20 μg/ml haemin. These data demonstrate that when haemin uptake is impaired, as in strain WS15, more haemin is required to elicit a change in the lipid A content. They also suggest that previously characterized haemin acquisition and detection proteins, such as Kgp and HmuR, contribute to the ability of *P. gingivalis* to 'sense' the environmental haemin concentration and alter the lipid A structural content accordingly. Consistent with the demonstrable effect of haemin on the lipid A structural content, two previous studies have shown that the haemin concentration can affect the potency (Champagne *et al.*, 1996) and composition (Cutler *et al.*, 1996) of *P. gingivalis* LPS. These studies were performed before MALDI-TOF analysis of lipid A structures was available, and it is highly likely that these changes in lipid A content contributed to the effects described in these publications.

The identification of haemin as a growth-medium component that can significantly alter the lipid A structural content of *P. gingivalis* has important implications for innate host responses. Specifically, it has recently been shown that the major penta-acylated lipid A structural forms found when *P. gingivalis* is grown at low haemin concentrations are TLR4 agonists and facilitate E-selectin activation on human endothelial cells (Reife *et al.*, 2006). In contrast, the tetra-acylated structural forms found at high haemin concentrations have been shown to be TLR4 antagonists (Coats *et al.*, 2005; Coats *et al.*, 2003). Therefore, increased haemin concentrations facilitate the production of lipid A structures that are TLR4 antagonists. The ability to significantly alter *P.*

gingivalis LPS interactions with TLR4 represents a form of immunomodulation (Portnoy, 2005) that may influence the homeostasis that exists between clinically healthy periodontal tissue and commensal oral bacteria.

Generation of innate host defence status in clinically healthy tissue

In the past decade, great progress has been made in further understanding host-microbe symbiosis (Wilson, 2005). Currently, it is believed that the interactions of microbes and host cells are complex and symbiotic microbe-host interactions may be more representative than pathological interactions. In fact, both humans and microbes benefit from the peaceful and dynamic co-existence and microbial colonization actually plays an obligatory role in human health (Tuomanen, 2005). We have proposed that the innate host defence status found in clinically healthy tissue occurs in response to the highly specific microbial consortium residing on the tooth surface (Roberts and Darveau, 2002). In the previous sections of this chapter, we have shown that the innate host response can display a remarkable specificity and that bacteria can capitalize upon that by altering their LPS structure to accommodate life in different host microenvironments. In the following section, we propose that this specificity has also contributed to the evolution of commensal bacterial species that interact with host periodontal tissue and elicit a 'beneficial' protective response. The best example of a beneficial protective response to commensal bacterial colonization comes from studies with germ-free mice where it has been shown that a normal functioning immune system does not develop histologically or molecularly in the absence of an intestinal commensal microbiota. Therefore, in this section, evidence that a beneficial homeostasis also exists between oral commensal bacteria and the host will be described.

Germ-free mice studies and intestinal and periodontal tissue development

A key contribution of germ-free (gnotobiotic) mice studies has been elucidating the participation of commensal bacteria in normal intestinal development (Gordon and Pesti, 1971; Hooper *et al.*, 2001; Macpherson and Harris, 2004; Umesaki and Setoyama, 2000; Xu and Gordon, 2003). Germ-free mice are completely devoid of bacteria. They are generated by sterile Caesarean section and raised aseptically in an isolator with sterile filtered air and are housed using sterile food, water, and bedding. Germ-free mice are distinct from specific-pathogen-free (SPF) mice which are devoid only of known mouse pathogens and harbour intestinal bacteria (Macpherson and Harris, 2004). Early studies in germ-free mice demonstrated that commensal bacteria have a direct impact on the morphology of the intestine (Duncan and Edberg, 1995; Falk *et al.*, 1998). Bacteria are responsible for degradation of mucus glycoproteins, and their absence results in an enlargement of the caecum. The villi of the small intestine are longer and the crypts are shorter and contain fewer cells in germ-free animals. In addition, commensal bacteria have effects on intestinal motility. Effects of bacteria can also be seen on the development and function of the intestinal immune system. Commensal bacteria are required for the complete development of Peyer's patches, the lamina propria, and the intraepithelial spaces, all three of the main immune elements found in the intestine (Duncan and Edberg, 1995; Falk *et al.*, 1998). More recently, studies in germ-free mice have revealed that the commensal bacteria induce angiogenesis, contributing to the development of the complex vascular beds found just underneath the mucosal surface (Stappenbeck *et al.*, 2002). Furthermore, it has been found that constitutive ICAM-1 expression in these vessels is also regulated by the presence of the commensal microbiota (Komatsu *et al.*, 2000). In fact the state of 'controlled' inflammation that normally exists in the intestine has been attributed to both the quality and quantity of intestinal commensal microorganisms (Cebra, 1999; Chadwick and Anderson, 1992).

In contrast, and surprisingly, we were unable to find any published studies examining the contribution of oral commensal bacteria on periodontal tissue development in germ-free mice. Therefore, our laboratory conducted a preliminary investigation comparing innate host-mediator expression in germ-free and conventionally reared mice (Dixon *et al.*, 2004). It was found that IL-1β levels were significantly higher in conventionally reared mice, consistent with the notion that commensal bacteria in the periodontium, as is the case in the intestine, actively participate in establishing the innate-mediator expression of clinically healthy or normal periodontal tissue. One of the reasons why IL-1β alone was identified as being differentially expressed in germ-free and conventionally reared mice may have been the limited number of samples examined in the pilot study. The reasons why this inflammatory mediator was elevated in healthy conventionally reared animals, compared to germ-free controls, may appear paradoxical since it has previously been associated with the development of periodontitis (Masada *et al.*, 1990). However, we suspect that the observed IL-1β expression represents a form of indirect activation of periodontal tissue by oral commensal bacteria. Indirect activation is defined as those effects which occur when a bacterium activates one cell type and then the product from that cell acts upon another cell or cell type. In contrast, direct activation occurs when the bacteria or bacterial components directly stimulate a cell to respond (e.g. monocytes, by the production of cytokines, chemokines, or cellular adhesion molecules). Both indirect and direct activation can potentially occur with both myeloid and non-myeloid cells (Gemmell and Seymour, 1993; Heath *et al.*, 1987; Lindemann *et al.*, 1995; Yamazaki *et al.*, 1992). Therefore, the presence of IL-1β in clinically healthy periodontal tissue may serve as a 'priming' mechanism for several different cell types found in the periodontium. It is therefore important to understand the global effects that oral commensal bacteria have on cells involved in maintaining the characteristic innate host defence mediator pattern observed in clinically healthy tissue.

Similarities between the mouse and human innate host defence status

Direct experimental comparisons between mouse and human innate defence mediator expression in clinically healthy periodontal tissue have not been made. However, there are several lines of indirect evidence suggesting that mice are similar to humans in that they maintain a balance between beneficial and destructive innate host activation of the periodontium. Firstly, germ-free mice have been used as a model for periodontal disease since it was found that these mice were more susceptible to bone loss than were conventionally reared animals (Crawford *et al.*, 1978). Similarly, the *P. gingivalis* gavage model, described by Baker *et al.* (Hart *et al.*, 2004), requires antibiotic administration to reduce the commensal bacterial load before *P. gingivalis* is administered. In both of these periodontitis disease models, the ecological competition for niche-occupation by the normal oral flora is absent or significantly reduced. It is likely that this has contributed to the success of these disease models. However, since we have shown that mouse oral commensal bacteria actively interact with mouse periodontal tissue as evidenced by increased IL-1β in conventionally reared mice (Dixon *et al.*, 2004), it is possible that removal of the commensal flora also reduces the protective function of periodontal tissue, further contributing to the success of these disease models. Secondly, studies in P- and E-selectin double-knockout mice (Niederman *et al.*, 2001) and IL-10 knockout mice (Al-Rasheed *et al.*, 2003) have shown that these animals spontaneously develop periodontitis. As described above, low level E-selectin is found in clinically healthy human periodontal tissue (Gemmell *et al.*, 1994; Moughal *et al.*, 1992; Nylander *et al.*, 1993; Tonetti, 1997; Tonetti *et al.*, 1994) and is believed to contribute to neutrophil transit though the periodontium (Tonetti, 1997). Periodontitis did not develop in P- and E-selectin knockout mice given continuous antibiotic administration (Niederman *et al.*, 2001), confirming the contribution of oral bacteria to disease. Therefore, spontaneous disease in this model is due to a disruption in the healthy balance between the oral commensal flora and periodontal tissue. Similarly, although in the reverse direction, the spontaneous disease observed in IL-10 knockout mice is due to the lack of this potent anti-inflammatory molecule. In this instance, the hypothesis is that the lack of the suppressive effects of IL-10 result in over-expression of innate host components made in response to oral bacteria, eventually leading to disease. Although systemic antibiotic administration is needed to test this hypothesis, the development of disease in the absence of IL-10 demonstrates that balance between pro- and anti-inflammatory mediators is needed to maintain periodontal health. Further evidence that increased host-mediated inflammation contributes to disease was obtained in an IL-1β over-producing transgenic mouse strain (Dayan *et al.*, 2004). In these mice, periodontitis develops with or without the continuous administration of systemic antibiotics (Dayan *et al.*, 2004), demonstrating that it is possible to develop periodontitis in the absence of a bacterial component during excessive host inflammatory mediator production.

Conclusions

The specificity of the innate host response is believed to contribute to the ability of bacteria to modify their PRR ligands to modulate innate host responses, facilitating the highly protective response observed in clinically healthy periodontal tissue.

References

Abreu, M.T., Vora, P., Faure, E., Thomas, L.S., Arnold, E.T., and Arditi, M. (2001) Decreased expression of Toll-like receptor-4 and MD-2 correlates with intestinal epithelial cell protection against dysregulated proinflammatory gene expression in response to bacterial lipopolysaccharide. J. Immunol. *167*, 1609–1616.

Al-Qutub, M.N., Braham, P.H., Karimi-Naser, L.M., Liu, X., Genco, C.A., and Darveau, R.P. (2006) Hemin-dependent modulation of the lipid A structure of *Porphyromonas gingivalis* lipopolysaccharide. Infect. Immun. *74*, 4474–4485.

Al-Rasheed, A., Scheerens, H., Rennick, D.M., Fletcher, H.M., and Tatakis, D.N. (2003) Accelerated alveolar bone loss in mice lacking interleukin-10. J. Dent. Res. *82*, 632–635.

Appelmelk, B.J., Shiberu, B., Trinks, C., Tapsi, N., Zheng, P.Y., Verboom, T., Maaskant, J., Hokke, C.H., Schiphorst, W.E., Blanchard, D., *et al.* (1998) Phase variation in *Helicobacter pylori* lipopolysaccharide. Infect. Immun. *66*, 70–76.

Attström, R., and Schroeder, H.E. (1979) Effect of experimental neutropenia on initial gingivitis in dogs. Scand. J. Dent. Res. *87*, 7–23.

Beamer, L.J., Carroll, S.F., and Eisenberg, D. (1999) The three-dimensional structure of human bactericidal/permeability-increasing protein: implications for understanding protein-lipopolysaccharide interactions. Biochem. Pharmacol. *57*, 225–229.

Bocker, U., Yezerskyy, O., Feick, P., Manigold, T., Panja, A., Kalina, U., Herweck, F., Rossol, S., and Singer, M.V. (2003) Responsiveness of intestinal epithelial cell lines to lipopolysaccharide is correlated with Toll-like receptor 4 but not Toll-like receptor 2 or CD14 expression. Int. J. Colorectal Dis. *18*, 25–32.

Boman, H.G. (2000) Innate immunity and the normal microflora. Immunol. Rev. *173*, 5–16.

Caroff, M., Tacken, A., and Szabó, L. (1988) Detergent-accelerated hydrolysis of bacterial endotoxins and determination of the anomeric configuration of the glycosyl phosphate present in the 'isolated lipid A' fragment of the *Bordetella pertussis* endotoxin. Carbohyd. Res. *175*, 273–282.

Carrassi, A., Abati, S., Santarelli, G., and Vogel, G. (1989) Periodontitis in a patient with chronic neutropenia. J. Periodontol. *60*, 352–357.

Cebra, J.J. (1999) Influences of microbiota on intestinal immune system development. Am. J. Clin. Nutr. *69*, 1046S-1051S.

Chadwick, V.S., and Anderson, R.P. (1992) Microorganisms and their products in inflammatory bowel disease., In Inflammatory Bowel Disease, R. P. MacDermott, and W. F. Stenson, eds. (Amsterdam: Elsevier Science), pp. 241–258.

Champagne, C.M., Holt, S.C., Van Dyke, T.E., Gordon, B.J., and Shapira, L. (1996) Lipopolysaccharide isolated from *Porphyromonas gingivalis* grown in hemin-limited chemostat conditions has a reduced capacity for human neutrophil priming. Oral Microbiol. Immunol. *11*, 319–325.

Chen, C., Coats, S.R., Bumgarner, R.E., and Darveau, R.P. (2007) Hierarchical gene expression profiles of HUVEC stimulated by different lipid A structures obtained from *Porphyromonas gingivalis* and *Escherichia coli*. Cell. Microbiol. 9, 1028–1038.

Chung, W.O., and Dale, B.A. (2004) Innate immune response of oral and foreskin keratinocytes: utilization of different signalling pathways by various bacterial species. Infect. Immun. *72*, 352–358.

Chung, W.O., Hansen, S.R., Rao, D., and Dale, B.A. (2004) Protease-activated receptor signalling increases epithelial antimicrobial peptide expression. J. Immunol. *173*, 5165–5170.

Coats, S.R., Do, C.T., Karimi-Naser, L.M., Braham, P.H., and Darveau, R.P. (2007) Antagonistic lipopolysaccharides block *E. coli* lipopolysaccharide function at human TLR4 via interaction with the human MD-2 lipopolysaccharide binding site. Cell. Microbiol. *9*, 1191–1202.

Coats, S.R., Pham, T.T., Bainbridge, B.W., Reife, R.A., and Darveau, R.P. (2005) MD-2 mediates the ability of tetra-acylated and penta-acylated lipopolysaccharides to antagonize *Escherichia coli* lipopolysaccharide at the TLR4 signalling complex. J. Immunol. *175*, 4490–4498.

Coats, S.R., Reife, R.A., Bainbridge, B.W., Pham, T.T., and Darveau, R.P. (2003) *Porphyromonas gingivalis* lipopolysaccharide antagonizes *Escherichia coli* lipopolysaccharide at toll-like receptor 4 in human endothelial cells. Infect. Immun. *71*, 6799–6807.

Cohen, D., and Slavkin, H. (2000) Periodontal disease and systemic disease., In Periodontal medicine, L. F. Rose, D. W. Cohen and R. J. Genco, eds. (Hamilton (Canada): B. C. Decker Inc), pp. 1–10.

Costerton, J.W., and Lewandowski, Z. (1997) The biofilm lifestyle. Adv. Dent. Res. *11*, 192–195.

Costerton, J.W., Lewandowski, Z., Caldwell, D.E., Korber, D.R., and Lappin-Scott, H.M. (1995) Microbial biofilms. Ann. Rev. Microbiol. *49*, 711–745.

Crawford, J.M. (1992) Distribution of ICAM-1, LFA-3 and HLA-DR in healthy and diseased gingival tissues. J. Periodont. Res. *27*, 291–298.

Crawford, J.M., Taubman, M.A., and Smith, D.J. (1978) The natural history of periodontal bone loss in germfree and gnotobiotic rats infected with periodontopathic microorganisms. J. Periodont. Res. *13*, 316–325.

Cunningham, M.D., Shapiro, R.A., Seachord, C., Ratcliffe, K., Cassiano, L., and Darveau, R.P. (2000) CD14 employs hydrophilic regions to 'capture' lipopolysaccharides. J. Immunol. *164*, 3255–3263.

Cutler, C.W., Eke, P.I., Genco, C.A., Van Dyke, T.E., and Arnold, R.R. (1996) Hemin-induced modifications of the antigenicity and hemin-binding capacity of *Porphyromonas gingivalis* lipopolysaccharide. Infect. Immun. *64*, 2282–2287.

Dale, B.A. (2002) Periodontal epithelium: a newly recognized role in health and disease. Periodontol. 2000 *30*, 70–78.

Dale, B.A., Kimball, J.R., Krisanaprakornkit, S., Roberts, F., Robinovitch, M., O'Neal, R., Valore, E.V., Ganz, T., Anderson, G.M., and Weinberg, A. (2001) Localized antimicrobial peptide expression in human gingiva. J Periodont. Res. *36*, 285–294.

Darveau, R.P., Belton, C.M., Reife, R.A., and Lamont, R.J. (1998) Local chemokine paralysis, a novel pathogenic mechanism for *Porphyromonas gingivalis*. Infect. Immun. *66*, 1660–1665.

Darveau, R.P., Tanner, A., and Page, R.C. (1997) The microbial challenge in periodontitis. Periodontol. 2000 *14*, 12–32.

Dayan, S., Stashenko, P., Niederman, R., and Kupper, T.S. (2004) Oral epithelial overexpression of IL-1alpha causes periodontal disease. J. Dent. Res. *83*, 786–790.

Dewhirst, F.E., Tamer, M.A., Ericson, R.E., Lau, C.N., Levanos, V.A., Boches, S.K., Galvin, J.L., and Paster, B.J. (2000) The diversity of periodontal spirochetes by 16S rRNA analysis. Oral Microbiol. Immunol. *15*, 196–202.

Dixon, D.R., Bainbridge, B.W., and Darveau, R.P. (2004) Modulation of the innate immune response within the periodontium. Periodontol. 2000 *35*, 53–74.

Dixon, D.R., and Darveau, R.P. (2005) Lipopolysaccharide heterogeneity: innate host responses to bacterial modification of lipid A structure. J. Dent. Res. *84*, 584–595.

Dixon, D.R., Reife, R.A., Cebra, J.J., and Darveau, R.P. (2004) Commensal bacteria influence innate status within gingival tissues: a pilot study. J. Periodontol. *75*, 1486–1492.

Duncan, H.E., and Edberg, S.C. (1995) Host-microbe interaction in the gastrointestinal tract. Crit. Rev. Microbiol. *21*, 85–100.

Erridge, C., Bennett-Guerrero, E., and Poxton, I.R. (2002) Structure and function of lipopolysaccharides. Microbes Infect. *4*, 837–851.

Falk, P.G., Hooper, L.V., Midtvedt, T., and Gordon, J.I. (1998) Creating and maintaining the gastrointestinal ecosystem: what we know and need to know from gnotobiology. Microbiol. Mol. Biol. Rev. *62*, 1157–1170.

Gangloff, S.C., Zahringer, U., Blondin, C., Guenounou, M., Silver, J., and Goyert, S.M. (2005) Influence of CD14 on ligand interactions between lipopolysaccharide and its receptor complex. J. Immunol. *175*, 3940–3945.

Gemmell, E., and Seymour, G.J. (1993) Interleukin 1, interleukin 6 and transforming growth factor-beta production by human gingival mononuclear cells following stimulation with *Porphyromonas gingivalis* and *Fusobacterium nucleatum*. J. Periodont. Res. *28*, 122–129.

Gemmell, E., Walsh, L.J., Savage, N.W., and Seymore, G.J. (1994) Adhesion molecule expression in chronic inflammatory periodontal disease tissue. J. Periodont. Res. *29*, 46–53.

Gordon, H.A., and Pesti, L. (1971) The gnotobiotic animal as a tool in the study of host microbial relationships. Bacteriol. Rev. *35*, 390–429.

Guo, L., Lim, K.B., Gunn, J.S., Bainbridge, B., Darveau, R.P., Hackett, M., and Miller, S.I. (1997) Regulation of lipid A modifications by *Salmonella typhimurium* virulence genes *phoP-phoQ*. Science *276*, 250–253.

Hajjar, A.M., Ernst, R.K., Tsai, J.H., Wilson, C.B., and Miller, S.I. (2002) Human Toll-like receptor 4 recognizes host-specific LPS modifications. Nat. Immunol. *3*, 354–359.

Han, J., Mathison, J.C., Ulevitch, R.J., and Tobias, P.S. (1994) Lipopolysaccharide (LPS) binding protein, truncated at Ile-197, binds LPS but does not transfer LPS to CD14. J. Biol. Chem. *269*, 8172–8175.

Hart, G.T., Shaffer, D.J., Akilesh, S., Brown, A.C., Moran, L., Roopenian, D.C., and Baker, P.J. (2004) Quantitative gene expression profiling implicates genes for susceptibility and resistance to alveolar bone loss. Infect. Immun. *72*, 4471–4479.

Hart, T.C., Shapira, L., and Van Dyke, T.E. (1994) Neutrophil defects as risk factors for periodontal diseases. J. Periodontol. *65*, 521–529.

Heath, J.K., Atkinson, S.J., Hembry, R.M., Reynolds, J.J., and Meikle, M.C. (1987) Bacterial antigens induce collagenase and prostaglandin E2 synthesis in human gingival fibroblasts through a primary effect on circulating mononuclear cells. Infect. Immun. *55*, 2148–2154.

Hedlund, M., Wachtler, C., Johansson, E., Hang, L., Somerville, J.E., Darveau, R.P., and Svanborg, C. (1999) P fimbriae-dependent, lipopolysaccharide-independent activation of epithelial cytokine responses. Mol. Microbiol. *33*, 693–703.

Hertz, C.J., Wu, Q., Porter, E.M., Zhang, Y.J., Weismuller, K.H., Godowski, P.J., Ganz, T., Randell, S.H., and Modlin, R.L. (2003) Activation of Toll-like receptor 2 on human tracheobronchial epithelial cells induces the antimicrobial peptide human beta defensin-2. J. Immunol. *171*, 6820–6826.

Hooper, L.V., Wong, M.H., Thelin, A., Hansson, L., Falk, P.G., and Gordon, J.I. (2001) Molecular analysis of commensal host-microbial relationships in the intestine. Science *291*, 881–884.

Huang, G.T., Kim, D., Lee, J.K., Kuramitsu, H.K., and Haake, S.K. (2001) Interleukin-8 and intercellular adhesion molecule 1 regulation in oral epithelial cells by selected periodontal bacteria: multiple effects of *Porphyromonas gingivalis* via antagonistic mechanisms. Infect. Immun. *69*, 1364–1372.

Janeway, C.J. (1992) The immune system evolved to discriminate infectious nonself from noninfectious self. Immunol. Today *13*, 11–16.

Jin, L. (2006) Research advances in periodontal etiopathology., In Research advances and clinical practice in periodontics: Bridging the gap., P. Bartold, I. Ishikawa, and N. Vergel de Dios, eds. (Adelaide: Asian Pacific Society of Periodontology), pp. 26–38.

Jin, L., and Darveau, R.P. (2001) Soluble CD14 levels in gingival crevicular fluid of subjects with untreated adult periodontitis. J. Periodontol. *72*, 634–640.

Jin, L., Lu, Q., Darveau, R., and Samaranayake, L. (2006) *P. gingivalis* LPS modulates antimicrobial peptide expression in gingival epithelia. J. Dent. Res. *85*, 898.

Jin, L., Ren, L., Leung, W.K., and Darveau, R.P. (2004) The *in vivo* expression of membrane-bound CD14 in periodontal health and disease. J. Periodontol. *75*, 578–585.

Kawahara, K., Tsukano, H., Watanabe, H., Lindner, B., and Matsuura, M. (2002) Modification of the structure and activity of lipid A in *Yersinia pestis* lipopolysaccharide by growth temperature. Infect. Immun. *70*, 4092–4098.

Kawai, T., and Akira, S. (2005) Pathogen recognition with Toll-like receptors. Curr. Opin. Immunol. *17*, 338–344.

Kim, J.I., Lee, C.J., Jin, M.S., Lee, C.H., Paik, S.G., Lee, H., and Lee, J.O. (2005) Crystal structure of CD14 and its implications for lipopolysaccharide signalling. J. Biol. Chem. *280*, 11347–11351.

Kobayashi, M., Saitoh, S., Tanimura, N., Takahashi, K., Kawasaki, K., Nishijima, M., Fujimoto, Y., Fukase, K., Akashi-Takamura, S., and Miyake, K. (2006) Regulatory roles for MD-2 and TLR4 in ligand-induced receptor clustering. J. Immunol. *176*, 6211–6218.

Komatsu, S., Berg, R.D., Russell, J.M., Nimura, Y., and Granger, D.N. (2000) Enteric microflora contribute to constitutive ICAM-1 expression on vascular endothelial cells. Am. J. Physiol. Gastrointest. Liver Physiol. *279*, G186–191.

Kornman, K.S., Page R.C., and Tonetti, M.S. (1997) The host response to the microbial challenge in periodontitis: assembling the players. Periodontol. 2000 *14*, 112–143.

Krisanaprakornkit, S., Kimball, J.R., Weinberg, A., Darveau, R.P., Bainbridge, B.W., and Dale, B.A. (2000) Inducible Expression of Human beta-Defensin 2 by *Fusobacterium nucleatum* in Oral Epithelial Cells: Multiple Signalling Pathways and Role of Commensal Bacteria in Innate Immunity and the Epithelial Barrier. Infect. Immun. *68*, 2907–2915.

Kroes, I., Lepp, P.W., and Relman, D.A. (1999) Bacterial diversity within the human subgingival crevice. Proc. Natl Acad. Sci. USA *96*, 14547–14552.

Kulik, E.M., Sandmeier, H., Hinni, K., and Meyer, J. (2001) Identification of archaeal rDNA from subgingival dental plaque by PCR amplification and sequence analysis. FEMS Microbiol. Lett. *196*, 129–133.

Kumada, H., Haishima, Y., Umemoto, T., and Tanamoto, K.-I. (1995) Structural study on the free lipid A isolated from lipopolysaccharide of *Porphyromonas gingivalis*. J. Bacteriol. *177*, 2098–2106.

Kumar, P.S., Griffen, A.L., Barton, J.A., Paster, B.J., Moeschberger, M.L., and Leys, E.J. (2003) New bacterial species associated with chronic periodontitis. J. Dent. Res. *82*, 338–344.

Lepp, P.W., Brinig, M.M., Ouverney, C.C., Palm, K., Armitage, G.C., and Relman, D.A. (2004) Methanogenic Archaea and human periodontal disease. Proc. Natl. Acad. Sci. USA *101*, 6176–6181.

Lindemann, R.A., Kjeldsen, M., and Cabret, M. (1995) Effect of whole oral bacteria and extracted lipopolysaccharides on peripheral blood leukocyte interleukin-2 receptor expression. J. Periodont. Res. *30*, 264–271.

Lu, Q., Jayatilake, J.A., Samaranayake, L.P., and Jin, L. (2006) Hyphal invasion of *Candida albicans* inhibits the expression of human beta-defensins in experimental oral candidiasis. J. Invest. Dermatol. *126*, 2049–2056.

Lu, Q., Jin, L., Darveau, R.P., and Samaranayake, L.P. (2004) Expression of human beta-defensins-1 and -2 peptides in unresolved chronic periodontitis. J. Periodont. Res. *39*, 221–227.

Lu, Q., Samaranayake, L.P., Darveau, R.P., and Jin, L. (2005) Expression of human beta-defensin-3 in gingival epithelia. J. Periodont. Res. *40*, 474–481.

MacFarlane, G.D., Herzberg, M.C., Wolff, L.F., and Hardie, N.A. (1992) Refractory periodontitis associated with abnormal polymorphonuclear leukocyte phagocytosis and cigarette smoking. J. Periodont. *63*, 908–913.

Mackenzie, I.C., Rittman, G., Gao, Z., Leigh, I., and Lane, E.B. (1991) Patterns of cytokeratin expression in human gingival epithelia. J. Periodont. Res. *26*, 468–478.

Macpherson, A.J., and Harris, N.L. (2004) Interactions between commensal intestinal bacteria and the immune system. Nat. Rev. Immunol. *4*, 478–485.

Madianos, P.N., Papapanou, P.N., and Sandros, J. (1997) *Porphyromonas gingivalis* infection of oral epithelium inhibits neutrophil transepithelial migration. Infect. Immun. *65*, 3983–3990.

Marsh, P.D., and Bradshaw, D.J. (1995) Dental plaque as a biofilm. J. Ind. Microbiol. *15*, 169–175.

Marsh, P.D., McDermid, A.S., McKee, A.S., and Baskerville, A. (1994) The effect of growth rate and haemin on the virulence and proteolytic activity of *Porphyromonas gingivalis* W50. Microbiology *140 (Pt 4)*, 861–865.

Masada, M.P., Persson, R., Kenney, J.S., Lee, S.W., Page, R.C., and Allison, A.C. (1990) Measurement of interleukin-1 alpha and -1 beta in gingival crevicular fluid: implications for the pathogenesis of periodontal disease. J. Periodont. Res. *25*, 156–163.

Mekalanos, J.J. (1992) Environmental signals controlling expression of virulence determinants in bacteria. J. Bacteriol. *174*, 1–7.

Melmed, G., Thomas, L.S., Lee, N., Tesfay, S.Y., Lukasek, K., Michelsen, K.S., Zhou, Y., Hu, B., Arditi, M., and Abreu, M.T. (2003) Human intestinal epithelial cells are broadly unresponsive to Toll-like receptor 2-dependent bacterial ligands: implications for host-microbial interactions in the gut. J. Immunol. *170*, 1406–1415.

Miller, S.I., Kukral, A.M., and Mekalanos, J.J. (1989) A two-component regulatory system (phoP phoQ) controls *Salmonella typhimurium* virulence. Proc. Natl. Acad. Sci. U S. A. *86*, 5054–5058.

Miyake, K. (2004) Innate recognition of lipopolysaccharide by Toll-like receptor 4-MD-2. Trends Microbiol. *12*, 186–192.

Moughal, N.A., Adonogianaki, E., Thornhill, M.K., and Kinane, D.F. (1992) Endothelial cell leukocyte adhesion molecule-1 (ELAM-1) and intercellular adhesion molecule-1 (ICAM-1) expression in gingival tissue during health and experimentally-induced gingivitis. J. Periodont. *27*, 623–630.

Murch, O., Collin, M., Hinds, C.J., and Thiemermann, C. (2007) Lipoproteins in inflammation and sepsis. I. Basic science. Intensive Care Med. *33*, 13–24.

Muroi, M., and Tanamoto, K. (2006) Structural regions of MD-2 that determine the agonist-antagonist activity of lipid IVa. J. Biol. Chem. *281*, 5484–5491.

Nagy, I., Pivarcsi, A., Koreck, A., Szell, M., Urban, E., and Kemeny, L. (2005) Distinct strains of *Propionibacterium acnes* induce selective human beta-defensin-2 and interleukin-8 expression in human keratinocytes through toll-like receptors. J. Invest. Dermatol. *124*, 931–938.

Nanci, A., and Bosshardt, D.D. (2006) Structure of periodontal tissues in health and disease. Periodontol. 2000 *40*, 11–28.

Niederman, R., Westernoff, T., Lee, C., Mark, L.L., Kawashima, N., Ullman-Culler, M., Dewhirst, F.E., Paster, B.J., Wagner, D.D., Mayadas, T., *et al.* (2001) Infection-mediated early-onset periodontal disease in P/E-selectin-deficient mice. J. Clin. Periodontol. *28*, 569–575.

Nishitani, C., Mitsuzawa, H., Sano, H., Shimizu, T., Matsushima, N., and Kuroki, Y. (2006) Toll-like receptor 4 region Glu24-Lys47 is a site for MD-2 binding: importance of CYS29 and CYS40. J. Biol. Chem. *281*, 38322–38329.

Nylander, K., Danielsen, B., Fejerskov, O., and Dabelsteen, E. (1993) Expression of the endothelial leukocyte adhesion molecule-1 (ELAM-1) on endothelial cells in experimental gingivitis in humans. J. Periodontol. *64*, 355–357.

Ogawa, T. (1993) Chemical structure of lipid A from *Porphyromonas (Bacteroides) gingivalis* lipopolysaccharide. FEBS Lett. *332*, 197–201.

Olczak, T., Simpson, W., Liu, X., and Genco, C.A. (2005) Iron and heme utilization in *Porphyromonas gingivalis*. FEMS Microbiol. Rev. *29*, 119–144.

Page, R.C., Beatty, P., and Waldrop, T.C. (1987) Molecular basis for the functional abnormality in neutrophils from patients with generalized prepubertal periodontitis. J. Periodont. Res. *22*, 182–183.

Page, R.C., and Schroeder, H.E. (1976) Pathogenesis of inflammatory periodontal disease. A summary of current work. [Review]. Lab. Invest. *33*, 235–249.

Paster, B.J., Boches, S.K., Galvin, J.L., Ericson, R.E., Lau, C.N., Levanos, V.A., Sahasrabudhe, A., and Dewhirst, F.E. (2001) Bacterial diversity in human subgingival plaque. J. Bacteriol. *183*, 3770–3783.

Portnoy, D.A. (2005) Manipulation of innate immunity by bacterial pathogens. Curr. Opin. Immunol. *17*, 1–4.

Rebeil, R., Ernst, R.K., Gowen, B.B., Miller, S.I., and Hinnebusch, B.J. (2004) Variation in lipid A structure in the pathogenic yersiniae. Mol. Microbiol. *52*, 1363–1373.

Reife, R.A., Coats, S.R., Al-Qutub, M., Dixon, D.R., Braham, P.A., Billharz, R.J., Howald, W.N., and Darveau, R.P. (2006) *Porphyromonas gingivalis* lipopolysaccharide lipid A heterogeneity: differential activities of tetra and penta-acylated lipid A structures on E-selectin expression and TLR4 recognition. Cell. Microbiol. *8*857–868.

Ren, L., Jin, L., and Leung, W.K. (2004) Local expression of lipopolysaccharide-binding protein in human gingival tissues. J. Periodont. Res. *39*, 242–248.

Ren, L., Leung, W.K., Darveau, R.P., and Jin, L. (2005) The expression profile of lipopolysaccharide-binding protein, membrane-bound CD14, and toll-like receptors 2 and 4 in chronic periodontitis. J. Periodontol. *76*, 1950–1959.

Ren, L., Leung, W.K., Loo, T.W., and Jin, L. (2005) Lipopolysaccharide-binding protein down-regulates the expression of interleukin-6 by human gingival fibroblast. J. Periodont. Res. *40*, 407–416.

Roberts, F.A., and Darveau, R.P. (2002) Beneficial bacteria of the periodontium. Periodontol. 2000 *30*, 40–50.

Roy, M.F., Lariviere, L., Wilkinson, R., Tam, M., Stevenson, M.M., and Malo, D. (2006) Incremental expression of Tlr4 correlates with mouse resistance to *Salmonella* infection and fine regulation of relevant immune genes. Genes Immun. *7*, 372–383.

Simpson, W., Olczak, T., and Genco, C.A. (2004) Lysine-specific gingipain K and heme/hemoglobin receptor HmuR are involved in heme utilization in *Porphyromonas gingivalis*. Acta Biochim. Pol. *51*, 253–262.

Smalley, J.W., Birss, A.J., McKee, A.S., and Marsh, P.D. (1996) Haemin binding as a factor in the virulence of *Porphyromonas gingivalis*. FEMS Microbiol. Lett. *141*, 65–70.

Socransky, S.S., and Haffajee, A.D. (1994) Evidence of bacterial etiology: a historical perspective. Periodontol. 2000 *5*, 7–25.

Stappenbeck, T.S., Hooper, L.V., and Gordon, J.I. (2002) Developmental regulation of intestinal angiogenesis by indigenous microbes via Paneth cells. Proc. Natl. Acad. Sci. USA *99*, 15451–15455.

Sugawara, Y., Uehara, A., Fujimoto, Y., Kusumoto, S., Fukase, K., Shibata, K., Sugawara, S., Sasano, T., and Takada, H. (2006) Toll-like receptors, NOD1, and NOD2 in oral epithelial cells. J. Dent. Res. *85*, 524–529.

Tatakis, D.N., and Kumar, P.S. (2005) Etiology and pathogenesis of periodontal diseases. Dent. Clin. North Am. *49*, 491–516.

Tonetti, M.S. (1997) Molecular factors associated with compartmentalization of gingival immune responses and transepithelial neutrophil migration. J. Periodont. Res. *32*, 104–109.

Tonetti, M.S., Imboden, M.A., Gerber, L., Lang, N.P., Laisue, J., and Mueller, C. (1994) Localized expression of mRNA for phagocyte-specific hemotactic cytokines in human periodontal infections. Infect. Immun. *62*, 4005–4014.

Tuomanen, E. (2005) Appreciating our usual guests. Science *308*, 635.

Uehara, A., Sugawara, Y., Kurata, S., Fujimoto, Y., Fukase, K., Kusumoto, S., Satta, Y., Sasano, T., Sugawara, S., and Takada, H. (2005) Chemically synthesized pathogen-associated molecular patterns increase the expression of peptidoglycan recognition proteins via toll-like receptors, NOD1 and NOD2 in human oral epithelial cells. Cell. Microbiol. *7*, 675–686.

Umesaki, Y., and Setoyama, H. (2000) Structure of the intestinal flora responsible for development of the gut immune system in a rodent model. Microbes Infect. *2*, 1343–1351.

Vankeerberghen, A., Nuytten, H., Dierickx, K., Quirynen, M., Cassiman, J.J., and Cuppens, H. (2005) Differential induction of human beta-defensin expression by periodontal commensals and pathogens in periodontal pocket epithelial cells. J. Periodontol. *76*, 1293–1303.

Visintin, A., Iliev, D.B., Monks, B.G., Halmen, K.A., and Golenbock, D.T. (2006) MD-2. Immunobiology *211*, 437–447.

Vora, P., Youdim, A., Thomas, L.S., Fukata, M., Tesfay, S.Y., Lukasek, K., Michelsen, K.S., Wada, A., Hirayama, T., Arditi, M., and Abreu, M.T. (2004) Beta-defensin-2 expression is regulated by TLR signalling in intestinal epithelial cells. J. Immunol. *173*, 5398–5405.

Wilson, M. (2005) Microbial inhabitants of humans. (New York: Cambridge University Press).

Wilson, M., Weightman, A., and Wade, W. (1997) Applications of molecular ecology in the characterisation of uncultured microorganisms associated with human disease. Rev. Med. Microbiol. *8*, 91–101.

Wurfel, M.M., Hailman, E., and Wright, S.D. (1995) Soluble CD14 acts as a shuttle in the neutralization of lipopolysaccharide (LPS) by LPS-binding protein and reconstituted high density lipoprotein. J. Exp. Med. *181*, 1743–1754.

Xu, J., and Gordon, J.I. (2003) Inaugural Article: Honor thy symbionts. Proc. Natl. Acad. Sci. USA *100*, 10452–10459.

Yamazaki, K., Ikarashi, F., Aoyagi, T., Takahashi, K., Nakajima, T., Hara, K., and Seymour, G.J. (1992) Direct and indirect effects of *Porphyromonas gingivalis* lipopolysaccharide on interleukin-6 production by human gingival fibroblasts. Oral Microbiol. Immunol. *7*, 218–224.

Yi, E.C., and Hackett, M. (2000) Rapid isolation method for lipopolysaccharide and lipid A from gram-negative bacteria. Analyst *125*, 651–656.

Yoshinari, N., Kameyama, Y., Aoyama, Y., Nishiyama, H., and Noguchi, T. (1994) Effect of long-term methotrexate-induced neutropenia on experimental periodontal lesion in rats. J. Periodont. Res. *29*, 393–400.

Zasloff, M. (2002) Antimicrobial peptides of multicellular organisms. Nature *415*, 389–395.

Zweigner, J., Schumann, R.R., and Weber, J.R. (2006) The role of lipopolysaccharide-binding protein in modulating the innate immune response. Microbes Infect. *8*, 946–952.

Molecular Approaches to Vaccination against Oral Infections

11

George Hajishengallis and Michael W. Russell

Abstract

Given that both dental caries and periodontitis have an infectious aetiology, immunization has been proposed as a means of controlling them. However, the approaches vary according to the nature of the bacteria involved and the mechanisms of pathogenesis for these two very different diseases. For dental caries, surface and secreted proteins involved in colonization of teeth by mutans streptococci, and specific functional domains derived from these, have been identified and demonstrated to serve as targets for antibodies that inhibit the cariogenic process. Strategies have been devised to induce salivary secretory IgA antibodies which are particularly well suited to fulfilling this function. Further progress will depend on applying the results from animal models to human trials. Periodontal vaccines are less well developed, but some antigenic targets of two principal agents, *Porphyromonas gingivalis* and *Aggregatibacter actinomycetemcomitans*, have been identified. Experiments in rodent and primate models have shown the ability of fimbrial antigens of either, and gingipains from the former, to elicit serum antibodies that inhibit colonization or the pathogenic mechanisms of these bacteria. However, questions remain about the role of host immune responses in the pathogenesis of periodontal disease, and whether current approaches to immunization can modulate these immunopathological processes.

Introduction and historical perspective of vaccination against oral diseases

Almost as soon as it was recognized that dental caries was due to bacterial infection in the 1920s, the concept arose of immunization against the responsible organisms, which at that time were thought to be lactobacilli. When, in the 1960s, dextran-forming acidogenic streptococci, especially *Streptococcus mutans*, which later became split into several species collectively known as 'mutans streptococci' including *S. sobrinus*, became recognized as the principal initiating agents of human caries, several groups took up the challenge of developing a vaccine against these organisms. Important factors contributing to this undertaking (Table 11.1), besides the rediscovery and characterization of *S. mutans* and its virulence factors, were the development of quantitative models of dental caries in experimental animals, and emerging comprehension of the mucosal immune system by which secretory IgA (S-IgA) antibodies are produced in the saliva and other secretions. Indeed, studies on vaccination against caries led to the development of several concepts in mucosal immunity, and the first demonstration that oral immunization of human subjects with a bacterial suspension could lead to the appearance of S-IgA antibodies in secretions such as saliva was performed using *S. mutans* (Mestecky *et al.*, 1978). During the 1970s also evidence accumulated that humans mounted antibody responses against mutans streptococci, and that these antibodies could be correlated positively or negatively with past experience of dental caries (as measured by indices of decayed, missing, and filled teeth), depending upon whether serum IgG or salivary IgA was being measured, and whether the subjects were currently free of active lesions or not (Challacombe and Lehner, 1976). Thus, it was possible to interpret the data in various ways to suggest that antibodies could confer protection, or conversely, that antibodies reflected previous exposure to mutans streptococci. Nevertheless, direct experimentation on rodents or monkeys demonstrated that either systemic or mucosal immunization regimens using mutans streptococcal antigen preparations could induce protective immunity that suppressed the development of dental caries following oral inoculation with mutans streptococci in the presence of a high-sucrose diet. Fears over the safety of immunization with whole mutans streptococci or ill-defined complex antigen preparations, which were reported to induce heart-cross-reactive antibodies, caused some concern, even though the molecular basis for such cross-reactivity has never been satisfactorily explained. However, the identification of important target antigens, particularly glucosyltransferases, glucan-binding proteins, and the AgI/II family of adhesins, has afforded the opportunity of developing vaccines that do not entail such adverse consequences. No molecularly defined antigen of mutans streptococci has been shown to elicit heart-reactive antibodies or to display serological cross-reactivity with human tissues (Russell and Wu, 1990).

The concept of vaccination against periodontal disease was inevitably slower to emerge, and not until specific organisms, such as *Porphyromonas*

Table 11.1 Historical milestones in caries immunity

'Mutans streptococci' as the principal cariogenic organisms
S-IgA as the major salivary form of antibody
Serotypes (now separate species) of mutans streptococci; cell wall antigens
Glucosyltransferases and sucrose in glucan formation
Development of animal models: rodents and primates
Identification of surface proteins on mutans streptococci; adhesins; AgI/II
Origins of salivary IgA within the common mucosal immune system
Mother-infant transmission of mutans streptococci; 'window of infectivity'

gingivalis, *Tannerella forsythia*, and *Treponema denticola* (the 'red complex'; Holt and Ebersole, 2005) and *Aggregatibacter (Actinobacillus) actinomycetemcomitans (A.a.)* were implicated as etiologic agents could this idea begin to be developed: the first essential for vaccination against any infection is to define the causative agent. However, it has become clear that, in contrast to dental caries which – perhaps uniquely – arises from the direct assault of bacteria and their products on a tissue, periodontal disease in common with many other infectious pathologies results as much from the undesirable side-effects of the host immune response in inflicting damage upon host tissues as from the activities of the causative organisms. This is particularly illustrated by the finding that severe combined immunodeficient mice, which lack both T and B cells and cannot mount specific immune responses, have substantially less *P. gingivalis*-induced bone loss than immunocompetent mice (Baker *et al.*, 1994). Mice deficient in MHC class II-restricted $CD4^+$ T cells also display greater resistance to *P. gingivalis*-induced bone loss than normal mice or mice deficient in MHC class I-restricted $CD8^+$ T cells (Baker *et al.*, 1999). Therefore, key to success in developing a vaccine against periodontal disease is an understanding of the immunopathological processes, albeit initiated by certain bacterial species, that result in inflammatory damage to the periodontium while generally failing to eliminate the pathogens.

Although caries and periodontal disease account for the great majority of dental diseases, other oral infections may also be amenable to immunological control by means of a vaccine. Oro-pharyngeal candidiasis is one potential candidate, but specific vaccination efforts are not nearly as advanced as those for caries or even periodontal disease. Current thinking seems to favour approaches that enhance innate defence mechanisms, such as defensins or histatins, that have been found to be effective against *Candida*. However, recent success in the development of vaccines against genital herpes virus and the release of a vaccine against carcinogenic types of human genital papillomavirus suggest that similar approaches may be feasible against their oral equivalents.

The mucosal immune system and oral immunity

The predominant isotype of antibody found in human saliva and to which therefore oral microorganisms are exposed is S-IgA, which consists of a polymeric, generally dimeric, form of IgA (pIgA) linked by the J chain and coupled to secretory component (SC; Mestecky *et al.*, 2005). S-IgA is the characteristic product of the mucosal immune system (Mestecky, 1987), which is largely independent of the circulatory immune system. Indeed, there are convincing arguments to show that the mucosal immune system is by far the largest component of the entire immune system, and that the production of IgA far outstrips the production of all other antibody isotypes. Daily production of S-IgA in adult humans has been estimated to amount to 3–8 g, and salivary output is in the range of 0.1–0.2 g/day (Russell, 2003). Studies conducted over many years have shown that S-IgA production is initiated largely at specialized inductive sites of the mucosal immune system (Fig. 11.1), such as the well-known intestinal Peyer's patches but also including similarly organized follicular lymphoid tissues in the large intestine and possibly in the bronchial tract, as well as the tonsils and adenoid in the pharynx (Waldeyer's ring; Brandtzaeg and Pabst, 2004). These tissues are covered by a specialized epithelium that contains 'M cells' which differ from the surrounding epithelial cells in possessing a microfold apical membrane that can ingest antigenic materials and pass them through, by a transcytotic vesicular mechanism, to the underlying antigen-presenting cells (APC). When stimulated by APC presenting processed antigenic peptides and antigen-sensitized T cells, the B cells in these mucosal inductive sites preferentially switch to express IgA and emigrate through the draining lymphatics into the circulation, while differentiating into polymeric IgA-secreting cells. Through the selective expression of mucosal homing receptors such as $\alpha4\beta7$ integrin which recognizes MAdCAM-1, and receptors that recognize certain mucosal chemokines, these IgA-secreting cells along with cognate T cells home to various effector sites that may be remote from the inductive site where they were originally stimulated (Youngman *et al.*, 2005). The B cells

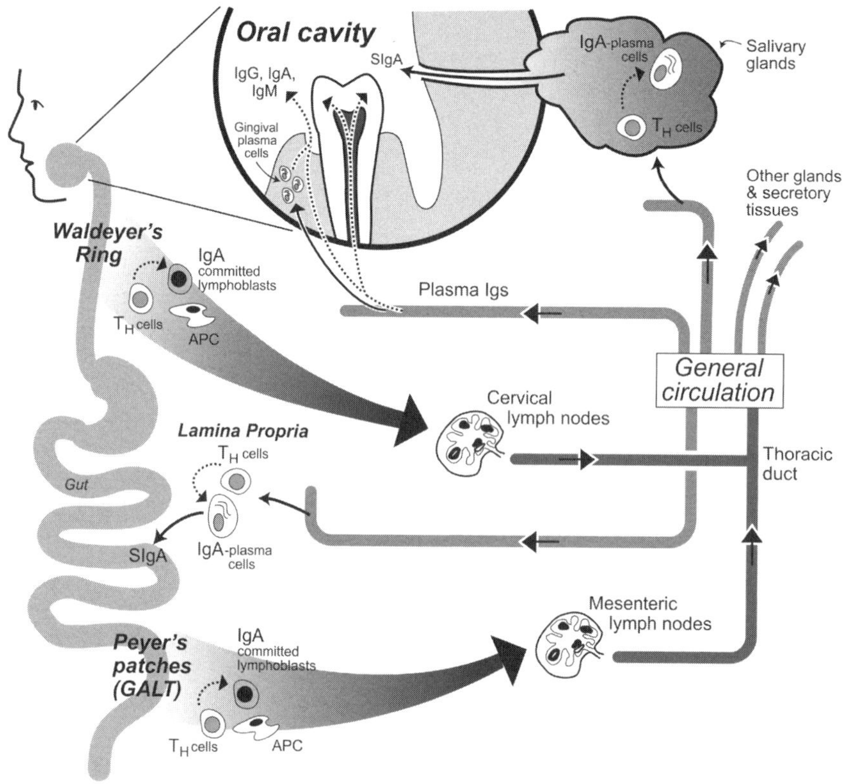

Figure 11.1 The mucosal immune system from the oral perspective, and the origins of salivary immunoglobulins Naïve mucosal B and T cells within the inductive sites represented by gut-associated lymphoid tissues, such as Peyer's patches in the small intestine, or the pharyngeal Waldeyer's ring in humans (nasal lymphoid tissue in rodents) are stimulated by antigens that penetrate the M cells of the follicular epithelium and are then processed by APC. The stimulated B cells undergo maturation in germinal centres and switch to IgA production either in the inductive sites or in the draining lymph nodes. B and T lymphoblasts migrate further through the lymphatics, enter the circulation, and home to mucosal effector sites such as the salivary glands where B cells terminally differentiate into polymeric IgA-secreting plasma cells. S-IgA is finally secreted by epithelial cells which transport the polymeric IgA using pIgR which becomes secretory component in the complete S-IgA. Other B lymphoblasts localize in the inflamed gingivae during periodontal disease and differentiate into plasma cells that secrete IgM, IgG, or monomeric IgA. Immunoglobulins derived from the circulation can also transude through the gingival crevices, or from the pulp though the dentinal tubules.

terminally differentiate into plasma cells which secrete large quantities of polymeric IgA close to the basolateral surfaces of the epithelia under which they reside. Cells in single-layered mucosal epithelia usually express the precursor of SC as a basolateral membrane receptor known as polymeric immunoglobulin receptor (pIgR) which binds J-chain-containing pIgA (and IgM, if available) and transports it by a vesicular transport mechanism to the apical surface (Kaetzel and Mostov, 2005). During transcytosis, the pIgR becomes covalently bonded through disulfide exchange to the pIgA, and it is also proteolytically cleaved from its transmembrane segment, so that it now becomes the SC of S-IgA which is released into the lumen. Although this mechanism of S-IgA production is often known as the 'common mucosal immune system' because it operates to distribute pIgA-secreting cells to sites remote from their induction, evidence has accumulated that it is not uniform, but that particular effector sites are selectively supplied with mucosal B and T cells induced in a limited set of inductive sites. In the case of the salivary glands, the inductive sites appear to lie largely (but not exclusively) in Waldeyer's ring or its equivalent in rodents, the nasal lymphoid tissue, and to a lesser extent in the intestine (Wu *et al.*, 1997). The precise molecular mechanisms responsible for this 'compartmentalization' of the common mucosal immune system (Moldoveanu *et al.*, 1995) are inadequately understood, but presumably depend upon the selective expression of homing addressins and chemokines, and their corresponding receptors on lymphocytes.

From the foregoing it will be apparent that vaccines intended to induce salivary IgA antibody responses will need to target mucosal inductive tissues, preferably those located in the pharynx or intestinal tract (Russell, 2003). This poses problems at several levels. The gastrointestinal tract is a hostile environment for many putative vaccine antigens on account of stomach acid and digestive enzymes, but there are pharmaceutical formulations available to circumvent or minimize the impact of these. It is also desirable to target the vaccine to the M cells in order to enhance its uptake. Micro- or nano-particles are more readily taken up than soluble molecules, and a variety of particulate formulations, including polymer particles and liposomes, have been devised with varying degrees of success. Another angle has been to develop molecules that may be able to recognize specific structures on the M cell membrane, and empirical approaches using lectins that have been found to bind to cell surfaces may work in this way. Alternatively, enterotoxins that bind gangliosides on cell membranes through their B subunits have been exploited as coupled delivery systems, although the precise mechanisms of their action are not clearly understood (Hajishengallis *et al.*, 2005a). More will be discussed about this approach under the section 'Novel Adjuvants'. A more challenging problem is that it has been known for a long time that soluble antigens especially, such as purified proteins, have the tendency to induce tolerance rather than active immunity when applied to mucosal surfaces of the intestine – the phenomenon known as 'oral tolerance', or more generally, 'mucosal tolerance' (Mestecky *et al.*, 2007). It has become clear over recent years that the default mode of response of the mucosal immune system, especially towards non-viable, non-threatening antigens including most free proteins, is tolerance, even though they may be immunologically 'foreign'. As the vast majority of such macromolecular material presented to the intestinal mucosa is food this is advantageous, and breakdown of tolerance leads to food hypersensitivities, allergies, or possibly more serious manifestations of inflammatory disease. The mucosal immune system has also co-evolved with a commensal microbiota that is generally beneficial to the host, and an immune response that would have the effect of eliminating such desirable organisms would be disadvantageous. Conversely, it is clear that the mucosal immune system can respond effectively to pathogenic stimuli, whether these are aggressive microorganisms or noxious toxins, and generate S-IgA antibodies or other responses against them. Thus a key feature of the mucosal immune system is that it can distinguish between the essentially harmless, even desirable, molecular and microbial constituents of the luminal environment, and dangerous pathogens and their products. The mechanisms by which this is accomplished, whether by recognition of so-called 'pathogen-associated molecular patterns' (Janeway and Medzhitov, 2002) or by responding to danger signals released from damaged cells and tissues, or both, have been hotly debated and reviewed elsewhere (Matzinger, 2007). New findings concerning the way that different classes of mucosal dendritic cells and other 'first-responder' cells at mucosal surfaces react to different types of stimuli are beginning to shed significant light on these mechanisms, but are beyond the scope of this review. The lesson, however, for mucosal vaccine development, is that effective vaccines will need to mimic these pathogenic stimuli in an appropriate manner. It has become abundantly clear that the general mode of action of molecular adjuvants, including several used parenterally in experimental animals, is to provide the 'second signal' to APC that causes them to express co-stimulatory receptors and to secrete cytokines which in turn elicit active responses from specific T and B cells. Moreover, different adjuvants can elicit different responses and lead to the selective generation of different classes of cell- or antibody-mediated immunity. Adjuvant development tailored to inducing the most appropriate response against a particular pathogen is therefore an important aspect of future vaccine design.

Caries vaccines

Objectives

Dental caries results from localized demineralization of tooth enamel by lactic acid secreted by cariogenic bacteria following fermentation of dietary sugars, and acid production at sufficiently destructive levels requires successful colonization and biofilm formation by mutans streptococci (Banas, 2004; Loesche, 1986). The colonization of tooth surfaces by mutans streptococci is dependent upon a two-step process consisting of a sucrose-independent initial adherence to the salivary pellicle-coated dental enamel, followed by sucrose-dependent microbial accumulation. Although this mechanism was proposed more than 25 years ago (Staat *et al.*, 1980), it remains fundamentally correct although enriched with significant molecular details derived from the cloning, sequencing, and functional characterization of key virulence factors. This knowledge is crucial for the selection of appropriate caries vaccine targets. The key colonization factors are the AgI/II salivary adhesins (Jenkinson and Demuth, 1997), named after the originally discovered member from *S. mutans* serotype c (Russell *et al.*, 1980a), a group of secreted or cell-associated glucosyltransferases (GTF) which produce adhesive glucans (Thurnheer *et al.*, 2006) and the glucan-binding proteins (Smith *et al.*, 1998). The molecular biology of these virulence factors is described in Chapter 5 by R.R.B. Russell.

The primary rational objectives for caries vaccine development have therefore been to induce salivary antibodies that disrupt these sucrose-independent or –dependent colonization mechanisms. However, other functional targets have been proposed over the years (Table 11.2) although none has progressed to a 'proof-of-concept' stage.

Table 11.2 Potential immunological intervention against prototypical cariogenic or periodontopathic organisms and candidate immunogens

Intervention targets	Anti-mutans streptococci strategies	Anti-*P. gingivalis* strategies
Colonization	Inhibition of binding to salivary pellicle or to early colonizers *Immunogen*: AgI/II (Hajishengallis *et al.*, 1992) Inhibition of glucan synthesis or glucan-mediated biofilm formation *Immunogens*: GTF, GBP (Jespersgaard *et al.*, 1999; Smith *et al.*, 2005)	Inhibition of binding to salivary pellicle, to early colonizers, or to gingival epithelium *Immunogen*: fimbriae (Sojar *et al.*, 1999; Sharma *et al.*, 2001)
Persistence (resistance to killing)	Opsonization and phagocytosis *Immunogen*: AgI/II (Scully *et al.*, 1980)	Opsonization and phagocytosis *Immunogens*: fimbriae, gingipains (Fan *et al.*, 2001; Nakagawa *et al.*, 2001)
Invasion	Inhibition of entry into dentinal tubules *Immunogen*: AgI/II (Love and Jenkinson, 2002; Hahn and Falkler, 1992)	Inhibition of invasion into epithelial cells *Immunogen*: fimbriae (Sojar *et al.*, 1999)
Immune evasion	Neutralization of IgA1 protease activity in the oral cavity *Immunogen*: IgA1 proteases of oral streptococci (not *S. mutans*)	Inhibition of degradation of complement, cytokines and innate immune receptors by gingipains *Immunogen*: gingipains (Grenier, 1992; Calkins *et al.*, 1998; Sugawara *et al.*, 2000)
Direct bacteria-induced tissue destruction	Inhibition of metabolism, sugar uptake, acid production (hypothetical) *Immunogens:* not determined	Inhibition of gingipain-mediated tissue degradation *Immunogen*: gingipains (Sorsa *et al.*, 1992)

Virulence factors of cariogenic streptococci as candidate vaccine immunogens

AgI/II protein adhesins

The gene encoding the 167-kDa AgI/II adhesin (also known as antigen B, P1, PAc, SpaP, or SR) is highly conserved between serotypes of mutans streptococci and shares significant homology with genes for surface proteins expressed by other streptococcal species (Brady *et al.*, 1991; Ma *et al.*, 1991; Okahashi *et al.*, 1989; Zhang *et al.*, 2006; Fig. 11.2). However, the functions of individual members of the AgI/II family vary, presumably reflecting species-specific adaptations for colonization of the oral cavity (Jenkinson and Demuth, 1997). Despite their structural similarities and an overall 66% amino acid sequence identity, *S. mutans* AgI/II and *S. sobrinus* SpaA (PAg) bind distinct salivary pellicle components (Brady *et al.*, 1992; Gibbons *et al.*, 1986; LaPolla *et al.*, 1991), although they cross-react serologically and may thus confer cross-protection following immunization with either antigen.

Kinetic analysis and comparative binding studies involving wild-type and AgI/II-deficient mutants of *S. mutans* have shown that AgI/II binds to saliva-coated hydroxyapatite (Hajishengallis *et al.*, 1994; Koga *et al.*, 1990; Lee *et al.*, 1989). The saliva-binding region (SBR) comprises residues 188–578 in the N-terminal one-third of AgI/II containing the alanine-rich repeat region (A-region) (Crowley *et al.*, 1993; Hajishengallis *et al.*, 1994; Moisset *et al.*, 1994; Fig. 11.2). An additional adherence domain occurs in the proline-rich repeat region (P-region; Munro *et al.*, 1993). These findings are consistent with the notion that AgI/II contains two saliva-interacting domains (Hajishengallis *et al.*, 1994; Peros and Gibbons, 1986), and that the A- and P-regions interact with each other (Seifert *et al.*, 2004; van Dolleweerd *et al.*, 2003; van Dolleweerd *et al.*, 2004). Crystallographic analysis of the intermediate or variable region which contains a putative sugar-binding domain (residues 464 to 840) reveals a β-sandwich structure that brings the A- and P-regions within close proximity (Troffer-Charlier *et al.*, 2002). The role of AgI/II as a virulence factor was established by comparing the cariogenic potential of wild-type and AgI/II-deficient *S. mutans* strains in the rat caries model in which wild-type *S. mutans* induced more carious lesions although not displaying a colonization advantage compared to the mutant strain (Crowley *et al.*, 1999). Although the function of AgI/II as a glucan-binding protein cannot be ruled out given the homology of its V-region with GbpC (Troffer-Charlier *et al.*, 2002), another plausible mechanism may involve the ability of AgI/II to bind collagen type I and thereby mediate the invasion of *S. mutans* and *S. sobrinus* into root dentinal tubules (Love and Jenkinson, 2002). The binding of AgI/II to several other extracellular matrix molecules (*e.g.* fibronectin and laminin; Sciotti *et al.*, 1997) through its A-region may contribute to the progression of dentinal carious

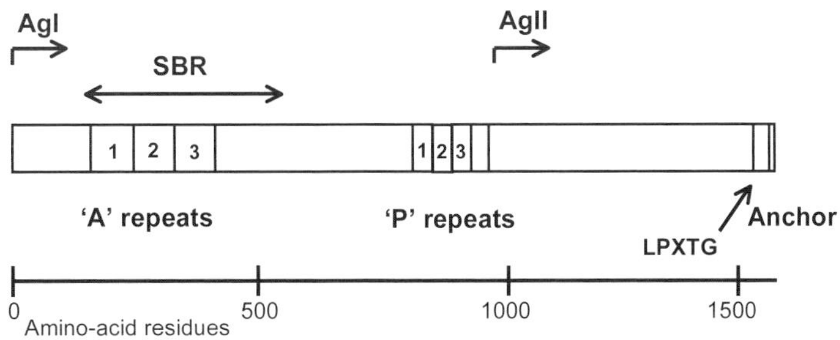

Figure 11.2 Schematic representation of AgI/II of mutans streptococci. AgI/II is approximately 167 kDa and contains >1500 amino acid residues. The A-region typically consists of three tandem repeats of 82 residues rich in alanine, and the P-region usually has three tandem repeats of 39 residues rich in proline. The 'saliva-binding region' (SBR; Hajishengallis *et al.*, 1995) is about 42 kDa within the N-terminal end, spanning the A-region and into the intermediate or variable region. Structural/functional studies suggest that the A- and P-regions interact to form a binding domain that recognizes proteins deposited in the dental pellicle. The C-terminal end consists of a typical Gram-positive wall anchor and membrane-spanning segment with an LPXTGE sortase recognition motif by which it becomes coupled to the peptidoglycan.

lesions and initiation of endodontic infections, and also facilitate the systemic dissemination of mutans streptococci under pathologic conditions that expose matrix molecules in the oral cavity. These putative virulence traits of AgI/II further underscore its suitability as a candidate for vaccine development.

Glucosyltransferases and glucan-binding proteins

GTFs synthesize the extracellular adhesive glucans that are essential for cell-to-cell adherence and the establishment of cohesive mutans streptococcal biofilms with increased cariogenic potential (Kuramitsu, 1993). *S. mutans* and *S. sobrinus* express several GTFs (140–160 kDa) which synthesize glucans varying in chain length, content of α-1,3 and α-1,6 glucosyl linkages, degree of branching, and solubility (Ferretti *et al.*, 1987; Honda *et al.*, 1990; Shiroza *et al.*, 1987). *S. mutans* expresses three genes for GTF: *gtf*B responsible for insoluble glucan synthesis (Shiroza *et al.*, 1987), *gtf*C for soluble and insoluble glucan synthesis (Ueda *et al.*, 1988), and *gtf*D for soluble glucan synthesis (Honda *et al.*, 1990). *S. sobrinus* expresses *gtf*I and *gtf*S, encoding enzymes for producing insoluble and soluble glucans, respectively (Gilmore *et al.*, 1990; Russell *et al.*, 1987). Epitope mapping studies have indicated that the catalytic region is located in the N-terminal one-third of GTF, whereas the glucan-binding region is in the C-terminal one-third (reviewed by Smith, 2002; Fig. 11.3). *S. mutans* expresses three cell-wall-associated glucan-binding proteins (GbpA, GbpB, and GbpC) and *S. sobrinus* produces five Gbps (reviewed by Smith *et al.*, 1998), which mediate cell-to-cell adherence by binding the glucans synthesized by GTF.

Animal studies with wild-type and isogenic mutants of *S. mutans* in which one or more *gtf* genes were inactivated, have established the role of GTF in cariogenicity and, moreover, show that insoluble glucan synthesis is more important than soluble glucan synthesis (Munro *et al.*, 1991; Yamashita *et al.*, 1993). Therefore, antibody-mediated interference with the enzymatic or glucan-binding activity of the *gtf*B-encoded enzyme may be a reasonable preventive approach against development of pathogenic mutans streptococcal biofilms. GTFs from *S. mutans* or *S. sobrinus* have been demonstrated to induce cross-neutralizing antibody responses, which moreover are protective against experimental caries induced by infection with either species (Smith and Taubman, 1997; Smith *et al.*, 1993; Taubman *et al.*, 1995). In contrast, there is no significant antigenic cross-reactivity between *S. mutans* and *S. sobrinus* Gbps (Smith *et al.*, 1998).

Other surface antigens

Early studies of caries immunity and vaccine development identified several other surface antigens of mutans streptococci (Russell, 1986; Russell *et al.*, 1986). Among these, antigen A (WapA, AgIII) emerged as a potential vaccine antigen (Hughes *et al.*, 1986; Russell *et al.*, 1983) which was developed for human trials that ultimately did not occur. The biological functions of WapA/AgIII were unclear but have recently been found to include binding to collagen and fibronectin (Han *et al.*, 2006; Russell *et al.*, 1995). However, as WapA/AgIII occurs only in *S. mutans* and *S. rattus*, but not *S. sobrinus*, a vaccine based on it would exclude protection against the last species.

The genome of *S. mutans* contains only six genes identified as encoding potential surface

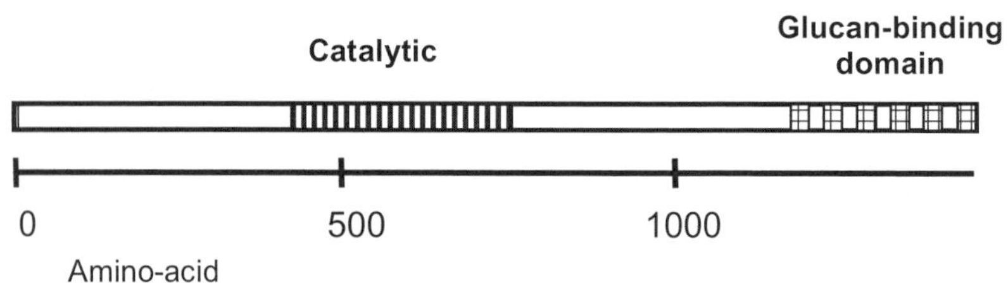

Figure 11.3 Schematic representation of glucosyltransferase of mutans streptococci. GTFs are approximately 150 kDa and contain >1300 amino acid residues. Epitope mapping, functional studies, and modelling based on amino acid sequences indicate that the catalytic regions of GTFs are located within the N-terminal halves of the molecules, and that glucan-binding domains occur as a series of repeats at the C-termini.

proteins containing a correctly located LPXTG sortase signal, including SpaP, WapA, GbpC, DexA, FruA, and an uncharacterized protein designated WapE (Ajdic *et al.*, 2002). Although dextranase and levansucrase were considered early on (Bahn *et al.*, 1977) they have not been pursued and WapE has not been evaluated as a potential vaccine antigen. Other cell surface components initially proposed as vaccine candidates include the cell wall polysaccharides responsible for serotype specificity, and lipoteichoic acid. However, the wall polysaccharides are weakly immunogenic, and lipoteichoic acid is widely distributed among oral streptococci and therefore an unlikely specific vaccine antigen.

Antibody to mutans streptococci: mechanisms of anti-cariogenic action

S-IgA antibodies are thought to protect against a variety of infections by blocking microbial adherence and by neutralizing viruses, toxins, or enzymes (Russell and Kilian, 2005). Thus blockade of AgI/II or Gbp and inhibition of GTF enzymatic function are likely mechanisms of protection against caries due to S-IgA antibodies induced by mucosal immunization.

Human S-IgA antibodies to AgI/II have been shown to inhibit the binding of isolated AgI/II and of *S. mutans* cells to saliva-coated hydroxyapatite in a dose-dependent fashion and with significantly greater potency compared to monomeric IgA or IgG antibodies (Hajishengallis *et al.*, 1992). Moreover, human salivary IgA antibodies neutralize the enzymatic activity of GTF (Smith *et al.*, 1985). Thus both sucrose-independent and sucrose-dependent stages of tooth colonization by mutans streptococci are amenable to immunological intervention. A potential limitation for a protective role of S-IgA antibodies arises from the ability of bacterial IgA1 proteases to cleave human IgA1 into intact Fab and Fc fragments (Kilian and Russell, 2005). These enzymes are produced by several species of oral streptococci including *S. gordonii*, *S. oralis*, and *S. mitis*, but not mutans streptococci, and thus may be found in saliva. Cleavage of S-IgA1 by IgA1 protease counteracts the *S. mutans* adherence-inhibiting effect of S-IgA1 antibodies (Hajishengallis *et al.*, 1992; Reinholdt and Kilian, 1987). However, because of the exquisite specificity of IgA1 proteases for human (and ape) IgA1, the biological significance of these observations is not readily testable in animal models. Although the generated Fab fragments retain antigen-binding activity, they fail to inhibit *S. mutans* adherence, probably because of their hydrophobic nature and propensity to interact with the salivary pellicle (Hajishengallis *et al.*, 1992; Reinholdt and Kilian, 1987). Whether Fab fragments of IgA1 antibodies to GTF retain neutralizing activity has not been determined.

The observation that *S. mutans* adherence can be inhibited in an AgI/II-specific fashion was initially made using rabbit IgG antibodies to antigen B (AgI/II; Douglas and Russell, 1984). Rabbit IgG antibodies to a PAc (AgI/II) segment within the SBR domain also inhibit *S. mutans* adherence to experimental salivary pellicles (Yu *et al.*, 1997). Moreover, IgG antibody to P1 (AgI/II) inhibits the saliva-mediated interbacterial adherence between *S. mutans* and early colonizers of the salivary pellicle (Lamont *et al.*, 1991). These observations suggest that induced antibodies have the potential to block both the direct and indirect anchoring of *S. mutans* to the tooth surface.

Independent studies have shown that rabbit IgG antibodies against the glucan-binding domain of *S. mutans* glucosyltransferase are potent inhibitors of glucan synthesis, although antibodies to the catalytic domain are less inhibitory (Jespersgaard *et al.*, 1999; Yu *et al.*, 1997). Rat IgG antibodies to synthetic peptides corresponding to sequences of the catalytic subdomains of GTF react with the native enzyme and partially inhibit its glucan synthesis activity (Smith *et al.*, 1999). Therefore, although access of the antibodies to the catalytic site may be restricted, such antibodies do display partial neutralization of catalytic activity. Modest enzymatic inhibition of *S. mutans* or *S. sobrinus* GTF has also been accomplished using rat antiserum to a conserved 22-amino acid residue sequence from the glucan-binding region (Taubman *et al.*, 1995). However, significantly higher inhibition of GTF enzymatic activity was achieved using antibodies to a diepitopic construct containing peptides from both the catalytic and the glucan-binding regions (Taubman *et al.*, 2001). Consistent with the proposed roles of AgI/II and GTF in sucrose-independent and sucrose-dependent adherence, respectively, rabbit IgG antibodies against a chimeric protein comprising the AgI/II SBR and the GTF glucan-binding domain, inhibit both the initial and the subsequent glucan-mediated adherence of *S. mutans* in an *in vitro* tooth surface model (Yu *et al.*, 1997).

Although S-IgA antibodies in saliva are thought to represent the primary defence mechanism against tooth colonization by mutans streptococci, circulating IgG antibodies may also be protective against caries at the gingival margin of the tooth, because crevicular fluid contains IgG derived by transudation from plasma (Challacombe *et al.*, 1978). IgG antibody can mediate opsonophagocytosis and killing of mutans streptococci by neutrophils (Scully *et al.*, 1980). This mechanism may be particularly relevant during tooth eruption when inflammation is common (Seward, 1971). In addition, IgG antibodies may confer protection by blocking the invasin function of AgI/II which directs *S. mutans* through dentinal tubules (Love and Jenkinson, 2002). IgG antibodies against *S. mutans* have been detected in the supernatant fluids from pulpal explant cultures (Hahn and Falkler, 1992). It is conceivable that neutralizing IgG antibodies exuding from the pulp could arrest the invasion of *S. mutans* through the dentine, but this intriguing hypothesis remains unproven.

Proof-of-principle induction of salivary immunity against caries

The ability of vaccines to induce immunity to dental caries in experimental animals is revealed by decreased colonization of tooth surfaces by implanted mutans streptococci and by reduction of subsequent carious lesion formation. Many studies, mostly in rats but also in monkeys, have demonstrated the potential of immunization for caries prevention, and have been extensively reviewed elsewhere (Hajishengallis and Michalek, 1999; Russell *et al.*, 2004; Smith, 2002).

Although the protective potential of intact AgI/II was demonstrated by systemic or mucosal (intranasal) immunization of rhesus monkeys and rats (Katz *et al.*, 1993; Lehner *et al.*, 1981), similar protective effects have been obtained using defined segments of it. Intranasal vaccination of mice with a CTB-conjugated 19-mer peptide from the SBR of AgI/II (residues 301–319) resulted in strong inhibition of *S. mutans* oral colonization, although the study did not address protection against caries (Takahashi *et al.*, 1991). The protective potential of SBR against experimental caries was established in a rat intranasal immunization study using the 42-kDa SBR linked to CTB and given with an adjuvant dose of CT (Hajishengallis *et al.*, 1995; Hajishengallis *et al.*, 1998a). The specificity of the protective antibody response was revealed by the finding that immunization with CTB-conjugated AgII (the C-terminal third of AgI/II; Fig. 11.2) was ineffective although both immunogens induced comparable levels of salivary IgA response (Hajishengallis *et al.*, 1998a). Interestingly, the salivary IgA response induced by AgI/II conjugated to CTB was primarily directed to the C-terminal AgII rather than to the SBR, as observed previously in monkeys immunized with *S. mutans* (Russell *et al.*, 1980b). This suggests the occurrence of intramolecular antigenic competition (Taussig, 1973), which might serve as a strategy whereby *S. mutans* diverts the host response to a part of the AgI/II molecule that is not involved in adherence. This putative evasion strategy becomes more meaningful given that AgII represents the protease-resistant region of AgI/II, in contrast to the N-terminal AgI region (including the SBR) which is quite sensitive to proteolytic digestion (Russell *et al.*, 1980a). Therefore, natural mucosal immunization upon swallowing *S. mutans* would probably result in inductive-site uptake of cells expressing truncated AgII molecules due to proteloytic digestion of the N-terminal AgI by enteric proteases with resulting IgA responses directed predominantly against the non-protective AgII region. Such considerations underscore the importance of active immunization using SBR or other protective AgI/II epitopes. B- and T-cell epitopes have been identified throughout AgI/II (Kelly *et al.*, 1995; Matsushita *et al.*, 1994) and thus emphasis should be placed on the selection of epitopes with functional significance in *S. mutans* colonization.

The SBR was moreover shown to confer protection against oral colonization of *S. mutans* and caries formation in mice when delivered intranasally by an attenuated *Salmonella typhimurium* vector (Huang *et al.*, 2001). However, SBR may not be the only protective immunogenic region of AgI/II, as intact AgI/II conferred improved protection compared to SBR, although the salivary IgA anti-SBR responses induced by the two immunogens were quite comparable (Hajishengallis *et al.*, 1998a). This suggests that additional protective epitopes are present in AgI, such as in the P-region which appears to define a second saliva-adherence domain (Munro *et al.*, 1993). Topical application of monoclonal antibodies which recognize the P-region prevents tooth colonization by *S. mutans* in primates and results in significant caries reduction (Lehner *et al.*, 1992).

Oral immunization studies in rodents using soluble or liposome-encapsulated intact GTF showed induction of salivary IgA antibodies that interfered with mutans streptococcal colonization and caries activity (Michalek *et al.*, 1986; Smith *et al.*, 1979). More recent studies using multiple antigenic peptides constructed from conserved peptide sequences corresponding to either the catalytic or the glucan-binding domains of GTF resulted in inhibition of *S. mutans-* or *S. sobrinus-*induced caries, presumably through inhibition of glucan synthesis (Smith and Taubman, 1997; Taubman *et al.*, 1995). Immunization of rats by subcutaneous injection (in the vicinity of salivary glands) with diepitopic constructs, including peptide epitopes from both the catalytic and glucan-binding regions of GTF, resulted in higher serum IgG and salivary IgA antibodies than did a mixture of both peptides, and conferred protection against caries (Taubman *et al.*, 2001). The induced antibodies also inhibited glucan formation by GTF *in vitro*. Additional diepitopic peptide constructs were synthesized to include GbpB peptides together with sequences from the catalytic (CAT) or glucan-binding (GLU) domains of GTF and used to immunize rats. GbpB-CAT was the most immunogenic construct which elicited significant levels of serum IgG and salivary IgA antibody to both components and induced protective immunity against *S. mutans-* or *S. sobrinus-*induced caries (Smith *et al.*, 2005).

Chimeric polypeptide immunogens comprising the SBR of AgI/II and a glucan-binding segment of GTF have been constructed with the objective of inducing immunity that might interfere with both the initial adherence of *S. mutans* to saliva-coated tooth surfaces and the subsequent sucrose-dependent accumulation of this pathogen (Yu *et al.*, 1997; Zhang *et al.*, 2002). The chimeric construct was found to be significantly more immunogenic than its constituent parts, giving rise to higher anti-SBR and anti-glucan-binding region antibody responses in serum and saliva and significantly reducing *S. mutans* colonization (Zhang *et al.*, 2002). Rats immunized subcutaneously with a fusion DNA vaccine encoding both the glucan-binding region of GTF and an AgI/II segment that encompassed both the SBR and the P-region generated serum and salivary antibody responses to both components and showed protection against caries (Guo *et al.*, 2004). However, the fusion construct was only slightly more effective than a similar DNA vaccine encoding only the AgI/II domains. A similar fusion DNA vaccine constructed by transferring the *S. mutans*-specific sequences into another vector suitable for use in humans was found to be immunogenic in mice by the intranasal route, eliciting antibody responses that persisted for at least 6 months (Xu *et al.*, 2006). The protective potential of this DNA vaccine was demonstrated in rats in which intranasal immunization resulted in specific salivary IgA antibody responses and reduction in caries caused by *S. mutans* or *S. sobrinus*, suggesting that the vaccine can induce cross-protective immunity (Xu *et al.*, 2007).

Only a limited number of caries immunization studies using subunit vaccines have been performed in adult human volunteers. Although useful for establishing safety and immunogenicity, these studies provide limited information with regard to the preventive potential of caries vaccination. Oral immunization with GTF and aluminium phosphate adjuvant induced a modest and short-lived salivary IgA response which, however, delayed the recolonization of indigenous mutans streptococci following dental prophylaxis, relative to non-immunized control subjects (Smith and Taubman, 1987). Salivary IgA responses to GTF were also induced by intranasal immunization using liposome-encapsulated GTF alone or adjuvanted with monophosphoryl lipid A (Childers *et al.*, 2000; Childers *et al.*, 1999). The effectiveness of antibodies to interfere with recolonization of tooth surfaces by mutans streptococci, following treatment with an antibacterial mouthwash, was also shown by topical application of antibodies to AgI/II (Ma and Lehner, 1990). No caries immunization studies have been performed yet in young children, although such trials would be essential for investigating the feasibility of preventing caries through vaccination.

Periodontal disease vaccines

Objectives

Periodontitis is an infection-driven chronic inflammatory disease that affects the integrity of the tooth-supporting tissues (Pihlstrom *et al.*, 2005). Periodontal tissue degradation results primarily from untoward inflammatory reactions arising from complex interactions between the host immune system and subgingival bacterial challenge (Baker, 2000; Taubman and Kawai, 2001). With the identification of relatively few organisms as major agents of human periodontal disease (the 'red complex'), especially *Porphyromonas gingivalis*, has come the concept of immunization against it. Most vaccine development efforts are aimed at conferring protection by antibody-mediated blockade of colonization factors, neutralization of virulence enzymes or other factors involved in microbial evasion of immunity, or opsonophagocytosis and killing of periodontal pathogens (Fan *et al.*, 2001). Given recent findings that some periodontal pathogens invade and reside within permissive cells (Lamont and Yilmaz, 2002), it is even conceivable that vaccination aimed at generating cell-mediated immunity might contribute to protection.

All these approaches share the feature of decreasing the bacterial challenge rather than modulating the host response. However, what constitutes protective immunity against periodontitis has not been adequately elucidated yet, and at least some aspects of a vaccine-induced host response may result in more harm than good. For example, immunization may result in formation of antibody-antigen complexes which can interact with Fcγ receptors and induce excessive production of TNF-α (Abrahams *et al.*, 2000), a cytokine implicated in periodontal bone resorption (Graves and Cochran, 2003). Furthermore, opsonophagocytic modes of defence against bacterial infections always involve the collateral release of inflammatory mediators, tissue-destructive enzymes, and toxic substances such as reactive oxygen intermediates from the activated phagocytes, as well as the generation of split products of complement activation which have chemoattractant and cell-activating properties. If the pathogens are completely eliminated, then the inflammation is resolved and tissue healing ensues. If not, then the stage is set for persistent chronic inflammation. However, a protective vaccine may be a feasible pursuit if appropriate immunogens and vaccination strategies that do not induce adverse immune reactions can be identified.

The greater complexity of pathogenic mechanisms in periodontal disease compared to caries, and the involvement of multiple putative periodontal pathogens render the development of a periodontitis vaccine a formidable challenge. However, the identification of protective epitopes from key virulence factors of the various pathogens could lead to a multivalent periodontitis vaccine, the success of which would have to be determined empirically. The functional characterization and mapping of virulence proteins and their use for inducing protective immunity in animal models have therefore been pursued for the development of a human periodontitis vaccine.

Target organisms: characterization of virulence factors and proof-of-principle immunization studies in rodents

The concept that vaccination against periodontal pathogens can protect against disease was shown in early studies using whole bacterial cells or sonicates as an immunogen. Rats monoinfected with *Eikenella corrodens* exhibit significant alveolar bone loss unless first vaccinated parenterally with sonicates of this pathogen, and protection against bone loss was attributed to increased serum antibody responses induced by immunization (Behling *et al.*, 1981). In another rat study, parenteral immunization with heat-killed cells of *P. gingivalis* resulted in specific serum and salivary antibodies which protected the rats against subsequent oral infection with the pathogen, as evidenced by reduced alveolar bone loss (Klausen *et al.*, 1991). The activities of tissue-degrading enzymes, gingival collagenase and cathepsins B and L, were also diminished in the immunized animals. A whole-cell vaccine based on *Porphyromonas gulae*, an animal periodontopathogen considered equivalent to human *P. gingivalis*, has been developed for potential use in dogs (Hardham *et al.*, 2005). When tested in a mouse model it induced high titres of specific antibodies and protected against alveolar bone loss caused by homologous and heterologous challenge with *Porphyromonas* species. However, parenteral immunization against *A.a.* resulted in increased bone loss in rats monoinfected with this bacterium (Taubman *et al.*, 1983). This study underscores the complexities associated with the development of a periodontal vaccine, since not all host responses are protective.

Because of problems relating to reactogenicity, many current vaccine development efforts focus on highly engineered strains or defined subunits of microorganisms designed to minimize harmful side-effects. In the case of periodontal disease, subunit vaccine approaches have concentrated on *P. gingivalis* virulence proteins, especially its fimbriae and cysteine proteinases. However, because of the number of organisms implicated, a successful vaccine against human periodontitis will likely require either the identification of key protective antigens from multiple periodontal pathogens, or the identification of shared, conserved antigenic determinants with cross-protective potential.

P. gingivalis *fimbriae*

The fimbriae form adhesive hair-like appendages on *P. gingivalis* cell surface and are of two types: 'major' fimbriae, 0.3 to 3.0 μm long, encoded by the *fimA* gene (Dickinson *et al.*, 1988) and 'minor' fimbriae of shorter length encoded by the *mfa1* gene (Umemoto and Hamada, 2003). Major fimbriae have been studied in considerably greater detail and their role in *P. gingivalis* virulence is well documented; FimA-deficient mutants are fairly avirulent compared to wild-type *P. gingivalis* in inducing periodontal bone loss (Gibson *et al.*, 2004; Malek *et al.*, 1994). The major fimbriae appear to have a multifaceted involvement in virulence, since they can interact with various dental or epithelial substrates, other bacteria, or innate immune cells (Lamont and Jenkinson, 1998). Their ability to mediate adherence of *P. gingivalis* in vitro to early tooth-colonizing bacteria (Lamont *et al.*, 1994; Lamont *et al.*, 1992), to saliva-coated tooth surfaces (Amano *et al.*, 1996) or to matrix proteins (Sojar *et al.*, 1995) and gingival epithelial cells (Yilmaz *et al.*, 2002) is likely crucial for *P. gingivalis* colonization of the subgingival region of the teeth (Noiri *et al.*, 2004). Fimbriae are moreover essential for *P. gingivalis* invasiveness and replication within epithelial and endothelial cells (Belanger *et al.*, 2006; Deshpande *et al.*, 1998; Lamont *et al.*, 1995), properties that are thought to contribute to periodontal disease pathogenesis. Peptide mapping studies indicate that the versatile binding capacity of *P. gingivalis* fimbriae is attributable to multiple adhesion epitopes within the 337-amino-acid-long structural subunit of FimA (reviewed in Hajishengallis, 2007; Fig. 11.4). Although discrete FimA peptides have been generally associated with binding to distinct host

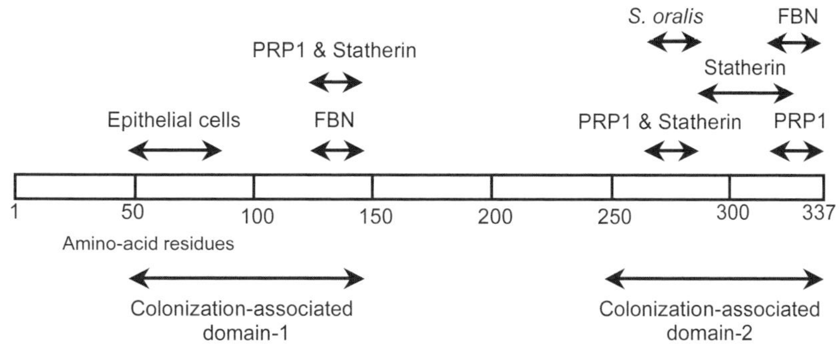

Figure 11.4 Schematic representation of *P. gingivalis* major fimbrial protein, FimA. *P. gingivalis* FimA (41 kDa; 337 amino acid residues) contains multiple adhesion epitopes located at the N-terminal end and C-terminal one-third. The adhesion epitopes mediate binding to salivary proteins (PRP1, proline-rich protein-1; Stath, statherin;), extracellular matrix proteins (FBN, fibronectin), to epithelial cells, and to the early colonizer, *Strep. oralis*.

or microbial receptors, binding activity can often be attributed to consensus amino-acid motifs shared by different FimA epitopes (Ogawa *et al.*, 1994a; Ogawa *et al.*, 1994b; Sojar *et al.*, 1995).

The concept that periodontal disease can be prevented by immunization with *P. gingivalis* fimbriae was first demonstrated in a rat model, in which protection was conferred against *P. gingivalis*-induced alveolar bone loss (Evans *et al.*, 1992). Subsequent studies focused on identification of immunodominant T and B cell epitopes of FimA, which in association with colonization epitopes, might be harnessed for the development of subunit vaccines against periodontitis. Fimbriae constitute an immunodominant antigen of *P. gingivalis* that readily reacts with sera from periodontitis patients (Boutsl *et al.*, 1996). B cell epitopes have been identified throughout the FimA sequence in animal studies, and epitopes defined by residues 31–50, 99–110, 212–231, and 312–331 appear to be immunodominant (Brant *et al.*, 1995; Ogawa, 1994; Ogawa *et al.*, 1991). Similarly, the segments corresponding to residues 61–70, 116–125, 151–160 and 316–325 have been identified as potent T cell epitopes, on the basis of their ability to stimulate *in vitro* production of interferon-γ, IL-4, IL-5 and IL-6 by T cells obtained from *P. gingivalis* fimbria-immunized mice (Ogawa *et al.*, 1995). The most potent T cell epitope in this study was found to be FimA 116–125 whereas another study identified an overlapping epitope (residues 103–122) as immunodominant. Strikingly, the ability of FimA 103–122 to stimulate primed mouse T cells was comparable to that of intact FimA (Deslauriers *et al.*, 1996).

FimA epitope mapping data have been used to develop mucosally delivered subunit vaccines against *P. gingivalis*-induced periodontal disease. FimA segments have been fused to the anchor region of a streptococcal surface protein and expressed on the surface of *Streptococcus gordonii* as a delivery vehicle for oral immunization (Sharma *et al.*, 1999; Sharma *et al.*, 1996; Sharma *et al.*, 1997). One FimA segment corresponds to residues 55 to 145 and contains the immunodominant T cell epitope (residues 103–122) as well as epitopes involved in binding to epithelial cells (residues 55–90) and to fibronectin (residues 126–145). A second FimA segment (residues 226 to 337) contains various B and T cell epitopes as well as epitopes involved in binding to early-colonizing bacteria or salivary proteins (Fig. 11.3). Oral immunization of rats with two *S. gordonii* vectors expressing FimA 55–145 or FimA 226–337 resulted in significant antibody responses against *P. gingivalis* in saliva (IgA antibodies) and serum (IgG and IgA antibodies) that were protective against subsequent *P. gingivalis*-induced periodontal bone loss (Sharma *et al.*, 2001).

Efforts to develop FimA-based vaccines also include the use of plasmid DNA immunization strategies. A FimA-encoding plasmid injected into the salivary glands of mice resulted in induction of anti-fimbrial IgA and IgG antibodies in saliva and serum and fimbria-specific cytotoxic T-cell activity in the spleen (Kawabata *et al.*, 1999). Intranasal immunization with a plasmid co-expressing FimA and IL-15, a cytokine that stimulates B cell proliferation and differentiation (Armitage *et al.*, 1995), resulted in significantly higher salivary IgA responses to FimA compared to a control plasmid expressing FimA alone (Guo *et al.*, 2006), although similar serum IgG antibody responses were induced by both plasmids. The protective potential of either DNA immunization approach remains to be established.

Another adjuvanted FimA vaccine comprising a C-terminal region of FimA (residues 266–337) fused to the B subunit of cholera toxin (CTB) has been expressed in potatoes but its immunogenicity has not yet been tested (Shin *et al.*, 2006).

Although the mechanisms of protection due to anti-fimbria antibody are not completely clear, it is likely that salivary IgA antibodies inhibit *P. gingivalis* colonization, either by blocking interbacterial interactions with early colonizers of the dental plaque, or by directly blocking its adherence to tooth surfaces or gingival epithelial cells. Salivary IgA antibodies can also inhibit the transmission of *P. gingivalis* between human contacts. As anti-fimbrial IgG antibody is highly opsonic, serum IgG antibodies might promote opsonophagocytosis and killing by neutrophils or macrophages (Fan *et al.*, 2001), although the finding that different genotypes of FimA generally do not share opsonic targets detracts from this possibility. *P. gingivalis* fimbriae have been classified into six genotypes (I to V and Ib) based on the sequence diversity of the *fimA* gene (reviewed in Amano *et al.*, 2004). *P. gingivalis* expressing Type II fimbriae is the most common with prevalence ranging from 37% to 66% of periodontitis patients, whereas the second most prevalent has been variably found to be Type IV, Ib, or I depending on the population studied (Amano *et al.*, 2000; Eick *et al.*, 2002; Missailidis *et al.*, 2004; Miura *et al.*, 2005; van der Ploeg *et al.*, 2004). Type I fimbriae display 97% amino acid identity with Type Ib, 76% identity with Type II, but only 46% identity with Type IV. Comparative studies have shown that *P. gingivalis* strains expressing Type II or Type I fimbriae are most potent in adhesive and invasive activities and perhaps include the most virulent strains (Nakagawa *et al.*, 2000; Nakagawa *et al.*, 2002; Nakagawa *et al.*, 2006). Therefore, selection of protective epitopes that are shared by *P. gingivalis* strains expressing fimbriae of Types I, Ib, or II would probably result in a vaccine that is active against most virulent *P. gingivalis* strains. Well conserved, extended FimA segments are shared between *P. gingivalis* strains expressing Type I or Type II fimbriae, and these bind equally well to saliva-coated hydroxyapatite (Nakagawa *et al.*, 2000). The saliva-binding domain defined by residues 266–337 displays 97% identity between Type I and Type II fimbriae, whereas epitopes interacting with proline-rich protein-1 residues 318–337) or statherin (residues 293–306 and 307–326) are 100% identical. In contrast, the internal region of Type I and Type II FimA, roughly defined by residues 100–250, is only 59% identical. The N-terminal one third of FimA is also well conserved between Type I and Type II (93% identity). Therefore the N-terminal and C-terminal one-third segments are appropriate for selecting protective epitopes that might induce colonization-blocking or opsonic antibodies against fimbriae present on the most pathogenic and prevalent strains of *P. gingivalis*.

P. gingivalis *cysteine proteinases and haemagglutinins*

P. gingivalis expresses Arg- and Lys-specific cysteine proteinases, known as gingipains, which readily degrade host proteins, including extracellular matrix and cytokines (Potempa *et al.*, 2000). The *rgpA* gene encodes a protein containing both a catalytic domain and an adhesion/haemagglutinin region, whereas *rgpB* encodes a protein with a highly homologous catalytic region that lacks a haemagglutinin domain, and *kgp* encodes a Lys-specific proteinase which includes a haemagglutinin region homologous to that of the RgpA (Curtis *et al.*, 1999; Fig. 11.5). The critical roles of RgpA, RgpB and Kgp in the virulence of *P. gingivalis* have been established in a mouse lesion model using isogenic mutants lacking each of the gingipains. All of the mutants displayed significantly reduced virulence compared with the wild-type parental strain (W50) and the order in which the gingipains contributed to virulence was Kgp >> RgpB ≥ RgpA (O'Brien-Simpson *et al.*, 2001).

Although the *P. gingivalis* RgpA/B and Kgp proteases degrade a variety of extracellular matrix proteins (*e.g.* fibronectin, laminin and several types of collagen), they can also induce host protein degradation indirectly through activation of latent host matrix metalloproteinases (Sorsa *et al.*, 1992). Moreover, the gingipains inhibit human complement-dependent bactericidal activity by

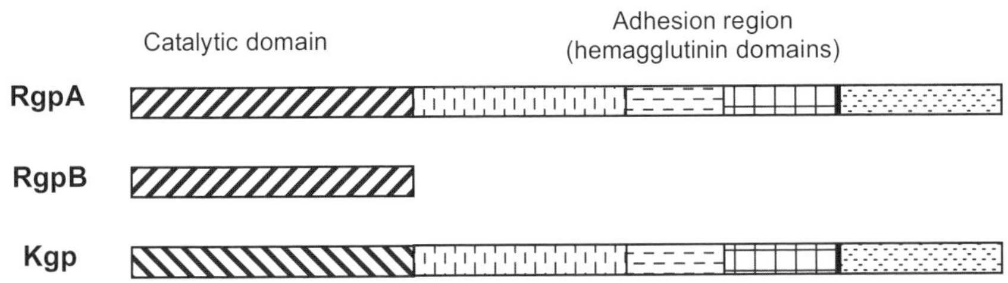

Figure 11.5 Schematic representation of *P. gingivalis* gingipains, RgpA, RgpB, and Kgp. The gingipains are a complex family of cysteine proteases, some of which are associated with haemagglutinins (Curtis *et al.*, 1999). RgpA and Kgp contain 1706 and 1723 amino acid residues, respectively (~190 kDa), and RgpB contains 736 residues (80 kDa). The catalytic domain of RgpA is largely homologous to RgpB, which lacks the haemagglutinin (adhesin) domains. The haemagglutinin domains of RgpA and Kgp are highly homologous.

degrading IgG and IgM as well as complement factor C3 (Grenier, 1992). CD14, which is an essential coreceptor for TLR-mediated innate immune recognition is degraded by gingipains (Sugawara *et al.*, 2000), and so also are numerous cytokines (Calkins *et al.*, 1998). Because gingipains have the potential to undermine the host immune response, neutralization of their enzymatic activity by antibodies induced through vaccination might not only suppress proteolytic destruction of periodontal tissue but also block the ability of *P. gingivalis* to evade host immunity (Table 11.2). RgpA and Kgp are also opsonic targets for IgG antibodies that significantly enhance neutrophil-mediated killing of *P. gingivalis* (Nakagawa *et al.*, 2001). Although the gingipains were isolated from strain HG66, antibody-mediated opsonization and killing was effective against four different strains representing distinct serotypes A, B, C, and D. Both the enzyme and the adhesin components of the gingipains are highly immunogenic (O'Brien-Simpson *et al.*, 2000), a property that can be exploited for vaccine development, especially as these enzymes are present in all isolates of *P. gingivalis*.

Rats immunized with a combination of RgpA/Kgp in incomplete Freund adjuvant elicited high-titre serum IgG2a antibody responses to the gingipains and were protected against *P. gingivalis*-induced periodontal bone loss (Rajapakse *et al.*, 2002). Although the control rats were positive for *P. gingivalis*, the pathogen was undetectable by DNA probe analysis of subgingival plaque samples from the RgpA/Kgp-immunized animals. In addition to possible neutralization of gingipains by antibody, the findings suggest that *P. gingivalis* may have been eliminated by IgG2a-mediated opsonization and facilitation of FcγRII-dependent phagocytosis and killing. It is also possible that specific IgG2a antibodies inhibited colonization by *P. gingivalis*.

The protective potential of anti-haemagglutinin antibody activity in the bone loss model is supported by independent studies, in which subcutaneous immunization of rats with recombinant haemagglutinin B (HagB) resulted in protection against *P. gingivalis*-induced periodontal bone loss (Katz *et al.*, 1999). Similarly, immunization with purified native *P. gingivalis* haemagglutinin resulted in reduced colonization of implanted *P. gingivalis* in hamsters with ligated teeth (Okuda *et al.*, 1988). Passively administered antibody was also protective, suggesting that antibody alone had a significant impact. *P. gingivalis* HagA and HagB have been expressed in avirulent *Salmonella* vectors for peroral administration which induced antibody and memory responses, although their protection against periodontitis was not assessed (Kohler *et al.*, 2000; Kozarov *et al.*, 2000).

Immunization with a peptide derived from the catalytic domain of RgpA, but not peptides from the haemagglutinin domain, induced a protective immune response against *P. gingivalis* infection in the mouse subcutaneous chamber model (Genco *et al.*, 1998). A protective immune response against *P. gingivalis*-induced periodontal bone loss was also induced in mice by immunization with synthetic peptides corresponding to repeated motifs within the adhesin domains of RgpA and Kgp that were conjugated to diphtheria toxic. Even more strongly protective were two similar diphtheria toxic-conjugated peptides corresponding to sequences of the catalytic sites of the RgpA and Kgp proteinases (O'Brien-Simpson *et al.*, 2005).

Immunization against RgpA has also been performed using an RgpA DNA plasmid vaccine administered intradermally in mice; this induced high levels of serum IgG antibodies against *P. gingivalis* which suppressed the proteolytic activity of both RgpA and RgpB and inhibited the binding of *P. gingivalis* to type I collagen as well as its haemagglutination. The immunized mice also exhibited resistance against *P. gingivalis* invasiveness in an abscess model (Yonezawa *et al.*, 2001).

Passive topical application in humans of a monoclonal antibody that recognized many laboratory strains and clinical isolates of *P. gingivalis* was reported to inhibit oral recolonization by *P. gingivalis* in sites with severe periodontitis for up to 9 months (Booth *et al.*, 1996). This protective mAb was subsequently mapped to the haemagglutinin domain of RgpA, and it inhibited haemagglutination by *P. gingivalis* or purified gingipains, a mechanism which may have contributed to the observed inhibition of recolonization (Booth and Lehner, 1997). It is known that RgpA and Kgp can bind and degrade haemoglobin and facilitate utilization of haemin as an iron source by *P. gingivalis* (Simpson *et al.*, 2004). IgG antibodies to Kgp induced by DNA immunization have been found to inhibit the binding of *P. gingivalis* to haemoglobin, restrict its *in vitro* growth when haemoglobin is the only iron source, and protect mice against lethal peritoneal infection by *P. gingivalis* (Kuboniwa *et al.*, 2001). Moreover, subcutaneous immunization of rats with the haemoglobin-binding domain of gingipain (HA2) resulted in specific IgG antibodies and protection against bone loss (DeCarlo *et al.*, 2003).

Whereas the protective effects of immunization with RgpA have been well documented, RgpB fails to induce protection against alveolar bone loss in mice, despite its ability to stimulate specific IgG antibodies (Gibson and Genco, 2001). Because RgpA and RgpB share similar catalytic domains, while RgpA additionally possesses a haemagglutinin domain, it may be concluded that protection is mediated by haemagglutinin domain-specific antibodies. However, as this conclusion is contradicted by the protective potential of anti-catalytic domain responses (Genco *et al.*, 1998; O'Brien-Simpson *et al.*, 2005), the reason why RgpB fails to induce protective immunity remains unclear. Nevertheless, RgpA/Kgp-based vaccines hold great potential for protection against *P. gingivalis* infection. Whether such vaccines will afford protection against periodontitis may depend in part on the extent to which other pathogens have a critical role in the human disease.

Cross-reactive proteins from periodontal pathogens

Since numerous pathogenic organisms contribute to human periodontitis, attempts to identify cross-protective immunogens present enormous interest. Heat-shock proteins (HSP) display a high degree of sequence homology among periodontopathogens (Hinode *et al.*, 1998), and have therefore been exploited for developing a cross-protective vaccine against multiple pathogens (Lee *et al.*, 2006). Antiserum to *P. gingivalis* HSP-60 was found to cross-react with HSPs from other species and, most importantly, to show strong opsonophagocytic function against all bacteria tested (*P. gingivalis*, *A.a.*, *P. intermedia*, *T. forsythia*, *F. nucleatum* and *T. denticola*; Choi *et al.*, 2005). Rats immunized with recombinant *P. gingivalis* HSP-60 and subsequently infected orally with *P. gingivalis* alone or in combination with *A.a.*, *P. intermedia*, *T. forsythia*, *F. nucleatum* and *T. denticola* revealed a strong inverse relationship between the induced anti-HSP-60 IgG levels and the degree of periodontal bone loss. Although these results support the protective potential of *P. gingivalis* HSP-60 as a universal periodontal immunogen, it is not clear whether such immunization would also affect commensal bacteria associated with periodontal health. This issue requires further study. Another concern involves possible cross-reactivity between bacterial and mammalian HSPs due to the conserved nature of these proteins (Munk *et al.*, 1989). As a bacterial HSP vaccine might lead to pathological reactions associated with autoimmunity to human HSPs, it is imperative to identify protective immunogenic epitopes that do not cross-react with human HSPs.

A.a. antigens

A.a. has received less attention than *P. gingivalis* with regard to vaccine development, but it has been extensively studied with respect to the role of T cell-mediated immunity in bone loss in rats. Whereas *A. a.*-sensitized Th1 clones adoptively transferred into naïve rats exacerbate periodontal bone resorption following oral infection with the pathogen, transfer of specific Th2 type clones results in induction of high levels of specific antibodies and protection against bone loss (Eastcott *et al.*, 1994; Eastcott *et al.*, 1990; Yamashita *et al.*, 1991; Yoshie *et al.*, 1987). It is thus possible that antibody responses, when directed to appropriate targets, may protect against this periodontopathogen.

Systemic or mucosal immunization of mice with an adjuvanted synthetic peptide derived from the sequence of *A.a.* fimbriae resulted in induction of specific salivary IgA and serum IgG antibodies (Honma *et al.*, 1999). Antibodies to the same peptide inhibit the adherence of fimbriated *A.a.* strains to experimental salivary pellicles (Harano *et al.*, 1995). Therefore, an anti-fimbrial vaccine may be able to interfere with *A.a.* colonization and subsequent induction of alveolar bone loss, although this has not yet been tested. Another study found that the mucosal immunogenicity of the capsule-like serotype b polysaccharide antigen of *A.a.* was significantly enhanced upon conjugation to bovine serum albumin and coadministration with CTB in mice. The resulting antibodies were highly opsonic for *A.a.* serotype b, which is perhaps the most pathogenic serotype of this species (Takamatsu-Matsushita *et al.*, 1996), but the *in vivo* protective potential of this vaccine was not studied.

Periodontal vaccination studies in non-human primates

Unlike immunization against caries, no periodontitis vaccine studies have been carried out in humans. However, several experiments have been performed in non-human primates, especially using the ligature-induced periodontitis model in cynomolgus monkeys (*Macaca fascicularis*). Immunization of monkeys with whole cells of *Bacteroides macacae*, a non-human primate equivalent of *P. gingivalis*, induced serum IgG antibodies and reduced subgingival colonization of *B. macacae*. However, comparable ligature-induced bone loss was observed in both immunized and control monkeys (Nisengard *et al.*, 1989). In another study, when monkeys were parenterally immunized with whole cells of *P. gingivalis* or *Prevotella intermedia*, approximately 2-log increases in specific serum IgG and IgA antibody responses were generated. Although immunization with *P. gingivalis* resulted in decreased levels of this organism in subgingival plaque, immunization with *P. intermedia* had no significant effect on colonization by this pathogen. Moreover, immunization failed to protect against periodontal bone loss (Ebersole *et al.*, 1991).

Immunization of monkeys with purified *P. gingivalis* cysteine proteinase in incomplete Freund's adjuvant elicited antibody responses against both the purified immunogen and whole cells of *P. gingivalis* but did not suppress the emergence of this pathogen in the oral flora, although the total number of gram-negative anaerobes was reduced in the immunized animals. Induction of bone loss appeared to be greater in the non-immunized group, although the differences did not reach statistical significance (Moritz *et al.*, 1998). Thus specific immunization against a known virulence factor displayed potential for protective immunity, although more studies are warranted to determine whether protection can be induced.

The choice of adjuvant may be an important factor in the success of vaccination. Immunization of monkeys with formalin-killed *P. gingivalis* in Syntex Adjuvant Formulation elicited specific IgG responses and conferred significant protection against bone loss, although the inhibitory effect against *P. gingivalis* colonization was less pronounced (Persson *et al.*, 1994). In a similar experiment, the levels of prostaglandin E2 (but not of IL-1β or TNF-α) in the gingival crevicular fluid were dramatically suppressed in immunized monkeys compared to controls, and were correlated with decreased bone loss scores (Roberts *et al.*, 2004). This study suggests that suppression of inflammatory mediators by immunization, possibly due to reduced pathogenic challenge as a result of antibody-dependent inhibition of colonization, constitutes a potential protective mechanism against periodontitis.

Novel adjuvants for oral vaccine applications

Mucosal administration of most soluble protein antigens, including those selected for immunization against oral infectious diseases, elicits little or no S-IgA response, for reasons discussed above in the section 'The mucosal immune system and oral immunity'. Several strategies have been developed to overcome the inherent unresponsiveness of the mucosal immune system to non-replicating protein antigens and thereby induce immune responses in mucosal secretions and in serum (Russell, 2003). Some of the most effective strategies include the use of heat-labile enterotoxins and derivatives thereof, either as adjuvants or as coupled delivery agents (Hajishengallis *et al.*, 2005a) or the expression of protein immunogens in attenuated live bacterial or viral vectors (Curtiss, 2005). Both of these strategies have been successfully used in proof-of-concept caries or periodontal vaccination studies in rodents (Hajishengallis *et al.*, 1998b; Huang *et al.*, 2001a; Katz *et al.*, 1993; Redman *et al.*, 1995; Sharma *et al.*, 2001) and their advantages and disadvantages, including potential toxicity, have been reviewed elsewhere (Hajishengallis *et al.*, 2005a; Hajishengallis and Michalek, 1999).

The discovery of Toll-like receptors (TLR) has not only resulted in a resurgent interest in innate immunity, but additionally suggested novel approaches to adjuvant development. The property of TLRs to function as adjuvant receptors is based on their ability to recognize and respond to conserved 'microbe-associated molecular patterns' (MAMP; Iwasaki and Medzhitov, 2004). MAMPs, which are targets of innate immune recognition by pattern-recognition receptors, effectively alert the immune system and stimulate APC. At the molecular level, TLR-stimulated APC provide costimulatory signals for T cell activation ultimately leading to induction of cell-mediated or humoral immunity (Iwasaki and Medzhitov, 2004). Several adjuvants that have been shown to help poorly immunogenic proteins to induce an immune response are substances of microbial origin. For example, muramyl dipeptide, one of the active ingredients of the Freund complete adjuvant, is derived from bacterial peptidoglycan and activates TLR2. Monophosphoryl lipid A is a chemically modified form of the lipid A from *Salmonella minnesota* LPS, which activates TLR4 (Cox *et al.*, 2006).

TLRs are central signalling receptors for triggering innate immunity and generally respond to different types of MAMPs, thereby endowing the innate immune response with a degree of specificity (Akira and Takeda, 2004; Iwasaki and Medzhitov, 2004). For example, TLR2 responds to agonists including lipoteichoic acid, bacterial lipoproteins, and peptidoglycan, TLR4 to enterobacterial LPS, TLR5 to flagellin, and TLR9 to bacterial DNA. Metaphorically, the TLRs read a 'bar code' on microbes, which then elicits the appropriate type of innate response (Aderem, 2003). Different TLR-interacting adjuvants may be appropriate for vaccination against different infectious diseases, depending on the responses required for effective immunity. TLR2 or TLR4 activation is thought to induce Th2- or Th1-biased responses, respectively (Agrawal *et al.*, 2003; Re and Strominger, 2001). It is therefore possible that immunization against dental caries, where protection is mediated by mucosal IgA antibody, may be preferentially enhanced with a TLR2 adjuvant. Conversely, immunization against intracellular bacteria, including some of the periodontal pathogens (Lamont and Yilmaz, 2002), may be most effectively induced by a TLR4 adjuvant that can stimulate Th1 cell-mediated immunity. However, it has become clear that T cell help for adaptive immune responses is not adequately describable in such simple dichotomous terms, and in reality most immune responses against pathogens or vaccines involve aspects of both Th1 and Th2 activity, as well as other modes of T cell help including the recently described Th17 axis which has recently been postulated to be involved in periodontal disease (Yu *et al.*, 2007).

One of the most actively researched areas of mucosal adjuvant development has focused on the gram-negative heat-labile enterotoxins of which cholera toxin (CT) and the closely related type I labile toxin (LT-I) of *Escherichia coli* are the best known examples. An extensive literature attests to the effectiveness of both CT and LT-I as potent adjuvants for the oral or intranasal delivery of experimental vaccines. Although most reports indicate that CT in particular preferentially induces Th2-biased responses, neither of these toxins has been found to activate TLR, and their mechanism of action has not been fully elucidated (Arce *et al.*, 2007). Indeed the toxicity of these adjuvants precludes their use in humans, and consequently considerable effort has been given to developing non-toxic mutants, or evaluating their non-toxic B units as adjuvants or as coupled delivery agents (Hajishengallis *et al.*, 2005a). The type II enterotoxins of *E. coli*, LT-IIa and LT-IIb, resemble the type I toxins as AB_5 heterohexamers in which the homologous A subunits have essentially the same ADP-ribosylating activity that upregulates adenyl cyclase, whereas their pentameric B subunits show little homology with those of type I enterotoxins and bind to different gangliosides. LT-IIa and LT-IIb have similar mucosal adjuvant activity to CT and LT-I, but appear to bias the immune response more towards Th1 or a mixed pattern of T cell help (Martin *et al.*, 2000). Interestingly, however, LT-IIa and LT-IIb exert their adjuvant effects in a different manner than the type I enterotoxins (Arce *et al.*, 2005). Moreover, their B pentamers activate APC through TLR2, in addition to their interaction with gangliosides (Hajishengallis *et al.*, 2005b; Liang *et al.*, 2007a). It was thought until recently that the immunomodulatory and adjuvant effects of the intact heat-labile enterotoxins depended upon their ganglioside-binding and catalytic-toxic activities. However, the catalytic activity of the A subunit may not be critical for adjuvant action, since the isolated non-toxic B pentamers can display immunomodulatory activity (reviewed in Hajishengallis *et al.*, 2005a). The ganglioside-binding activity of the B pentamers is important (Guidry *et al.*, 1997; Nashar *et al.*, 1996), but apparently not sufficient since certain point mutations that do not preclude ganglioside binding result in defective immunomodulatory action (Aman *et al.*, 2001; Fraser *et al.*, 2003). These findings suggest that additional receptors may complement immunomodulatory signalling. We have shown that the B pentamer of LT-IIb binds both ganglioside GD1a and TLR2 in a functional signalling complex (Liang *et al.*, 2007a; Liang *et al.*, 2007b), and may thereby promote adaptive immune responses through GD1a/TLR2-dependent induction of costimulatory molecules and immunoenhancing cytokines by APC (Iwasaki and Medzhitov, 2004). The existence of an array of TLRs with both overlapping and unique immunostimulatory signalling may help in their exploitation for the development of novel adjuvants for use in both caries and periodontal vaccination.

Concluding remarks and future directions

From the above it should be evident that it is technically feasible to develop a vaccine against dental caries. Proof-of-principle has been demonstrated in rodent and primate models that immunization can elicit antibodies in serum and saliva against key antigenic targets of mutans streptococci, and can suppress colonization by these organisms with subsequently diminished development of caries lesions. *In vitro* studies further illustrate plausible mechanisms of action of these antibodies in inhibiting steps in caries pathogenesis. Further progress, however, will to an increasing extent require translation of the findings from animal models into human trials. This raises important ethical issues that are intertwined with economic factors, which have been aired recently elsewhere (Taubman and Nash, 2006). The incidence of dental caries has declined considerably in western nations over the past several decades, especially among the more affluent sectors of the population, for reasons that usually include the use of fluoride, improved access to high-quality dental care, and public awareness and education leading to better eating and oral hygiene habits. However, dental caries remains prevalent among underserved populations both within affluent western nations and elsewhere. A safe and effective vaccine could have significant impact among these groups where modern professional dental care is infrequently available. It is unlikely that a caries vaccine would attain 100% effectiveness either in terms of preventing all caries within a vaccinated subject, or in protecting all subjects, but even 50% effectiveness (by either criterion) could have a substantial impact in improving quality of life (less morbidity, pain, suffering, and consequential effects) as well as on the financial costs of treating caries.

Studies in rodents and non-human primates suggest the feasibility of developing a vaccine for the control of periodontal disease, although major challenges remain. Proof-of-principle studies in rodents discussed above show that immunization with functional domains of *P. gingivalis* gingipains represents a promising approach to developing a vaccine against periodontal infection with *P. gingivalis*. Whether a gingipain-based vaccine will have a significant protective impact on periodontal disease in humans remains to be determined. Given the polymicrobial aetiology of periodontitis, a cross-protective or a multivalent vaccine will probably be required. Another formidable challenge in periodontal vaccine development involves our limited understanding of the immune regulatory mechanisms operating in periodontal disease pathogenesis. Whereas specific immunity has been strongly implicated in protection, it has also become clear that the immune system itself plays an important role in the immunopathogenesis of periodontal disease and that some aspects of the adaptive immune response contribute to periodontal tissue destruction. The finding that immunodeficient mice display resistance to *P. gingivalis*-induced bone loss can also be interpreted to suggest that innate immunity has a protective role against periodontitis (Baker *et al.*, 1994). Another recent study using IL-17-receptor-knockout mice implies an important role for innate responses and the so-called 'Th17 axis of immunity' in the control of *P. gingivalis* infection and associated bone loss (Yu *et al.*, 2007).

In contrast to caries vaccines which are implicitly designed for prophylactic use, it is possible that periodontal vaccines could be more appropriate for therapeutic application in individuals already affected by the disease. This has the potential advantage that the principal target etiological agent(s) can be identified in each case. However, most experimental work in animal models so far has been designed to test periodontal vaccines in a prophylactic setting. Therapeutic models may offer a novel and fruitful direction to explore.

It may also be advantageous to design vaccines, including the selection of adjuvants, that, besides enhancing specific immunity to the target periodontopathogens, also elicit appropriate non-inflammatory modes of immune response that diminish tissue damage and promote restoration, and maybe also stimulate innate immune mechanisms. For example, TLR-interacting adjuvants may be capable of stimulating the microbicidal activity of macrophages thereby enhancing defence against periodontal pathogens. However, this approach might also enhance the destructive aspects of cell-mediated immunity, by inducing excessive levels of bone-resorbing cytokines, such as TNF-α, or activating Th1 cells that promote RANKL-dependent osteoclastogenesis (Taubman and Kawai, 2001; Teng, 2006). These considerations highlight the importance of better understanding host-pathogen interactions in the periodontium and elucidating the host response signalling pathways. It remains a challenge in vaccine development to fine-tune the host response to maximize protection and minimize the destructive aspects.

References

Abrahams, V.M., Cambridge, G., Lydyard, P.M., and Edwards, J.C. (2000) Induction of tumor necrosis factor alpha production by adhered human monocytes: a key role for Fcγ receptor type IIIa in rheumatoid arthritis. Arthritis Rheum. *43*, 608–616.

Aderem, A. (2003) Phagocytosis and the inflammatory response. J. Infect. Dis. *187 Suppl. 2*, S340–345.

Agrawal, S., Agrawal, A., Doughty, B., Gerwitz, A., Blenis, J., Van Dyke, T., and Pulendran, B. (2003) Cutting edge: different Toll-like receptor agonists instruct dendritic cells to induce distinct Th responses via differential modulation of extracellular signal-regulated kinase-mitogen-activated protein kinase and c-Fos. J. Immunol. *171*, 4984–4989.

Ajdic, D., McShan, W.M., McLaughlin, R.E., Savic, G., Chang, J., Carson, M.B., Primeaux, C., Tian, R.Y., Kenton, S., Jia, H.G., *et al.* (2002) Genome sequence of *Streptococcus mutans* UA159, a cariogenic dental pathogen. Proc. Natl. Acad. Sci., USA. *99*, 14434–14439.

Akira, S., and Takeda, K. (2004) Toll-like receptor signalling. Nat. Rev. Immunol. *4*, 499–511.

Aman, A.T., Fraser, S., Merritt, E.A., Rodigherio, C., Kenny, M., Ahn, M., Hol, W. G.J., Williams, N.A., Lencer, W.I., and Hirst, T.R. (2001) A mutant cholera toxin B subunit that binds GM1-ganglioside but lacks immunomodulatory or toxic activity. Proc. Natl. Acad. Sci. USA. *98*, 8536–8541.

Amano, A., Kuboniwa, M., Nakagawa, I., Akiyama, S., Morisaki, I., and Hamada, S. (2000) Prevalence of specific genotypes of *Porphyromonas gingivalis* fimA and periodontal health status. J. Dent. Res. *79*, 1664–1668.

Amano, A., Nakagawa, I., Okahashi, N., and Hamada, N. (2004) Variations of *Porphyromonas gingivalis* fimbriae in relation to microbial pathogenesis. J. Periodont. Res. *39*, 136–142.

Amano, A., Sharma, A., Lee, J.Y., Sojar, H.T., Raj, P.A., and Genco, R.J. (1996) Structural domains of *Porphyromonas gingivalis* recombinant fimbrillin that mediate binding to salivary proline-rich protein and statherin. Infect. Immun. *64*, 1631–1637.

Arce, S., Nawar, H., Russell, M.W., and Connell, T.D. (2005) Differential binding of *Esherichia coli* enterotoxins LT-IIa, LT-IIb and of cholera toxin elicits differences in apoptosis, proliferation, and activation in lymphoid cells. Infect. Immun. *73*, 2718–2727.

Arce, S., Nawar, H.F., Muehlinghaus, G., Russell, M.W., and Connell, T.D. (2007) In vitro induction of immunoglobulin A (IgA)- and IgM-secreting plasma blasts by cholera toxin depends on T-cell help and is mediated by CD154 up-regulation and inhibition of IFN-γ synthesis. Infect. Immun. *75*, 1413–1423.

Armitage, R.J., Macduff, B.M., Eisenman, J., Paxton, R., and Grabstein, K.H. (1995) IL-15 has stimulatory activity for the induction of B cell proliferation and differentiation. J. Immunol. *154*, 483–490.

Bahn, A.H., Shklair, I.L., and Hayashi, J.A. (1977) Immunization with dextransucrases, levansucrases, and glycosidic hydrolases from oral streptococci II. Immunization with dextransucrases, levansucrases, and glycosidic hydrolases from oral streptococci in monkeys. J Dent. Res. *56*, 1586–1598.

Baker, P.J. (2000) The role of immune responses in bone loss during periodontal disease. Microbes Infect. *2*, 1181–1192.

Baker, P.J., Dixon, M., Evans, R.T., Dufour, L., Johnson, E., and Roopenian, D.C. (1999) CD4(+) T cells and the proinflammatory cytokines gamma interferon and interleukin-6 contribute to alveolar bone loss in mice. Infect. Immun. *67*, 2804–2809.

Baker, P.J., Evans, R.T., and Roopenian, D.C. (1994) Oral infection with *Porphyromonas gingivalis* and induced alveolar bone loss in immunocompetent and severe combined immunodeficient mice. Arch Oral. Biol. *39*, 1035–1040.

Banas, J.A. (2004) Virulence properties of *Streptococcus mutans*. Front. Biosci. *9*, 1267–1277.

Behling, U.H., Sallay, C., Sanavi, F., Pham, P.H., and Nowotny, A. (1981) Humoral immunity and reduced periodontal bone loss in *Eikenella corrodens*-monoassociated rats. Infect. Immun. *33*, 801–805.

Belanger, M., Rodrigues, P.H., Dunn, W.A., Jr., and Progulske-Fox, A. (2006) Autophagy: a highway for *Porphyromonas gingivalis* in endothelial cells. Autophagy *2*, 165–170.

Booth, V., Ashley, F.P., and Lehner, T. (1996) Passive immunization with monoclonal antibodies against *Porphyromonas gingivalis* in patients with periodontitis. Infect. Immun. *64*, 422–427.

Booth, V., and Lehner, T. (1997) Characterization of the *Porphyromonas gingivalis* antigen recognized by a monoclonal antibody which prevents colonization by the organism. J. Periodont. Res. *32*, 54–60.

Boutsl, E.A., Koseki, T., Nishihara, T., and Ishikawa, I. (1996) Characterization of the immunodominant antigens of *Porphyromonas gingivalis* 381 in high-responder patients. Oral Microbiol. Immunol. *11*, 236–241.

Brady, L.J., Crowley, P.J., Ma, K.C., Kelly, C., Lee, S.F., Lehner, T., and Bleiweis, A.S. (1991) Restriction fragment length polymorphisms and sequence variation within the *spaP* gene of *Streptococcus mutans* serotype c isolates. Infect. Immun. *59*, 1803–1810.

Brady, L.J., Piacentini, D.A., Crowley, P.J., Oyston, P.C., and Bleiweis, A.S. (1992) Differentiation of salivary agglutinin-mediated adherence and aggregation of mutans streptococci by use of monoclonal antibodies against the major surface adhesin P1. Infect. Immun. *60*, 1008–1017.

Brandtzaeg, P., and Pabst, R. (2004) Let's go mucosal: communication on a slippery ground. Trends Immunol. *25*, 570–577.

Brant, E.E., Sojar, H.T., Sharma, A., Bedi, G.S., Genco, R.J., and De Nardin, E. (1995) Identification of linear antigenic sites on the *Porphyromonas gingivalis* 43-kDa fimbrillin subunit. Oral Microbiol. Immunol. *10*, 146–150.

Calkins, C.C., Platt, K., Potempa, J., and Travis, J. (1998) Inactivation of tumor necrosis factor-α by proteinases (gingipains) from the periodontal pathogen, *Porphyromonas gingivalis*. Implications of immune evasion. J. Biol. Chem. *273*, 6611–6614.

Challacombe, S.J., and Lehner, T. (1976) Serum and salivary antibodies to cariogenic bacteria in man. J. Dent. Res. 55 *(Special Issue C)*, C139-C148.

Challacombe, S.J., Russell, M.W., Hawkes, J.E., Bergmeier, L.A., and Lehner, T. (1978) Passage of immunoglobulins from plasma to the oral cavity in rhesus monkeys. Immunology *35*, 923–931.

Childers, N.K., Miller, K.L., Tong, G., Llarena, J.C., Greenway, T., Ulrich, J.T., and Michalek, S.M. (2000) Adjuvant activity of monophosphoryl lipid A for nasal and oral immunization with soluble or liposome-associated antigen. Infect. Immun. *68*, 5509–5516.

Childers, N.K., Tong, G., Mitchell, S., Kirk, K., Russell, M.W., and Michalek, S.M. (1999) A controlled clinical study of the effect of nasal immunization with a *Streptococcus mutans* antigen alone or incorporated into liposomes on induction of immune responses. Infect. Immun. *67*, 618–623.

Choi, J.I., Choi, K.S., Yi, N.N., Kim, U.S., Choi, J.S., and Kim, S.J. (2005) Recognition and phagocytosis of multiple periodontopathogenic bacteria by anti-*Porphyromonas gingivalis* heat-shock protein 60 antisera. Oral Microbiol. Immunol. *20*, 51–55.

Cox, E., Verdonck, F., Vanrompay, D., and Goddeeris, B. (2006) Adjuvants modulating mucosal immune responses or directing systemic responses towards the mucosa. Veterin. Res. *37*, 511–539.

Crowley, P.J., Brady, L.J., Michalek, S.M., and Bleiweis, A.S. (1999) Virulence of a spaP mutant of *Streptococcus mutans* in a gnotobiotic rat model. Infect. Immun. *67*, 1201–1206.

Crowley, P.J., Brady, L.J., Piacentini, D.A., and Bleiweis, A.S. (1993) Identification of a salivary agglutinin-binding domain within cell surface adhesin P1 of *Streptococcus mutans*. Infect. Immun. *61*, 1547–1552.

Curtis, M.A., Kuramitsu, H.K., Lantz, M., Macrina, F.L., Nakayama, K., Potempa, J., Reynolds, E.C., and Aduse-Opoku, J. (1999) Molecular genetics and nomenclature of proteases of *Porphyromonas gingivalis*. J. Periodont. Res. *34*, 464–472.

Curtiss, R. (2005) Antigen delivery systems II: Development of live recombinant attenuated bacterial antigen and DNA vaccine delivery vector vaccines. In Mucosal Immunology, 3rd Edn, J. Mestecky, J. Bienenstock, M. E. Lamm, L. Mayer, W. Strober, and J. R. McGhee, eds. (Amsterdam, Elsevier/Academic Press), pp. 1009–1038.

DeCarlo, A.A., Huang, Y., Collyer, C.A., Langley, D.B., and Katz, J. (2003) Feasibility of an HA2 domain-based periodontitis vaccine. Infect. Immun. *71*, 562–566.

Deshpande, R.G., Khan, M.B., and Genco, C.A. (1998) Invasion of aortic and heart endothelial cells by *Porphyromonas gingivalis*. Infect. Immun. *66*, 5337–5343.

Deslauriers, M., Haque, S., and Flood, P.M. (1996) Identification of murine protective epitopes on the *Porphyromonas gingivalis* fimbrillin molecule. Infect. Immun. *64*, 434–440.

Dickinson, D.P., Kubiniec, M.A., Yoshimura, F., and Genco, R.J. (1988) Molecular cloning and sequencing of the gene encoding the fimbrial subunit protein of *Bacteroides gingivalis*. J. Bacteriol. *170*, 1658–1665.

Douglas, C. W.I., and Russell, R. R.B. (1984) Effect of specific antisera upon *Streptococcus mutans* adherence to saliva-coated hydroxylapatite. FEMS Microbiol. Lett. *25*, 211–214.

Eastcott, J.W., Yamashita, K., Taubman, M.A., Harada, Y., and Smith, D.J. (1994) Adoptive transfer of cloned T helper cells ameliorates periodontal disease in nude rats. Oral Microbiol. Immunol. *9*, 284–289.

Eastcott, J.W., Yamashita, K., Taubman, M.A., and Smith, D.J. (1990) Characterization of rat T-cell clones with bacterial specificity. Immunology *71*, 120–126.

Ebersole, J.L., Brunsvold, M., Steffensen, B., Wood, R., and Holt, S.C. (1991) Effects of immunization with *Porphyromonas gingivalis* and *Prevotella intermedia* on progression of ligature-induced periodontitis in the nonhuman primate *Macaca fascicularis*. Infect. Immun. *59*, 3351–3359.

Eick, S., Rodel, J., Einax, J.W., and Pfister, W. (2002) Interaction of *Porphyromonas gingivalis* with KB cells: comparison of different clinical isolates. Oral Microbiol. Immunol. *17*, 201–208.

Evans, R.T., Klausen, B., Sojar, H.T., Bedi, G.S., Sfintescu, C., Ramamurthy, N.S., Golub, L.M., and Genco, R.J. (1992) Immunization with *Porphyromonas (Bacteroides) gingivalis* fimbriae protects against periodontal destruction. Infect. Immun. *60*, 2926–2935.

Fan, Q., Sims, T., Sojar, H., Genco, R., and Page, R.C. (2001) Fimbriae of *Porphyromonas gingivalis* induce opsonic antibodies that significantly enhance phagocytosis and killing by human polymorphonuclear leukocytes. Oral Microbiol. Immunol. *16*, 144–152.

Ferretti, J.J., Gilpin, M.L., and Russell, R.R. (1987) Nucleotide sequence of a glucosyltransferase gene from *Streptococcus sobrinus* MFe28. J. Bacteriol. *169*, 4271–4278.

Fraser, S.A., de Haan, L., Hearn, A.R., Bone, H.K., Salmond, R.J., Rivett, A.J., Williams, N.A., and Hirst, T.R. (2003) Mutant *Escherichia coli* heat-labile toxin B subunit that separates toxoid-mediated signaling and immunomodulatory action from trafficking and delivery functions. Infect. Immun. *71*, 1527–1537.

Genco, C.A., Odusanya, B.M., Potempa, J., Mikolajczyk-Pawlinska, J., and Travis, J. (1998) A peptide domain on gingipain R which confers immunity against *Porphyromonas gingivalis* infection in mice. Infect. Immun. *66*, 4108–4114.

Gibbons, R.J., Cohen, L., and Hay, D.I. (1986) Strains of *Streptococcus mutans* and *Streptococcus sobrinus* attach to different pellicle receptors. Infect. Immun. *52*, 555–561.

Gibson, F.C., 3rd, and Genco, C.A. (2001) Prevention of *Porphyromonas gingivalis*-induced oral bone loss following immunization with gingipain R1. Infect. Immun. *69*, 7959–7963.

Gibson, F.C., 3rd, Hong, C., Chou, H.H., Yumoto, H., Chen, J., Lien, E., Wong, J., and Genco, C.A. (2004) Innate immune recognition of invasive bacteria accelerates atherosclerosis in apolipoprotein E-deficient mice. Circulation *109*, 2801–2806.

Gilmore, K.S., Russell, R.R., and Ferretti, J.J. (1990) Analysis of the *Streptococcus downei* gtfS gene, which specifies a glucosyltransferase that synthesizes soluble glucans. Infect. Immun. *58*, 2452–2458.

Graves, D.T., and Cochran, D. (2003) The contribution of interleukin-1 and tumor necrosis factor to periodontal tissue destruction. J. Periodontol. *74*, 391–401.

Grenier, D. (1992) Inactivation of human serum bactericidal activity by a trypsinlike protease isolated from Porphyromonas gingivalis. Infect. Immun. *60*, 1854–1857.

Guidry, J.J., Cardenas, L., Cheng, E., and Clements, J.D. (1997) Role of receptor binding in toxicity, immunogenicity, and adjuvanticity of *Escherichia coli* heat-labile enterotoxin. Infect. Immun. *65*, 4943–4950.

Guo, H., Wang, X., Jiang, G., and Yang, P. (2006) Construction of a sIgA-enhancing anti-*Porphyromonas gingivalis* FimA vaccine and nasal immunization in mice. Immunol. Lett. *107*, 71–75.

Guo, J.H., Jia, R., Fan, M.W., Bian, Z., Chen, Z., and Peng, B. (2004) Construction and immunogenic characterization of a fusion anti-caries DNA vaccine against PAc and glucosyltransferase I of *Streptococcus mutans*. J. Dent. Res. *83*, 266–270.

Hahn, C.L., and Falkler, W.A., Jr. (1992) Antibodies in normal and diseased pulps reactive with microorganisms isolated from deep caries. J. Endodontol. *18*, 28–31.

Hajishengallis, G. (2007) Peptide mapping of a functionally versatile fimbrial adhesin from *Porphyromonas gingivalis*. Int J Pept Res Ther *In press*, Epub ahead of print; DOI:10.1007/s10989-007-9084-1.

Hajishengallis, G., Arce, S., Gockel, C.M., Connell, T.D., and Russell, M.W. (2005a). Immunomodulation with enterotoxins for the generation of secretory immunity or tolerance: applications for oral infections. J. Dent. Res. *84*, 1104–1116.

Hajishengallis, G., Hollingshead, S.K., Koga, T., and Russell, M.W. (1995) Mucosal immunization with a bacterial protein antigen genetically coupled to cholera toxin A2/B subunits. J. Immunol. *154*, 4322–4332.

Hajishengallis, G., Koga, T., and Russell, M.W. (1994) Affinity and specificity of the interactions between *Streptococcus mutans* antigen I/II and salivary components. J. Dent. Res. *73*, 1493–1502.

Hajishengallis, G., and Michalek, S.M. (1999) Current status of a mucosal vaccine against dental caries. Oral Microbiol. Immunol. *14*, 1–20.

Hajishengallis, G., Nikolova, E., and Russell, M.W. (1992) Inhibition of *Streptococcus mutans* adherence to saliva-coated hydroxyapatite by human secretory immunoglobulin A antibodies to the cell surface protein antigen I/II: Reversal by IgA1 protease cleavage. Infect. Immun. *60*, 5057–5064.

Hajishengallis, G., Russell, M.W., and Michalek, S.M. (1998a). Comparison of an adherence domain and a structural region of *Streptococcus mutans* antigen I/II in protective immunity against dental caries in rats after intranasal immunization. Infect. Immun. *66*, 1740–1743.

Hajishengallis, G., Russell, M.W., and Michalek, S.M. (1998b). Effectiveness of an adherence domain in comparison to a structural domain of *Streptococcus mutans* antigen I/II in protection against dental caries in rats after intranasal immunization. Infect. Immun. *66*, 1740–1743.

Hajishengallis, G., Tapping, R.I., Martin, M.H., Nawar, H., Lyle, E.A., Russell, M.W., and Connell, T.D. (2005b). Toll-like receptor 2 mediates cellular activation by the B subunits of type II heat-labile enterotoxins. Infect. Immun. *73*, 1343–1349.

Han, T.K., Zhang, C., and Dao, M.L. (2006) Identification and characterization of collagen-binding activity in *Streptococcus mutans* wall-associated protein: a possible implication in dental root caries and endocarditis. Biochem. Biophys. Res. Commun. *343*, 787–792.

Harano, K., Yamanaka, A., and Okuda, K. (1995) An antiserum to a synthetic fimbrial peptide of

Actinobacillus actinomycetemcomitans blocked adhesion of the microorganism. FEMS Microbiol. Lett. *130*, 279–285.

Hardham, J., Reed, M., Wong, J., King, K., Laurinat, B., Sfintescu, C., and Evans, R.T. (2005) Evaluation of a monovalent companion animal periodontal disease vaccine in an experimental mouse periodontitis model. Vaccine *23*, 3148–3156.

Hinode, D., Nakamura, R., Grenier, D., and Mayrand, D. (1998) Cross-reactivity of specific antibodies directed to heat shock proteins from periodontopathogenic bacteria of human origin. Oral Microbiol. Immunol. *13*, 55–58.

Holt, S.C., and Ebersole, J.L. (2005) *Porphyromonas gingivalis, Treponema denticola,* and *Tannerella forsythia*: the 'red complex', a prototype polybacterial pathogenic consortium in periodontitis. Periodontol. 2000 *38*, 72–122.

Honda, O., Kato, C., and Kuramitsu, H.K. (1990) Nucleotide sequence of the *Streptococcus mutans gtfD* gene encoding the glucosyltransferase-S enzyme. J. Gen. Microbiol. *136*, 2099–2105.

Honma, K., Kato, T., and Okuda, K. (1999) Salivary immunoglobulin A production against a synthetic oligopeptide antigen of *Actinobacillus actinomycetemcomitans* fimbriae. Oral Microbiol. Immunol. *14*, 288–292.

Huang, Y., Hajishengallis, G., and Michalek, S.M. (2001) Induction of protective immunity against *Streptococcus mutans* colonization after mucosal immunization with attenuated *Salmonella enterica* serovar typhimurium expressing an *S. mutans* adhesin under the control of *in vivo*-inducible nirB promoter. Infect. Immun. *69*, 2154–2161.

Hughes, M., Machardy, S., Sheppard, A., Langford, D., and Shepherd, W. (1986) Experiences in the development of a dental caries vaccine. In Molecular Microbiology and Immunobiology of *Streptococcus mutans,* S. Hamada, S.M. Michalek, H. Kiyono, L. Menaker, and J.R. McGhee, eds. (Amsterdam, The Netherlands, Elsevier Science Publishers), pp. 349–357.

Iwasaki, A., and Medzhitov, R. (2004) Toll-like receptor control of the adaptive immune responses. Nat. Immunol. 5, 987–995.

Janeway, C.A., and Medzhitov, R. (2002) Innate immune recognition. Ann. Rev. Immunol. *20*, 197–216.

Jenkinson, H.F., and Demuth, D.R. (1997) Structure, function and immunogenicity of streptococcal antigen I/II polypeptides. Mol Microbiol 23, 183–190.

Jespersgaard, C., Hajishengallis, G., Greenway, T.E., Smith, D.J., Russell, M.W., and Michalek, S.M. (1999) Functional and immunogenic characterization of two cloned regions of *Streptococcus mutans* glucosyltransferase I. Infect. Immun. *67*, 810–816.

Kaetzel, C.S., and Mostov, K. (2005) Immunoglobulin transport and the polymeric immunoglobulin receptor. In Mucosal Immunology, 3rd edn, J. Mestecky, M.E. Lamm, W. Strober, J. Bienenstock, J.R. McGhee, and L. Mayer, eds. (Amsterdam, Elsevier/Academic Press), pp. 211–250.

Katz, J., Black, K.P., and Michalek, S.M. (1999) Host responses to recombinant hemagglutinin B of *Porphyromonas gingivalis* in an experimental rat model. Infect. Immun. *67*, 4352–4359.

Katz, J., Harmon, C.C., Buckner, G.P., Richardson, G.J., Russell, M.W., and Michalek, S.M. (1993) Protective salivary immunoglobulin A responses against *Streptococcus mutans* infection after intranasal immunization with *S. mutans* antigen I/II coupled to the B subunit of cholera toxin. Infect. Immun. *61*, 1964–1971.

Kawabata, S., Terao, Y., Fujiwara, T., Nakagawa, I., and Hamada, S. (1999) Targeted salivary gland immunization with plasmid DNA elicits specific salivary immunoglobulin A and G antibodies and serum immunoglobulin G antibodies in mice. Infect. Immun. *67*, 5863–5868.

Kelly, C.G., Todryk, S., Kendal, H.L., Munro, G.H., and Lehner, T. (1995) T-cell, adhesion, and B-cell epitopes of the cell surface *Streptococcus mutans* protein antigen I/II. Infect. Immun. 63, 3649–3658.

Kilian, M., and Russell, M.W. (2005) Microbial evasion of IgA functions. In Mucosal Immunology, J. Mestecky, M.E. Lamm, W. Strober, J. Bienenstock, J.R. McGhee, and L. Mayer, eds. (Elsevier Academic Press), pp. 291–304.

Klausen, B., Evans, R.T., Ramamurthy, N.S., Golub, L.M., Sfintescu, C., Lee, J.Y., Bedi, G., Zambon, J.J., and Genco, R.J. (1991) Periodontal bone level and gingival proteinase activity in gnotobiotic rats immunized with *Bacteroides gingivalis.* Oral Microbiol. Immunol. 6, 193–201.

Koga, T., Okahashi, N., Takahashi, I., Kanamoto, T., Asakawa, H., and Iwaki, M. (1990) Surface hydrophobicity, adherence, and aggregation of cell surface protein antigen mutants of *Streptococcus mutans* serotype c. Infect. Immun. 58, 289–296.

Kohler, J.J., Pathangey, L., Hasona, A., Progulske-Fox, A., and Brown, T.A. (2000) Long-term immunological memory induced by recombinant oral *Salmonella* vaccine vectors. Infect. Immun. 68, 4370–4373.

Kozarov, E., Miyashita, N., Burks, J., Cerveny, K., Brown, T.A., McArthur, W.P., and Progulske-Fox, A. (2000) Expression and immunogenicity of hemagglutinin A from *Porphyromonas gingivalis* in an avirulent *Salmonella enterica* serovar *typhimurium* vaccine strain. Infect. Immun. 68, 732–739.

Kuboniwa, M., Amano, A., Shizukuishi, S., Nakagawa, I., and Hamada, S. (2001) Specific antibodies to *Porphyromonas gingivalis* Lys-gingipain by DNA vaccination inhibit bacterial binding to hemoglobin and protect mice from infection. Infect. Immun. 69, 2972–2979.

Kuramitsu, H.K. (1993) Virulence factors of mutans streptococci: Role of molecular genetics. Crit. Rev. Oral Biol. Med. 4, 159–176.

Lamont, R.J., Chan, A., Belton, C.M., Izutsu, K.T., Vasel, D., and Weinberg, A. (1995) *Porphyromonas gingivalis* invasion of gingival epithelial cells. Infect. Immun. 63, 3878–3885.

Lamont, R.J., Demuth, D.R., Davis, C.A., Malamud, D., and Rosan, B. (1991) Salivary-agglutinin mediated adherence of *Streptococcus mutans* to early plaque bacteria. Infect. Immun. 59, 3446–3450.

Lamont, R.J., Gil, S., Demuth, D.R., Malamud, D., and Rosan, B. (1994) Molecules of *Streptococcus gordonii* that bind to *Porphyromonas gingivalis*. Microbiology *140*, 867–872.

Lamont, R.J., Hersey, S.G., and Rosan, B. (1992) Characterization of the adherence of *Porphyromonas gingivalis* to oral streptococci. Oral Microbiol. Immunol. *7*, 193–197.

Lamont, R.J., and Jenkinson, H.F. (1998) Life below the gum line: Pathogenic mechanisms of *Porphyromonas gingivalis*. Microbiol. Mol. Biol. Rev. *62*, 1244–1263.

Lamont, R.J., and Yilmaz, O. (2002) In or out: the invasiveness of oral bacteria. Periodontol. 2000 *30*, 61–69.

LaPolla, R.J., Haron, J.A., Kelly, C.G., Taylor, W.R., Bohart, C., Hendricks, M., Pyati, J., Graff, R.T., Ma, J. K.-C., and Lehner, T. (1991) Sequence and structural analysis of protein antigen I/II (SpaA) of *Streptococcus sobrinus*. Infect. Immun. *59*, 2677–2685.

Lee, J.Y., Yi, N.N., Kim, U.S., Choi, J.S., Kim, S.J., and Choi, J.I. (2006) *Porphyromonas gingivalis* heat shock protein vaccine reduces the alveolar bone loss induced by multiple periodontopathogenic bacteria. J. Periodont. Res. *41*, 10–14.

Lee, S.F., Progulske-Fox, A., Erdos, G.W., Piacentini, D.A., Ayakawa, G.Y., Crowley, P.J., and Bleiweis, A.S. (1989) Construction and characterization of isogenic mutants of *Streptococcus mutans* deficient in major surface protein antigen P1 (Ag I/II). Infect. Immun. *57*, 3306–3313.

Lehner, T., Ma, J.K., and Kelly, C.G. (1992) A mechanism of passive immunization with monoclonal antibodies to a 185000 M(r) streptococcal antigen. Adv. Exp. Med. Biol. *327*, 151–163.

Lehner, T., Russell, M.W., Caldwell, J., and Smith, R. (1981) Immunization with purified protein antigens from *Streptococcus mutans* against dental caries in rhesus monkeys. Infect. Immun. *34*, 407–415.

Liang, S., Wang, M., Tapping, R.I., Stepensky, V., Nawar, H.F., Triantafilou, M., Triantafilou, K., Connell, T.D., and Hajishengallis, G. (2007a). Ganglioside GD1a is an essential coreceptor for toll-like receptor 2 signaling in response to the B subunit of Type IIB enterotoxin. J. Biol. Chem. *282*, 7532–7542.

Liang, S., Wang, M., Triantafilou, K., Triantafilou, M., Nawar, H.F., Russell, M.W., Connell, T.D., and Hajishengallis, G. (2007b). The A subunit of Type IIb enterotoxin (LT-IIb) suppresses the proinflammatory potential of the B subunit and its ability to recruit and interact with TLR2. J. Immunol. *178*, 4811–4819.

Loesche, W.J. (1986) Role of *Streptococcus mutans* in human dental decay. Microbiol. Rev. *50*, 353–380.

Love, R.M., and Jenkinson, H.F. (2002) Invasion of dentinal tubules by oral bacteria. Crit. Rev. Oral Biol. Med. *13*, 171–183.

Ma, J.K.-C., Kelly, C.G., Whiley, R.A., and Lehner, T. (1991) Conservation of the gene encoding streptococcal antigen I/II in oral streptococci. Infect. Immun. 59, 2686–2694.

Ma, J.K.-C., and Lehner, T. (1990) Prevention of colonization of *Streptococcus mutans* by topical application of monoclonal antibodies in human subjects. Arch. Oral Biol. 35 *(Supplement)*, 115s-122s.

Malek, R., Fisher, J.G., Caleca, A., Stinson, M., van Oss, C.J., Lee, J.Y., Cho, M.I., Genco, R.J., Evans, R.T., and Dyer, D.W. (1994) Inactivation of *Porphyromonas gingivalis* fimA gene blocks periodontal damage in gnotobiotic rats. J. Bacteriol. *176*, 1052–1059.

Martin, M.H., Metzger, D.J., Michalek, S.M., Connell, T.D., and Russell, M.W. (2000) Comparative analysis of the mucosal adjuvanticity of the type II heat-labile enterotoxins, LT-IIa and LT-IIb. Infect. Immun. *68*, 281–287.

Matsushita, K., Nisizawa, T., Nagaoka, S., Kawagoe, M., and Koga, T. (1994) Identification of antigenic epitopes in a surface protein antigen of *Streptococcus mutans* in humans. Infect. Immun. *62*, 4034–4042.

Matzinger, P. (2007) Friendly and dangerous signals: is the tissue in control? Nat. Immunol. *8*, 11–13.

Mestecky, J. (1987) The common mucosal immune system and current strategies for induction of immune response in external secretions. J. Clin. Immunol. *7*, 265–276.

Mestecky, J., McGhee, J.R., Arnold, R.R., Michalek, S.M., Prince, S.J., and Babb, J.L. (1978) Selective induction of an immune response in human external secretions by ingestion of bacterial antigen. J. Clin. Invest. *61*, 731–737.

Mestecky, J., Moro, I., Kerr, M.A., and Woof, J.M. (2005) Mucosal immunoglobulins. In Mucosal Immunology, 3rd Edn, J. Mestecky, J. Bienenstock, M. E. Lamm, L. Mayer, W. Strober, and J. R. McGhee, eds. (Amsterdam, Elsevier/Academic Press), pp. 153–181.

Mestecky, J., Russell, M.W., and Elson, C.O. (2007) Perspectives on oral vaccines: is oral tolerance a barrier? J. Immunol. *(in press)*.

Michalek, S.M., Morisaki, I., Kimura, S., Hamada, S., Curtiss III, R., and McGhee, J.R. (1986) Effective immunity to *Streptococcus mutans*-induced dental caries. In Molecular Microbiology and Immunobiology of *Streptococcus mutans*, S. Hamada, S.M. Michalek, H. Kiyono, L. Menaker, and J.R. McGhee, eds. (Amsterdam, Elsevier Science Publishers), pp. 269–278.

Missailidis, C.G., Umeda, J.E., Ota-Tsuzuki, C., Anzai, D., and Mayer, M.P. (2004) Distribution of fimA genotypes of *Porphyromonas gingivalis* in subjects with various periodontal conditions. Oral Microbiol. Immunol. *19*, 224–229.

Miura, M., Hamachi, T., Fujise, O., and Maeda, K. (2005) The prevalence and pathogenic differences of *Porphyromonas gingivalis* fimA genotypes in patients with aggressive periodontitis. J. Periodont. Res. *40*, 147–152.

Moisset, A., Schatz, N., Lepoivre, Y., Amadio, S., Wachsmann, D., Schöller, M., and Klein, J.-P. (1994) Conservation of salivary glycoprotein-interacting and human immunoglobulin G-cross-reactive domains of antigen I/II in oral streptococci. Infect. Immun. *62*, 184–193.

Moldoveanu, Z., Russell, M.W., Wu, H.-Y., and Mestecky, J. (1995) Compartmentalization within the mucosal immune system. Adv. Exp. Med. Biol. *371A*, 97–101.

Moritz, A.J., Cappelli, D., Lantz, M.S., Holt, S.C., and Ebersole, J.L. (1998) Immunization with *Porphyromonas gingivalis* cysteine protease: effects on experimental gingivitis and ligature-induced periodontitis in *Macaca fascicularis*. J. Periodontol. *69*, 686–697.

Munk, M.E., Schoel, B., Modrow, S., Karr, R.W., Young, R.A., and Kaufmann, S.H. (1989) T lymphocytes from healthy individuals with specificity to self-epitopes shared by the mycobacterial and human 65-kilodalton heat shock protein. J. Immunol. *143*, 2844–2849.

Munro, C., Michalek, S.M., and Macrina, F.L. (1991) Cariogenicity of *Streptococcus mutans* V403 glucosyltransferase and fructosyltransferase mutants constructed by allelic exchange. Infect. Immun. 59, 2316–2323.

Munro, G.H., Evans, P., Todryk, S., Buckett, P., Kelly, C.G., and Lehner, T. (1993) A protein fragment of streptococcal cell surface antigen I/II which prevents adhesion of *Streptococcus mutans*. Infect. Immun. *61*, 4590–4598.

Nakagawa, I., Amano, A., Kimura, R.K., Nakamura, T., Kawabata, S., and Hamada, S. (2000) Distribution and molecular characterization of *Porphyromonas gingivalis* carrying a new type of fimA gene. J. Clin. Microbiol. *38*, 1909–1914.

Nakagawa, I., Amano, A., Kuboniwa, M., Nakamura, T., Kawabata, S., and Hamada, S. (2002) Functional differences among FimA variants of *Porphyromonas gingivalis* and their effects on adhesion to and invasion of human epithelial cells. Infect. Immun. *70*, 277–285.

Nakagawa, I., Inaba, H., Yamamura, T., Kato, T., Kawai, S., Ooshima, T., and Amano, A. (2006) Invasion of epithelial cells and proteolysis of cellular focal adhesion components by distinct types of *Porphyromonas gingivalis* fimbriae. Infect. Immun. *74*, 3773–3782.

Nakagawa, T., Sims, T., Fan, Q., Potempa, J., Travis, J., Houston, L., and Page, R.C. (2001) Functional characteristics of antibodies induced by Arg-gingipain (HRgpA) and Lys-gingipain (Kgp) from Porphyromonas gingivalis. Oral Microbiol. Immunol. *16*, 202–211.

Nashar, T.O., Webb, H.M., Eaglestone, S., Williams, N.A., and Hirst, T.R. (1996) Potent immunogenicity of the B subunits of *Escherichia coli* heat-labile enterotoxin: receptor binding is essential and induces differential modulation of lymphocyte subsets. Proc. Natl. Acad. Sci. USA. *93*, 226–230.

Nisengard, R., Blann, D., Zelonis, L., McHenry, K., Reynolds, H., and Zambon, J. (1989) Effects of immunization with *B. macacae* on induced periodontitis--preliminary findings. Immunol. Investig. *18*, 225–237.

Noiri, Y., Li, L., Yoshimura, F., and Ebisu, S. (2004) Localization of *Porphyromonas gingivalis*-carrying fimbriae in situ in human periodontal pockets. J. Dent. Res. *83*, 941–945.

O'Brien-Simpson, N.M., Black, C.L., Bhogal, P.S., Cleal, S.M., Slakeski, N., Higgins, T.J., and Reynolds, E.C. (2000) Serum immunoglobulin G (IgG) and IgG subclass responses to the RgpA-Kgp proteinase-adhesin complex of *Porphyromonas gingivalis* in adult periodontitis. Infect. Immun. *68*, 2704–2712.

O'Brien-Simpson, N.M., Paolini, R.A., Hoffmann, B., Slakeski, N., Dashper, S.G., and Reynolds, E.C. (2001) Role of RgpA, RgpB, and Kgp proteinases in virulence of *Porphyromonas gingivalis* W50 in a murine lesion model. Infect. Immun. *69*, 7527–7534.

O'Brien-Simpson, N.M., Pathirana, R.D., Paolini, R.A., Chen, Y.-Y., Veith, P.D., Tam, V., Ally, N., Pike, R.N., and Reynolds, E.C. (2005) An immune response directed to proteinase and adhesin functional epitopes protects against *Porphyromonas gingivalis*-induced periodontal bone loss. J. Immunol. *175*, 3980–3989.

Ogawa, T. (1994) The potential protective immune responses to synthetic peptides containing conserved epitopes of *Porphyromonas gingivalis* fimbrial protein. J. Med. Microbiol. *41*, 349–358.

Ogawa, T., Kusumoto, Y., Uchida, H., Nagashima, S., Ogo, H., and Hamada, S. (1991) Immunobiological activities of synthetic peptide segments of fimbrial protein from *Porphyromonas gingivalis*. Biochem. Biophys. Res. Commun. *180*, 1335–1341.

Ogawa, T., Ogo, H., and Hamada, S. (1994a). Chemotaxis of human monocytes by synthetic peptides that mimic segments of *Porphyromonas gingivalis* fimbrial protein. Oral Microbiol. Immunol. *9*, 257–261.

Ogawa, T., Uchida, H., and Hamada, S. (1994b). *Porphyromonas gingivalis* fimbriae and their synthetic peptides induce proinflammatory cytokines in human peripheral blood monocyte cultures. FEMS Microbiol. Lett. *116*, 237–242.

Ogawa, T., Uchida, H., and Yasuda, K. (1995) Mapping of murine Th1 and Th2 helper T-cell epitopes on fimbriae from *Porphyromonas gingivalis*. J. Med. Microbiol. *42*, 165–170.

Okahashi, N., Sasakawa, C., Yoshikawa, M., Hamada, S., and Koga, T. (1989) Molecular characterization of a surface protein antigen gene from serotype *c* *Streptococcus mutans*, implicated in dental caries. Mol. Microbiol. *3*, 673–678.

Okuda, K., Kato, T., Naito, Y., Takazoe, I., Kikuchi, Y., Nakamura, T., Kiyoshige, T., and Sasaki, S. (1988) Protective efficacy of active and passive immunizations against experimental infection with Bacteroides gingivalis in ligated hamsters. J. Dent. Res. *67*, 807–811.

Peros, W.J., and Gibbons, R.J. (1986) Evidence suggesting multiple binding sites in experimental pellicles for *Streptococcus mutans* JBP. J. Dent. Res. *65*, 1332–1334.

Persson, G.R., Engel, D., Whitney, C., Darveau, R., Weinberg, A., Brunsvold, M., and Page, R.C. (1994) Immunization against *Porphyromonas gingivalis* inhibits progression of experimental periodontitis in nonhuman primates. Infect. Immun. *62*, 1026–1031.

Pihlstrom, B.L., Michalowicz, B.S., and Johnson, N.W. (2005) Periodontal diseases. Lancet *366*, 1809–1820.

Potempa, J., Banbula, A., and Travis, J. (2000) Role of bacterial proteinases in matrix destruction and modulation of host responses. Periodontol. 2000 *24*, 153–192.

Rajapakse, P.S., O'Brien-Simpson, N.M., Slakeski, N., Hoffmann, B., and Reynolds, E.C. (2002) Immunization with the RgpA-Kgp proteinase-adhesin complexes of Porphyromonas gingivalis protects against periodontal bone loss in the rat periodontitis model. Infect. Immun. *70*, 2480–2486.

Re, F., and Strominger, J.L. (2001) Toll-like receptor 2 (TLR2) and TLR4 differentially activate human dendritic cells. J. Biol. Chem. *276*, 37692–37699.

Redman, T.K., Harmon, C.C., Lallone, R.L., and Michalek, S.M. (1995) Oral immunization with recombinant *Salmonella typhimurium* expressing surface protein antigen A of *Streptococcus sobrinus*: Dose response and induction of protective humoral responses in rats. Infect. Immun. 63, 2004–2011.

Reinholdt, J., and Kilian, M. (1987) Interference of IgA protease with the effect of secretory IgA on adherence of oral streptococci to saliva-coated hydroxyapatite. J. Dent. Res. 66, 492–497.

Roberts, F.A., Houston, L.S., Lukehart, S.A., Mancl, L.A., Persson, G.R., and Page, R.C. (2004) Periodontitis vaccine decreases local prostaglandin E2 levels in a primate model. Infect. Immun. *72*, 1166–1168.

Russell, M.W. (1986) Protein antigens of *Streptococcus mutans*. In Molecular Microbiology and Immunobiology of *Streptococcus mutans*, S. Hamada, S.M. Michalek, H. Kiyono, L. Menaker, and J.R. McGhee, eds. (Amsterdam, Elsevier), pp. 51–59.

Russell, M.W. (2003) Mucosal immunity. In New Bacterial Vaccines, R.W. Ellis, and B.R. Brodeur, eds. (Georgetown TX and New York, Eurekah. com and Kluwer Academic/Plenum), pp. 63–79.

Russell, M.W., Bergmeier, L.A., Zanders, E.D., and Lehner, T. (1980a). Protein antigens of *Streptococcus mutans*: purification and properties of a double antigen and its protease-resistant component. Infect. Immun. *28*, 486–493.

Russell, M.W., Challacombe, S.J., and Lehner, T. (1980b). Specificity of antibodies induced by *Streptococcus mutans* during immunization against dental caries. Immunology *40*, 97–106.

Russell, M.W., Childers, N.K., Michalek, S.M., Smith, D.J., and Taubman, M.A. (2004) A caries vaccine? The state of the science of immunization against dental caries. Caries Res. *38*, 230–235.

Russell, M.W., Harrington, D.J., and Russell, R.R. (1995) Identity of *Streptococcus mutans* surface protein antigen III and wall-associated protein antigen A. Infect. Immun. 63, 733–735.

Russell, M.W., and Kilian, M. (2005) Biological activities of IgA. In Mucosal Immunology, J. Mestecky, M.E. Lamm, W. Strober, J. Bienenstock, J.R. McGhee, and L. Mayer, eds. (Elsevier Academic Press), pp. 267–290.

Russell, M.W., and Wu, H. (1990) *Streptococcus mutans* and the problem of heart cross-reactivity. Crit Rev Oral Biol Med *1*, 191–205.

Russell, R.R., Gilpin, M.L., Mukasa, H., and Dougan, G. (1987) Characterization of glucosyltransferase expressed from a *Streptococcus sobrinus* gene cloned in *Escherichia coli*. J. Gen. Microbiol. *133*, 935–944.

Russell, R.R., Peach, S.L., Colman, G., and Cohen, B. (1983) Antibody responses to antigens of *Streptococcus mutans* in monkeys (*Macaca fascicularis*) immunized against dental caries. J. Gen. Microbiol. *129*, 865–875.

Russell, R.R.B., Abdulla, E., Gilpin, M.L., and Smith, K. (1986) Characterization of *Streptococcus mutans* surface antigens. In Molecular Biology and Immunobiology of *Streptococcus mutans*, S. Hamada, S.M. Michalek, H. Kiyono, L. Menaker, and J.R. McGhee, eds. (Amsterdam, Elsevier), pp. 61–70.

Sciotti, M.-A., Yamodo, I., Klein, J.-P., and Ogier, J.A. (1997) The N-terminal half part of the oral streptococcal antigen I/IIf contains two distinct binding domains. FEMS Microbiol. Lett. *153*, 439–445.

Scully, C.M., Russell, M.W., and Lehner, T. (1980) Specificity of opsonizing antibodies to antigens of *Streptococcus mutans*. Immunology *41*, 467–473.

Seifert, T.B., Bleiweis, A.S., and Brady, L.J. (2004) Contribution of the alanine-rich region of *Streptococcus mutans* P1 to antigenicity, surface expression, and interaction with the proline-rich repeat domain. Infect. Immun. *72*, 4699–4706.

Seward, M. (1971) Local disturbances attributed to eruption of the human primary dentition. Br. Dent. J. *130*, 72–77.

Sharma, A., Honma, K., Evans, R.T., Hruby, D.E., and Genco, R.J. (2001) Oral immunization with recombinant *Streptococcus gordonii* expressing *Porphyromonas gingivalis* FimA domains. Infect. Immun. 69, 2928–2934.

Sharma, A., Honma, K., Sojar, H.T., Hruby, D.E., Kuramitsu, H.K., and Genco, R.J. (1999) Expression of saliva-binding epitopes of the *Porphyromonas gingivalis* FimA protein on the surface of *Streptococcus gordonii*. Biochem. Biophys. Res. Commun. *258*, 222–226.

Sharma, A., Nagata, H., Hamada, N., Sojar, H.T., Hruby, D.E., Kuramitsu, H.K., and Genco, R.J. (1996) Expression of functional *Porphyromonas gingivalis* fimbrillin polypeptide domains on the surface of *Streptococcus gordonii*. Appl. Environ. Microbiol. *62*, 3933–3938.

Sharma, A., Sojar, H.T., Hruby, D.E., Kuramitsu, H.K., and Genco, R.J. (1997) Secretion of *Porphyromonas gingivalis* fimbrillin polypeptides by recombinant *Streptococcus gordonii*. Biochem. Biophys. Res. Commun. *238*, 313–316.

Shin, E.A., Lee, J.Y., Kim, T.G., Park, Y.K., and Langridge, W.H. (2006) Synthesis and assembly of an adjuvanted *Porphyromonas gingivalis* fimbrial antigen fusion protein in plants. Protein Expr. Purif. *47*, 99–109.

Shiroza, T., Ueda, S., and Kuramitsu, H.K. (1987) Sequence analysis of the *gtfB* gene from *Streptococcus mutans*. J. Bacteriol. *169*, 4263–4270.

Simpson, W., Olczak, T., and Genco, C.A. (2004) Lysine-specific gingipain K and heme/hemoglobin receptor HmuR are involved in heme utilization in *Porphyromonas gingivalis*. Acta Biochim. Pol. *51*, 253–262.

Smith, D.J. (2002) Dental caries vaccines: prospects and concerns. Crit. Rev. Oral Biol. Med. *13*, 335–349.

Smith, D.J., Heschel, R.L., King, W.F., and Taubman, M.A. (1999) Antibody to glucosyltransferase induced by synthetic peptides associated with catalytic regions of alpha-amylases. Infect. Immun. *67*, 2638–2642.

Smith, D.J., King, W.F., Rivero, J., and Taubman, M.A. (2005) Immunological and protective effects of diepitopic subunit dental caries vaccines. Infect. Immun. *73*, 2797–2804.

Smith, D.J., King, W.F., Wu, C.D., Shen, B.I., and Taubman, M.A. (1998) Structural and antigenic characteristics of *Streptococcus sobrinus* glucan binding proteins. Infect. Immun. *66*, 5565–5569.

Smith, D.J., and Taubman, M.A. (1987) Oral immunization of humans with *Streptococcus sobrinus* glucosyltransferase. Infect. Immun. 55, 2562–2569.

Smith, D.J., and Taubman, M.A. (1997) Vaccines against dental caries infection. In New Generation Vaccines, M.M. Levine, G. C. Woodrow, J.B. Kaper, and G.S. Cobon, eds. (New York, Marcel Dekker), pp. 913–930.

Smith, D.J., Taubman, M.A., and Ebersole, J.L. (1979) Effect of oral administration of glucosyltransferase antigens on experimental dental caries. Infect. Immun. *26*, 82–89.

Smith, D.J., Taubman, M.A., and Ebersole, J.L. (1985) Salivary IgA antibody to glucosyltransferase in man. Clin Exp Immunol *61*, 416–424.

Smith, D.J., Taubman, M.A., Holmberg, C.F., Eastcott, J., King, W.F., and Ali-Salaam, P. (1993) Antigenicity and immunogenicity of a synthetic peptide derived from a glucan-binding domain of mutans streptococcal glucosyltransferase. Infect. Immun. *61*, 2899–2905.

Sojar, H.T., Lee, J.-Y., and Genco, R.J. (1995) Fibronectin binding domain of *P. gingivalis* fimbriae. Biochem. Biophys. Res. Commun. *216*, 785–792.

Sorsa, T., Ingman, T., Suomalainen, K., Haapasalo, M., Konttinen, Y.T., Lindy, O., Saari, H., and Uitto, V.J. (1992) Identification of proteases from periodontopathogenic bacteria as activators of latent human neutrophil and fibroblast-type interstitial collagenases. Infect. Immun. *60*, 4491–4495.

Staat, R.H., Langley, S.D., and Doyle, R.J. (1980) *Streptococcus mutans* adherence: presumptive evidence for protein-mediated attachment followed by glucan-dependent cellular accumulation. Infect. Immun. *27*, 675–681.

Sugawara, S., Nemoto, E., Tada, H., Miyake, K., Imamura, T., and Takada, H. (2000) Proteolysis of human monocyte CD14 by cysteine proteinases (gingipains) from *Porphyromonas gingivalis* leading to lipopolysaccharide hyporesponsiveness. J. Immunol. *165*, 411–418.

Takahashi, I., Okahashi, N., Matsushita, K., Tokuda, M., Kanamoto, T., Munekata, E., Russell, M.W., and Koga, T. (1991) Immunogenicity and protective effect against oral colonization by *Streptococcus mutans* of synthetic peptides of a streptococcal surface protein antigen. J. Immunol. *146*, 332–336.

Takamatsu-Matsushita, N., Yamaguchi, N., Kawasaki, M., Yamashita, Y., Takehara, T., and Koga, T. (1996) Immunogenicity of *Actinobacillus actinomycetemcomitans* serotype b-specific polysaccharide-protein conjugate. Oral Microbiol. Immunol. *11*, 220–225.

Taubman, M.A., Holmberg, C.J., and Smith, D.J. (1995) Immunization of rats with synthetic peptide constructs from the glucan-binding or catalytic region of mutans streptococcal glucosyltransferase protects against dental caries. Infect. Immun. 63, 3088–3093.

Taubman, M.A., Holmberg, C.J., and Smith, D.J. (2001) Diepitopic construct of functionally and epitopically complementary peptides enhances immunogenicity, reactivity with glucosyltransferase, and protection from dental caries. Infect. Immun. *69*, 4210–4216.

Taubman, M.A., and Kawai, T. (2001) Involvement of T-lymphocytes in periodontal disease and in direct and indirect induction of bone resorption. Crit. Rev. Oral Biol. Med. *12*, 125–135.

Taubman, M.A., and Nash, D.A. (2006) The scientific and public-health imperative for a vaccine against dental caries. Nat. Rev. Immunol. *6*, 555–563.

Taubman, M.A., Yoshie, H., Wetherell, J.R., Jr., Ebersole, J.L., and Smith, D.J. (1983) Immune response and periodontal bone loss in germfree rats immunized and infected with *Actinobacillus actinomycetemcomitans* J. Periodont. Res. *18*, 393–401.

Taussig, M.J. (1973) Antigenic competition. Curr Top Microbiol Immunol *60*, 125–174.

Teng, Y.T. (2006) Protective and destructive immunity in the periodontium: Part 1-Innate and humoral immunity and the periodontium. J. Dent. Res. *85*, 198–208.

Thurnheer, T., van der Ploeg, J.R., Giertsen, E., and Guggenheim, B. (2006) Effects of *Streptococcus mutans* gtfC deficiency on mixed oral biofilms *in vitro*. Caries Res. *40*, 163–171.

Troffer-Charlier, N., Ogier, J., Moras, D., and Cavarelli, J. (2002) Crystal structure of the V-region of *Streptococcus mutans* antigen I/II at 2.4 Å resolution suggests a sugar preformed binding site. J. Mol. Biol. *318*, 179–188.

Ueda, S., Shiroza, T., and Kuramitsu, H.K. (1988) Sequence analysis of the *gtfC* gene from *Streptococcus mutans* GS-5. Gene 69, 101–109.

Umemoto, T., and Hamada, N. (2003) Characterization of biologically active cell surface components of a periodontal pathogen. The roles of major and minor fimbriae of *Porphyromonas gingivalis*. J. Periodontol. *74*, 119–122.

van der Ploeg, J.R., Giertsen, E., Ludin, B., Morgeli, C., Zinkernagel, A.S., and Gmur, R. (2004) Quantitative detection of *Porphyromonas gingivalis* fimA genotypes in dental plaque. FEMS Microbiol. Lett. *232*, 31–37.

van Dolleweerd, C.J., Chargelegue, D., and Ma, J.K.C. (2003) Characterization of the conformational epitope of Guy's 13, a monoclonal antibody that prevents *Streptococcus mutans* colonization in human. Infect. Immun. *71*, 754–765.

van Dolleweerd, C.J., Kelly, C.G., Chargelegue, D., and Ma, J.K. (2004) Peptide mapping of a novel discontinuous epitope of the major surface adhesin from *Streptococcus mutans*. J. Biol. Chem. *279*, 22198–22203.

Wu, H.-Y., Nguyen, H., and Russell, M.W. (1997) Nasal lymphoid tissue (NALT) as a mucosal immune inductive site. Scand. J. Immunol. *46*, 506–513.

Xu, Q.A., Yu, F., Fan, M.W., Bian, Z., Chen, Z., Fan, B., Jia, R., and Guo, J.H. (2006) Immunogenicity and persistence of a targeted anti-caries DNA vaccine. J. Dent. Res. *85*, 915–918.

Xu, Q.A., Yu, F., Fan, M.W., Bian, Z., Chen, Z., Peng, B., Jia, R., and Guo, J.H. (2007) Protective efficacy of a targeted anti-caries DNA plasmid against cariogenic bacteria infections. Vaccine 25, 1191–1195.

Yamashita, K., Eastcott, J.W., Taubman, M.A., Smith, D.J., and Cox, D.S. (1991) Effect of adoptive transfer of cloned *Actinobacillus actinomycetemcomitans*-specific T helper cells on periodontal disease. Infect. Immun. 59, 1529–1534.

Yamashita, Y., Bowen, W.H., Burne, R.A., and Kuramitsu, H.K. (1993) Role of *Streptococcus mutans* gtf genes in caries induction in the specific-pathogen-free rat model. Infect. Immun. *61*, 3811–3817.

Yilmaz, O., Watanabe, K., and Lamont, R.J. (2002) Involvement of integrins in fimbriae-mediated binding and invasion by *Porphyromonas gingivalis*. Cell Microbiol. *4*, 305–314.

Yonezawa, H., Ishihara, K., and Okuda, K. (2001) Arg-gingipain a DNA vaccine induces protective immunity against infection by *Porphyromonas gingivalis* in a murine model. Infect. Immun. 69, 2858–2864.

Yoshie, H., Taubman, M.A., Olson, C.L., Ebersole, J.L., and Smith, D.J. (1987) Periodontal bone loss and immune characteristics after adoptive transfer of *Actinobacillus*-sensitized T cells to rats. J. Periodont. Res. 22, 499–505.

Youngman, K.R., Lazarus, N.H., and Butcher, E.C. (2005) Lymphocytic homing: chemokines and adhesion molecules in T cell and IgA plasma cell localization in the mucosal immune system. In Mucosal Immunology, 3rd edn, J. Mestecky, M. E. Lamm, W. Strober, J. Bienenstock, J.R. McGhee, and L. Mayer, eds. (Amsterdam, Elsevier/Academic Press), pp. 667–680.

Yu, H., Nakano, Y., Yamashita, Y., Oho, T., and Koga, T. (1997) Effects of antibodies against cell surface protein antigen PAc-glucosyltransferase fusion proteins on glucan synthesis and cell adhesion of *Streptococcus mutans*. Infect. Immun. 65, 2292–2298.

Yu, J.J., Ruddy, M.J., Wong, G.C., Sfintescu, C., Baker, P.J., Smith, J.B., Evans, R.T., and Gaffen, S.L. (2007) An essential role for IL-17 in preventing pathogen-initiated bone destruction: recruitment of neutrophils to inflamed bone requires IL-17 receptor-dependent signals. Blood 37, 3794–3802.

Zhang, P., Jespersgaard, C., Lamberty-Mallory, L., Katz, J., Huang, Y., Hajishengallis, G., and Michalek, S.M. (2002) Enhanced immunogenicity of a genetic chimeric protein consisting of two virulence antigens of *Streptococcus mutans* and protection against infection. Infect. Immun. *70*, 6779–6787.

Zhang, S., Green, N.M., Sitkiewicz, I., Lefebvre, R.B., and Musser, J.M. (2006) Identification and characterization of an antigen I/II family protein produced by group A *Streptococcus*. Infect. Immun. *74*, 4200–4213.

Abbreviations

AP-PCR	Arbitrarily primed polymerase chain reaction
CRISPR	Clustered regularly inter-spaced short palindrome repeats
CTD	C-terminal domains
DGGE	Denaturing gradient gel electrophoresis
DGGEGE	Denaturing gradient gel electrophoresis gel expansion
ECAM	Endothelial cell adhesion molecule
EPEC	Enteropathogenic *E. coli*
ERIC	Enterobacterial repetitive intergenic consensus
FISH	Fluorescence *in situ* hybridization
HGR	Habitat gene reservoir
HGT	Horizontal gene transfer
HSP	Heat shock protein
ICAM	Intercellular adhesion molecule
ICAT	Isotope-coded affinity tag
Ig CAM	Immunoglobulin superfamily cell adhesion molecule
IEF	Iso-electric focusing
IP	Intracellular polyglucose
IPG	Immobilized pH gradient
IVET	*In vivo* expression technology
IVIAT	*In vivo* induced antigen technology
LAP	Localized aggressive periodontitis
LC-MS	Liquid chromatography–mass spectrometry
LOPIT	Localization of organelle proteins by isotope tagging
LPB	Lipopolysaccharide binding protein
LTA	Lipoteichoic acid
MALDI-TOF	Matrix-assisted laser desorptive ionization time-of-flight
MAMP	Microbial-associated molecular patterns
MAP	Mitogen-activated protein
MLST	Multilocus sequence typing
MMP	Matrix metalloprotease
MS	Mass spectrometry
Omp	Outer membrane protein
ORF	Open reading frame
PAMP	Pathogen-associated molecular pattern
PAGE	Polyacrylamide gel electrophoresis

PCR	Polymerase chain reaction
PFGE	Pulsed field gel electrophoresis
PMF	Peptide mass fingerprinting
qRT-PCR	Real-time PCR
RAPD/AP-PCR	Random amplification of polymorphic DNA/arbitrarily primed PCR
Rep-PCR	Repetitive element-based PCR
RFLP	Restriction fragment length polymorphism
RISA	Ribosomal intergenic spacer analysis
RISSC	Ribosomal intergenic spacer sequence collection
SILAC	Stable isotope labelling amino acid cell culture
STM	Signature-tagged mutagenesis
TIGR	The Institute for Genomic Research
T-RFLP	Terminal restriction fragment length polymorphism
TLR	Toll-like receptor
TTSS	Type III secretion system
VCAM	Vascular cell-adhesion molecule
WCI	Widespread colonization island of *A. actinomycetemcomitans*

Index

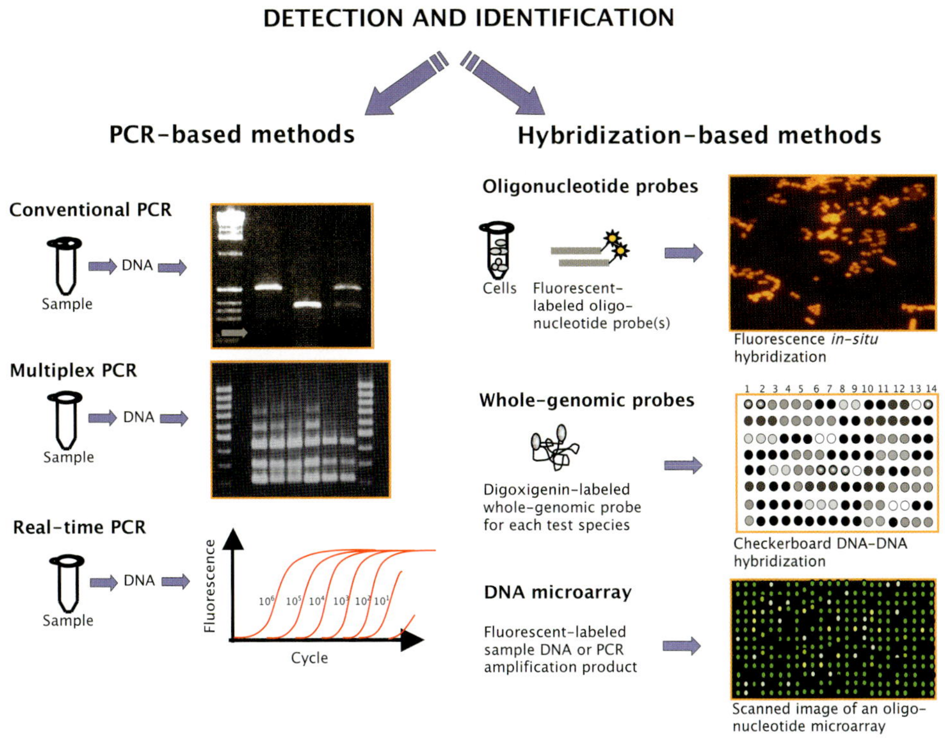

Plate 1.1 Examples of molecular techniques used for the detection and identification of oral bacteria from clinical samples.

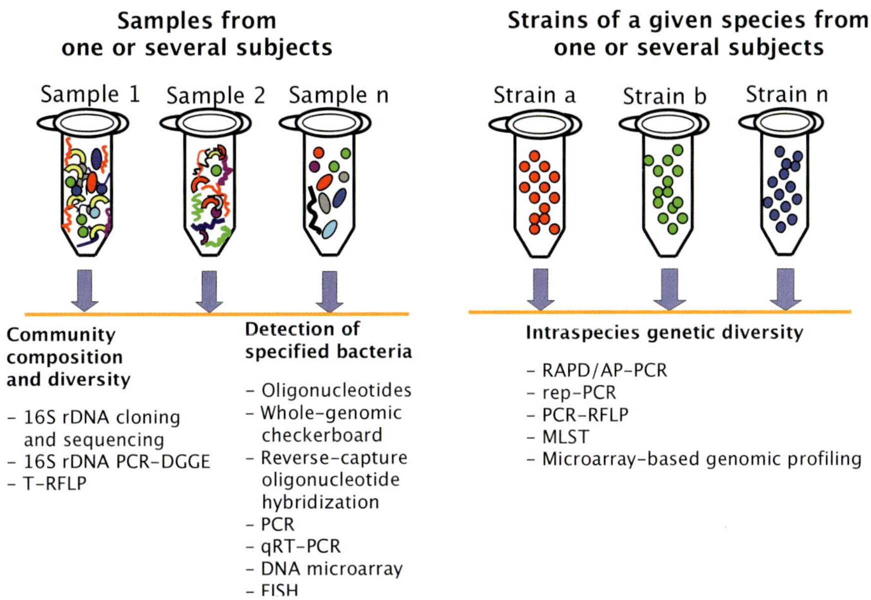

Plate 1.2 Examples of molecular techniques used for studying the composition and diversity of oral bacterial community samples; detection of specified bacteria from clinical samples; and comparison of different strains of the same species. For abbreviations, see Table 1.2.

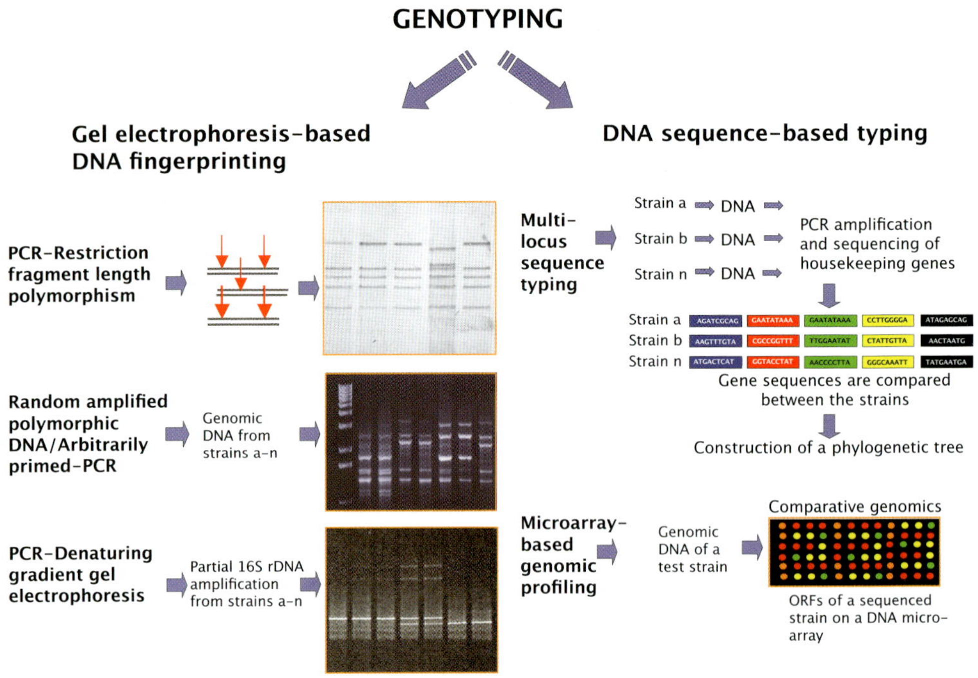

Plate 1.3 Examples of molecular techniques used for the genotyping of bacterial strains.

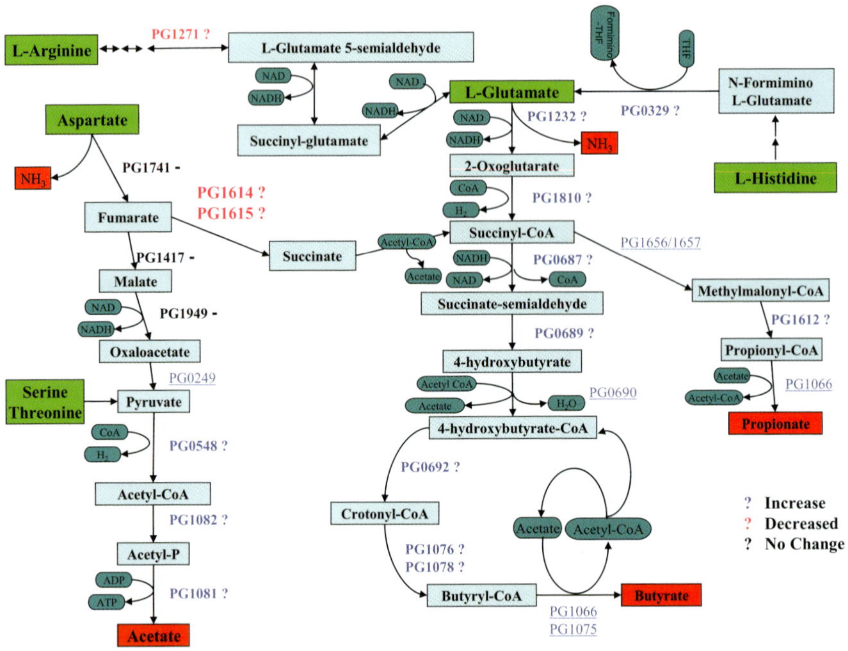

Plate 2.7 Summary of the effects of haem-limitation on the protein abundances of enzymes in the major glutamate and aspartate catabolic pathways of P. gingivalis. Bacteria were grown in continuous culture, with a mean generation time of 6.9 h, in brain heart infusion medium supplemented with 5 mg/mL filter-sterilized cysteine hydrochloride and either 5.0 µg/mL haemin (haem-excess) or 0.1 µg/mL haemin (haem-limitation). Substrates of the pathways are shown in green, intermediates are shown in light blue and end-products in red. P. gingivalis enzymes catalysing reactions are represented by their TIGR accession number. There was a co-ordinated increase in the abundance of enzymes in the glutamate pathway during haem-limited growth and a decrease in the level of the haem-containing fumarate reductase enzymes (PG1614-1615). (Unpublished data).

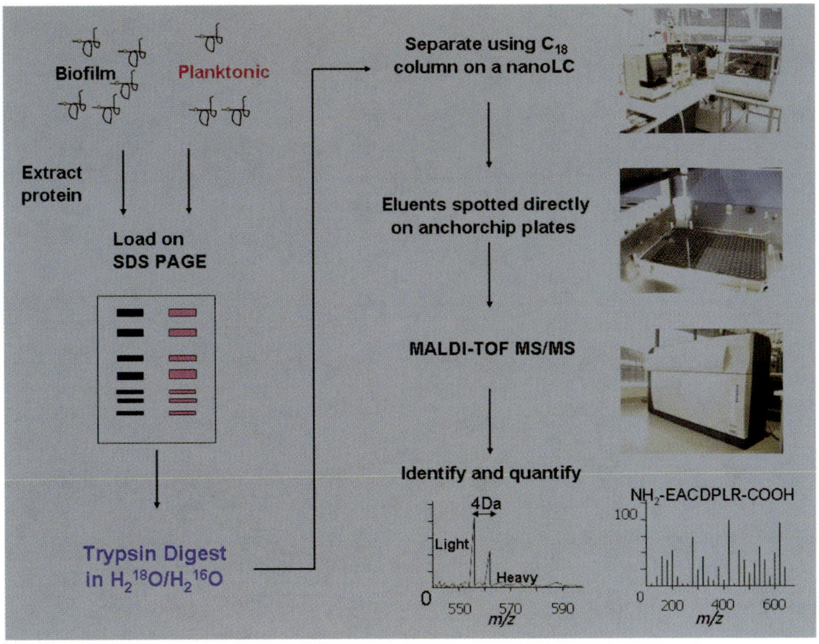

Plate 2.9 A typical scheme of performing $^{18}O^{16}O$ proteolytic labelling, coupled with mass spectrometry, for the study of bacterial proteome changes under two different growth conditions.

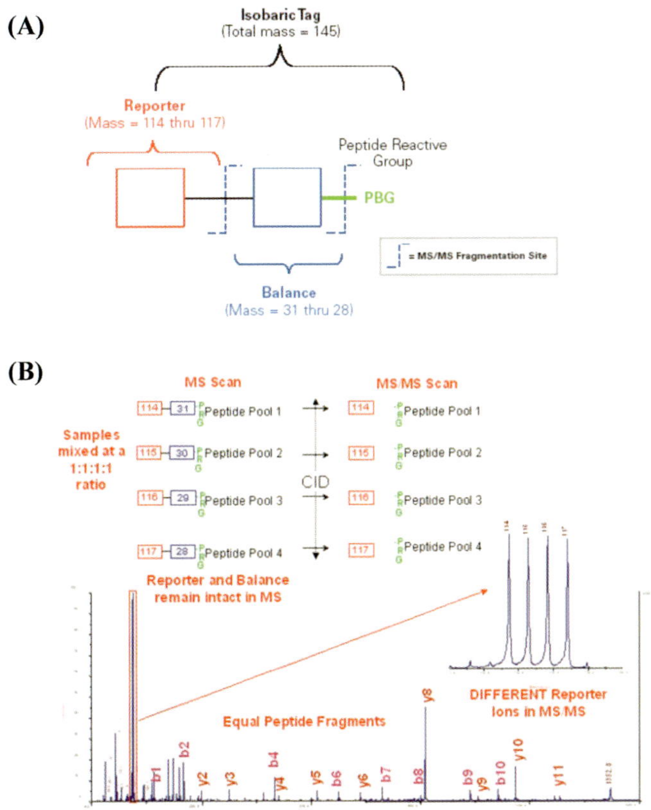

Plate 2.10 (A) The iTRAQ reagent consisting of the reporter, balance and peptide reactive groups. (B) Spectra showing the MS/MS spectra after CID and the different reporter ions (114–117 mz) at the 'quiet region' of the MS spectra which are used for quantitation. (http://docs.appliedbiosystems.com/pebiodocs/00113379.pdf).

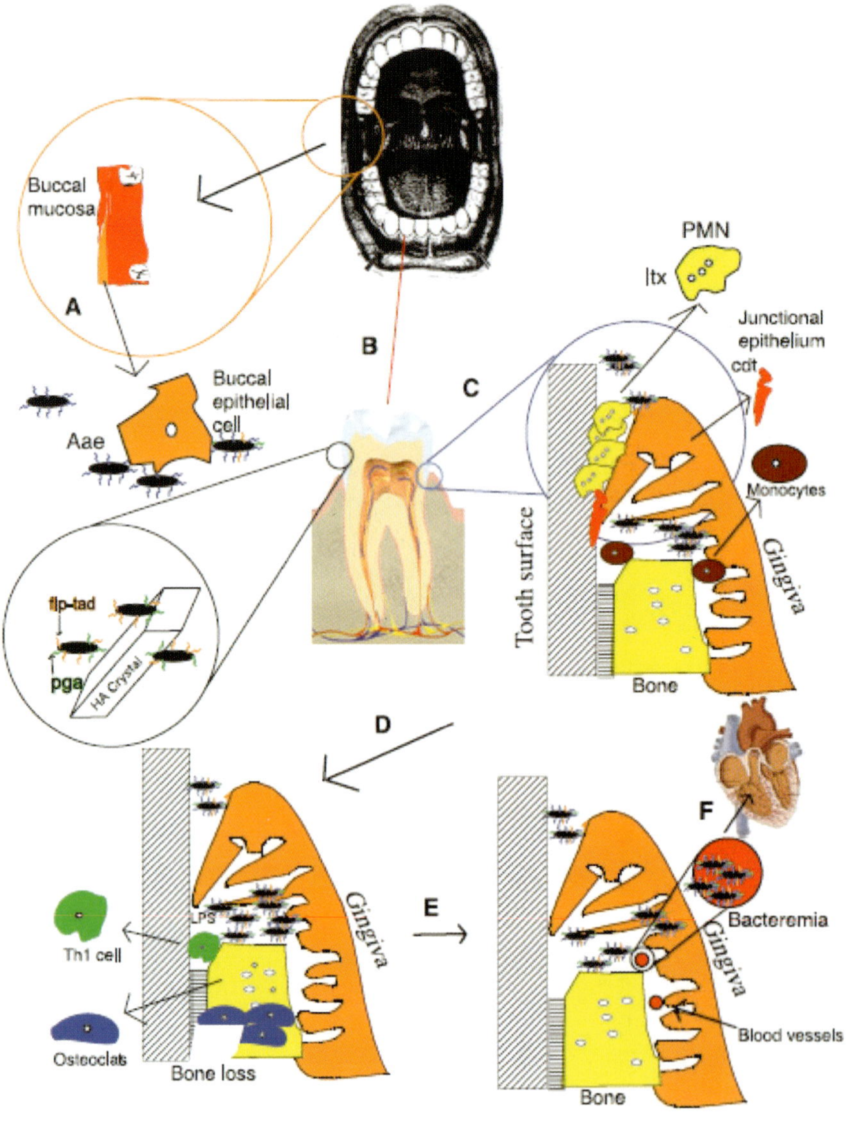

Plate 6.6 Illustration of the relationship of *A.a.* to oral and systemic disease. Panel A shows a proposed way in which *A.a.* first colonizes the mouth via the BEC. Panel B shows how Flp-Tad and PGA all provide *A.a.* with its ability to move from the BEC to the tooth where it encounters a non-shedding surface for a more permanent colonization in the oral cavity. Panel C shows how *A.a.* migrates into the sulcular space and is confronted by PMNs. It is proposed that *A.a.* Ltx can neutralize the PMNs' ability to destroy this foreign invader. This panel also illustrates how Cdt (red arrow-like symbol) is in a position to interact with monocytes in the sub-epithelial space. Then, in conjunction with *A.a.* LPS, is in a position to affect lymphocytes which can then stimulate (as seen in panel D) affected lymphocytes to stimulate precursor cells, initiating bone loss and impairing bone repair. In panel E, it is proposed that *A.a.* can gain access to the bloodstream, collect on heart valves and contribute to an increased risk for heart disease in susceptible subjects. (Reprinted from Fine *et al.*, 2005.)

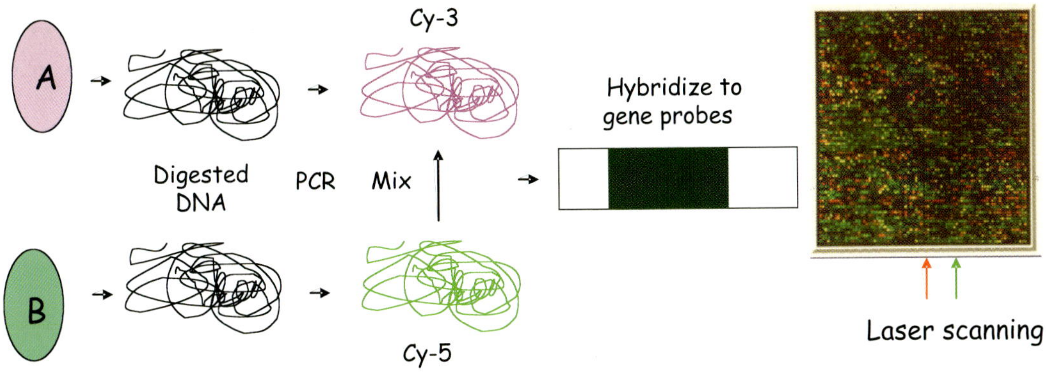

Plate 7.2 Comparison of strains by 'genomotyping'. To compare the gene content and diversity between strains A and B, fragments of their genomic DNA are differentially labelled with Cy3 andCy5 dyes, mixed and hybridized to microarrays. Fluorescent signals are scanned, normalized and scored for divergence.

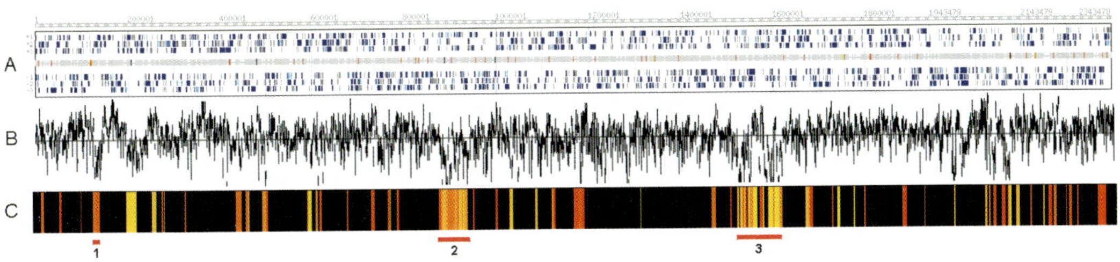

Plate 7.3 Distribution of divergent genes in the genome of strain W83. A. W83 genome and six-frame distribution of predicted genes. B. GC distribution based on G+C content of 500-bp windows across the genome. C. Simulated head map of the distribution of genes classified as slightly (yellow) to highly (red) divergent between strains W83 and ATCC 33277. The black background indicates areas where genes are present in both genomes. Printed with permission (Chen *et al.*, 2004).

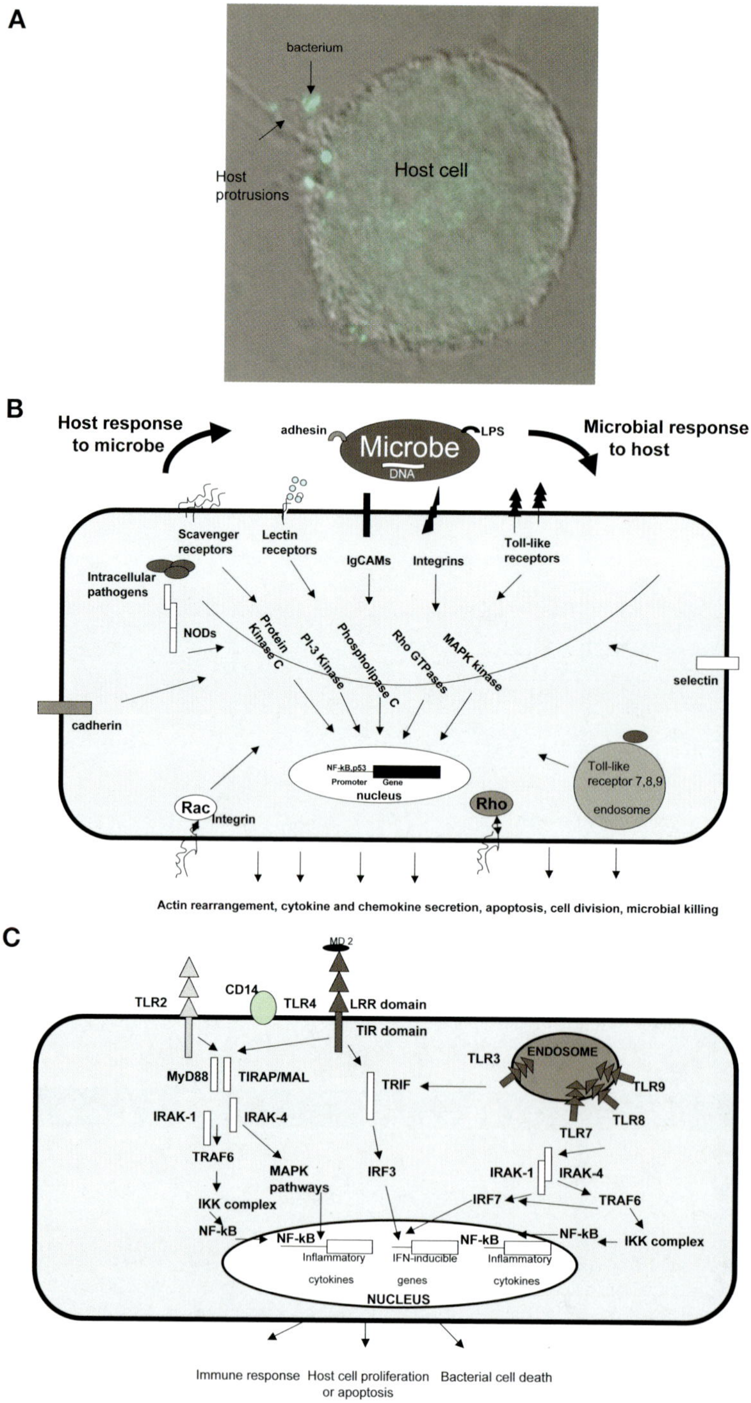

Plate 9.3 Receptor and signalling cascades during host-pathogen interaction. A. Co-culture of *P. gingivalis* with human umbilical vein endothelial cells (HUVECs). Upon exposure to *P. gingivalis* HUVEC forms fingerlike cell protrusions interacting with the microorganism. B. Receptors and signalling cascades are engaged upon microbial contact with the host cell. Multiple receptors can simultaneously recognize microbial components. The binding to receptor triggers intracellular signals that, through various pathways engaging multiple molecules, deliver the message to the nucleus and alter gene transcription in the host cell. This ultimately leads to a host cell response manifesting itself in actin rearrangement, cytokine/chemokine secretion, immune response, apoptosis, cell division, and microbial killing. The host-microbe contact results not only in host response but also in microbial response, thus indicating that this is a highly interactive process. C. TLR pathways.

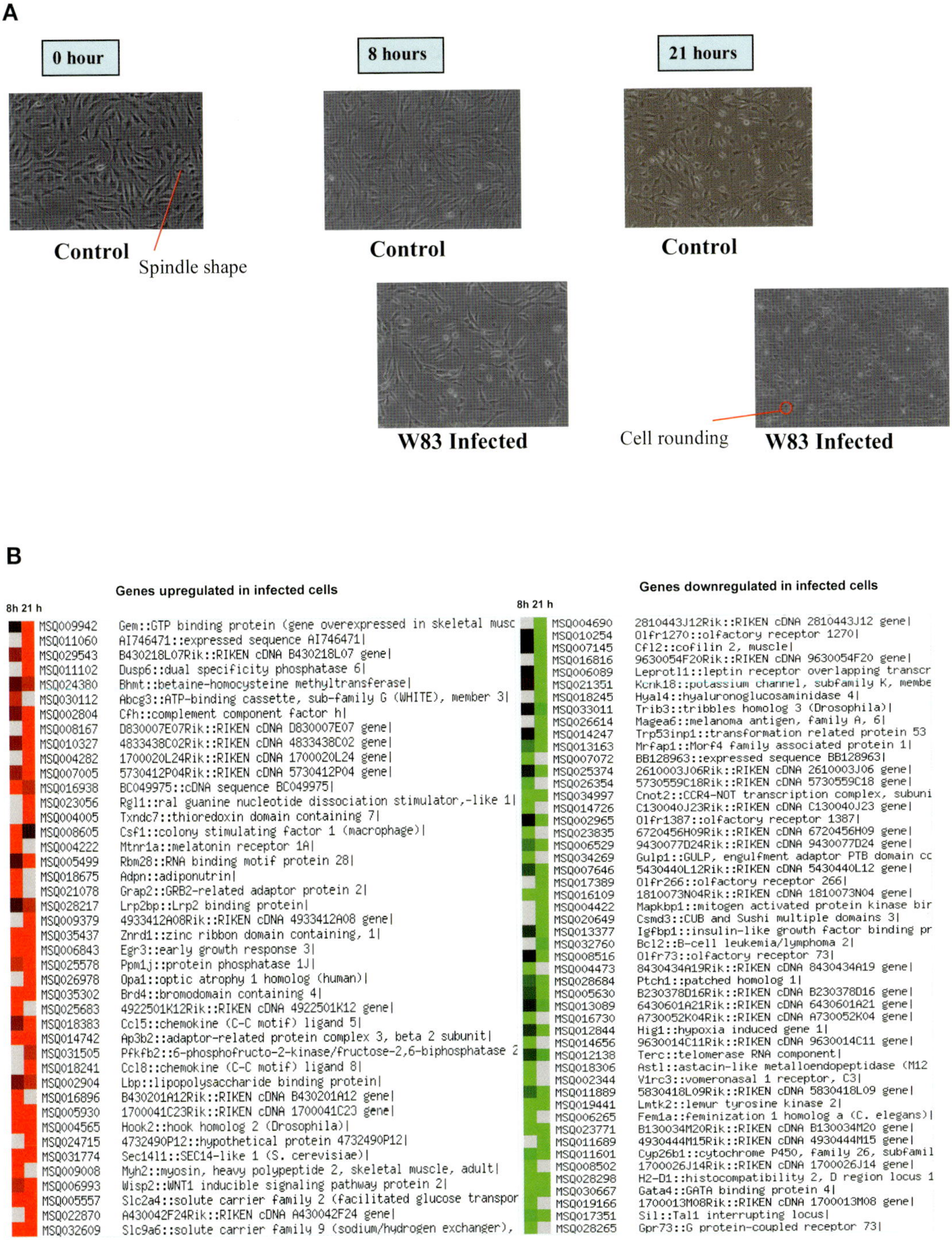

Plate 9.4 Host response to oral bacteria gleaned from genomic approaches. A. Morphological changes occurring in fibroblasts infected with *P. gingivalis* W83 at various times post-infection (8 hours and 21 hours). (Continued overleaf.) B. Transcriptional response of the infected fibroblasts at various times post-infection. At 8 hours, some 23 000 genes were detected, of which 182 were regulated. At 21 hours, some 20 000 genes were detected, of which 568 were regulated, at a 1.8-fold change. Among the regulated genes were pro-inflammatory cytokines or cytokine receptors; e.g. Cxcl5, which has recently been shown to play a role in oral carcinoma development. This study sets the stage for determining the *P. gingivalis*-activated transcriptional networks that may play a role not only in the development of periodontal diseases but also in the aetiology of cancer. The red and green colours represent up and down gene regulation, respectively. (Continued overleaf.)

C

P. gingivalis-induced anti-apoptosis.

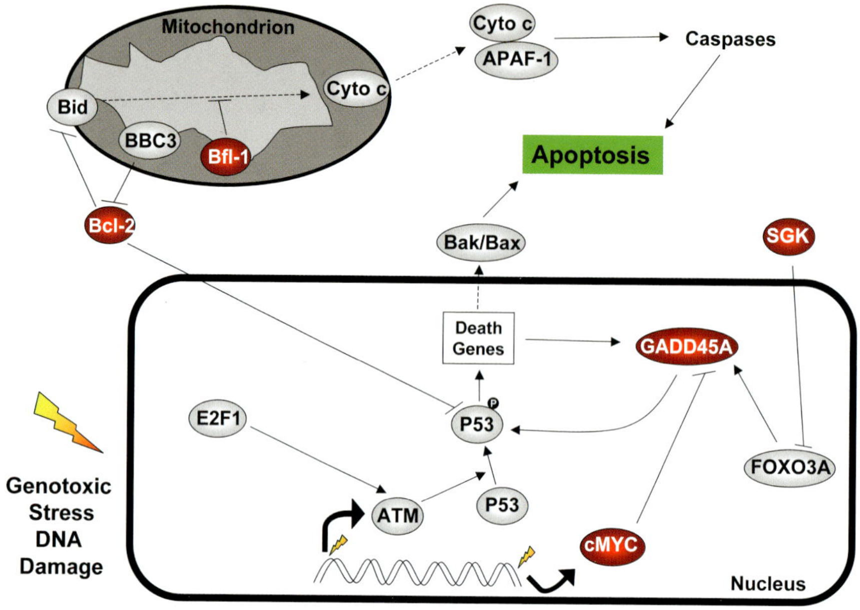

A. actinomycetemcomitans-induced apoptosis.

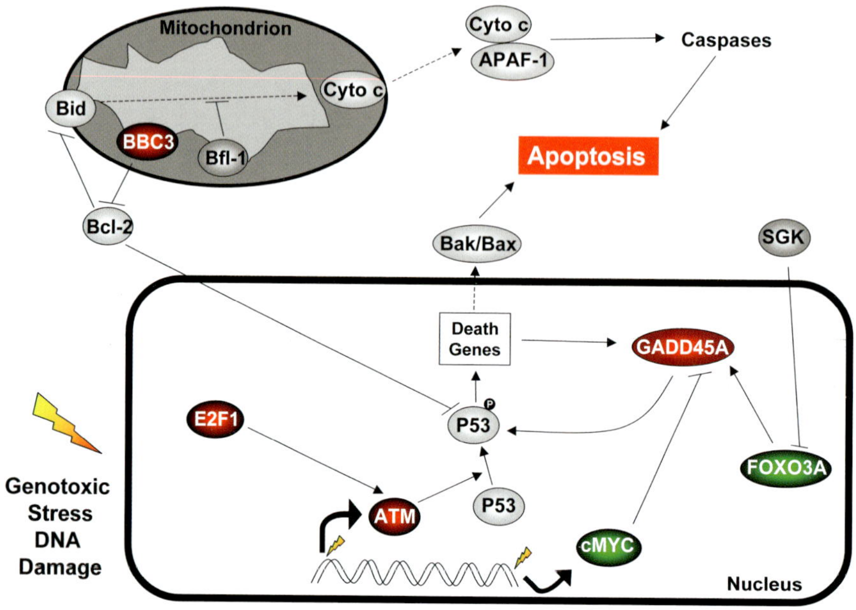

Plate 9.4 Continued. C. Apoptotic pathways affected by *A.a.* and *P. gingivalis*. D. Epithelial cell response to commensal (*S. gordonii*) and pathogenic (*P. gingivalis*) microorganisms. E. Cell-line dependent (IHGK and HeLa) host response to *A.a.* Figures adapted from *Cell. Microbiol.* (2005) *7*, 811–823, (C); *Infect Immun.* (2007) *75*, 2540–2547 (D); *Oral Microbiol. Immunol.* (2006), 21261–267 (E) and reprinted by permission.

D

F. nucleatum-infected

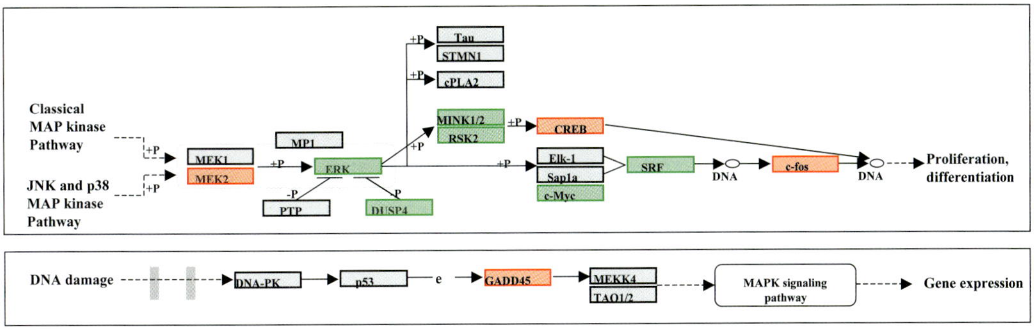

S. gordonii-infected

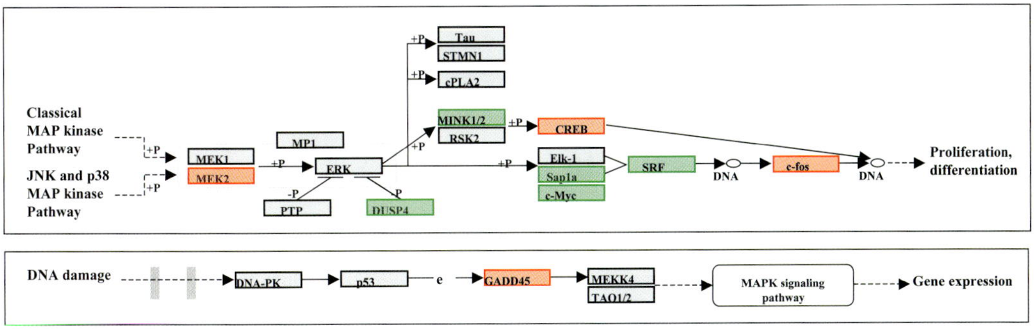

E

IHGK cells

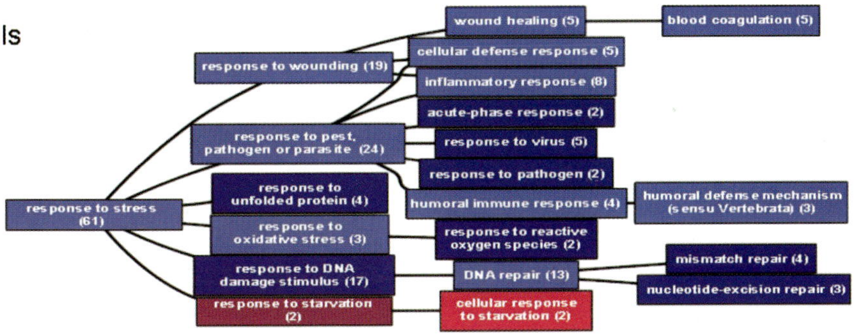

HeLa cells

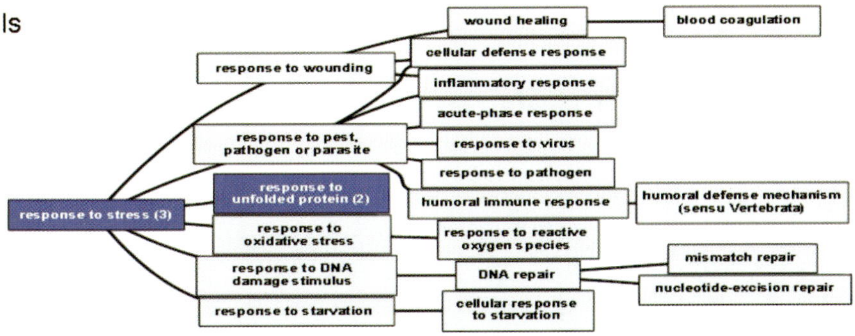

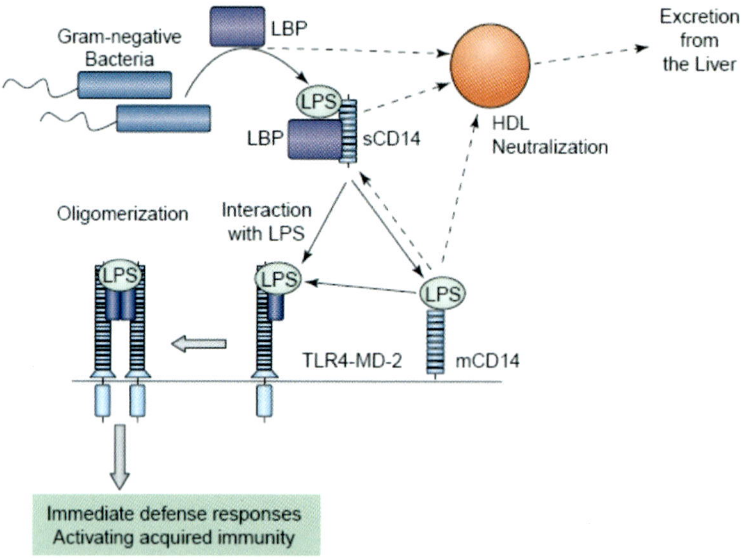

Plate 10.1 Lipopolysaccharide (LPS) processing, signalling and clearance. LPS in the outer membrane of Gram-negative bacteria has to be processed by LPS-binding protein (LBP) and CD14. LBP removes LPS from the outer membrane and forms complexes consisting of LPS, LBP and soluble CD14 (sCD14). These complexes deliver LPS to membrane CD14 for LPS signalling. LPS processing for signalling is depicted by solid lines. LPS is neutralized by its transfer to high-density lipoprotein (HDL) from LBP or CD14. LPS clearance is shown by dotted lines. LPS bound to CD14 is transferred to Toll like receptor (TLR)4-MD-2, which undergoes oligomerization and triggers a signal. The signal via TLR activates immediate and acquired immune responses. (Taken from Miyake, 2004, with permission.)